SPRAY-FREEZE-DRYING OF FOODS AND BIOPRODUCTS

SPRAY-FREEZE-DRYING OF FOODS AND BIOPRODUCTS

Theory, Applications and Perspectives

S. Padma Ishwarya

CRC Press
Taylor & Francis Group
Boca Raton London New York

CRC Press is an imprint of the
Taylor & Francis Group, an **informa** business

Cover image credit:
Coffee image:
Food photo created by azerbaijan_stockers -
HYPERLINK www.freepik.com
Images of drug & phase diagram:
https://imgbin.com
Image of microstructure
Ishwarya, S.P., & Anandharamakrishnan, C. (2015). Spray-Freeze-Drying approach for soluble coffee processing and its effect on quality characteristics. *Journal of Food Engineering*, 149, 171–180.

First edition published 2022
by CRC Press
6000 Broken Sound Parkway NW, Suite 300, Boca Raton, FL 33487-2742

and by CRC Press
4 Park Square, Milton Park, Abingdon, Oxon, OX14 4RN

CRC Press is an imprint of Taylor & Francis Group, LLC

Library of Congress Cataloging-in-Publication Data
Names: Ishwarya, S. Padma, 1988- author.
Title: Spray-freeze-drying of foods and bioproducts : theory, applications, and perspectives /
S. Padma Ishwarya.
Description: First edition. | Boca Raton : CRC Press, 2022. | Includes bibliographical references and index.
Identifiers: LCCN 2021051245 (print) | LCCN 2021051246 (ebook) | ISBN 9780367435608 (hbk) |
ISBN 9781032207759 (pbk) | ISBN 9781003019312 (ebk)
Subjects: LCSH: Food–Drying. | Food--Preservation. | Spray drying.
Classification: LCC TP371.66 .I84 2022 (print) | LCC TP371.66 (ebook) | DDC 664/.028--dc23/eng/20211230
LC record available at https://lccn.loc.gov/2021051245
LC ebook record available at https://lccn.loc.gov/2021051246

ISBN: 978-0-367-43560-8 (hbk)
ISBN: 978-1-032-20775-9 (pbk)
ISBN: 978-1-003-01931-2 (ebk)

DOI: 10.1201/9781003019312

Typeset in Times
by MPS Limited, Dehradun

Preface

The key driving force behind most inventions is 'necessity'. Conversely, the concept of *spray-freeze-drying* was born in 1959 when Meryman first showed that vacuum is not necessary for freeze-drying and it is possible to freeze-dry products at atmospheric pressure. Since then, this synergistic technology between spray-drying and freeze-drying has evolved across six decades. Today, it is well-known as the best alternative to its parent techniques for the production of powdered products with unique structure and superior quality. Spray-freeze-drying (SFD) provides the signature advantages of spray drying and freeze-drying, whilst overcoming the limitations of both.

Foods hold the credit of being the pioneering product of spray-freeze-drying, with the apple juice and egg albumin dried using this technique as early as 1970. To mention a few merits, SFD has demonstrated excellent retention of volatiles in aromatic products, higher viability of probiotics and improved oxidative stability of polyunsaturated fatty acids with nutraceutical potential. SFD for the production of biological products such as pharmaceuticals was proposed only in the end of the 20th century, but the advancements have indeed been rapid in this field. At present, spray-freeze-drying is considered as a potential technology for bulk-powder manufacturing of biologicals. SFD is capable of producing homogeneous and free-flowing bulk powders of biologics, which exhibit reproducible particle size and shape, which are ready-to-fill in any type of container. Consequently, studies on the feasibility and usefulness of spray-freeze-drying in food and bioproducts processing are on the rise. The core application areas of SFD include the lyophilization of foods and biologicals, encapsulation of food bioactives and active pharmaceutical ingredients, drug delivery (oral, pulmonary and transdermal) and nanoparticle production. At present, automation of spray-freeze-dryers is the major focus of the Industry and Academia, which is progressing well as evidenced from the development of lab-scale and pilot-scale spray-freeze-drying equipment.

This title, *Spray-Freeze-Drying of Foods and Bioproducts: Theory, Applications, and Perspectives* aims at presenting to the readers, a comprehensive review of this interesting drying technology. The scope of this book comprising 12 chapters is divided into four parts. The first three chapters provide a thorough appreciation of the theory and principle of each unit operation comprising the spray-freeze-drying process: atomization, spray-freezing and freeze-drying. The fourth chapter explains the morphology of spray-freeze-dried products, which is considered typical of the process leading to various functional properties. The chapters through five to eight discuss various facets of SFD applications with respect to foods and biologics. The 21st century is certainly the *era of nanotechnology* and spray-freeze-drying is no exception to this! Chapter 9 elucidates the applications of SFD in preparing drug-loaded nanoparticles and nanocomposite carriers for targeted delivery. Spray-freeze-dried products are preferred over the spray dried and freeze-dried counterparts mainly due to their characteristic functional properties, which are explained in Chapter 10 along with the methods of characterization. The final two chapters deal respectively with the computational modeling and the challenges and future directions of research pertaining to the SFD process. In the relevant chapters and sections, this book will also shed light upon recent advancements in the field of spray-freeze-drying.

The aim of the book is to enlighten the readers with substantial information and deeper insights on this key technology. Understanding the nuances of spray-freeze-drying would aid in designing science-based trials rather than following a hit-and-trial approach.

S. Padma Ishwarya

Acknowledgements

"A teacher affects eternity; he can never tell where his influence stops"

- Henry B. Adams

With profound gratitude, I dedicate this book to my respected mentor, Dr. C. Anandharamakrishnan (Director, National Institute of Food Technology Entrepreneurship and Management - Thanjavur (NIFTEM-T), India). The content that I have presented in this book is the essence of knowledge that I gained from him, during my tenure as a doctoral student under his guidance.

****

"Behind every child who believes in himself is a parent who believed first."

- Matthew L. Jacobson

I am extremely thankful to my parents, Mr. K. Shankaran and Mrs. Indrani Shankaran, for their boundless love and the confidence they have in me and my efforts. My heartfelt remembrance for my grandfather, Late. Mr. S. Kandaswami 'Thuraivan' (Former Deputy Director General (East zone), All India Radio), whom I have always admired for his profound knowledge, sincerity, discipline and magnanimity.

****

I thank Mr. Stephen Zollo, Senior Editor, CRC Press, Taylor & Francis Group, for his zeal in coordinating this project through different stages from proposal to production.

S. Padma Ishwarya

About the Author

Dr. S. Padma Ishwarya is presently an Institute postdoctoral Fellow at the Polymer Engineering and Colloid Science Laboratory of the Department of Chemical Engineering, Indian Institute of Technology (IIT) Madras, Chennai, India. Before joining the IIT Madras, she was a SERB-National postdoctoral fellow at the CSIR-National Institute for Interdisciplinary Science and Technology (NIIST), Thiruvananthapuram, India.

Dr. Ishwarya completed B.Tech in Food Technology from the A.C. College of Technology, Anna University, Chennai, India. Then, she pursued M.Sc. and Ph.D. in Food Technology, from the CSIR – Central Food Technological Research Institute (CFTRI), Mysuru, India. She carried out her doctoral research under the prestigious Prime Minister's Fellowship Scheme for Doctoral Research of the Science and Engineering Research Board, Government of India. Her thesis was entitled *Development of a combined experimental and computational modeling approach to investigate the influence of bran addition on the volume and structural development during the breadmaking process.* Padma has a complementing experience in the food industry. She was a quality assurance officer at the Coffee & Beverages division of Nestlé India Limited (2011–2013). Later, her doctoral research was done in collaboration with General Mills India Pvt. Ltd. (2014–2017).

Dr. Ishwarya's research focuses upon soft materials in foods, including foams, emulsions and gels. She is also well-versed in the areas of soluble coffee processing and quality aspects; valorization of food industry by-products and wastes to produce surfactants and emulsifiers; food structure and functionality; spray drying; spray-freeze-drying; and encapsulation and bakery technology. She has more than 15 peer-reviewed publications in reputed international journals, two books, eight book chapters and two popular articles to her credit.

Dr. Ishwarya is the recipient of several awards and fellowships in recognition of her distinguished academic record. To cite a few, in 2020, she won the 'Young Scientist Award' instituted by the Association of Food Scientists & Technologists (India) (AFST[I]). In 2021, her popular science article was selected for the 'AWSAR (Augmenting Writing Skills for Articulating Research) Award' founded by the Department of Science and Technology (DST), Government of India. She serves as a Reviewer for the *International Journal of Food Engineering*, published by De Gruyter, Germany and as a review editor in the Editorial Board of Food Chemistry, a specialty section of Frontiers in Chemistry and Frontiers in Nutrition.

CHAPTER 1

The Inception, Evolution and Theory of Spray-freeze-drying

Spray-freeze-drying is a hybrid drying technology that overcomes the limitations and imbibes the signature merits of spray-drying and freeze-drying techniques (Figure 1.1). The concept of spray-freeze-drying (SFD) was conceived in an attempt to avoid the longer drying time of freeze-drying. Accordingly, during SFD, a liquid feed is atomized to increase its surface area for enhanced heat and mass transfer. Subsequently, the atomized feed droplets undergo rapid freezing in the presence of a cold gas or a cryogenic liquid at ultra-low temperature to form solid particles, which are then freeze-dried. Alternatively, a fluidized bed freeze-drier can be employed to further reduce the drying time by forced convection heat and mass transfer (Al-Hakim, 2004).

The competitive edge of SFD over conventional freeze-drying can be substantiated with the help of Planck's equation (Figure 1.2), which shows that drying times vary approximately with the square of the sample thickness. For a sample of dimension 0.5 cm, a conventional freeze-drying process would require 21 hours of drying time. In contrast, atomization of the bulk feed solution into fine droplets of size 50 μm would reduce the freeze-drying time to 7 seconds (Ishwarya, Anandharamakrishnan, & Stapley, 2017). Thus, a 100-fold reduction in sample dimension shortens the residence time by over 10000 folds (Figure 1.2).

A customized *Google Scholar* search for the keyword *spray-freeze-drying* during the period from 2010 to 2020 shows a progressive increase in the number of publications including articles and patents (Figure 1.3). Hence, the growing research interest in this field of study is evident. In this introductory chapter, the readers would be taken through the seven-decade journey of *spray-freeze-drying* technique since its inception in the early 20th century. Sidney W. Benson and David A. Ellis from the Chemistry Department of the University of Southern California and H. T. Meryman from the Biophysics Division of National Naval Medical Center, Maryland, were the innovators behind this state-of-the-art drying technology. The next section would present a timeline of developments in the spray-freeze-drying process since its advent.

1.1 THE ADVENT OF SPRAY-FREEZE-DRYING

Drying involves the removal of water from a material via its vapor form. The preservative effect of drying is based on the reduction of water available for chemical reactions, microbial growth and enzyme activity, all of which deteriorate a product's quality with time. Consequently, drying leads to shelf-stable products, which retain their quality attributes even when stored under ambient conditions for several months to years (Ishwarya, Anandharamakrishnan, & Stapley, 2015). Thus, drying is of relevance to the bulk manufacture and supply of food powders and biologicals. Industrial scale production of dry food powders and biologicals with long-term storage

DOI: 10.1201/9781003019312-1

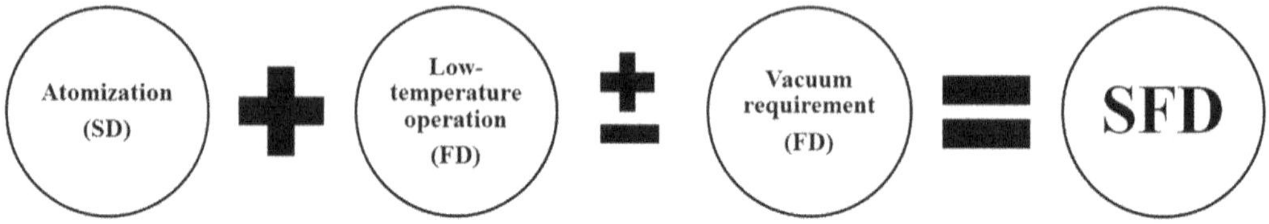

Figure 1.1 *Spray-freeze-drying* – A synergistic drying process.

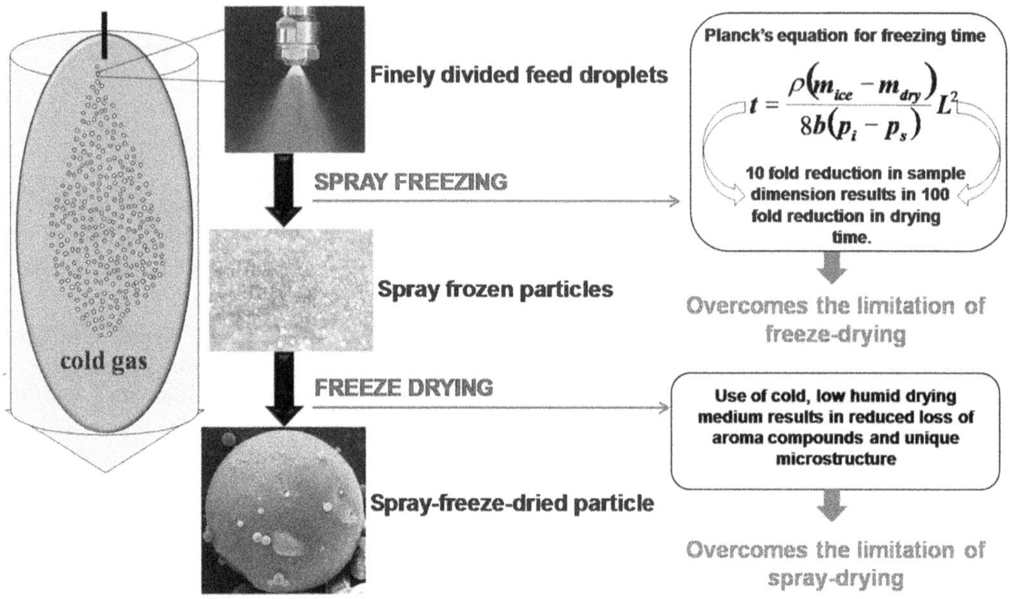

Figure 1.2 Schematic diagram of the spray-freeze-drying process (Ishwarya, Anandharamakrishnan, & Stapley, 2017).

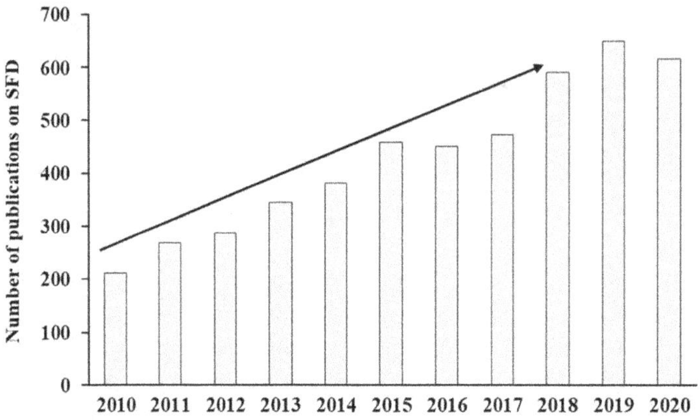

Figure 1.3 Increasing trend of publications on spray-freeze-drying during the last decade.

stability at room temperature can cater to the needs of developing markets without the requirement for a complex cold chain infrastructure.

Spray drying is an important industrial drying process, which leverages the principle of atomization to produce free-flowing powders with a homogeneous and controlled particle size distribution, at a rapid drying rate. During spray drying, hot air flows inside a cylindroconical chamber to evaporate water from the atomized feed droplets and convert the aqueous feed into a dry powdered product. The particle size of spray dried powders can be precisely controlled by tuning the atomization parameters – i.e. liquid and gas pressure (hydraulic or twin-fluid nozzle), wheel speed (rotary atomizer) or frequency of vibration (ultrasonic atomizer) (Ishwarya, Anandharamakrishnan, & Stapley, 2015). Therefore, an atomizer is considered the heart of a spray dryer. The minute droplet size of atomized feed solution is significant in deciding the residence time of product within the spray drying chamber, which is typically in the order of seconds or tens of seconds (Woo, 2017). Residence time is typically the total time during which the product is under treatment (Thuse et al., 1968). Despite the reduced particle residence time, the high-temperature operation during spray drying leads to thermal degradation of heat sensitive products and loss of volatile substances (Bhandari, Patel, & Chen, 2008). Nevertheless, the aforementioned challenge is overcome by the freeze-drying process.

During conventional freeze-drying, a slab of material is frozen (stage I: freezing) and the ice crystals thus formed are sublimed off to the vapor phase in a vacuum chamber (stage II: primary drying) (Liu, Zhao, & Feng, 2008). Then, the water that remains occluded in the freeze-concentrated non-ice phase is removed by desorption (stage III: secondary drying) (Liapis & Bruttini, 2009). The porous microstructure created by the sublimation of ice crystals leads to excellent rehydration property of the powdered products. Freeze-drying results in products with better color, flavor and texture and biologicals with better stability and viability, for instance, the probiotics (biologicals refer to substances of biological or natural origin that are used as drugs, vaccines, pesticides and so on). More specifically, the low-temperature operation of freeze-drying preserves the volatile compounds and enhances the aroma and flavor quality of premium products such as instant coffee and fruit juice powders. Nevertheless, freeze-drying is more expensive than spray drying per unit weight of water removed. High cost of processing is due to prolonged drying times (low drying rate), energy-intensive batch operation, requirement of high vacuum and low capacity because of low drying rate. All the aforementioned shortcomings limit the commercial applications of freeze-drying (Lopez-Quiroga, Antelo, & Alonso, 2012; Malecki et al., 1970; Quast & Karel, 1968; Ratti, 2008). Besides, the freeze-drying process does not yield homogeneous and free-flowing powders (Lowe et al., 2018), due to lack of precise control over the particle size.

The major components of a freeze-dryer include (McHugh, 2018):

1. ***Refrigeration unit:*** Cools the shelves in the product chamber to freeze the product and also the condenser located inside the freeze-dryer.
2. ***Product chamber:*** Could be a manifold attached with flasks or a larger chamber equipped with shelves or trays on which the product is spread and dried.
3. ***Vacuum pump:*** Removes non-condensable gases from the system to create the low-pressure environment required for sublimation (Stapley, 2008).
4. ***Condenser:*** Collects the vapors that are sublimed off the product and condenses them back into solid form (ice).

Of the aforementioned, vacuum pump is the bottleneck component that consumes more electrical energy during the freeze-drying process (Figure 1.4).

From the given discussion, it is obvious that a desirable drying temperature, residence time and energy efficiency often cannot be attained simultaneously. Because, lower drying temperature results in longer drying time so as to assure the same level of drying (Fu et al., 2018). An approach

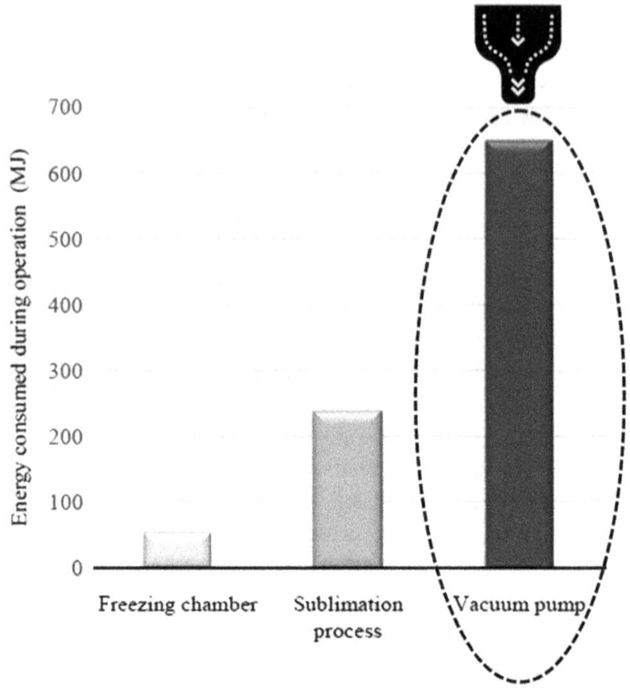

Figure 1.4 Vacuum pump – The *bottleneck* component in a freeze-dryer with respect to energy consumed during the operation (freeze-drying of 100 kg of raspberries) (Kešelj et al., 2017).

to increase the drying rates is to diminish the material dimensions, for which atomization is the apt operation. In addition, eliminating the vacuum operation would aid in improving the energy efficiency of a low-temperature drying process such as freeze-drying.

In their pioneering study conducted in 1948, Benson and Ellis determined the surface area of protein particles that were produced by atomizing different protein solutions directly into a bath of cryogenic liquid (liquid nitrogen). Upon freezing, the atomized protein feed formed a large amount of white precipitate from which the liquid nitrogen was evaporated. Then, the resultant precipitate was vacuum-dried at around –30°C for 8 to 12 hours to yield a fine white powder. Unlike in spray drying, the protein particles were produced with no requirement of heat, which prevented their thermal denaturation (Benson & Ellis, 1948). Thus, the concept of including atomization in a freeze-drying process, namely, *spray freezing* was disclosed in this study. Nevertheless, a complete elucidation of the concept of spray-freeze-drying happened only in 1959 while unveiling the mysterious function of a *void*, i.e. the *vacuum* applied during the freeze-drying process.

As mentioned earlier, vacuum pump is a vital component of the conventional freeze-dryer. Contrary to the belief that its function is to reduce the total pressure of the freeze-dryer system, the role of a vacuum pump is to remove the non-condensable gases (i.e. water vapor) and facilitate their transfer to the vapor trap. However, it was not realized until H. T. Meryman demonstrated for the first time in 1959 that vacuum is not mandatory for the freeze-drying process (Stapley & Rielly, 2009). He proposed that, *'the drying rate of a material undergoing freeze-drying is a function of the ice temperature and the vapor pressure gradient between the site of water vapor formation and the drying medium rather than the total pressure in the drying chamber'*. Meryman was the first to demonstrate that freeze-drying can be carried out under atmospheric conditions by circulating a convective freeze-drying medium such as cold air stream that is held dry by a molecular sieve desiccant or by a refrigerated condenser. Based on his experiments, he established that freeze-drying can be carried out even in the absence of a vacuum pump, provided the vapor pressure of water at the sample surface is lower than the saturation vapor pressure of ice at the temperature of the sample being dried.

A closer look at the mechanism of freeze-drying from Meryman's perspective provides a convincing proof that the vacuum system can be totally eliminated from a freeze-dryer without jeopardizing the drying rate. When a biological sample is frozen, most of the water in the solution is removed, isolated and concentrated in the form of ice crystals. Faster the freezing, smaller the size of these ice crystals and vice versa. The purpose of the subsequent freeze-drying step is to remove water from these ice crystals without allowing much changes in the sample. The rate at which the water molecules are removed from the ice crystals is exclusively based on their temperature. Further, the proportion of water molecules that return to the crystal is dependent on the concentration of the surrounding vapor. Therefore, the net removal of water from the ice crystals depends on the efficiency with which the vapor is prohibited from returning to the crystal (Meryman 1959). In turn, the water vapor removal is hindered by two factors: (1) resistance to diffusion of water vapor imposed by the previously dried shell of the sample and (2) impeded transfer of water vapor molecules due to its collision with the air molecules during its transit from the sample surface to the vapor trap. However, the former factor has a stronger influence on the water vapor removal than the latter.

The factors controlling the rate of diffusion of water vapor through the dried shell are the rate of vapor production (which is a function of sample temperature), resistance offered by the dried layer and water vapor pressure at the sample surface. As the sample temperature and dimensions are fixed, the only governing factor in freeze-drying is the water vapor pressure at the sample surface rather than the total gas pressure. The difference in vapor pressure between the sample surface and its surrounding promotes an effective removal of water vapor once it reaches the sample surface. Thus, during freeze-drying, the passage of water vapor through the dried sample is a function of the vapor pressure gradient rather than the total pressure of the system.

In his experiment, Meryman created the vapor pressure gradient by blowing cold and dry air past the sample to carry away the water vapor molecules as and when they reached the sample surface. In other words, the dew point of air was maintained at a level that was lesser than the temperature of food to be dried (Stapley & Rielly, 2009). This increased the rate of air transfer past the specimen to reduce the thickness of the boundary layer that is rich in water vapor. Dry air was generated by passing the air through a desiccant before blowing it across the sample. The air temperature was reduced by placing the drying set-up within a cold chamber or by wrapping the system inside a tube and cooling the same by a conventional refrigeration system. This atmospheric freeze-drying system achieved drying of a 2 mm cube of mouse kidney in 8 hours (Meryman, 1959).

1.2 PHASE DIAGRAM OF WATER

The concept of eliminating vacuum pump from a freeze-dryer as explained earlier can be appreciated further by reading the *phase diagram of water*. Phase is a physically unique and chemically homogeneous part of a system, which is identified by its specific chemical composition and structure. Pure water can exist in solid, liquid or gaseous phase. Under specific conditions of temperature and pressure, the phase or state of water can be mapped from a phase diagram (Figure 1.5). The temperature (0.01°C) and pressure (0.6 kPa) at which the solid, liquid and vapor phases of water coexist in equilibrium is termed as the *triple point*. In the phase diagram, the zone of freeze-drying is below the triple point (Figure 1.5). The pre-requisite for sublimation is that both the vapor pressure and temperature are below the triple point of water (Mumenthaler & Leuenberger, 1991). In other words, ice sublimes off to vapor at temperature below the triple point,

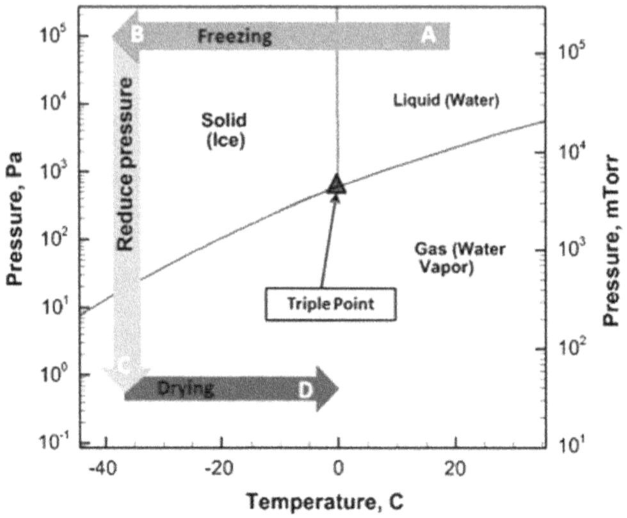

Figure 1.5 Phase diagram of water (Ganguly, Nail, & Alexeenko, 2012; Wagner, Saul, & Pruss, 1994).

provided, the partial pressure of water vapor is lower than the partial pressure of ice (Shivanand & Mukhopadhayaya, 2017). The chart on vapor pressure over ice at different temperatures (Table 1.1) reveals the relevance of circulating cold gas to achieve the vapor pressure gradient, as specified in Meryman's theory (1959).

A best instance of atmospheric freeze-drying phenomenon amidst us is the condition that prevails in Antarctica, wherein sublimation of ice occurs on a very large scale at ambient pressure. Strong winds blow off and loosen the surface ice to render it airborne and eventually sublime (Mann, 1998). Another daily life example could be of the clothes put out to dry on a frosty day that freeze like boards and yet dry perfectly (Anandharamakrishnan, 2008). The given findings put forth by Meryman led to the inception of spray-freeze-drying process. However, the term *spray-freeze-drying* was not mentioned in the first publication of Meryman. Instead, the drying process was referred to as *sublimation freeze-drying without vacuum* or *atmospheric pressure freeze-drying* and the drying apparatus was denoted as *dry-air device*. Indeed, Meryman's concept of eliminating the vacuum pump from a freeze dryer proved to be advantageous in terms of mechanical simplicity and possibility to dry multiple samples simultaneously. The aforementioned are not possible in vacuum drying as multiple samples impose physical obstacles and impede the straight-lined passage of vapor to the trap, which reduces the efficiency of the vacuum system.

1.3 EVOLUTION OF THE SPRAY-FREEZE-DRYING PROCESS

After Benson, Ellis and Meryman demonstrated the proof of concepts that led to the advent of spray-freeze-drying technology, the process has undergone substantial changes in terms of its instrumentation and applications. Across more than six decades, the developments in the SFD process have been mainly aimed at reducing the processing cost and drying time so as to improve its industry-friendliness and commercial viability. Different approaches proposed by various researchers are discussed subsequently.

In consensus with Meryman's concept, Lewin & Mateles (1962) also demonstrated the feasibility of conducting freeze-drying at atmospheric pressure by maintaining a sufficiently low partial pressure of water in the drying chamber. The need for vacuum pump and interlock systems was eliminated,

Table 1.1 Vapor pressure over ice chart (Shivanand & Mukhopadhayaya, 2017)

Temperature (°C)	Vapor Pressure (mbar)	Temperature (°C)	Vapor Pressure (mbar)
0	6.11148	–76	0.00104
–2	5.176893	–78	0.00076
–4	4.374295	–80	0.000547
–6	3.686353	–82	0.000387
–8	3.099737	–84	0.00028
–10	2.598446	–86	0.0002
–12	2.173149	–88	0.000133
–14	1.811846	–90	0.000096
–16	1.506539	–92	0.000065
–18	1.24896	–94	0.000045
–20	1.032446	–96	0.000031
–22	0.850861	–98	0.00002
–32	0.30824		
–34	0.249045		
–36	0.200383		
–38	0.160786		
–40	0.128389		
–42	0.102258		
–44	0.08106		
–46	0.063995		
–48	0.050262		
–50	0.03933		
–52	0.030664		
–54	0.023865		
–56	0.018398		
–58	0.014132		
–60	0.010799		
–62	0.008213		
–64	0.006213		
–66	0.00468		
–68	0.003506		
–70	0.002613		
–72	0.001933		
–74	0.001413		

which facilitated continuous freeze-drying operation. Nevertheless, the given advantages were obtained only at the expense of prolonged drying times. In addition, the operational cost of atmospheric freeze-drying (AFD) was on par with the conventional freeze-drying method (Woodward, 1963; Noyes, 1968). Particularly, the AFD was not suitable for puree or bulk liquid because of the very slow drying rate. In view of addressing the aforementioned bottleneck, Woodward (1963) made an attempt to identify the root causes for the increase in drying time in the absence of vacuum. He concluded that product dimensions and gas temperature are the key factors that stretched the total drying time, rather than the gas flow rate. These findings emphasized the importance of including an atomization step in the freeze-drying process to reduce the sample dimensions. In 1968, Quast & Karel noticed the formation of an impermeable layer over the frozen liquid surface that resisted the mass flow rate. The authors showed that, intermittent removal of the surface layer by manual agitation could reduce the surface resistance and increase the sublimation rates. Interestingly, this study involved food samples like coffee solutions (20% and 30% solids), 10% glucose, 10% microcrystalline cellulose and 2% potato starch as model systems to validate the aforementioned concept. This shows the longstanding association between foods and the SFD process.

Later, researchers shifted their focus upon the freezing and freeze-drying mechanisms to reduce the drying time of SFD process. Freezing mechanisms were studied by microscopic examination of structural changes in the sample. In this context, Malecki et al. (1970) attempted to reduce the drying time by atomizing liquid foods (apple juice, liquid egg and orange juice) into a cryogenic medium (liquid nitrogen), followed by sublimation of the frozen particles in a fluidized bed at atmospheric pressure. The reduced droplet size substantially increased the sublimation rate, as the drying progressed in an unrestricted manner from each frozen droplet. Also, the resistance to water-vapor diffusion through both dried and drying media were found to be the rate-controlling factors during the freeze-drying process, which varied proportionately with the diffusion path permeability of the dried material. Therefore, reduced sample dimensions accelerated the rate of atmospheric sublimation by shortening the diffusion path of vapor molecules (Malecki et al., 1970).

In 1974, Heldman & Hohner arrived at a mathematical model for the simultaneous heat and mass transfer occurring during the atmospheric freeze-drying process. The model predicted the effect of sample size on atmospheric freeze-drying. Results showed that a rapid reduction in drying time could be achieved at reduced sample dimensions and the consequent increase in surface mass transfer coefficient. The time to dry a sample to any given dimensionless moisture content should vary with the ratio of the square of the sample thickness. Hence, reducing the dimensions of the material was confirmed as the favorable approach for reducing both freezing and freeze-drying times. Boeh-Ocansey (1983) studied the sublimation rate during the fluidized bed freeze-drying of differently-sized carrot pieces at –5°C and –10°C. The objective of their investigation was to prove the effect of shortened transit path of vapor molecules to the vapor trap on the drying time. Results showed the increase in drying time with the increase in sample thickness. Samples of thickness below a certain critical value exhibited higher drying rates and substantially shorter drying time in an atmospheric freeze-dryer relative to a conventional vacuum freeze-dryer.

Later, researchers came up with an alternative approach of contacting the atomized feed with cold, dry gas in a fluidized bed at atmospheric pressure (Mumenthaler & Leuenberger, 1991) or sub-atmospheric pressure (Anandharamakrishnan et al., 2008; Anandharamakrishnan, Rielly, & Stapley, 2010). Fluidization was adopted to apply uniform heat to the frozen particles during the freeze-drying process, by leveraging the small sample dimensions, thereby reducing the drying time without the risk of particle melting and collapse. However, this is not possible to obtain with the conventional freeze-drying, wherein the latent heat of sublimation is supplied by conductive or radiative heating. Moreover, fluidized bed has the advantage of excellent heat and mass transfer due to good contact between the particles and fluids. This renders the SFD suitable for large-scale operations with good mixing between particles that lead to fast drying at a low capital cost of construction. This marked the beginning of atmospheric spray fluidized bed freeze-drying (ASFBFD) and vacuum spray fluidized bed freeze-drying (VSFBFD) (Ishwarya et al., 2015).

1.4 PROCESS AND EQUIPMENT OF SPRAY-FREEZE-DRYING: AN OVERVIEW

Based on the discussion in the previous section, spray-freeze-drying can be defined as *'a technique which involves a liquid or solution being atomized into droplets, solidified by contact with a cold fluid and ice subliming at low temperature and pressure'* (Leuenberger, 2002). It is a three-step batch process comprising the atomization, freezing and freeze-drying phases. SFD has also been described as a two-step semi-continuous process, including the spray-freezing and freeze-drying operations, according to which, it has been defined as the '*solidification of an atomized liquid by contacting it with an inert cooling medium below the freezing point of the liquid*'. Liquid nitrogen, cold air and dry ice are the commonly used cooling media (Al-Hakim et al., 2006). During the atomization step, the feed solution or emulsion containing solids is finely divided into droplets using high precision and low-shear frequency nozzles. These droplets fall under gravity into a freezing chamber and congeal to form the frozen spheres. The freezing chamber is supplied with a cryogen, which could be a liquid or dry gas that is capable of maintaining the internal air temperature of the freezing chamber below –110°C. This leads to flash-freezing of the droplets into distinct spheres of diameter in the range of 250–1000 μm, depending on the nozzle diameter (Luy, 2016). The aforesaid step is analogous to the spray drying process, except that the hot air is replaced by cold gas. Depending on whether the atomized feed is frozen by a cryogenic gas, liquid or vapor over the cryogenic liquid, SFD can be classified as follows (Ishwarya et al., 2015):

i. ***Spray-freezing into Vapor over Liquid (SFV/L):*** In this case, the feed solution is atomized through a nozzle positioned at a small distance above a boiling cryogenic liquid (Figure 1.6[a]). The droplets begin to solidify during their transit through the vapor gap and are completely frozen on contacting the cryogenic liquid (Adams, Beck, & Menson, 1982; Briggs & Maxwell, 1976; Buxton & Peach, 1984; Dunn, Masavage, & Sauer, 1972; Sauer, 1969).
ii. ***Spray-freezing into Vapor (SFV):*** Here, the atomized feed droplets are frozen on contact with cold, dry gas (Figure 1.6[b, d & e]).
iii. ***Spray-freezing into Liquid (SFL):*** In this mode of spray-freezing, the nozzle is immersed below the surface of a cryogenic liquid. SFL is based on the liquid-liquid impingement between the pressurized feed liquid emanating from a nozzle and the cryogenic liquid. In SFL, the droplet freezing is instantaneous. To avoid aggregation of particles, the cryogen is stirred by an impeller inserted into the vessel (Figure 1.6[c]).

Subsequently, the spray frozen particles are collected and transferred to a freeze-drying unit to obtain the dried particles. Thus, the second step in the SFD process is freeze-drying. During freeze-drying, the frozen microspheres are dried under low temperature and low pressure conditions to form a homogeneous powdered product with narrow particle size in the range of 200–800 μm and superior flowability. The required latent heat of sublimation is supplied by radiation and temperature-controlled surfaces (Luy, 2016). The freeze-drying step can be carried out in any of the following modes:

i. Conventional vacuum freeze-Drying (VFD) (Figure 1.6[a]);
ii. Atmospheric freeze-drying (AFD) (Figure 1.6[b];
iii. Atmospheric fluidized bed freeze-drying (AFBFD) (Figure 1.6[d]);
iv. Sub-atmospheric (or) vacuum fluidized bed freeze-drying (SAFBFD or VFBFD) (Figure 1.6[e])

Subsequently, the freeze-dried product is transferred to a collection vessel which is connected to the sterile feeders of the powder filling line (Lowe et al., 2018).

Possibility of conducting spray-freeze-drying as a one-step continuous process has also been realized. In this case, the solids-containing feed liquid is continuously dried into instantly soluble powdered product by spraying the extract into a directly condensing vacuum freezing chamber. The atomized droplets freeze in transit and the frozen particles are conducted through a directly condensing vacuum drying chamber equipped with a heat source. Finally, the dried product is collected from the drying chamber (Thuse et al., 1968).

The instrumentation of spray-freeze-drying process is the one which has truly evolved across years since its inception. One of the early SFD systems used an atomizer with 1-mm orifice to generate droplets of diameter 0.1 mm. The vacuum drying assembly was the simplest, comprising a 500 mL round-bottom flask with a ground joint to hold the frozen material. This assembly was

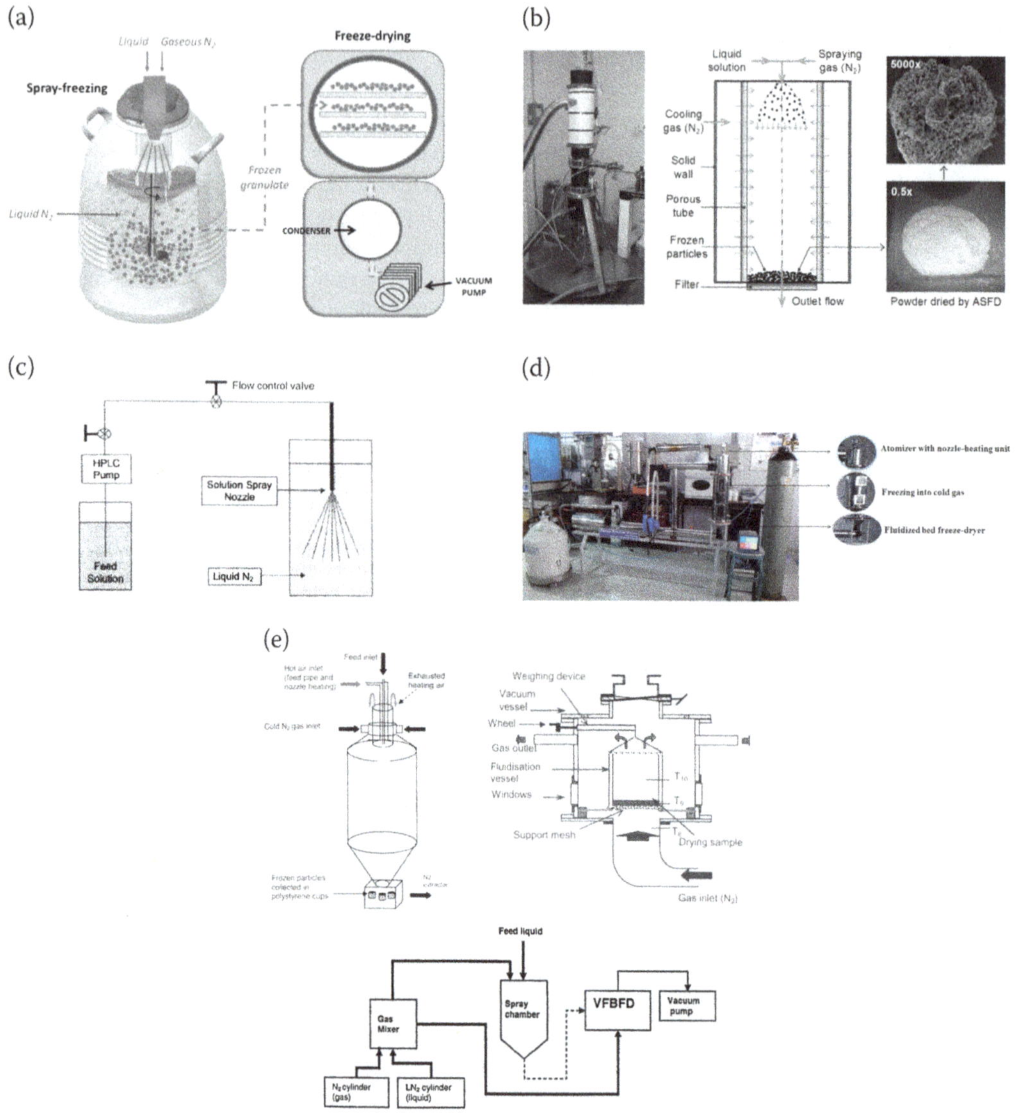

Figure 1.6 Schematic representation of (a) SFV/L process followed by conventional VFD (Bhatta, Stevanovic Janezic, & Ratti, 2020); (b) SFV process followed by atmospheric freeze-drying (Sebastião et al., 2017); (c) SFL process (Rogers et al., 2002); (d) atmospheric spray fluidized bed freeze-drying apparatus (Reproduced with permission from C. Anandharamakrishnan); (e) SFV followed by VFBFD (Anandharamakrishnan, 2008; Anandharamakrishnan et al., 2010).

connected to a 500 mL trap which was dipped in a dry ice-ether-bath to collect the water, which in turn was connected to a vacuum pump (Benson & Ellis, 1948; Flosdorf & Mudd, 1935). On the other hand, Meryman's device for atmospheric freeze-drying was essentially a cylinder with a central tube. The screens contained the desiccant, which filled the annular space surrounding the central tube. A blower driven by a motor through a sealed bearing functioned to draw air through the central tube across the specimen and back through the desiccant. The seal bearing ensured that the system is air-tight upon assembling. The sample was placed in a basket that was inserted through a port into the central tube. After placing the sample, the system was sealed by a stopper. The assembly as described earlier was either placed in a cold chamber or wrapped in a tube and cooled by a conventional refrigeration system. The air temperature of the aforementioned system was accurately controlled by a thermocouple (Meryman, 1959).

From the spray-freeze-dryers described earlier, the modern-day equipment (Figure 1.7) have come a long way in terms of their automation, controls and ease of operation. Depicted in Figure 1.7 are the spray-freezing and spray-freeze-drying rigs located at the Lougborough University, United Kingdom (Figure 1.7[a]) and the Indian Institute of Food Processing Technology (IIFPT), India (Figure 1.7[b]). Essentially, these systems are an assembly of two units, namely, the spray-freezing chamber and the freeze-dryer. In the system shown in Figure 1.7(a), the spray-freezing chamber is cooled by introducing liquid nitrogen from a dewar. Before passing the liquid nitrogen, the chamber is initially purged with nitrogen gas from a cylinder. Subsequently, the wall temperature of spray chamber and the gas temperature are adjusted to –85°C (±2°C) by regulating a needle valve at the gas mixer. A hydraulic nozzle is used to atomize a liquid feed from a pressurized feed tank. A unique feature in these SFD rigs is the circulation of heated air around the nozzle feed pipe to maintain the nozzle body at 20°C. This is to prevent freezing of liquid feed within the nozzle. Finally, the frozen particles are collected from the outlet of the chamber in a polystyrene cup kept in a polystyrene box (Anandharamakrishnan, 2008). Then, the frozen particles are collected and transferred to a freeze-drying unit (vacuum fluidized bed freeze-dryer), followed by the collection of dried particles.

On the other hand, in the set-up shown in Figure 1.7(b), spray-freezing is carried out in a stainless steel vessel containing liquid nitrogen and freeze-drying of the frozen particles is carried out using a conventional vacuum freeze-dryer. An impeller is included to stir the contents of the stainless steel vessel to prevent agglomeration of particles during freezing. The detailed descriptions of these advanced present-day spray-freeze-dryers will be explained in the forthcoming chapters of this book.

1.5 ADVANTAGES OF SPRAY-FREEZE-DRYING OVER SPRAY-DRYING AND FREEZE-DRYING

The advantages of spray-freeze-drying over the conventional drying techniques can be explained under five major headers:

i. Short drying time
ii. Homogeneous and reproducible particle size distribution
iii. Superior flow characteristics
iv. Instant reconstitution
v. Oxidative stability

1.5.1 Short drying time

Compared to the conventional freeze-drying process, a 30% reduction in drying time has been achieved with SFD (Karthik & Anandharamakrishnan, 2013; Ishwarya & Anandharamakrishnan, 2015). The key benefits of short drying time are the higher aroma retention and improved oxidative stability of aromatic products such as coffee and specialty oils, respectively. Short drying time of SFD is also advantageous in preserving the biological activity and stability of proteins and other active pharmaceutical ingredients of similar kind.

1.5.2 Homogeneous and reproducible particle size distribution

Particle size plays a key role in deciding the solubility and flow characteristics of food powders. It bears a direct relationship with the nozzle diameter (of atomizer). An increase in nozzle diameter from 300 to 400 μm increased the particle size of SFD powder from 400–600 μm to

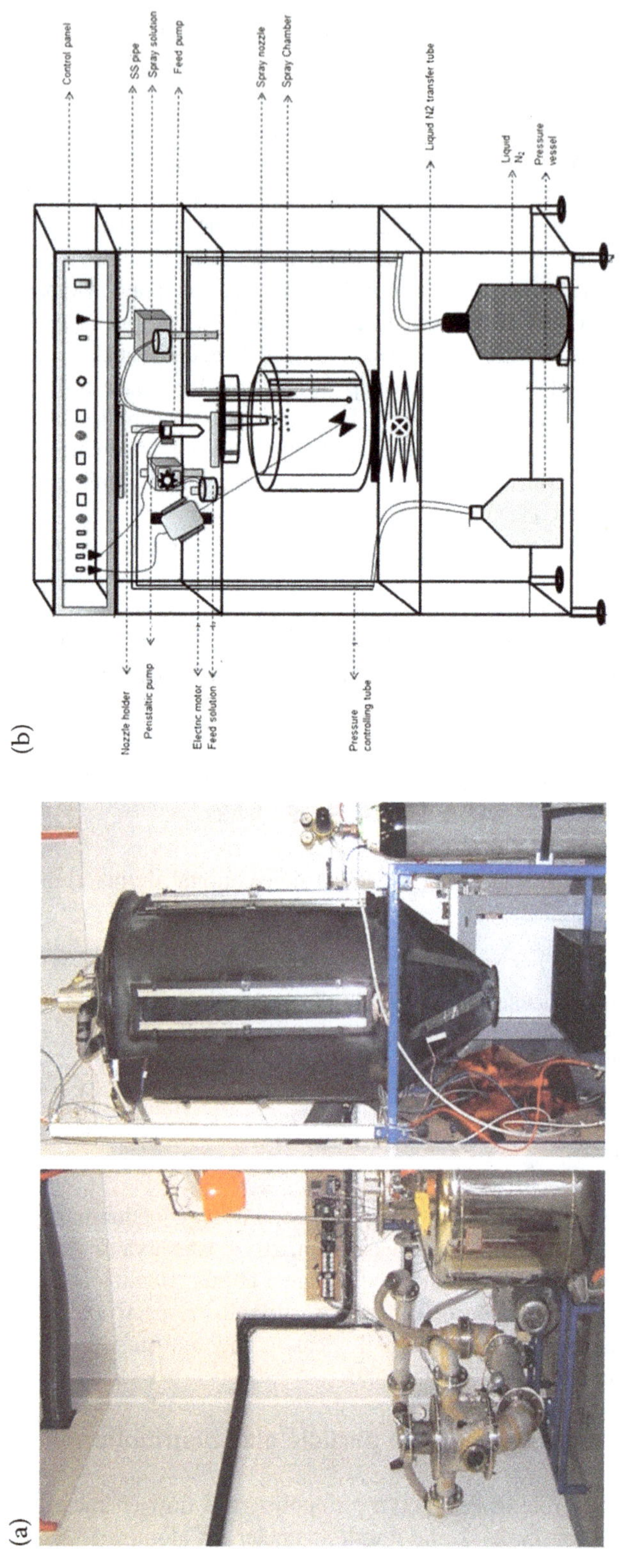

Figure 1.7 The modern-day spray-freeze-drying rigs integrated with (a) vacuum fluidized bed freeze-dryer (Anandharamakrishnan, 2008); (b) conventional freeze-dryer (Reproduced with permission from C. Anandharamakrishnan).

600–800 μm (Lowe et al., 2018). Particle size analysis of spray-freeze-dried powders by laser-light scattering revealed a monomodal distribution of particles. This was different from the bimodal size distribution of powder produced by the conventional freeze-drying process (Figure 1.8). The mean particle diameter of SFD product was found to be lower than that of the freeze-dried product, attributed to the precise control of particle size obtained by regulating the atomization parameters. However, this precision over particle size is not attainable with the conventional freeze-drying process, wherein, the frozen coffee slab is merely ground at a low temperature (Ishwarya & Anandharamakrishnan, 2015). Further, SFD results in a narrow and reproducible particle size distribution with virtually no fines below 10 μm (Figure 1.9) (Lowe et al., 2018).

1.5.3 Superior flow characteristics

1Spray-freeze-drying facilitates the aseptic production of free-flowing bulk powder, which is homogeneous and ready-to-fill into a variety of container types. Based on the flow-function test plots carried out using a shear-cell apparatus – a method commonly used to characterize powder-flow

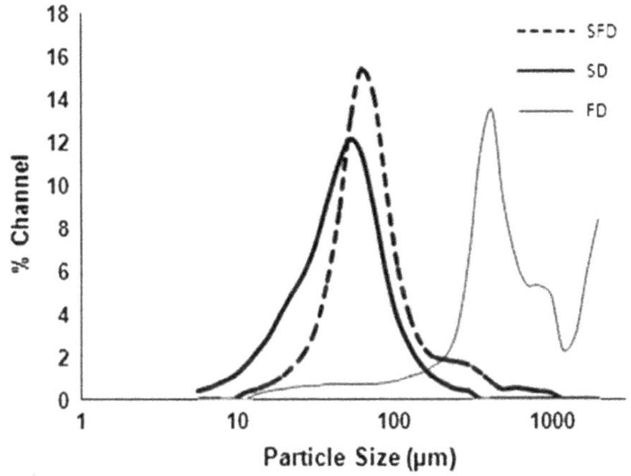

Figure 1.8 Particle size distribution – Comparison plot of spray-freeze-dried (SFD), freeze dried (FD) and spray dried (SD) coffee samples (Ishwarya & Anandharamakrishnan, 2015).

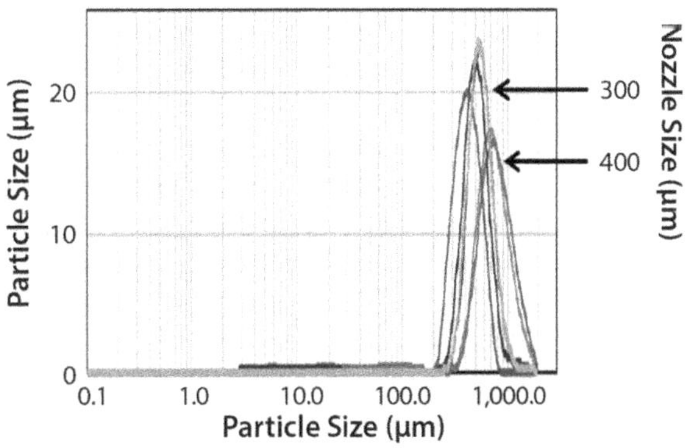

Figure 1.9 Particle-size distributions of different batches of spray-freeze-dried powder produced using lab-scale and pilot-scale dryers using two different nozzle sizes (Lowe et al., 2018).

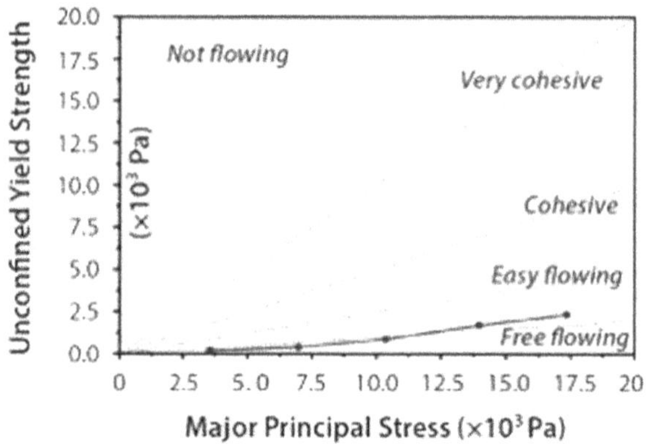

Figure 1.10 Flow-function test of a powder batch of monoclonal antibodies produced by spray-freeze-drying using a shear cell apparatus; dotted lines indicate the designation of powder flow behavior (Lowe et al., 2018).

behavior, powders are classified as either *free flowing, easy flowing, cohesive, very cohesive* or *not flowing*. Analysis has shown that flow behavior of spray-freeze-dried powders is either free-flowing or easy-flowing (Figure 1.10) (Lowe et al., 2018). Further, the free flow and tapped bulk densities of spray-freeze-dried powders were high compared to their spray-dried and freeze-dried counterparts (Ishwarya & Anandharamakrishnan, 2015). This can be attributed to their non-cohesive nature (Geldart, 1986) relative to the spray-dried powder and smaller particle size than the freeze-dried powder. Smaller particle size leads to reduced number of interparticle voids with higher contact surface areas per unit volume, which result in higher bulk density (Caparino et al., 2012). Further, the lower drying temperature may also be responsible for a simultaneous increase in the bulk density and solubility of spray-freeze-dried products (Hassen & Al-Kahtani, 1990). High bulk density of SFD products is advantageous for long distance transportation owing to the lesser cost of packaging and transit compared to higher bulk density products resultant from conventional drying processes (Bhandari, Patel, & Chen, 2008).

1.5.4 Instant reconstitution properties

Instantaneous reconstitution is an additional advantage of bulk powders produced by SFD technology. The spherical shape and porous microstructure of spray-freeze-dried products (Figure 1.11) result in superior rehydration properties, implied by shorter rehydration time than conventionally dried powders. For instance, the solubility time for spray-freeze-dried DHA and coffee powders was significantly less when compared to their spray-dried (SD) and freeze-dried (FD) counterparts (Table 1.2).

On rehydration, the porous structure of spray-freeze-dried products (Figure 1.11) promotes capillary imbibition of water (Saguy et al., 2005). This is expected of a SFD product as the occurrence of porous microstructure is the outcome of ice crystal formation during the freezing step, which sublimes during the subsequent freeze-drying step to form the internal pores (Stapley, 2008; Anandharamakrishnan et al., 2010). Further, the smaller particle size of SFD powders compared to the FD product increases their specific surface area that adds to the surface energy of the particle and thereby improves solubility (Kaptay, 2012). Owing to their high surface area and porosity, rapid wetting and dissolution of the SFD powders can be achieved with different volumes of water for injection (WFI) as a diluent (Figure 1.12). Irrespective of the particle size, the time for complete reconstitution was constantly 2–4 minutes. Size-exclusion chromatography (SEC) was used to confirm that the aggregation levels of SFD powders remained unaffected after reconstitution regardless of the reconstitution volume (Lowe et al., 2018) (Figure 1.12).

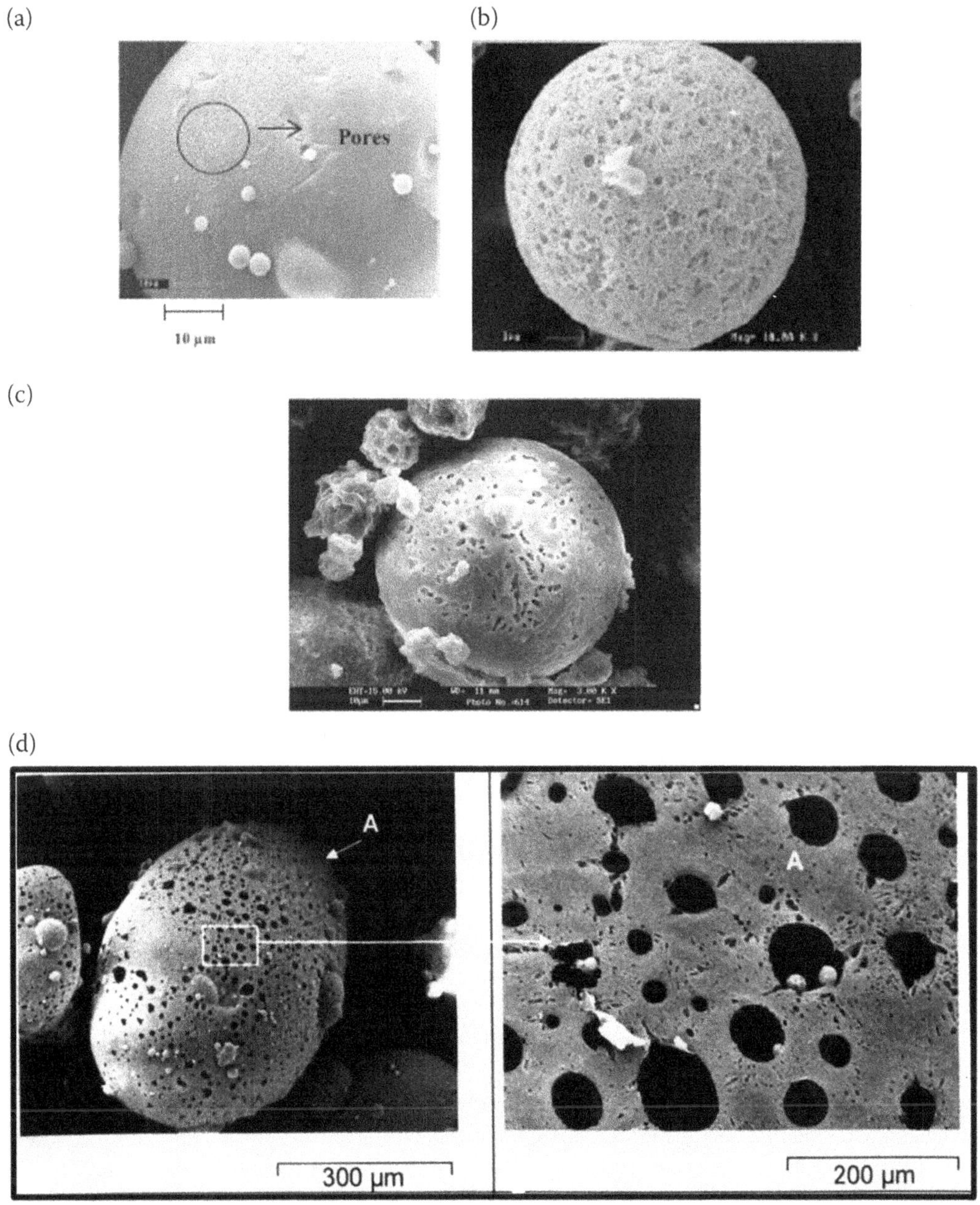

Figure 1.11 Porous microstructure of spray-freeze-dried (a) soluble coffee (Ishwarya & Anandharamakrishnan, 2015); (b) vanillin (Hundre et al., 2015); (c) DHA (Karthik & Anandharamakrishnan, 2013); (d) whey protein (Anandharamakrishnan, 2008).

1.5.5 Oxidative stability

Yet another advantage of spray-freeze-drying is the low degree of oxidation during drying. For instance, with respect to DHA microencapsulates, spray drying and freeze-dying resulted in doubled oxidation levels compared to the SFD process due to its high-temperature operation and longer drying time, respectively (Karthik & Anandharamakrishnan, 2013).

Table 1.2 Reconstitution time of spray-freeze-dried products

Product	Reconstitution time (seconds)			Reference
	SFD	FD	SD	
DHA	30–50	36–48	89–102	Karthik & Anandharamakrishnan (2013)
Coffee	10–12	21–23	19–21	Ishwarya & Anandharamakrishnan (2015)

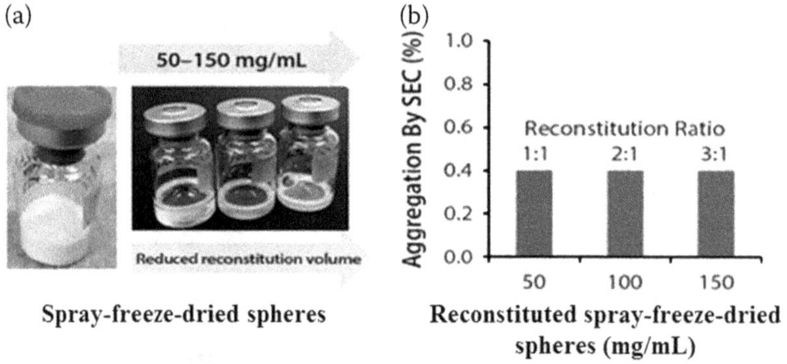

Figure 1.12 (a) Reconstitution of spray-freeze-dried monoclonal antibodies using different volumes of water for injection (WFI) diluent; (b) aggregation levels of spray-freeze-dried monoclonal antibodies determined by size-exclusion chromatography (SEC) after reconstitution with different volumes of WFI diluent (Lowe et al., 2018).

1.6 APPLICATIONS OF SPRAY-FREEZE-DRYING IN FOOD AND BIOPRODUCTS PROCESSING: AN OVERVIEW

Spray-freeze-drying outweighs the conventional drying techniques in terms of its versatile and high-value applications. The unique surface morphology, superior reconstitution properties and the improved stability and bioavailability of bioactive components resultant from SFD render it a suitable technique for the bulk production of powders in both the food and pharmaceutical sectors. Time and cost savings can be expected from an optimally designed spray-freeze-drying process. Furthermore, the ability of SFD to encapsulate poorly water soluble drugs (Leuenberger, 2002) and the unique aerodynamic qualities of the porous particles produced (D'Addio et al., 2012; Sweeney et al., 2005; Wang et al., 2012) has rendered it a particularly attractive process for producing particles for pulmonary delivery. As a result, spray-freeze-dried powders exhibit utility in drug delivery through the nasal and colonic routes and in the development of needle-free intradermal injection system (Wanning, Süverkrüp & Lamprecht, 2015). So far, volatile substances, thermally labile food bioactives and poorly water-soluble drugs are the candidate products that have been successfully produced by leveraging the technical advantages of spray-freeze-drying process. In the last few decades, nanotechnology and nanoparticle production have gained increasing attention in almost all the fields. In this context, SFD has been used for the production of food ingredient nanoparticles and nanocomposite micro-carriers that are inhalable and capable of colonic delivery.

Thus, it is evident that the applications of SFD can be grouped under four categories:

I. lyophilization of foods and biologicals;
II. encapsulation of food bioactives and active pharmaceutical ingredients;

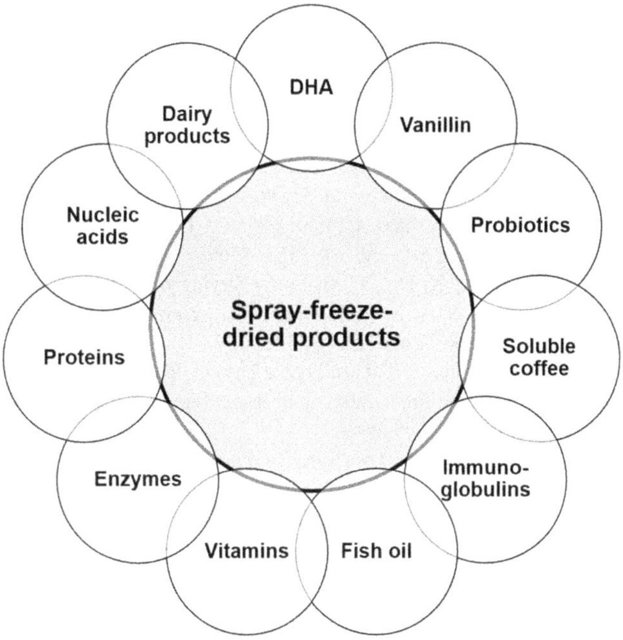

Figure 1.13 Foods and biologicals produced by the spray-freeze-drying process.

III. drug delivery (oral, pulmonary and transdermal) and
IV. nanoparticle production.

These widely varied applications of SFD in the production of food powders and biologicals (Figure 1.13) would be detailed with pertinent case studies in the forthcoming chapters of this book.

CONCLUSIONS

Thus, this introductory chapter provided a microscopic view of the spray-freeze-drying process right from its advent through its development across these years to its applications and advantages. The upcoming chapters will present a more detailed, macroscopic perspective of each aspect discussed in this chapter. The nuances of spray-freeze-drying process in terms of its constituent unit operations, characterization of product attributes and its applications in the manufacture of various food and biological products will be discussed. The concluding chapters will deliberate upon the recent developments and challenges involved in industrializing the spray-freeze-drying process, besides highlighting the future opportunities pertaining to this innovative drying technology.

REFERENCES

Adams, T. H., Beck, J. P., & Menson, R. C. (1982). *Novel Particulate Compositions*. US Patent 4,323,478.

Al-Hakim, K. (2004). *An Investigation of Spray-Freezing and Spray-Freeze-Dryings*. PhD Thesis. Loughborough University, United Kingdom.

Al-Hakim, K., Wigley, G., & Stapley, A. G. F. (2006). Phase Doppler Anemometry Studies of Spray Freezing. Trans IChemE, Part A, *Chemical Engineering Research and Design*, 84(A12), 1142–1151.

Anandharamakrishnan, C. (2008). *Experimental and Computational Fluid Dynamics Studies on Spray-freeze-drying and Spray-drying of proteins*. PhD Thesis. Loughborough University, United Kingdom.

Anandharamakrishnan, C., Gimbun, J., Stapley, A. G. F., & Rielly, C. D. (2008). Application of Computational Fluid Dynamics (CFD) Simulations to Spray-freeze Drying Operations. *Proceedings of the 16th International Drying Symposium (IDS 2008)*, pp. 537–545.

Anandharamakrishnan, C., Rielly, C. D., & Stapley, A. G. F. (2010). Spray-freeze-Drying of Whey Proteins at Sub-atmospheric Pressures. *Dairy Science and Technology*, 90, 321–334.

Benson, S. W., & Ellis, D. A. (1948). Surface Areas of Proteins I. Surface Areas and Heat of Adsorption. *Journal of the American Chemical Society*, 70, 3563–3569.

Bhandari, B. R., Patel, K. C., & Chen, X. D. (2008). Spray Drying of Food Materials Process and Product Characteristics. In: X.-D. Chen & A. S. Mujumdar (Eds.), *Drying Technologies in Food Processing*. Blackwell Publishing Ltd., United Kingdom, pp. 113–159.

Bhatta, S., Stevanovic Janezic, T., & Ratti, C. (2020). Freeze-Drying of Plant-based Foods. *Foods*, 9(1), 87, 22.

Boeh-Ocansey, O. (1983). A Study of the Freeze-drying of Some Liquid Foods in Vacuo and at Atmospheric-pressure. *Drying Technology*, 2(3), 389–405.

Briggs, A. R., & Maxwell, T. J. (1976). *Method of Preparation of Lyophilized Biological Products*. US Patent 3,932,943.

Buxton, I. R., & Peach, J. M. (1984). *Process and Apparatus for Freezing a Liquid Medium*. US Patent 4,470,202.

Caparino, A., Tang, J., Nindo, C. I., Sablani, S. S., Powers, J. R., & Fellman, J. K. (2012). Effect of Drying Methods on the Physical Properties and Microstructures of Mango (Philippine 'Carbao' var.) Powder. *Journal of Food Engineering*, 111, 135–148.

D'Addio, S. M., Chan, J. G. Y., Kwok, P. C. L., Prud'homme, R. K., & Chan, H.-K. (2012). Constant Size, Variable Density Aerosol Particles by Ultrasonic Spray Freeze Drying. *International Journal of Pharmaceutics*, 427, 185–191.

Dunn, D. B., Masavage, G. J., & Sauer, H. A. (1972). *Method of Freezing Solution Droplets and the like Using Immiscible Refrigerants of Differing Densities*. US Patent 3,653,222.

Flosdorf, E. W., & Mudd, S. (1935). Procedure and Apparatus for Preservation in "Lyophile" form of Serum and Other Biological Substances. *The Journal of Immunology*, 29, 389–425.

Fu, N., Huang, S., Xiao, J., & Chen, X. D. (2018). Producing Powders Containing Active Dry Probiotics with the Aid of Spray Drying. In: F. Toldrá (Ed.), *Advances in Food and Nutrition Research* (Volume 85). Elsevier Science, Netherlands, pp. 211–262.

Ganguly, A., Nail, S. L., & Alexeenko, A. A. (2012). Rarefied Gas Dynamics Aspects of Pharmaceutical Freeze-drying. *Vacuum*, 86(11), 1739–1747.

Geldart, D. (1986). *Gas Fluidization Technology*. John Wiley & Sons Ltd., Chichester, UK.

Hassen, A., Al-Kahtani, B. H. H., 1990. Spray Drying of Roselle (*Hibiscus sabdariffa* L.) Extract. *Journal of Food Science*, 55 (4), 1073–1076.

Heldman, D. R., & Hohner, G. A. (1974). An Analysis of Atmospheric Freeze Drying. *Journal of Food Science*, 39(1), 147–155.

Hundre, S. Y., Karthik, P., & Anandharamakrishnan, C. (2015). Effect of Whey Protein Isolate and β-cyclodextrin Wall Systems on Stability of Microencapsulated Vanillin by Spray–freeze Drying Method. *Food Chemistry*, 174, 16–24.

Ishwarya, S. P., & Anandharamakrishnan, C. (2015). Spray-Freeze-Drying Approach for Soluble Coffee Processing and Its Effect on Quality Characteristics. *Journal of Food Engineering*, 149, 171–180.

Ishwarya, S. P., Anandharamakrishnan, C., & Stapley, A. G. F. (2015). Spray-Freeze-drying: A Novel Process for the Drying of Foods and Bioproducts. *Trends in Food Science & Technology*, 41, 161–181.

Ishwarya, S. P., Anandharamakrishnan, C., & Stapley, A. G. F. (2017). Spray-freeze-drying of Dairy Products. In: C. Anandharamakrishnan (Ed.), *Handbook of Drying for Dairy Products*. John Wiley & Sons, Oxford, pp. 123–148.

Kaptay, G., 2012. On the Size and Shape Dependence of the Solubility of Nanoparticles in Solutions. *International Journal of Pharmaceutics*, 430, 253–257.

Karthik, P., & Anandharamakrishnan, C. (2013). Microencapsulation of Docosahexaenoic Acid by Spray-freeze-drying Method and Comparison of its Stability with Spray-drying and Freeze-drying Methods. *Food and Bioprocess Technology*, 6, 2780–2790.

Kešelj, K., Pavkov, I., Radojčin, M., & Stamenković, Z. (2017). Comparison of Energy Consumption in the Convective and Freeze Drying of Raspberries. *Journal on Processing and Energy in Agriculture*, 21(4), 192–196.

Leuenberger, H. (2002). Spray Freeze Drying the Process of Choice for Low Water Soluble Drugs? *Journal of Nanoparticle Research*, 4, 111–119.

Lewin, L. M., & Mateles, R. I. (1962). Freeze Drying without Vacuum, a Preliminary Investigation. *Food Technology*, 16(1), 94.

Liapis, A. I., & Bruttini, R. (2009). A Mathematical Model for the Spray Freeze Drying Process: The Drying of Frozen Particles in Trays and in Vials on Trays. *International Journal of Heat and Mass Transfer*, 52, 100–111.

Liu, Y., Zhao, Y., & Feng, X. (2008). Energy Analysis for a Freeze-drying Process. *Applied Thermal Engineering*, 28, 675–690.

Lopez-Quiroga, E., Antelo, L. T., & Alonso, A. A. (2012). Time Scale Modeling and Optimal Control of Freeze Drying. *Journal of Food Engineering*, 111, 655–666.

Lowe, D., Mehta, M., Govindan, G., & Gupta, K. (2018). Spray Freeze-Drying Technology: Enabling Flexibility of Supply Chain and Drug-Product Presentation for Biologics. *Bioprocess International*. Available from: https://bioprocessintl.com/manufacturing/supply-chain/spray-freeze-drying-technology-enabling-flexibility-of-supply-chain-and-drug-product-presentation-for-biologics/ (Accessed 6 September 2021).

Luy, B. (2016). Innovative Bulk Drying of Frozen Microspheres by Spray Freeze Drying. Proceedings of the AAPS National Biotechnology Conference, Boston, MA.

Malecki, G. J., Shinde, P., Morgan, A. I., & Farkas, D. F. (1970). Atmospheric Fluidized Bed Freeze Drying. *Food Technology*, 24, 601–603.

Mann, G. W. (1998). Surface Heat and Water Vapour Budgets over Antarctica. The Environment Centre, The University of Leeds, July 1998, Leeds – UK.

McHugh, T. (2018). Freeze-Drying Fundamentals. *IFT Food Technology Magazine*. Available from: https://www.ift.org/news-and-publications/food-technology-magazine/issues/2018/february/columns/processing-freeze-drying-foods (Accessed 6 September 2021).

Meryman, H. T. (1959). Sublimation Freeze-drying without Vacuum. *Science*, 130(3376), 628–629.

Mumenthaler, M., & Leuenberger, H. (1991). Atmospheric Spray Freeze-drying: A Suitable Alternative in Freeze Drying Technology. *International Journal of Pharmaceutics*, 72, 97–110.

Noyes, R. (1968). *Freeze-drying Foods and Biologicals*. Noyes Development Corporation.

Quast, D. G., & Karel, M. (1968). Dry Layer Permeability and Freeze-drying Rates in Concentrated Fluid Systems. *Journal of Food Science*, 33(2), 170–175.

Ratti, C. (2008). Freeze and Vacuum Drying of Foods. In: X. D. Chen & A. S. Mujumdar (Eds.), *Drying Technologies in Food Processing*. Blackwell Publishing, Singapore, pp. 225–251.

Rogers, T. L., Hu, J., Hu, Z., Johnston, K. P., & Williams, R. O. (2002). A Novel Particle Engineering Technology: Spray-freezing into Liquid. *International Journal of Pharmaceutics*, 242, 93–100.

Saguy, S. I., Marabi, A., & Wallach, R., 2005. Liquid imbibitions during rehydration of dry porous foods. *Innovative Food Science and Emerging Technologies*, 6(1), 37–43.

Sauer, H. A. (1969). *Method and Apparatus for Freeze-drying*. US Patent 3,484,946.

Sebastião, B. I., Robinson, T. D., & Alexeenko, A. (2017). Atmospheric Spray Freeze-drying: Numerical Modeling and Comparison with Experimental Measurements. *Journal of Pharmaceutical Sciences*, 106(1), 183–192.

Shivanand, & Mukhopadhayaya, S. (2017). A Review on Lyophilization: A Technique to Improve Stability of Hygroscopic, Thermolabile substances. *PharmaTutor*, 5(11), 28–39.

Stapley, A. G. F., & Rielly, C. D. (2009). Developments in Fluidised Bed Freeze Drying. *Food Science and Technology*, 23(2), 19–21.

Stapley, A. (2008). Freeze Drying. In: J. A. Evans (Ed.), *Frozen Food Science and Technology*. Blackwell Publishing Ltd., Oxford, UK, pp. 248–275.

Sweeney, L. G., Wang, Z., Loebenberg, R., Wong, J. P., Lange, C. F., & Finlay, W. H. (2005). Spray-freeze-dried Liposomal Ciprofloxacin Powder for Inhaled Aerosol Drug Delivery. *International Journal of Pharmaceutics*, 305, 180–185.

Thuse, E., Jose, S., Ginnette, L. F., & Derby, R. R. (1968). *Spray Freeze Drying System*. US Patent 3362835A.

Wagner, W., Saul, A., & Pruss, A. (1994). International Equations for the Pressure Along the Melting and Along the Sublimation Curve of Ordinary Water Substance. *Journal of Physical and Chemical Reference Data*, 23(3), 515–527.

Wang, Y., Kho, K., Cheow, W. S., & Hadinoto, K. (2012). A Comparison Between Spray Drying and Spray-freeze Drying for Dry Powder Inhaler Formulation of Drug-loaded Lipid Polymer Hybrid Nanoparticles. *International Journal of Pharmaceutics*, 424, 98–106.

Wanning, S., Süverkrüp, R., & Lamprecht, A. (2015). Pharmaceutical Spray Freeze Drying. *International Journal of Pharmaceutics*, 488, 136–153.

Woo, M. (2017). Recent Advances in the Drying of Dairy Products. In: C. Anandharamakrishnan (Ed.), *Handbook of Drying for Dairy Products*. John Wiley & Sons, Chichester, UK, pp. 249–268.

Woodward, H. T. (1963). Freeze Drying without Vacuum. *Food Engineering*, 35, 96–97.

CHAPTER 2

Understanding the Nuances of Spray-freezing Process

Atomization and spray-freezing are often intricate to be distinguished as two distinct steps due to their extremely short time-scales. Thus, in this chapter, these two stages of the spray-freeze-drying process would be dealt as an integrated operation. A precise definition for spray-freezing was given by Windhab (1999) as, *'the solidification of a liquid by atomization into a relatively cold atmosphere'*. Likewise, Al-Hakim, Wigley, & Stapley (2006) described spray-freezing as, *'the solidification of an atomized liquid by contacting it with an inert cooling medium below the freezing point of the liquid'*. Certainly, spray-freezing is a critical step of the SFD process, due to its control over the number, shape, size, and size distribution of droplets and the location of ice crystals within the microstructure of frozen droplets. These crystals sublime later to leave behind pores that constitute the porous microstructure of SFD-particles. Moreover, due to their smaller Biot numbers for heat transfer, tiny droplets provide a more uniform temperature field for freezing to facilitate homogeneous nucleation and microstructure development (MacLeod et al., 2006).

Apart from particle size and porosity, the spray-freezing step also influences the kinetics of primary sublimation during the subsequent freeze-drying process. Generally, during atomization, 1 m^3 of bulk feed liquid is converted to around 2×10^{12} droplets having a uniform size of about 100 μm. This generates a total surface area of more than 60000 m^2 (Masters, 2002). The fine size of droplets enhances the sublimation rate (Pham, 1986) and surface mass-transfer coefficient during the freezing and freeze drying stages, respectively. Sublimation is faster when there are few interconnected pores that are large compared to multiple isolated pores that are small (Al-Hakim, 2004).

For any unit operation, it is imperative to understand its phenomenology and identify the parameters that are associated with the characteristics of final product (Hindmarsh, Russell, & Chen, 2007). Accordingly, the spray-freezing process is controlled by several factors such as the atomizer type, atomization energy, nozzle geometry, and mode of freezing. These parameters influence the surface composition of final powders, which determines the type of component that preferentially migrates to their surface during particle formation. In turn, the surface composition is related to functional properties of powders such as dispersibility, flowability and stickiness (Kim, Chen, & Pearce, 2003).

In the above background, this chapter aims at explaining the theory, classification, instrumentation, and influential process parameters of spray-freezing. Understanding the nuances of spray-freezing would aid in science-based standardization of the SFD process for any product, rather than adopting a trial-and-error approach that is often tedious, time-consuming and expensive.

2.1 THE MECHANISM AND CLASSIFICATION OF SPRAY-FREEZING

Spray-freezing is a particle engineering approach, which encompasses the spray operation/ atomization and freezing. The spray operation or atomization is central to the spray-freeze-drying

DOI: 10.1201/9781003019312-2

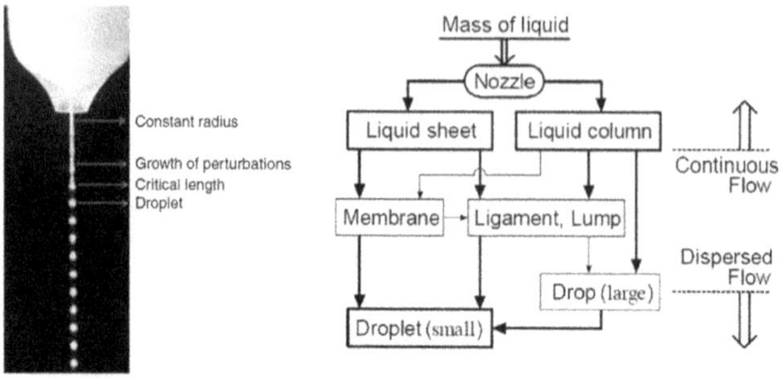

Figure 2.1 Mechanism of spray operation: (a) Plateau's mechanism (Wu et al., 2014); (b) Rayleigh's liquid jet theory (Yao et al., 2014).

process, which is based on the liquid disintegration phenomenon (Anandharamakrishnan & Ishwarya, 2015). It involves the fragmentation of a liquid mass into a fine mist of droplets, which collectively constitute the *'spray'* (Yao et al., 2014). Joseph Plateau (1873) and Lord Rayleigh (1878) were the pioneers to elucidate the mechanism of liquid instability that governs atomization. As stated by Plateau, liquid jet having a constant radius initially, would fall vertically under gravity. Subsequently, when the liquid length extends and attains a critical value at which the surface tension of the liquid drops, the jet loses its cylindrical shape and disintegrates into a mist of droplets (Figure 2.1[a]). Thus, Plateau attributed the liquid instability solely to the change in surface tension of the liquid. But, according to Rayleigh's *'liquid jet theory'*, instability and disintegration are caused by the aerodynamic forces. The feed liquid flows through the nozzle and the orifice edge before evolving into a liquid sheet or liquid column. In the beginning, the liquid sheet breaks up into membrane or elongated ligaments that are more or less cylindrical, and later divides into droplets (Figure 2.1[b]). Though Rayleigh's theory included the influence of surface tension and inertial forces, it did not consider the effects of viscosity, atomization gas and the medium to which the liquid is released.

In the above background, the explanation provided by Ohnesorge (1936) is known for its clarity. The relationship put forth by him comprised all the relevant factors responsible for the spray formation, which is numerically expressed by the dimensionless Ohnesorge number (*Oh*) (Eq. 2.1).

$$Oh = \frac{\sqrt{We}}{Re} = \frac{\mu}{\sqrt{\rho\sigma L}} = \frac{Viscous\ \ forces}{\sqrt{(inertia\ \times\ surface\ \ tension)}} \tag{2.1}$$

In the above equation, *We*: Weber number; *Re*: Reynolds number; μ, ρ *and* σ: viscosity, density and surface tension of the feed droplet, respectively; *L*: characteristic dimension of the feed droplet (i.e. volume per unit area). According to Ohnesorge's relationship, the tendency of a liquid jet to disintegrate depends on its viscosity, density, surface tension and jet size. The feed liquid disintegrates at the tip of the atomizer due to the turbulence in the emanating liquid jet and the action of air forces. Nevertheless, the disintegration is resisted by the viscous and surface tension forces in the liquid. Once the droplet is airborne, the realignment of shear stresses within the liquid causes the droplet fission and atomization.

The spray operation functions as a tool for size control, acceleration of heat and mass transfer and shortening of drying time. It forms fine droplets of the feed material that are rapidly frozen as smaller droplets have considerably larger surface area-to-volume ratio and thereby offer minimal resistance to heat transfer (Al-Hakim, 2004; Demarco & Renzi, 2018). The significance of spray-freezing step is apparent due to its role in determining the quality of final particles, especially, their

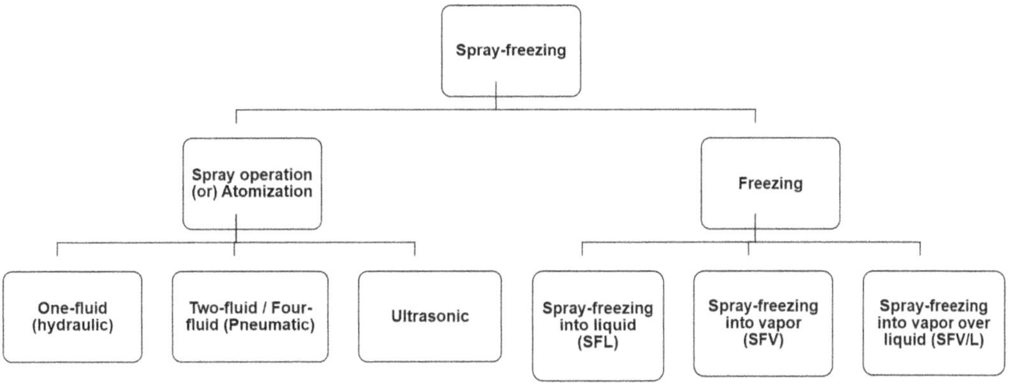

Figure 2.2 Classification of spray-freezing process based on its constituent operations.

unique porous microstructure. The events of spray operation, i.e., disintegration of liquid into droplets is controlled by the nozzle type, size, and geometry, viscosity and surface tension of the feed liquid, and the properties of the medium into which the liquid stream is discharged (Chin & Lefebvre, 1993). The above factors form the basis for the classification of spray operation (Figure 2.2). The classification based on nozzle type is governed by the type of energy (pressure energy, kinetic energy or ultrasonic energy) utilized to shear a fluid into small drops. The droplet size and size distribution vary with the nozzle type. Similarly, the classification based on freezing mode depends on the medium into which the liquid stream is discharged after atomization: cryogenic liquid, cold gas or the vapor above a cryogenic liquid. Notably, the freezing mode controls the ice morphology, the resistance to water vapor removal during sublimation and also the quality of final product (Pikal et al., 2002).

The forthcoming sections would discuss in detail about the different types of atomizers and freezing modes. The rationale of selection, merits and demerits of each atomizer and freezing mode would be elaborated.

2.2 CLASSIFICATION OF ATOMIZERS USED IN SPRAY-FREEZING

2.2.1 One-fluid or hydraulic nozzle

In hydraulic nozzles, conversion of pressure energy into kinetic energy governs the spray formation. Pressure energy can be defined as the energy in a fluid due to the applied pressure (force per unit area). Within the hydraulic nozzle (Figure 2.3), the feed liquid is discharged under pressure through an orifice using the energy provided by a pressurized gas (usually nitrogen or air). The pressure energy required for atomization is provided by a feed pump, typically a piston-type positive displacement or a diaphragm pump. Pump selection depends on the liquid viscosity. The operating pressure of these pumps is in the range of 0.3 to 6.9 MPa (Al-Hakim, 2004; Perry et al., 1997). Energized by gas pressure, the feed liquid emerges from the orifice in the form of a high-speed film and instantly disintegrates into a mist of droplets (Figure 2.3). The size of resultant droplets is directly proportional to the flow rate and viscosity of the feed and inversely related to the spraying pressure, which is usually in the range of 250–10,000 psi (Adali et al., 2020; Anandharamakrishnan & Ishwarya, 2015). While feed solution with low viscosities (1.7–3.9 cP) lead to particles of diameter less than 100 μm, high feed viscosity (>10 cP) do not produce particles of similar size (Barron et al., 2003).

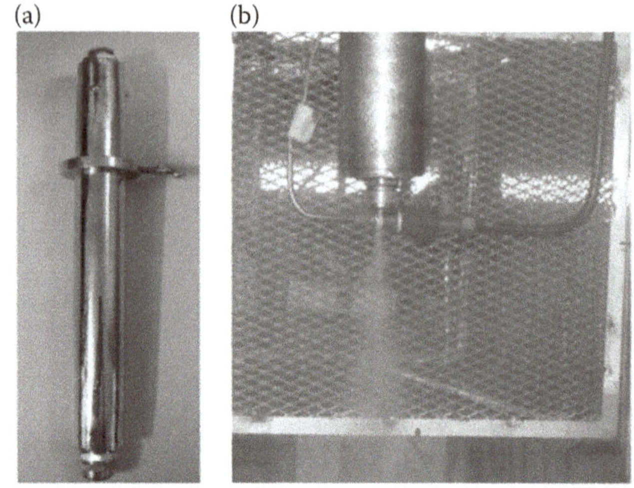

Figure 2.3 (a) Hydraulic nozzle (Anandharamakrishnan & Ishwarya, 2015); (b) Spray emerging from a hydraulic nozzle (Ishwarya, Anandharamakrishnan, & Stapley, 2017).

2.2.2 Two-fluid nozzle atomizer

In two-fluid nozzles, the feed liquid and compressed gas (air or nitrogen) are co-injected into the nozzle inlet and pressurized through the same orifice (Figure 2.4). Spray formation is driven by the energy of compressed gas flow, which interacts with the feed liquid, generates a shear field and results in a polydisperse size distribution (Adali et al., 2020). The interaction between feed liquid and compressed gas can take place either inside or outside the nozzle body, referred to as the internal and external mixing, respectively (Figure 2.5) (Marshall, 1954; Masters, 1991; Perry et al., 1997). The two fluid nozzles form a narrow spray angle of 17-22° and produce fine droplet sizes with a narrow particle size distribution. Moreover, any type of feed pump (Figure 2.4) can be used to transfer the feed liquid to the twin-fluid nozzle; but, it draws the feed by the negative pressure caused by the gas

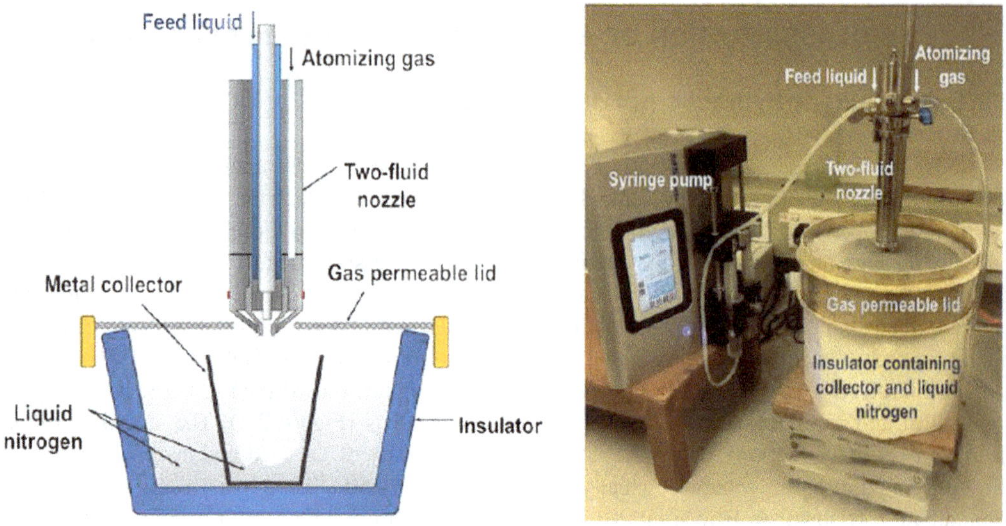

Figure 2.4 Construction of a two-fluid nozzle and its application in spray-freeze-drying process (Liang et al., 2018).

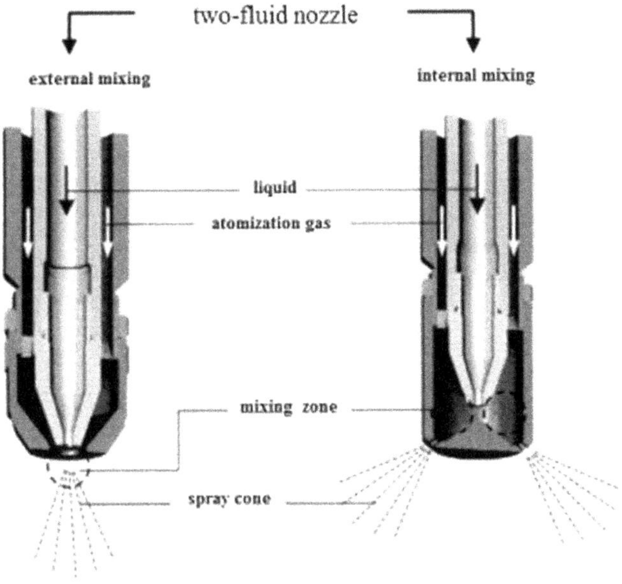

Figure 2.5 Designs of two-fluid nozzle: (a) external mixing nozzle; (b) internal mixing nozzle (Tabeei, Samimi, & Mohebbi-Kalhori, 2020).

flow. Usually, a tall form chamber is used in combination with the pneumatic nozzles. A twin-fluid atomizer is preferred over other nozzle types for spray-freezing due to the absence of moving parts and the resultant ease of heating it to prevent freezing of feed within the nozzle. It requires a high gas-to-feed liquid ratio of about 2:1 to cause sufficient liquid disintegration to generate droplets. In this context, nitrogen is the ideal atomization gas in order to maintain a constant gas (say, nitrogen) composition inside the spraying chamber (Al-Hakim, 2004).

As the two-fluid nozzles have a single liquid supply line, its use for encapsulation applications is occasionally challenged by the requirement for a common solvent that solubilizes both the active pharmaceutical ingredient (API) and the wall material (Adali et al., 2020). The above limitation can be overcome by transforming the configuration of two-fluid nozzles to three-fluid or four-fluid nozzles, determined by the number of gas passages. Three-fluid nozzle is a configuration that permits the supply of two independent feeds in conjunction with one gas channel. But, use of three-fluid nozzle for SFD application is yet to be explored. On the other hand, four-fluid nozzle – a recent breakthrough with respect to atomization in SFD, can handle two independent liquid feed streams atomized by two different air streams. These streams impinge on each other at the nozzle tip (a focal spot) to generate a thin film of liquid with a high-velocity gas flow. The liquid disintegration and generation of a spray of small droplets occur as a result of the shock waves generated at the nozzle tip (Cal & Sollohub, 2010) (Figure 2.6).

Studies that used four-fluid nozzles for SFD demonstrated the fine and porous structure and large specific surface area of the final particles. Four-fluid nozzles lead to particles with reasonably narrower size distribution than the two-fluid counterpart (Ozeki et al., 2005). Also, they are suitable for poorly water-soluble APIs when used in conjuction with either aqueous or organic solvents (Niwa, Shimabara, Kondo, & Danjo, 2009; Niwa, Shimabara, & Danjo, 2010). However, the porosity and specific surface area of particles from organic solvent-based feed are lower than that obtained from an aqueous feed system (Niwa, Shimabara, & Danjo, 2010). Not limited to the above, four-fluid nozzles are capable of producing polymeric particles with prolonged or controlled release (Cal & Sollohub, 2010; Niwa et al., 2009; Niwa et al., 2010).

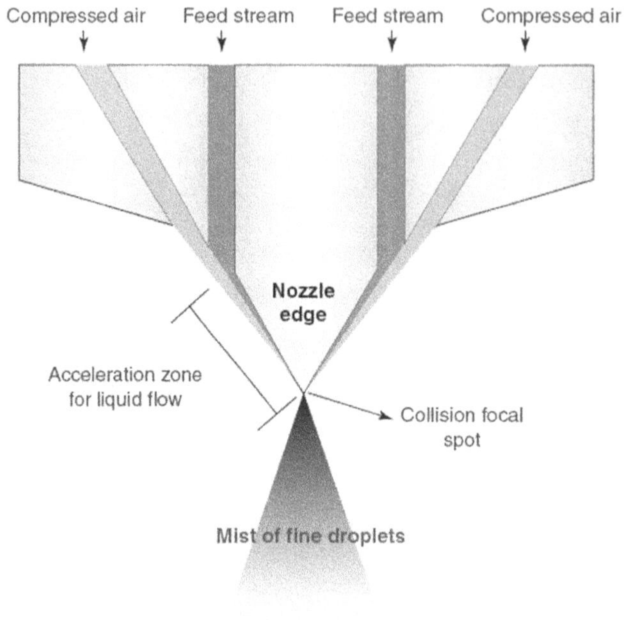

Figure 2.6 Four-fluid nozzle (Adapted from Mizoe et al., 2008).

Generally, with pneumatic nozzles, the droplet size is directly related to the feed flow rate and fluid cap diameter and inversely proportional to the atomization gas flow rate and pressure (Anandharamakrishnan & Ishwarya, 2015; Costantino et al., 2000). Nevertheless, a disadvantage associated with both hydraulic and twin-fluid nozzles is the polydisperse droplet size distribution, which affects the freezing kinetics in the next step. This led to the advent of ultrasonic nozzles and monodisperse droplet generators for spray-freezing application.

2.2.3 Ultrasonic atomizer

Ultrasonic atomization of a feed liquid is achieved by vibrating it through cavitation at the tip of an ultrasonic nozzle. It is driven by the conversion of high-frequency electric signal into mechanical energy. This energy conversion is accomplished using a pair of disc-shaped piezoelectric transducers that are positioned between a mechanical amplifying element and a support element and housed inside a syringe plunger. The resultant mechanical expansion and contraction of the transducers cause ultrasonic vibrations of high frequency (Adali et al., 2020), in the magnitude of kilohertz (kHz). These vibrations are sent down the nozzle's titanium horn that vibrates ultrasonically at the nozzle's atomizing tip. The vibrational energy causes the feed liquid moving down the nozzle's center to form capillary waves. The liquid attains its critical wave amplitude on reaching the atomizing surface. Then, the ultrasonic energy concentrated at the atomizing surface breaks it into a spray of fine droplets with an extremely narrow droplet size distribution (Figure 2.7) (D'Addio et al., 2012). Thus, similar to the pressure energy of pneumatic and hydraulic nozzles, the nozzle vibration frequency causes the droplet fission in ultrasonic atomizers. Besides, the amplitude of vibration and the area of vibrating surface also play a role in ultrasonic atomization (Lixin et al., 2004).

In contrast to the pressure nozzles that convey a high initial velocity to the droplets to result in a wider droplet size distribution, the ultrasonic nozzles impart velocity that is one to two orders of magnitude lower than the former. The low velocity of droplets results in a homogeneous droplet

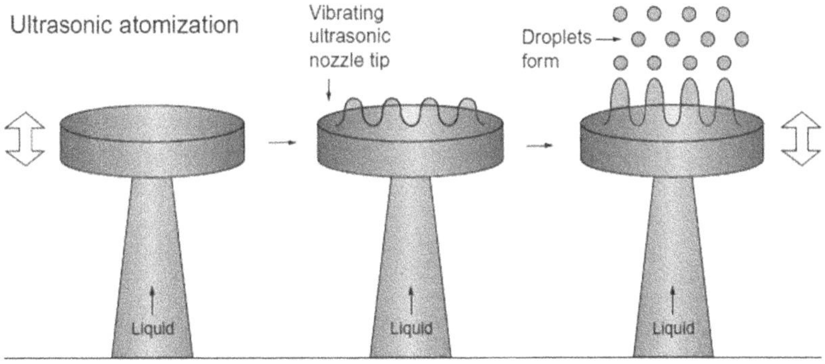

Figure 2.7 Schematic diagram of ultrasonic atomization (Gogate, 2015).

size distribution (Lixin et al., 2004). The droplet size bears an inverse relationship with the ultrasonic frequency. Typically, a 60 kHz spraying nozzle generates droplets of size in the range of 20-80 μm (Maa, Prestrelski, & Burkoth, 2003). Thus, compared to pressure and pneumatic nozzles, ultrasonic atomizers have a greater control over the geometric size of final particles by tuning their vibration frequency to regulate the droplet size. Compared to twin-fluid nozzles, the droplets generated by an ultrasonic atomizer are comparatively easy to be captured effectively in a cryogenic liquid that is placed at a short distance beneath the nozzle (D'Addio et al., 2012).

2.2.4 Monodisperse droplet generator

A monodisperse droplet generator (MDG) overcomes the limitations of hydraulic and twin-fluid nozzles with respect to their wider droplet size distribution. It produces droplets with a well-defined trajectory and a narrow size distribution (Figure 2.8). The above aspect is relevant from the perspectives of available space and distance in the spray-freezing chamber. Nozzles that lead to a narrow coverage with reasonable trajectory inside the chamber and a significant extent of atomization uniformity are advantageous. Here, the droplet formation is controlled by shearing the feed liquid with a second fluid (Xu et al., 2006) or using electrostatic repulsive force (Rosell-Llompart & Fernández de la Mora, 1994). Alternatively, direct pressurization of feed liquid is also carried out (Ishwarya, Anandharamakrishnan, & Stapley, 2015; Rogers et al., 2008). As the liquid is pumped through a narrow orifice (100–150 μm), a frequency generator imparts a pressure pulse (1500–2000 Hz) to the feed reservoir. The sinusoidal perturbations on the liquid jet surface, whose frequency is controlled by the frequency of pressure pulse, grow till they ultimately disrupt the jet into droplets. Consequently, in accordance with the Rayleigh's theory (1945), the liquid jet fragments into individual droplets before exiting through the orifice. The size of droplets emerging from a monodisperse droplet generator can be regulated by changing the frequency, orifice size and feed characteristics. But, the thin liquid threads that connected the drops before breakup can either coalesce with the main droplet or form smaller satellite droplets. The above can be prevented by altering the frequency to such values that would stabilize the droplet formation (Rogers et al., 2008).

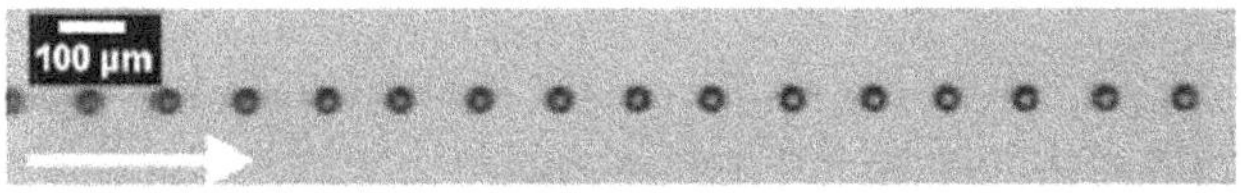

Figure 2.8 Typical controlled jet breakup into droplets in a monodisperse droplet generator (Frommhold et al., 2014).

Generally, the low-temperature operation of spray-freezing can lead to solidification of feed within the nozzle and its subsequent blockage. In addition, it may also cause spray turbulence inside the spraying chamber and affect the physical properties of feed liquid, which can impede the atomization level of sprays (Al-Hakim, 2004). The advantages, disadvantages and other salient features of different atomizers used in the spray-freeze-drying process are compiled in Table 2.1.

2.3 PARAMETERS OF ATOMIZATION

Atomization energy and the viscosity, surface tension and flow rate (flow pressure) of feed are the four major parameters of atomization that influence the droplet size and eventually the final particle size. With the same type of atomizer and feed liquid, an increase in atomization pressure reduces the droplet size (Masters, 1991). On the other hand, increasing the feed flow rate at constant atomization pressure, increases the volume of liquid to be atomized and thereby results in larger droplet size. Similarly, higher feed viscosity increases the droplet size, as the applied atomization energy has to combat the larger viscous forces to attain smaller droplet sizes. A feed liquid having high surface tension poses challenges during atomization, as the atomizer must overcome the surface tension of feed liquid to create more surface area. Therefore, emulsification of feed liquid with a suitable emulsifier followed by homogenization is recommended before spray-freezing (Ishwarya, Anandharamakrishnan, & Stapley, 2015).

Specifically, in two-fluid nozzles, apart from the above parameters, the final particle size is directly proportional to the fluid cap inner diameter and inversely related to the atomization gas (air or nitrogen) pressure or flow rate. In addition, the mass flow ratio, i.e., the ratio between atomization gas and liquid feed, is indicative of particle size, as increasing this ratio decreases the particle size (Costantino et al., 2000). Likewise, in a hydraulic nozzle, high-viscous feed formulations impede atomization at low feed flow rate and produce droplets of size greater than 100 μm. This is because, high viscosity hinders the destabilization phenomenon in the emanating jet of feed liquid and postpones the onset of disintegration. Consequently, atomization occurs in the lower velocity regions. Nevertheless, droplet size less than 100 μm is achieved at low viscosity and increased flow rate of the feed liquid due to intense atomization (Barron et al., 2003). The influence of various atomization parameters on the final particle size are summarized in Table 2.2.

Along with the spray nozzles, deflector jets are used occasionally in the SFD system, which are placed above the cooling jacket of the spray-freezing chamber. These jets instigate gas flow and promote dispersion of droplets to reduce the negative impact of droplet-droplet interactions on their heat release. When the gas flow rate is not regulated, the jets can cause a turbulent flow that leads to droplet-droplet and droplet-wall collisions and impact the powder size distribution (Sebastião et al., 2019). Yet another phenomenon that occurs during atomization is the freezing of feed within the nozzle due to the low temperature used in spray freezing. Generally, nozzle heaters (Figure 2.9) or plastic nozzles are used to combat the above event (Ishwarya, Anandharamakrishnan, & Stapley, 2015).

2.4 FREEZING OF SPRAYED DROPLETS: PRINCIPLE AND CLASSIFICATION

Freezing operation of the spray-freeze-drying process is a rapid phenomenon unlike its prolonged duration in a conventional freeze-drying process for bulk liquids. The freezing step is responsible for the unique porous microstructure of final product, attributed to the high rate at which it cools the atomized droplets (Gwie et al., 2006). Spray-freezing into either cold gas or cryogenic liquid results in interesting microstructures from both solutions (Hindmarsh, Russell, & Chen, 2004; MacLeod et al., 2006) and emulsions (Hindmarsh et al., 2004). The different phases of spray-freezing can be explained based on the inferences from studies on single-droplet freezing

Table 2.1 Lists of different types of atomizers used in SFD (Modified from Adali et al., 2020)

Atomizer type	Atomization energy	Type of spray	Droplet size (μm)	Advantages	Limitations
Hydraulic (Pressure)	Pressure energy converted to kinetic energy	Coarse and less homogeneous	120–250	• Reasonably narrow spray angle that helps reducing the capital cost of spray chamber. • Product powder with higher density with excellent flow properties. • Depending on the desired final product, it produces particles of a relatively large size. • Atomizer duplication: Pressure nozzles can be integrated in multiple nozzle arrangements to obtain an increased amount of flow rate and particle size flexibility. • Pressure nozzles result in particles with less occluded air when compared to twin-fluid atomizers. As a result, the powdered product is of higher density, with good flow characteristics.	• Risk of blockage at low feeding pressures. • Production of spray with a wide droplet size distribution. • The broadness of droplet size distribution increases the energy requirement and the particle size distribution of final product. • Requires routine changing of its internal parts, due to abrasion on the nozzle design caused by feed pressurization through the orifice.
Pneumatic (Two/ Four-fluid)	Kinetic energy	Coarse and less homogeneous	5–100	• Suitable to handle highly viscous solutions. • Better control over particle size than in the hydraulic nozzle.	• High consumption of compressed gas which increases the operational cost. • Introduction of a significant quantity of warm gas into the spray could affect the cooling rates during spray-freezing. • The temperature gradient caused by gas affects the freezing time of particles. • Requires periodic changing of the air and liquid caps.
Ultrasonic	Electrical energy converted to mechanical energy	Fine with narrow size distribution	18–68	• More uniform droplet size distribution • The larger outlet for feed droplet and the absence of any moving parts prevent clogging of the nozzle • Easy maintenance and operation	• Only relevant for low viscosity Newtonian fluids. • A higher intensity of vibrational energy utilized for ultrasonic atomization or the recirculation of feed solution may heat the solution, and lead to degradation of heat-labile drugs.
Monodisperse droplet generator	Shear force, electrostatic repulsive force or pressure energy	Fine with narrow size distribution	20–200	• Relatively easier process optimization due to higher reproducibility • Effective process control to optimal product quality • Spherical particles with a narrow particle size distribution • Better prediction of process parameters and energy requirements • High particle collection efficiency from the product chamber due to uniform characteristics of the dried particles.	• Reducing the orifice diameter of the nozzle to produce droplets of smaller size increases the risk of clogging during the run and lead to poor process economy. • The low productivity and high operating costs constrain their lab-scale applications.

Table 2.2 Atomization parameters and their influence on particle size (Modified from Ishwarya, Anandharamakrishnan, & Stapley, 2015)

Product	Cryogen	Nozzle type	Feed rate	Air flow rate/air pressure	Particle size	Reference
Whey protein	Nitrogen gas at inlet gas temperature, i. −10ºC ii. −15ºC iii. −30ºC	Hydraulic nozzle	0.0125 $kg.s^{-1}$	8 bar	i. 480±53 µm ii. 393±75 µm iii. 412±4 µm	Anandharamakrishnan et al. (2010)
Insulin	LN2	PEEK* i. 63.5µm I.D, 10 cm in length ii. 127 µm I.D, 15 cm in length	NA	5000 psi (34.5 Mpa)	3 µm	Rogers et al. (2002)
Inulin stabilized influenza subunit vaccine	LN2	Two fluid Nozzle	NA	700 In/h	10–11 µm	Amorij et al. (2007)
Darbepoetin Alfa	Frozen ethanol with a LN2 overlay	Ultrasonic Atomization probe (6 mm diameter; 20 kHz)	0.5 mL/min	NA	29±1 µm	Burke et al. (2004)
Polymeric nanoparticle aggregates	LN2	Two fluid Nozzle	0.24 L/h	240 L/h	1500–2500 nm	Cheow et al. (2011)
Kanamycin 2%,10%,15%	LN2	Two-fluid nozzle	25 mL/min	100k Pa	15.1 µm, 13.5 µm, 14.9 µm	Her et al. (2010)
Bovine Serum Albumin (BSA), Mannitol, Lysozyme at feed concentration: 40 mg/cm3	LN2	Ultrasonic Atomization probe (40 kHz)	0.5 mL/min	NA	30.2 µm (BSA) 17.2±3.6 µm (Mannitol) 34.8 µm (Lysozyme)	D'Addio et al. (2012)
Protein BSA (5 mg/mL) Lysozyme (5 mg/mL) Lysozyme (50 mg/mL) Lysozyme (100 mg/mL)	LN2	PEEK* nozzle	10 mL/min	17.2 MPa	0.05–1 µm 0.05–1 µm 4.0–12 µm 6.0–40 µm	Engstrom et al. (2007)

Protein Lysozyme (50 mg/mL) Lysozyme (100 mg/mL)	Isopentane	PEEK nozzle	10 mL/min	17.2 MPa	0.05–1 μm 0.05–1 μm	Engstrom et al. (2007)
i. Excipient free anti IgE antibody ii. Anti-IgE antibody: Trehalose 60:40	LN2 LN2	Two-fluid nozzle Ultrasonic nozzle Two fluid nozzle-do-	15 mL/min 5 mL/min 15 mL/min 15 mL/min	1050 L/h NA 600 L/h 1050 L/h	7.0 μm 5.9 μm 32 μm 19 μm 5.9 μm	Maa et al. (1999)
Trehalose	LN2	Ultrasonic nozzle (120 kHz)	3 mL/min	NA	20–90 μm	Sonner et al. (2002)
Darbepoetin Alfa	LN2	Ultrasonic nozzle 20 kHz	0.5 mL/min	NA	29 ± 1 μm	Nguyen et al. (2004)
Liposomal Ciprofloxacin,	LN2	Two fluid nozzle	–	–	2.8 μm	Sweeney et al. (2005)
Influenza subunit vaccine powder	LN_2	Two-fluid nozzle, 0.5 mm orifice	5 mL/ min	700 L/h	23% of SFD powder particles between 1–5 μm	Saluja et al. (2010)
Danazol Carbamazepine	LN_2	PEEK* nozzle, 63.5 μm I.D,	11 mL/min	4000 PSI	7.1 μm 7.11 μm	Hu et al. (2002)
Bovine serum albumin	LN_2	PEEK* nozzle, 63.5 μm I.D	12 mL/min	5000 PSI	<500 nm	Yu et al. (2004)
Ovalbumin (white albumin of chicken egg)	LN_2	Two-fluid nozzle	25 mL/min.	Air pressure of 100 kPa	11.8 μm	Yeom & Song (2010)
Encapsulation of *Lactobacillus paracasei*	LN_2	Pneumatic nozzle	a. 0.15 b. 0.3 c. 0.8 mL/min	2.12 3.08 4,52 L/min	400–1800 μm	Semyonov et al. (2010)

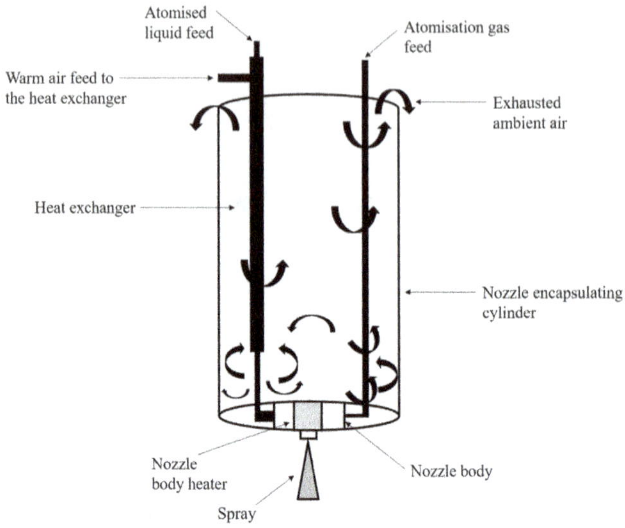

Figure 2.9 Heating arrangement for a twin-fluid nozzle (Redrawn from Al-Hakim, 2004).

(Hindmarsh, Russell, & Chen, 2003; MacLeod et al., 2006) and a plot that depicts the evolution of droplet temperature and solute concentration with time (Figure 2.10). In the plot shown in Figure 2.10, the relationship between droplet temperature and time is referred to as the *'freezing curve'*. A freezing curve provides insight to the amount of ice at any given temperature, which is a function of the freezing point depression that in turn varies with the solute concentration (Goff, *n.d.*). It explains the following four phases.

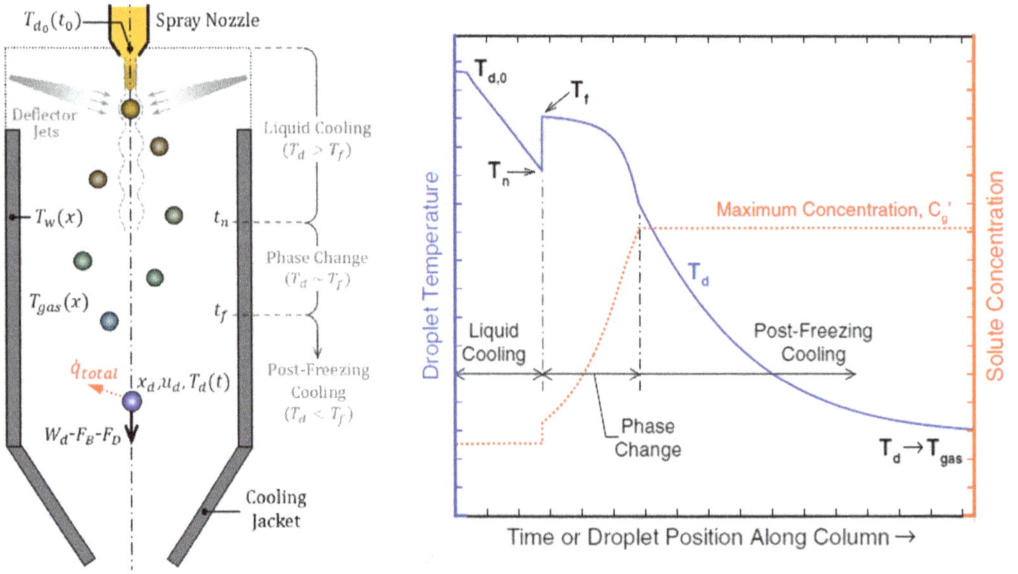

Figure 2.10 Schematic diagram of the typical temperature and solute concentration evolutions during the spray-freezing of a single droplet (Sebastião et al., 2019).

1. ***Initial cooling (or) supercooling to below the normal freezing temperature:*** The liquid droplets are sprayed at a temperature of T_{d_0}. Subsequently, the heat transfer from droplet into dry air via convection, radiation, and evaporation cools these droplets to the nucleation temperature T_n (Figure 2.10).
2. ***Nucleation and recalescence:*** After being cooled to the nucleation temperature, the water molecules in feed solution aggregate spontaneously to form a template. Subsequently, the other water molecules also adhere to the template and form an ice crystal. Thus, nucleation marks the beginning of the freezing phase of the feed solution. With time, as the temperature falls below the equilibrium freezing point, the likelihood of nucleation increases on account of the thermodynamic driving force. Later, it decreases due to the kinetic limitations caused by high viscosity (Pikal, Rambhatla, & Ramot, 2002). Also, nucleation depends on the distribution of nuclei or nucleation promoters, which instigates subsequent events such as ice crystal growth and recalescence (a temporary rise in temperature). This is because, the mass transfer is significantly suppressed with the build-up of a thin layer of less-volatile solutes at the droplet surface.
3. ***Freezing:*** After the above events, the temperature of super-cooled liquid abruptly rises to the equilibrium freezing temperature, T_f (Figure 2.10). Phase change or solidification happens due to the release of latent heat. After ice formation, the solute diffuses toward the liquid phase till the *'glass transition point'*, which is the stage of its maximum freeze concentration (C_g).
4. ***Final cooling of frozen droplet to gas temperature:*** After the glass transition point, the droplet enters the glassy state, which is then cooled to the gas temperature (T_{gas}) inside the freezing chamber (Figure 2.10) (MacLeod et al., 2006; Sebastião et al., 2019).

The important parameters of spray-freezing that can be obtained from the freezing curve are the cooling rate, nucleation temperature (indicated by a point of inflection in the freezing curve), freezing temperature, freezing rate, and the length of recalescence stage. All the above parameters, along with the feed composition or solute content play an important role in determining the microstructure of frozen droplets (MacLeod et al., 2006). This effect is further carried over to the microstructure of freeze-dried particles after moisture removal (Ishwarya et al., 2015). Due to the presence of fewer nucleation sites, which is proportional to the cube of droplet diameter (d^3), smaller droplets formed at a lower cooling rate would demand a higher degree of supercooling before nucleation. Therefore, the slow-cooled small droplets lead to crystals with larger mean hydraulic radius and polydisperse size distribution, than those formed at a faster cooling rate. Thus, the density of nucleation sites increases with rise in the melt cooling rate (Hindmarsh, 2003). In this manner, freezing rate exerts an indirect influence on the porous microstructure of final dried particles as the pores are nothing but the voids left behind after the sublimation of ice crystals during freeze-drying. After all, the distribution and connectivity of pores constitute the unique porous microstructure of spray-freeze-dried particles, besides controlling the drying rate (Al-Hakim, 2004). Similarly, at a lower concentration of the solute, the degree of freeze concentration is less. As a result, the inhibition of crystal growth is less, which leads to larger crystal sizes with markedly thinner walls (MacLeod et al., 2006).

2.4.1 Classification of freezing operation during spray-freeze-drying

Spray-freezing is classified based on the physical state of the cryogen into which the feed liquid is atomized. It could be the gas headspace above the surface of a cryogen liquid, a pool of cryogenic liquid, or cold gas referred to as, spray-freezing-into-vapor over liquid (SFV/L) (Figure 2.11[a]), spray-freezing-into-liquid (SFL) (Figure 2.11[b]), and spray-freezing-into-vapor (SFV) (Figure 2.11[c]), respectively. Freezing rate varies with the mode of spray-freezing and the rapidity decreases in the order of SFL (150°C per unit time) > SFV/L > SFV (10's of °C per unit time) (Al-Hakim, 2004; Gao, Smith, & Sego, 2000; Liao and Ng, 1990). SFV/L is the predominantly used freezing mode in general and SFL is specifically preferred in the spray-freeze-

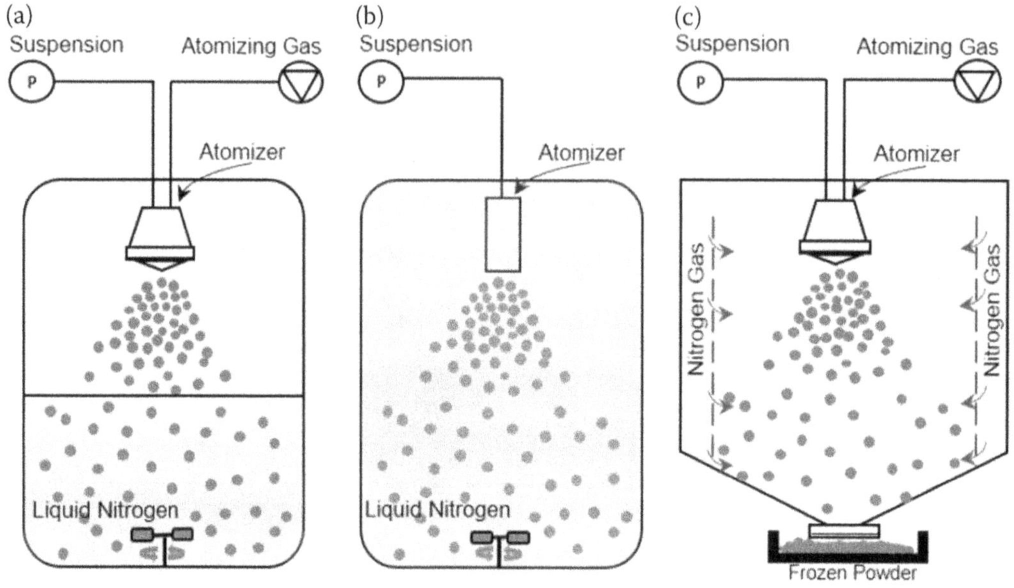

Figure 2.11 Schematic diagrams of different spray-freezing techniques: (a) Spray-freezing-into-vapor over liquid (SFV/L); (b) Spray-freezing-into-liquid (SFL); (c) Spray-freezing-into-vapor (SFV) (Adali et al., 2020).

drying of pharmaceuticals and biologicals (Chang and Baust, 1991; Maa et al., 1999; Sonner, 2002). The direct impingement of feed liquid beneath the surface of cryogenic liquid in SFL preserves the stability of active ingredients. On the other hand, SFV is preferred for food processing applications as it eliminates the need to separate frozen solids from the cryogenic liquid, which in turn must essentially be food-grade and non-contaminating. Both SFL and SFV/L require constant stirring of cryogenic liquid to alleviate the incidence of powder lumping inside the cryogenic bath. Each of the abovementioned freezing processes would be elaborated in the forthcoming sections.

2.4.1.1 Spray-freezing into vapor (SFV)

During SFV, the atomized liquid is frozen upon contact with the cold dry gas contained in a chamber. Before spraying the feed liquid, the cold gas is purged into the drying chamber, such that it attains the required temperature for subliming the frozen solvent. Later, the vapor is transported into a recycling chamber, wherein it condenses on a cold surface. The recycled gas returns to the chamber for further sublimation. Instead of purging the spray chamber with cold gas, the feed solution can be sprayed through a piezoelectric droplet generator nozzle into a jacketed chamber cooled by liquid nitrogen that creates the milieu for freezing and subsequent drying by sublimation. The above approach circumvents the direct contact between the product and cryogenic liquid or vapor, such that it is easy to recover the frozen product after spray-freezing (Eggerstedt et al., 2012).

The major advantages of SFV method have been realized in food products with respect to shorter drying time, enhanced aroma retention, and fine powders with good flowability. However, a major apprehension with the SFV mode of spray-freezing is encountered when operated in the counter-current flow configuration. The collection efficiency and particle elutriation were found to be affected. For instance, when SFV was used at atmospheric pressure to obtain an instant water-soluble drug of size 10–30 μm, difficulties were observed while capturing the particles with a gas filter (Leuenberger, 2002). Conversely, when a co-current flow configuration was employed to transport the frozen powder to the exit filter, the above challenge was resolved (Wang et al., 2006). Here, cold nitrogen gas was supplied to the chamber via its lateral porous walls, which are the sites of frozen particle formation. The drying was then carried out on an exit filter disc at atmospheric pressure. The above approach was found to be successful in obtaining free-flowing porous powders of bovine serum albumin (BSA) and skim milk with optimal retention of their relevant biological properties. Nevertheless, due to the cost of cold dry gas, the SFV process is generally preferred for high-value products such as pharmaceuticals, biologicals or functional materials (Moritz & Nagy, 2002).

2.4.1.2 Spray-freezing into vapor over liquid (SFV/L)

The mode of spray-freezing is referred to as *'spray-freezing-into-vapor over liquid (SFV/L)'*, when the feed solution is atomized through a nozzle that is placed at a defined distance above the boiling cryogenic liquid. The solidification of droplets commence during their transit through the headspace vapor, after which they freeze completely upon contacting the cryogenic liquid (Adams, Beck, & Menson, 1982; Briggs & Maxwell, 1976; Buxton & Peach, 1984; Dunn, Masavage, & Sauer, 1972; Sauer, 1969). Then, the suspended frozen particles are collected by sieving or after evaporating the cryogen. Subsequently, the frozen particles are transferred to a freeze-dryer to obtain dry powder (Ferrati et al., 2018). During their flight towards the cryogenic liquid, the droplets may collide and coalesce and the solutes might precipitate and grow in the unfrozen zones of the droplets (Hu et al., 2002).

The above events have an undesirable influence on the size distribution of atomized droplets and the particle morphology is fixed only when the droplets are completely solidified after contacting the cryogen. Hence, the major disadvantage associated with the SFV/L method is its negative impact on the stability of active pharmaceutical ingredients and other protein-based biologicals due to protein adsorption at the ice-water interface (Costantino et al., 2000 & 2002; Gombotz et al., 1990; Gusman & Johnson, 1990; Webb et al., 2002). Protein denaturation is more likely to occur during SFV/L, due to the following reasons:

- cold denaturation (Tsai, Maizel & Nussinov, 2002);
- freeze-concentration effect;
- pH changes (Lam et al., 1996; Pikal-Cleland & Carpenter, 2001; Pikal-Cleland et al., 2003);
- adsorption of proteins at the air-liquid interface or ice-water interface (Costantino et al., 2000; Hsu et al., 1995);
- Shear associated with the phase separation during the spray-freezing process: the phase separation between ice- and freeze-concentrated phases can subject the proteins to freezing-related stress. This can potentially end up in denaturation or aggregation of proteins during drying and/or reconstitution.
- loss of hydrogen bonding upon removal of water can cause the structural rearrangement of proteins (Flanders Health Blog, 2019).

The protein stability is affected by their adsorption at the air/gas-liquid interface, which is the major site for the denaturation of proteins and peptides (Niven, 1996). Air/gas corresponds to that from the two-fluid nozzle during the atomization or vapor phase of the cryogen during spray-freezing, and water is available in the feed droplet phase. The rate of protein denaturation increases with the freezing rate and interfacial area (Maa & Hsu, 1996; MacRichie, 1978; Maa, Nguyen & Hsu, 1998; Thurow & Geisen, 1984). In the SFV/L process, the transit of droplets through the vapor phase is unavoidable. Hence, before the droplets reach the surface of liquid nitrogen, the majority of protein aggregation would have already occurred (Yu, Johnston, & Williams, 2006).

To evade the above possibilities, Costantino et al. (2000 & 2002) adopted an alternative approach to contact the atomized droplets with liquid nitrogen spray directed from four nozzles, after which the suspended frozen particles were separated from the liquid nitrogen. The aforementioned demerits of SFV/L led to the advent of an alternative spray-freezing approach, namely, the spray-freezing-into-liquid (SFL).

2.4.1.3 Spray-freezing into liquid (SFL)

Spray-freezing-into liquid (SFL) is a cryogenic particle engineering approach, which involves the atomization of a feed solution (containing active pharmaceutical ingredients (APIs) or excipients) through an insulated nozzle that is placed directly underneath the cryogenic liquid, to produce frozen nanostructured particles (Barron et al., 2003). Consequently, the SFL process leads to a relatively lower degree of protein aggregation and lesser loss of enzyme activity, than SFV/L. The frozen particles are separated from the cryogenic liquid by sieve filtration or other methods and then freeze-dried (Hu, Johnston, & Williams, 2004; Yu et al., 2002). Liquid nitrogen, argon, hydrofluoroether, and pentane are the commonly used cryogenic liquids, which can be used at atmospheric pressure. On the other hand, a pressurized system could be used with liquid CO_2, propane or ethane as the cryogenic liquids. SFL causes an intense atomization to produce tiny droplets, due to the liquid-liquid impingement between the pressurized feed solution exiting the nozzle and the cryogenic liquid, besides the high-velocity feed liquid emanating through the nozzle's orifice. When two liquids collide under the above conditions, the high Reynolds and Weber numbers cause intense liquid disintegration into micron-scale droplets (Hu et al., 2002; Rogers et al., 2002). Moreover, the higher viscosity and density of the cryogenic liquid compared to vapor/gas implies that much smaller droplets can be produced by SFL than those obtained from SFV or SFV/L (Ishwarya, Anandharamakrishnan, & Stapley, 2015).

Further, as mentioned earlier, SFL leads to ultra-rapid freezing rates due to the extremely low boiling point of the cryogenic liquid (e.g. −196°C for liquid nitrogen), high specific surface area of the droplets and high heat transfer coefficients. The rapid freezing rates reduce the probability of phase separation, change in pH of dissolved substances, and crystalline growth of water (Rogers et al., 2003; Yu et al., 2004). In other words, the freezing rates are rapid enough to capture the API in an amorphous state by not allowing the crystallization time. This lessens the occurrence of protein denaturation during spray-freezing (Engstrom et al., 2007; Hu, Johnston, & Williams, 2004; Rogers et al., 2003). After freeze-drying, the final microparticles retain the shape of initial atomized droplets with a highly porous microstructure due to the channels created when the solvent(s) were removed during sublimation. As a result, SFL-processed powders have a large surface area. In general, either heated nozzles or capillary nozzles made of a material with very low thermal conductivity (ex. polyether–ether ketone or PEEK) are used for SFL to prevent clogging of the nozzle by ice formation.

The advent of SFL technology was intended to minimize the stability loss of active components and enhance the wetting and dissolution properties of poorly water soluble APIs for drug delivery applications (Hu et al., 2002; Rogers et al., 2001, 2002, & 2003; Yu et al., 2002; Yu et al., 2004). Accordingly, its key advantages are the uniform dispersion of API molecules throughout the solidified carrier matrix of the frozen microparticle that includes numerous pores after freeze-drying. The larger surface area of SFL particles has been attributed to the high degree of supersaturation and the resultant rapid nucleation rates of dissolved substances. The restricted growth after crystallization produce very small submicron domains (Hu et al., 2002; Rogers et al., 2002).

When the SFL-processed microparticles are placed in an aqueous medium for dissolution, their porous channels are penetrated through and filled by the medium. As a result, the particles undergo instantaneous wetting and dissolution. The large surface area and amorphous nature of SFL microparticles also play a role in their dissolution. This aspect of SFL is advantageous to improve the dissolution of a lipophilic API by encapsulating it within hydrophilic excipients. SFL is considered as the most appropriate mode of spray-freezing for protein based APIs. It confers protection upon the proteins against processing-induced instability (Wang, 2000) such as denaturation, which occurs during spray drying, conventional freeze-drying and even spray-freeze-drying processes that employ the SFV and SFV/L modes of freezing.

Nevertheless, during SFL, boiling-off of cryogen can occur due to the heat introduced by the relatively warm feed jet from heated nozzles. The above phenomenon is termed as the *Leidenfrost Effect*, which causes some droplets to encounter lower cooling rates owing to the insulating effect of boiled-off vapor around them. This effect prolongs the phase of ice crystal formation due to low thermal conductivity, resulting in low freezing rate (Engstrom et al., 2007; Wanning, Süverkrüp, & Lamprecht, 2015). Leidenfrost effect was found to be prominent while using liquid nitrogen as the cryogenic liquid. But, the same was not observed with iso-pentane due to its significantly higher boiling point (27°C) and larger heat of vaporization and thus the low tendency to evaporate (Engstrom et al., 2007). Thus, iso-pentane is a more appropriate cryogenic liquid for SFL as it can be cooled down to −160°C before it freezes (Ishwarya, Anandharamakrishnan, & Stapley, 2015).

2.5 GUIDELINES FOR DESIGNING A SPRAY-FREEZING CHAMBER

An important aspect in the spray-freezing process is the design of the spray-freezing chamber, which is supposedly more intricate and demand adequate control over the temperature and environment. A thesis by Al-Hakim (2004) shares ample insights to the designing of a spray-freezing rig for the SFV mode, the key interpretations of which are discussed in this section. Firstly, the size of the chamber must be sufficiently large such that the droplets are solidified before contacting the chamber wall. This curtails the sticking of particles to the chamber wall. Prevention of stickiness solves the concerns of corrosion and contamination and reduces the process downtime and loss of productivity due to requirement for periodic cleaning of the chamber. Typically, the drop freezing has been observed to complete within 1 second. The following factors must be taken into account before finalizing the dimensions of the spraying chamber:

1. Droplet size;
2. Trajectory path of the droplets inside the chamber, and,
3. Freezing rate.

As discussed in Section 2.2.2, tall-form spray chambers are predominantly used for spray-freeze-drying applications, owing to their ability to accommodate heated twin-fluid nozzles and occupying less floor space. According to Masters (1991), a chamber that accommodates a spray nozzle must have a minimum diameter-to-length ratio of 3:1. Moreover, the design of spray-freezing chamber should be flexible enough to permit either of the two gas flow configurations, i.e., counter or co-current and allow variability in the positioning of the spraying point at the top of the chamber, apart from being exactly vertical. In accordance with the above guidelines, a SFV chamber comprises three sections:

1. a top cone housing to accommodate the atomizer and the freezing gas inlet or outlet (positioned at 180° from each other) depending on whether the gas flow is counter-current or co-current. The atomization gas is supplied to the nozzle from a compressed nitrogen gas cylinder, the flow and pressure of which are regulated by a solenoid valve and pneumatic regulator, respectively. The feed liquid is transported from the feed tank into the spray nozzle using a diaphragm pump and its pressure is controlled by changing the power supply to the pump;
2. a cylindrical section that may be fitted with see-through glass windows to monitor the spray-freezing process, and,
3. a bottom cone that accommodates the gas inlet/outlet and a particle collection box attached to its lower opening. The formed particles are directed into the box and separated from the gas stream, which is vented to the exhausted ventilation extractor (Al-Hakim, 2004).

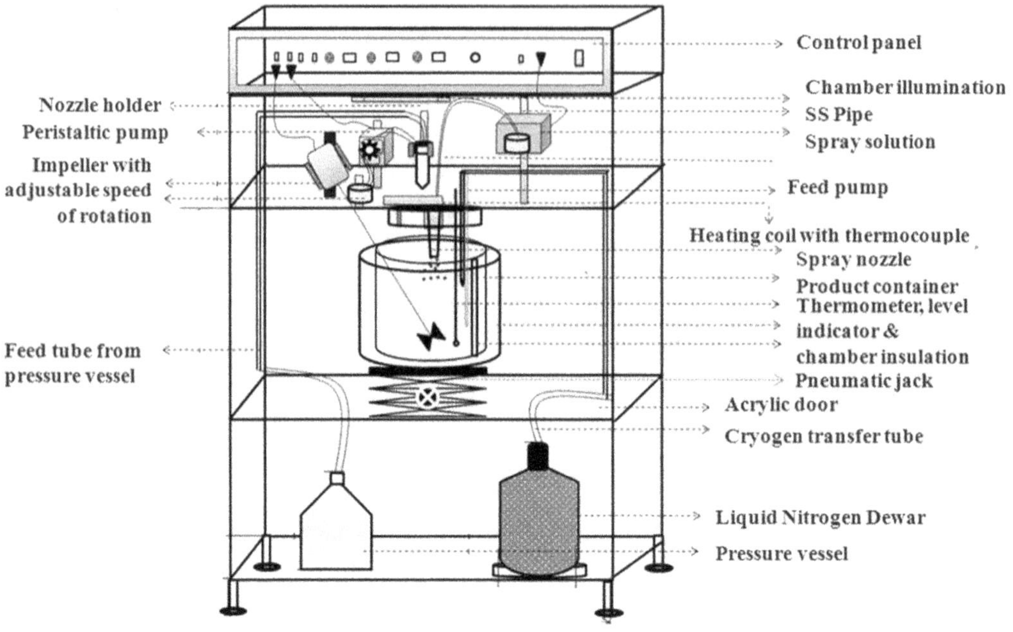

Figure 2.12 Design diagram of a SFV/L apparatus. (Drawn and presented with permission from C. Anandharamakrishnan)

The typical design of a SFV/L rig is shown in Figure 2.12 The rig is an arrangement comprising mainly of a twin fluid nozzle placed within a nozzle holder along with the heating arrangement, peristaltic pump or pressure vessel and a stainless steel product container (SS 304 or SS 316 grade) connected to a liquid nitrogen dewar. The contents of the container are stirred by an impeller with adjustable speed of rotation. The distance between the nozzle and surface of liquid nitrogen is adjusted by a pneumatic jack connected to the container. The cryogenic liquid is transferred from a pressurized liquid nitrogen dewar through a transfer tube into the product container.

A typical apparatus for the spray freezing into liquid (SFL) process and its components (Figure 2.13) was proposed by Barron et al. (2003). The feed solution is stored in a stainless reservoir [A] (inner diameter (ID): 11/16 inch; outer diameter (OD): 1 inch) fitted with a stainless steel piston and sealed with o-rings. A digital syringe pump [B] drives the piston by virtue of the CO_2 pressure. A PEEK nozzle [C] integrated with the stainless steel precipitation cell [D] (ID: 4 inch; length: 6 inch) is connected to the feed reservoir. The flow of feed solution from the reservoir through the nozzle and into the precipitation cell is regulated by the high pressure valves placed between the ports. Other ports that enter into the precipitation cell are the inlet for the cryogenic liquid [E], a release valve, a digital pressure transducer and display, a temperature probe and digital display say. The liquid CO_2 is pumped from a compressed gas tank through a stainless steel tubing (OD: 1/8-inch; ID: 0.060-inch id) [F] into the precipitation cell. This tubing is submerged in a bath containing chilled ethanol and dry ice. In addition, a pressure gauge [G] and high-pressure regulator [H] are available to monitor the pressure and control the flow rate of liquid CO_2 from the gas tank to the precipitation cell, respectively. The outlet port [I] is a stainless steel tubing (OD: ¼ inch; ID: 1/8 inch) dipped in a warm water bath coupled to a flow meter [J]. Similar to that in the SFV/L rig, the precipitation cell of the SFL apparatus is fitted with a paddle stirrer rotating about a shaft with a cylindrical magnet at the top connected to an outer drive motor. When the SFL process is on, the precipitation cell remains immersed in an ethanol and dry ice bath (Barron et al., 2003).

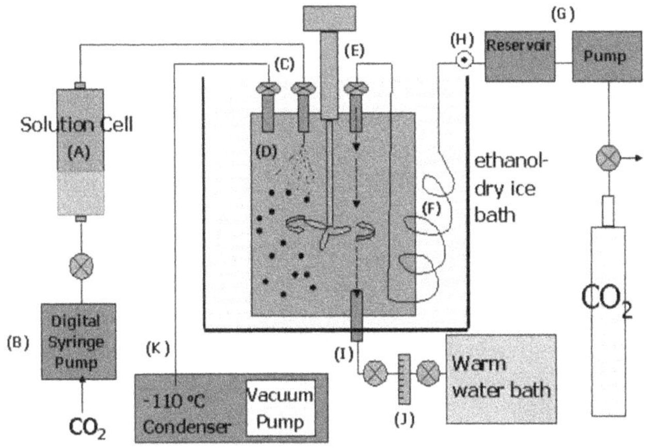

Figure 2.13 Design diagram of a SFL apparatus (Barron et al., 2003).

2.6 CHARACTERIZATION OF SPRAY-FREEZING PARAMETERS

2.6.1 Measurement of droplet characteristics (axial velocity, size distribution)

During spray-freezing, complex microstructures are created within the droplets through the formation of ice crystals. The size and number of ice crystals are governed by the cooling conditions including slip velocity. Further, measuring the droplet diameter and velocity of the spray-freezing system is useful in predicting or estimating the following (Al-Hakim, Wigley, & Stapley, 2006):

a. heat transfer coefficients by the calculation of slip velocities
b. average 'time of flight' (t) between the time at which the droplets of various sizes theoretically leave the nozzle and reach each measurement point, given by the following equations,

$$V_d = \frac{ds}{dt} \tag{2.2}$$

$$\therefore t = \int_{s1}^{s2} \frac{1}{V_d} ds \tag{2.3}$$

Where V_d is the velocity of droplet (m s^{-1}) and s is the distance (m).

c. input data to a simple *'heat transfer model of drop freezing'* to compute the freezing time.

In the above background, Al-Hakim, Wigley, & Stapley (2006) employed a high power and high resolution phase Doppler anemometer (PDA) system to quantify the size and axial velocity of discrete droplets inside a spray-freezing chamber. PDA is a single-point and non-invasive optical diagnostic technique, which is used to measure the instantaneous velocity and diameter of hundreds and thousands of individual drops or particles transiting through a small control volume within an extremely short duration in the order of one second (Albrecht, 2003; Bachalo & Houser, 1984). The above study investigated the vertical velocity component and diameter of droplets at various locations along the centerline of the spray under the nozzle for both twin-fluid and hydraulic atomizers (60, 100, 160 and 220 mm for twin-fluid nozzle, and 38 and 108 mm for hydraulic nozzle). In addition, the effect of different chamber gas temperatures (20°C, −22°C, −42°C and −62°C) on the spray characteristics was also explored.

For the stated purpose, spray freezing trials were performed in a specially designed cylindrical chamber (height: 1.5 m; diameter: 0.8 m) with plane windows to enable PDA measurements (Figure 2.14[a & b]). Before atomization, the chamber was purged with dry nitrogen gas from a cylinder to dehumidify the chamber and then cooled using a liquid nitrogen supply. The different chamber temperatures were achieved by regulating the flows of liquid nitrogen and nitrogen cylinder gas to the chamber. The temperature at the spray point was recorded using a thermocouple in close vicinity, which was found to be less than 2°C of the preferred set point. The PDA receiver was positioned at a scattering angle of 70° (Figure 2.14[a]) to assure that the collected scattered light was due to first order refraction only. It was observed that a maximum of 200000 drops could be measured within 3 seconds for the pneumatic nozzle, with a characteristic mean diameter value (d_{10}: the proportion of particles with diameter lesser than this value is 10%) and Sauter mean diameter (d_{32}: diameter of a sphere with its volume-to-surface area ratio equivalent to that of a particle under consideration) of 12 μm and 15 μm, respectively. Whereas, for the hydraulic nozzle, merely 5000 droplets could be measured in the same duration, owing to a smaller population of significantly larger drops having corresponding d_{10} and d_{32} values of 50 μm and 105 μm.

Overall, the results of Al-Hakim et al. (2006) revealed the following:

i. With respect to axial velocity of the droplets, the sprays emerging from pneumatic nozzle showed decelerating droplets with only a slight difference in drop size at each distance. Exceptionally, the 20-micron drops showed marginally higher velocities at 60 mm and 100 mm from the nozzle. Contrastingly, hydraulic nozzles resulted in droplets showing substantial deceleration with increase in drop size. But the velocity did not vary significantly with the distance from the nozzle. Owing to their higher inertia, larger drops are less influenced by deviations in gas velocity.
ii. For a pneumatic nozzle, irrespective of the distance of measurement, at ambient temperature, there were no significant changes in the drop size distribution (DSD) (Figure 2.15[a]), as the Weber numbers (a dimensionless number obtained from the ratio between drag force and cohesion force, applicable in fluid flow situations involving an interface between two fluids) were far below the

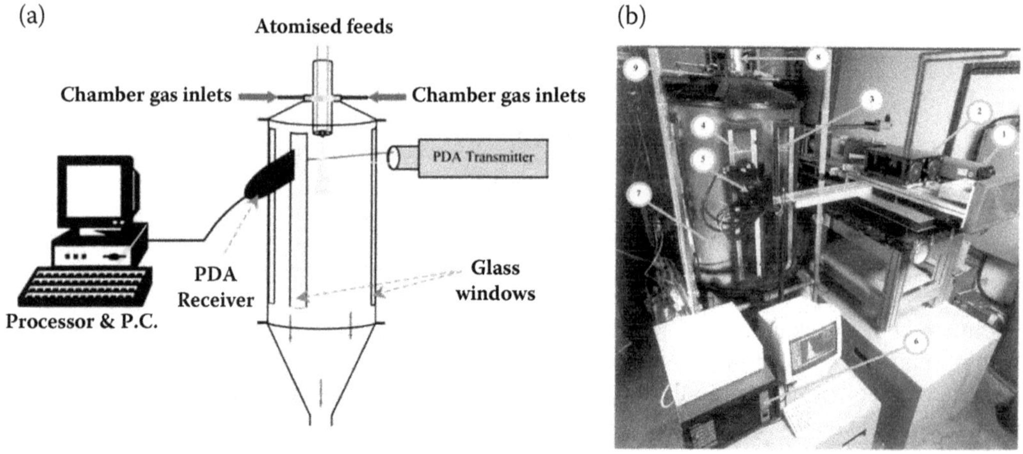

Figure 2.14 (a) Schematic of spray-freezing rig and PDA apparatus; (b) Photograph of spray-freezing rig and PDA apparatus during measurements [(1) laser source; (2) laser transmitter; (3) input laser beam; (4) measurement volume; (5) PDA receiver; (6) signal processor; (7) spraying chamber; (8) spray nozzle inside a holder, and (9) chamber gas inlets] (Modified from Al-Hakim, Wigley & Stapley, 2006).

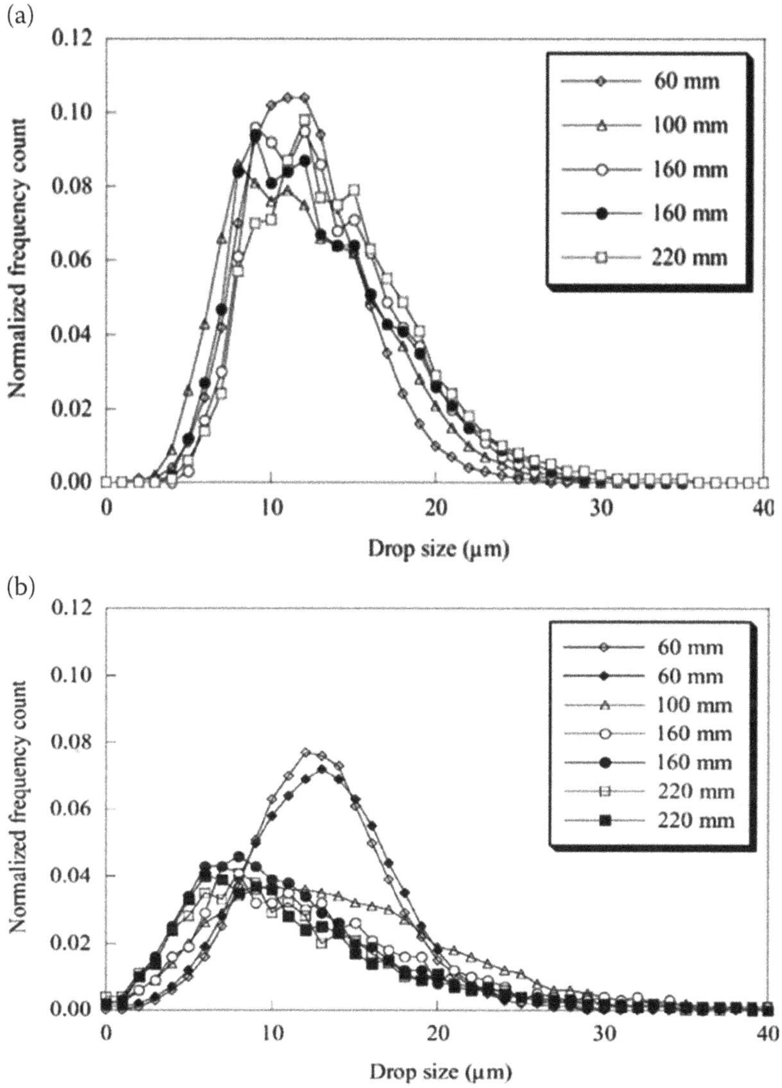

Figure 2.15 Drop size distribution resultant from spray-freezing using pneumatic nozzle as a function of distance from the nozzle: (a) at ambient temperature; (b) at –42°C (Al-Hakim et al., 2006).

critical value for droplet breakup. However, there was a slight increase in the d_{32} with increase in distance, implying that the larger droplets are capable of penetration.

iii. Contrary to spray-freezing at ambient temperature, the influence of distance on DSD of pneumatic nozzles was prominent at sub-zero chamber temperature (–42°C). At greater distances, both the peak and peak areas of DSD plot were skewed to lower values (Figure 2.15[b]). But, the values of d_{32} increased due to the presence of very large droplets. Due to higher surface tension, Weber number is much smaller for sub-ambient droplets than for ambient droplets and hence the droplet break up is improbable.

iv. In contrast to pneumatic nozzles, the hydraulic nozzles (Figure 2.16[a & b]) showed a substantial reduction in the measurable droplets with decrease in chamber temperature from ambient (20°C) to

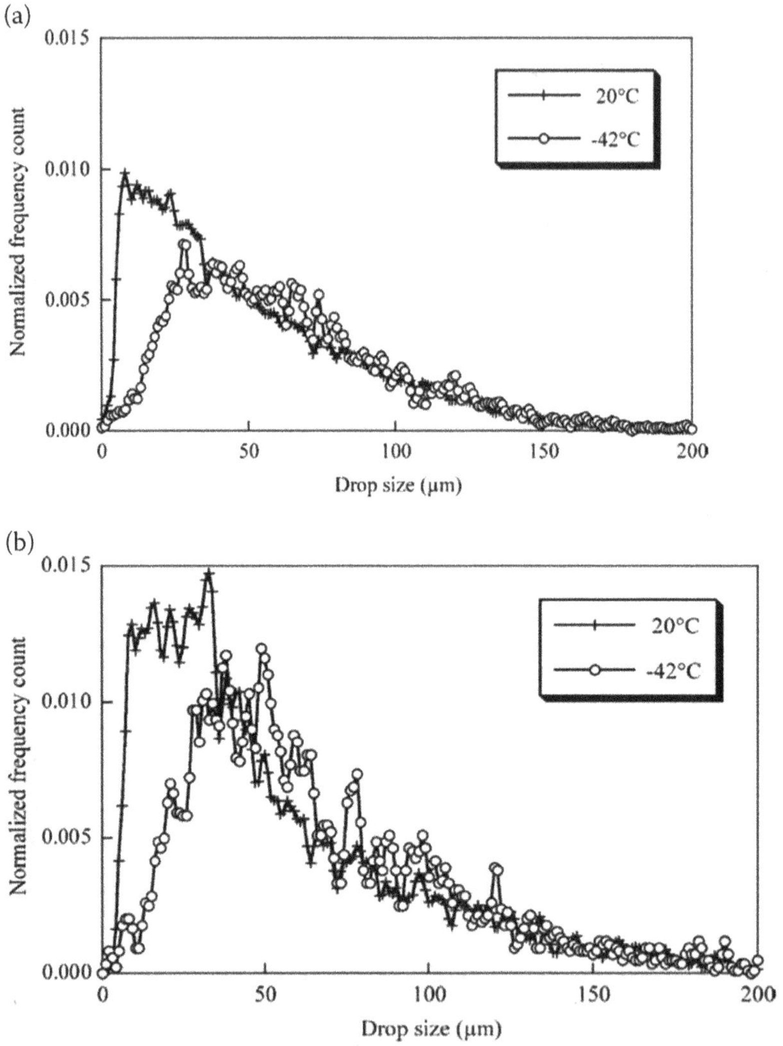

Figure 2.16 Drop size distribution resultant from spray-freezing using hydraulic nozzle as a function of chamber temperature: (a) at 38 mm from the nozzle; (b) at 108 mm from the nozzle (Al-Hakim et al., 2006).

−42°C. In general, the drop sizes (d_{32}) from hydraulic nozzle were significantly larger (105 μm) than those obtained from the pneumatic nozzles (14-18 μm). But, increase in distance did not affect the measured DSD except for a minor increase in d_{32}.

v. Based on the Hindmarsh freezing model (Eq. 2.4) and PDA data, it was found that at a constant slip velocity (2 or 10 m s^{-1}), the time for nucleation and solidification increased with increase in drop size (from 2 to 150 μm). Irrespective of the drop size, slip velocity showed an inverse relationship with the nucleation and solidification times.

$$mc_p \frac{dT_d}{dt} = hA(T_d - T_{gas}) \tag{2.4}$$

Where m is the droplet mass (kg), c_p is the specific heat capacity of gas (J kg^{-1} K^{-1}), T_d and T_{gas} are the temperatures of droplet and gas (K), respectively, h is the heat transfer coefficient (W m^{-2} K^{-1}) and A ($=\pi d^2/4$; d = droplet diameter [m]) is the frontal area of droplet (m^2).

Nevertheless, the use of Phase Doppler Anemometry to measure droplet size exhibited certain limitations. The measurements could be performed only at a maximum axial distance of 0.2 m below the nozzle. At distances beyond 0.2 m, measurements were challenged due to reduction in the refractive scatter caused by droplet freezing and smaller number of droplets entering the measurement volume caused by spreading of the spray. Moreover, greater interference of droplets were not in the measurement volume owing to intense ''fog'' at a greater depth from the nozzle (Anandharamakrishnan et al., 2010).

2.6.2 Measurement of gas temperature during spray-freezing

Measuring the temperature of cold gas inside the spray-freezing chamber is complex as the frozen particles tend to deposit on the thermocouples. As a preventive step to avoid measurement errors, a simple shield can be employed to protect the thermocouples from the direct impact of particles (Anandharamakrishnan, 2008; Papadakis & King, 1988). A schematic representation of the above approach is depicted in Figure 2.17. For the measurement of cold gas temperature, a defined number of thermocouples are placed at uniformly-spaced distances (say about 10 cm) along a plastic rod traversing a region from the centerline to the wall (Figure 2.17). A plastic cap as shown in the figure is used to protect the thermocouple from particle impact. To enable online

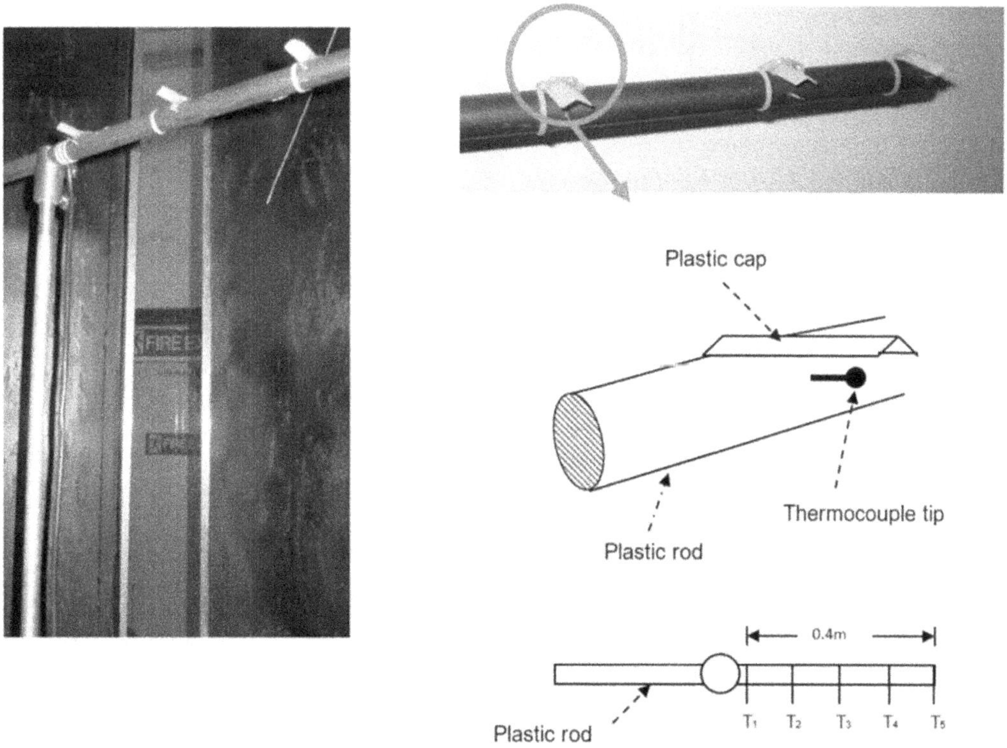

Figure 2.17 Shielded thermocouple used for temperature measurements (Anandharama krishnan, 2008).

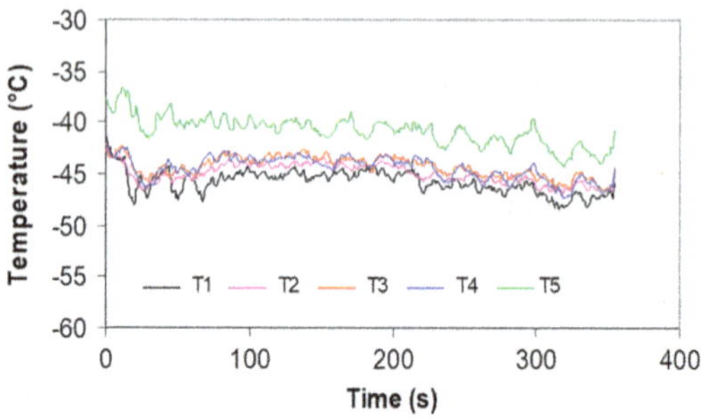

Figure 2.18 Temperature measurements during spray-freezing at –42°C chamber temperature at an axial distance z = 1.23 m below the nozzle (Anandharamakrishnan, 2008).

temperature measurements, all the thermocouples can be logged to data acquisition hardware on a desktop or laptop. Figure 2.18 shows a model data for the gas temperature measurements inside a spray-freezing chamber at –42°C and at an axial distance of z = 1.23 m from the nozzle, recorded from five thermocouples (Anandharamakrishnan, 2008).

CONCLUSIONS

The information presented in this chapter established that the spray-freezing step of SFD process is indeed complex with several phenomena occurring simultaneously. Spray-freezing is the key to the production of high-quality particles with good pore size distribution by preventing solute migration and preserving the stability of active ingredient at the air-water interface. Understanding the nuances of spray-freezing operation and an appropriate choice of atomizer type and mode of freezing would aid in the control of final particle size and microstructure. Though imaging techniques have been applied to visualize and study the spontaneous spraying and freezing operations, the scope still prevails for a comprehensive spray characterization including the droplet size distribution from different nozzle types, droplet trajectories within the cryogenic liquid during the SFL process, evolution kinetics of droplet microstructure and stability of APIs at the air-water interface. With the advent of cutting-edge cryo-imaging systems and image processing software, a wealth of information is envisaged to arise from the future investigations on spray-freezing.

REFERENCES

Adali, M. B., Barresi, A. A., Boccardo, G., & Pisano, R. (2020). Spray Freeze-Drying as a Solution to Continuous Manufacturing of Pharmaceutical Products in Bulk. *Processes*, 8(6), 709.

Adams, T. H., Beck, J. P., & Menson, R. C. (1982). *Novel particulate compositions*. US Patent 4,323,478.

Albrecht, H. E. (2003). *Laser doppler and phase doppler measurement techniques*, Springer, Berlin; London.

Al-Hakim, K. (2004). *An Investigation of Spray-Freezing and Spray-Freeze-Dryings*. PhD Thesis, Loughborough University, United Kingdom.

Al-Hakim, K., Wigley, G., & Stapley, A. G. F. (2006). Phase Doppler anemometry studies of spray freezing. Trans. ChemE, Part A, *Chemical Engineering Research and Design*, 84, 1142–1151.

Amorij, J. P., Saluja, V., Petersen, A. H., Hinrichs, W. L. J., Huckriede, A., & Frijlink, H. W. (2007). Pulmonary delivery of an inulin-stabilized influenza subunit vaccine prepared by spray freeze drying induces systemic, mucosal humoral as well as cell mediated immune responses in BALB/c mice. *Vaccine*, 25, 8707–8717.

Anandharamakrishnan, C. & Ishwarya, S. P. (2015). *Spray Drying Techniques for Food Ingredient Encapsulation* (Chapter 1), John Wiley & Sons, Ltd., Chichester, UK, and the Institute of Food Technologists, Chicago, IL, pp. 1–36.

Anandharamakrishnan, C. (2008). *Experimental and computational fluid dynamics studies on spray-freeze-drying and spray-drying of proteins*. Ph.D. thesis. Loughborough University, UK.
Anandharamakrishnan, C., Rielly, C. D., & Stapley, A. G. F. (2010). Spray-freeze-drying of whey proteins at sub-atmospheric pressures. *Dairy Science and Technology*, 90, 321–334.
Bachalo, W. D., & Houser, M. J. (1984). Phase/doppler spray analyzer for simultaneous measurement of drop size and velocity distributions. *Optical Engineering*, 23, 583–590.
Barron, M. K., Young, T. J., Johnston, K. P., & Williams, R. O. (2003). Investigation of processing parameters of spray freezing into liquid to prepare polyethylene glycol polymeric particles for drug delivery. *AAPS Pharm Sci Tech*, 4(2), 1–13, Article 12.
Briggs, A. R., & Maxwell, T. J. (1976). Method *of preparation of lyophilized biological products*. US Patent 3,932,943.
Burke, P. A., Klumb, L. A., Herberger, J. D., Nguyen, X. C., Harrell, R. A., & Monica, Z. (2004). Poly (Lactide-Co-Glycolide) Microsphere formulations of Darbepoetin Alfa: Spray drying is an alternative to encapsulation by spray-freeze drying. *Pharmaceutical Research*, 21(3), 500–506.
Buxton, I. R., & Peach, J. M. (1984). *Process and apparatus for freezing a liquid medium*. US Patent 4,470,202.
Cal, K., & Sollohub, K. (2010). Spray drying technique. I: hardware and process parameters. *Journal of Pharmaceutical Sciences*, 99, 575–586.
Chang, Z., & Baust, J. (1991). Ultra-rapid freezing by spraying/plunging: Precooling in the cold gaseous layer. *Journal of Microscopy*, 161, 435–444.
Cheow, W. S., Ng, M. L. L., Kho, K., & Hadinoto, K. (2011). Spray freeze drying production of thermally sensitive polymeric nanoparticle aggregates for inhaled drug delivery: effect of freeze drying adjuvants. *International Journal of Pharmaceutics*, 404, 289–300.
Chin, J. S., & Lefebvre, A. H. (1993). Flow patterns in internal-mixing, twin-fluid atomizers. *Atomization and Sprays*, 3, 463–475.
Costantino, H. R., Firouzabadian, L., Hogeland, K., Wu, C. C., Beganski, C., Carrasquillo, K. G., Cordova, M., Griebenow, K., Zale, S. E., & Trace, (2000). Protein spray-freeze drying. Effect of atomisation conditions on particle size and stability. *Pharmaceutical Research*, 17(11), 1374–1383.
Costantino, H. R., Firouzabadian, L., Wu, C., Hogeland, K. C., Carrasquillo, K. G., Cordova, M., et al. (2002). Protein Spray freeze drying. 2. Effect of formulation variables on particle size and stability. *Journal of Pharmaceutical Sciences*, 91, 388–395.
D'Addio, S. M., Chan, J. G. Y., Kwok, P. C. L., Prud'homme, R. K., & Chan, H.-K. (2012). Constant size, variable density aerosol particles by ultrasonic spray freeze drying. *International Journal of Pharmaceutics*, 427, 185–191.
Demarco, F. W., & Renzi, E. (2018). Bulk freeze-drying by spray freezing and agitated drying. Spanish Patent ES2649045T3.
Dunn, D. B., Masavage, G. J., & Sauer, H. A. (1972). *Method of freezing solution droplets and the like using immiscible refrigerants of differing densities*. US Patent 3,653,222.
Eggerstedt, S. N., Dietzel, M., Sommerfeld, M., Süverkrüp, R., & Lamprecht, A. (2012). Protein spheres prepared by drop jet freeze drying. *International Journal of Pharmaceutics*, 438, 160–166.
Engstrom, J. D., Simpson, D. T., Lai, E. S., Williams, R. O., & Johnston, K. P. (2007). Morphology of protein particles produced by spray freezing of concentrated solutions. *European Journal of Pharmaceutics and Biopharmaceutics*, 65, 149–162.
Ferrati, S., Wu, T., Fuentes, O., Brunaugh, A. D., Kanapuram, S. R., & Smyth, H. D. C. (2018). Influence of formulation factors on the aerosol performance and stability of lysozyme powders: A systematic approach. *AAPS PharmSciTech*, 19, 2755–2766.
Flanders Health Blog (2019). Available from: https://www.flandershealth.us/therapeutic-proteins/spray-freeze-drying.html (Accessed 15 August 2021).
Frommhold, P. E., Lippert, A., Holsteyns, F. L., & Mettin, R. (2014). High-speed monodisperse droplet generation by ultrasonically controlled micro-jet breakup. *Experiments in Fluids*, 55, 1716.
Gao, W., Smith, D. W., & Sego, D. C. (2000). Freezing temperatures of freely falling industrial wastewater droplets. *Journal of Cold Regions Engineering*, 14, 101–118.
Goff (*n.d.*). Theoretical Aspects of the Freezing Process. Available from: https://www.uoguelph.ca/foodscience/book/export/html/1881 (Accessed on 20 August 2021).

Gogate, P. R. (2015). The use of ultrasonic atomization for encapsulation and other processes in food and pharmaceutical manufacturing. In: J. A. Gallego-Juárez & K. F. Graff (Eds.), *Power Ultrasonics Applications of High-Intensity Ultrasound*, Woodhead Publishing, Cambridge, UK, pp. 911–935.

Gombotz, W. R., Healy, M. S., Brown, L. R., & Auer, H. E. (1990). *Process for producing small particles of biologically active pharmaceuticals*. WO/1990/013285.

Gusman, M. I., & Johnson, S. M. (1990). *Cryochemical method of preparing ultrafine particles of high-purity superconducting oxides*. Application (pp. 13). US, (SRI International, USA), US.

Gwie, C. G., Griffiths, R. J., Cooney, D. T., Johns, M. L., & Wilson, D. I. (2006). Microstructures formed by spray freezing of food fats. *Journal of the American Oil Chemists' Society*, 83, 1053–1062.

Her, J.-Y., Song, C.-S., Lee, S. J., & Lee, K.-G. (2010). Preparation of kanamycin powder by an optimized spray freeze-drying method. *Powder Technology*, 199, 159–164.

Hindmarsh, J. P. (2003). *Fundamentals of spray crystallisation of food solutions*. Ph.D. dissertation. The University of Auckland, New Zealand.

Hindmarsh, J. P., Russell, A. B., & Chen, X. D. (2003). Experimental and numerical analysis of the temperature transition of a suspended freezing water droplet. *International Journal of Heat and Mass Transfer*, 46, 1199–1213.

Hindmarsh, J. P., Russell, A. B., & Chen, X. D. (2004). Experimental and numerical analysis of the temperature transition of a suspended freezing food solution droplet. *Chemical Engineering Science*, 59, 2503–2515.

Hindmarsh, J. P., Russell, A. B., & Chen, X. D. (2007). Fundamentals of the Spray freezing of foods e microstructure of frozen droplets. *Journal of Food Engineering*, 78, 136–150.

Hsu, C. C., Nguyen, H. M., Yeung, D. A., Brooks, D. A., Koe, G. S., Bewley, T. A., & Pearlman, R. (1995). Surface denaturation at solid-void interface-a possible pathway by which opalescent particulates form during the storage of lyophilized tissue-type plasminogen activator at high temperatures. *Pharmaceutical Research*, 12, 69–77.

Hu, J., Johnston, K. P., & Williams, R. O. (2004). Rapid dissolving high potency danazol powders produced by spray freezing into liquid process. *International Journal of Pharmaceutics*, 271, 145–154.

Hu, J., Rogers, T. L., Brown, J., Young, T., Johnston, K. P., & Williams, R. O. (2002). Improvement of dissolution rates of poorly water soluble APIs using novel spray freezing into liquid technology. *Pharmaceutical Research*, 19, 1278–1284.

Ishwarya, S. P., Anandharamakrishnan, C., & Stapley, A. G. F. (2015). Spray-freeze-drying: A novel process for the drying of foods and bioproducts. *Trends in Food Science & Technology*, 41, 161–181.

Ishwarya, S. P., Anandharamakrishnan, C., & Stapley, A. G. F. (2017). Spray-freeze-drying of dairy products. In: Anandharamakrishnan, C. (Ed.), *Handbook of Drying for Dairy Products*. John Wiley & Sons, Oxford, pp. 123–148.

Kim, E. H. J., Chen, X. D., & Pearce, D. (2003). On the mechanisms of surface formation and the surface composition of industrial milk powders. *Drying Technology*, 21, 265–278.

Lam, X. M., Costantino, H. R., Overcashier, D. E., Nguyen, T. H., & Hsu, C. C. (1996). Replacing succinate with glycolate buffer improves the stability of lyophilized interferon-γ. *International Journal of Pharmaceutics*, 142, 85–95.

Leuenberger, H. (2002). Spray Freeze drying e the process of choice for low water soluble drugs? *Journal of Nanoparticle Research*, 4, 111–119.

Liang, W., Chow, M. Y. T., Chow, S. F., Chan, H.-K., Kwok, P. C. L., & Lam, J. K. W. (2018). Using two-fluid nozzle for spray freeze drying to produce porous powder formulation of Naked Sirna for inhalation. *International Journal of Pharmaceutics*, 552, 67–75.

Liao, J. and Ng, K. C. (1990). Effect of ice nucleators on snow making and spray freezing. *Industrial Engineering Chemistry Research*, 29, 361–366.

Lixin, H., Kumar, K., & Mujumdar, A. S. (2004). Simulation of spray evaporation using pressure and ultrasonic atomizer – A comparative analysis. *Russia TSTU Trans.* (English Version), 10, 83–100.

Lord Rayleigh (1878). On the instability of jets. *Proceedings of the London Mathematical Society*, 10, 4–13.

Maa, Y.-F., & Hsu, C. C. (1996). Effect of high shear on proteins. *Biotechnology and Bioengineering*, 51, 458–465.

Maa, Y.-F., Nguyen, P.-A., Sweeney, T., Shire, S. J., & Hsu, C. C. (1999). Protein inhalation powders: spray drying vs spray freeze drying. *Pharmaceutical Research*, 16, 249–254.

Maa, Y.-F., Nguyen, P.-A., & Hsu, C. C. (1998). Spray drying of air-sensitive recombinant human growth hormone. *Journal of Pharmaceutical Sciences* 87, 152–159.

Maa, Y.-F., Prestrelski, S. J., & Burkoth, T. L. (2003). *Spray freeze-dried compositions.* United States Patent US7229645B2.

MacLeod, C. S., McKittrick, J. A., Hindmarsh, J. P., Johns, M. L., & Wilson, D. I. (2006). Fundamentals of spray freezing of instant coffee. *Journal of Food Engineering*, 74, 451–461.

MacRichie, F. (1978). Proteins at interfaces. *Advances in Protein Chemistry*, 32, 283–311.

Marshall, W. R. (1954). Atomization and spray drying. *Chemical Engineering Progress Monograph Series*, 50, 50–56.

Masters, K. (1991). *Spray drying handbook* (5th Edition). Longman Scientific & Technical, Harlow.

Masters, K. (2002). *Spray Drying in Practice.* Spray Dry Consult, Denmark, pp. 1–35.

Mizoe, T., Ozeki, T., & Okada, H. (2008). Application of a four-fluid nozzle spray drier to prepare inhalable rifampicin-containing mannitol microparticles. *AAPS PharmSciTech*, 9, 755–761.

Moritz, T., & Nagy, A. (2002). Preparation of super soft granules from nanosized ceramic powders by spray freezing. *Journal of Nanoparticle Research*, 4, 439–448.

Nguyen, X. C., Herberger, J. D., & Burke, P. A. (2004). Protein powders for encapsulation: a comparison of spray-freeze drying and spray drying of Darbepoetin Alfa. *Pharmaceutical Research*, 21(3), 507–514.

Niven, R. (1996). Protein nebulization II. Stabilization of G-CSF to air-jet nebulization and the role of protectants. *International Journal of Pharmaceutics*, 127, 191–201.

Niwa, T., Shimabara, H., & Danjo, K. (2010). Novel spray freeze drying technique using four fluid nozzle e development of organic solvent system to expand its application to poorly water soluble drugs. *Chemical & Pharmaceutical Bulletin*, 58, 195–200.

Niwa, T., Shimabara, H., Kondo, M., & Danjo, K. (2009). Design of porous microparticles with single micron size by novel spray freeze drying technique using four fluid nozzle. *International Journal of Pharmaceutics*, 382, 88–97.

Ohnesorge, W. V. (1936). Die Bildung von Tropfen an Düsen und die Auflösung flüssiger Strahlen. *Zeitschrift für Angewandte Mathematik und Mechanik*, 16, 355–358.

Ozeki, T., Beppu, S., Mizoe, T., Takashima, Y., Yuasa, H., & Okada, H. (2005). Preparation of two-drug composite microparticles to improve the dissolution of insoluble drug in water for use with a 4-fluid nozzle spray drier. *Journal of Controlled Release*, 107, 387–394.

Papadakis, S. E., & King, C. J. (1988). Air temperature and humidity profiles in spray drying. 1. Features predicted by the particle source in cell model. *Industrial Engineering Chemistry Research*, 27, 2111–2116.

Perry, R. H., Green, D. W., & Maloney, J. O. (1997). Perry's Chemical Engineers' Handbook. McGraw-Hill, New York; London.

Pham, Q. T. (1986). Simplified equation for predicting the freezing time of foodstuffs. *International Journal of Food Science & Technology*, 21, 209–219.

Pikal, M. J., Rambhatla, S., & Ramot, R. (2002). The impact of the freezing stage in lyophilization: Effects of the ice nucleation temperature on process design and product quality. *American Pharmaceutical Review*, 5, 48–52.

Pikal-Cleland, K. A. & Carpenter, J. F. (2001). Lyophilization-induced protein denaturation in phosphate buffer systems: monomeric and tetrameric beta-galactosidase. *Journal of Pharmaceutical Sciences*, 90, 1255–1268.

Pikal-Cleland, K. A., Cleland, J. L., Anchordoquy, T. J., & Carpenter, J. F. (2003). Effect of glycine on pH changes and protein stability during freeze-thawing in phosphate buffer systems. *Journal of Pharmaceutical Sciences*, 91, 1969–1979.

Plateau, J. (1873). Experimental and theoretical statics of liquids subject to molecular forces only. Gauthier-Villars, Paris 1, 4–13.

Rogers, S., Wu, W. D., Saunders, J., & Chen, X. D. (2008). Characteristics of milk powders produced by spray freeze drying. *Drying Technology*, 26, 404–412.

Rogers, T. L., Hu, J., Yu, Z., Johnston, K. P., & Williams, R. O. (2002). A novel particle engineering technology: spray-freezing into liquid. *International Journal of Pharmaceutics*, 242, 93–100.

Rogers, T. L., Johnston, K. P., & Williams, R. O. (2001). Solution-based particle formation of pharmaceutical powders by supercritical or compressed fluid CO_2 and cryogenic spray-freezing technologies. *Drug Development and Industrial Pharmacy*, 27, 1003–1015.

Rogers, T. L., Nelsen, A. C., Sarkari, M., Young, T. J., Johnston, K. P., & Williams, R. O. (2003). Enhanced aqueous dissolution of a poorly water soluble drug by novel particle engineering technology: spray-freezing into liquid with atmospheric freeze-drying. *Pharmaceutical Research*, 20, 485–493.

Rosell-Llompart, J., & Fernández de la Mora, J. (1994). Generation of monodisperse droplets 0.3 to 4 μm in diameter from electrified cone-jets of highly conducting and viscous liquids. *Journal of Aerosol Science*, 25, 1093–1119.

Salman, A. D., Hounslow, M. J., & Seville, J. P. K. (2007). *Handbook of Powder Technology, Granulation*. Elsevier Publishing, Amsterdam.

Saluja, V., Amorij, J.-P., Kapteyn, J. C., de Boer, A. H., Frijlink, H. W., & Hinrichs, W. L. J. (2010). A comparison between spray drying and spray freeze drying to produce an influenza subunit vaccine powder for inhalation. *Journal of Controlled Release*, 144(2), 127–133.

Sauer, H. A. (1969). *Method and apparatus for freeze-freeze drying*. United States Patent US3484946.

Sebastião, I. B., Bhatnagar, B., Tchessalov, S., Ohtake, S., Plitzko, M., Luy, B., & Alexeenko, A. (2019). Bulk dynamic spray freeze-drying part 1: Modeling of droplet cooling and phase change. *Journal of Pharmaceutical Sciences*, 108(6), 2063–2074.

Semyonov, D., Ramon,O., Kaplun, Z., Levin-Brener, L., Gurevich, N., & Shimoni, E. (2010). Microencapsulation of *Lactobacillus paracasei* by spray freeze drying. *Food Research International*, 43, 193–202.

Sonner, C. (2002). *Protein loaded powder produced by spray freeze-drying*. Thesis. Institute for Pharmacy and Food Chemistry, University of Erlangen-Nuremberg, Erlangen, Germany.

Sonner, C., Maa, Y.-F., & Lee, G. (2002). Spray Freeze Drying for protein powder preparation. Journal of Pharmaceutical Sciences, 91(10), 2122–2139.

Sweeney, L. G., Wang, Z., Loebenberg, R., Wong, J. P., Lange, C. F., & Finlay, W. H. (2005). Spray-freeze-dried liposomal ciprofloxacin powder for inhaled aerosol drug delivery. *International Journal of Pharmaceutics*, 305, 180–185.

Tabeei, A., Samimi, A., & Mohebbi-Kalhori, D. (2020). CFD modeling of an industrial scale two-fluid nozzle fluidized bed granulator. *Chemical Engineering Research and Design*, 159, 605–614.

Thurow, H., & Geisen, K. (1984) Stabilization of dissolved proteins against denaturation at hydrophobic interfaces. *Diabetologia*, 27, 212–218.

Tsai, C.-J., Maizel, J. V., & Nussinov, R. (2002). The hydrophobic effect: A new insight from cold denaturation and a two-state water structure. *Critical Reviews in Biochemistry and Molecular Biology*, 37, 55–69.

Wang, W. (2000). Lyophilization and development of solid protein pharmaceuticals. *International Journal of Pharmaceutics*, 203, 1–60.

Wang, Z. L. et al. (2006). Powder Formation by atmospheric spray-freeze-drying. *Powder Technology*, 170, 45–52.

Wanning, S., Süverkrüp, R., & Lamprecht, A. (2015). Pharmaceutical spray freeze drying. *International Journal of Pharmaceutics*, 488, 136–153.

Webb, S. D., Golledge, S. L., Cleland, J. L., Carpenter, J. F., & Randolph, T. W. (2002). Surface adsorption of recombinant human interferon-gamma in lyophilized and spray-lyophilized formulations. *Journal of Pharmaceutical Sciences*, 91, 1474–1487.

Windhab, E. J. (1999). New developments in crystallization processing. *Journal of Thermal Analysis and Calorimetry*, 57, 171–180.

Wu, W. D., Liu, W., Gengenbach, T., Woo, M. W., Selomulya, C., Chen, X. D., & Weeks, M. (2014). Towards spray drying of high solids dairy liquid: Effects of feed solid content on particle structure and functionality. *Journal of Food Engineering*, 123, 130–135.

Xu, J. H. , Li, S. W. , Tan, J. , Wang, Y.J. , & Luo, G. S. (2006). Preparation of highly monodisperse droplet in a T-junction microfluidic device. *AIChE Journal*, 52, 3005–3010.

Yao, J., Furusawa, S., Kawahara, A., & Sadatomi, M. (2014). Influence of some geometrical parameters on the characteristics of prefilming twin-fluid atomization. *Transactions of the Canadian Society for Mechanical Engineering*, 38, 391–404.

Yeom, G.-S. , & Song, C.-S. (2010). Experimental and numerical investigation of the characteristics of spray-freeze drying for various parameters: effects of product height, heating plate temperature, and wall temperature. *Drying Technology*, 28(8), 165–179.

Yu, Z., Garcia, A. S., Johnston, K. P., & Williams, R. O. (2004). Spray freezing into liquid nitrogen for highly stable protein nanostructured microparticles. *European Journal of Pharmaceutics and Biopharmaceutics*, 58, 529–537.

Yu, Z., Rogers, T. L., Hu, J., Johnston, K. P., & Williams III, R. O. (2002). Preparation and characterization of microparticles containing peptide produced by a novel process: spray freezing into liquid. *European Journal of Pharmaceutics and Biopharmaceutics*, 54, 221–228.

Yu, Z., Johnston, K. P., & Williams, R. O. (2006). Spray freezing into liquid versus spray-freeze drying: Influence of atomization on protein aggregation and biological activity. *European Journal of Pharmaceutical Sciences*, 27, 9–18.

CHAPTER 3

Freeze-drying

Freeze-drying is the final step of the spray-freeze-drying (SFD) process. Generally, freeze-drying occurs in three stages: freezing, primary drying (sublimation drying), and secondary drying (evaporative or desorption drying) (Figure 3.1). However, unlike conventional freeze-drying, in SFD, freezing is considered as a distinct unit operation. As a comprehensive discussion on the spray-freezing operation has already been presented in the previous chapter, this chapter would focus on the primary and secondary drying stages. After spray freezing, the frozen feed droplets suspended in the cryogenic liquid are sieved and collected or separated after allowing the cryogen to boil off. Then, the separated frozen particles are freeze-dried in a drying module, under vacuum, atmospheric, or fluidized conditions, depending on which the SFD process is classified as:

i. Vacuum spray-freeze-drying (VSFD)
ii. Atmospheric spray-freeze-drying (ASFD)
iii. Atmospheric fluidized bed spray-freeze-drying (AFBSFD)
iv. Vacuum or sub-atmospheric fluidized bed spray-freeze-drying (VFBSFD or SAFBSFD)

Freeze-drying is an equally crucial step in the SFD process, owing to its influence on the product microstructure and instant reconstitution behavior. Also, it is a factor that determines whether the SFD process can be operated on a batch or continuous mode. This chapter intends to explain the principle of operation and instrumentation of the abovementioned modes of freeze-drying during SFD.

3.1 VACUUM SPRAY-FREEZE-DRYING

Vacuum freeze-drying (VFD), also referred to as *'lyophilization,'* is a well-known dehydration process for the production of high-value and premier-quality foods and pharmaceuticals. Coffee, milk powder and infant formula are popular amongst the vacuum freeze-dried food products. The bioprocessing applications include freeze-drying of living cells, vaccines, enzymes, and biological media (Chow et al., 2008; Fonseca et al., 2004; Nasirpour et al., 2007; Pardo et al., 2002; Thomas et al., 2004; Yang et al., 2010). Essentially, primary drying involves the sublimation of ice crystals formed during the spray-freezing step. It is conducted in a low-temperature and low-pressure environment. The voids left behind after the sublimation of ice crystals lead to the porous microstructure of the final product.

Freezing and secondary drying stages are usually completed within hours. But, completion of primary drying takes hours to even days, depending on the shelf temperature, the critical temperature of the product, and its solid concentration (Scoffin 2014). The long duration of primary drying is responsible for the prolonged processing time and high manufacturing cost of vacuum freeze-drying. While sublimation accounts for nearly 45% of the total energy consumption of

DOI: 10.1201/9781003019312-3

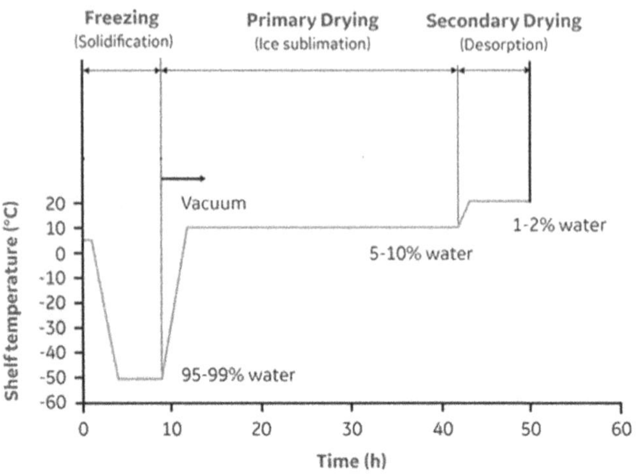

Figure 3.1 Schematic diagram of the freeze-drying process comprising three stages that remove water from a product by freezing it and placing it under specific vacuum pressures and temperatures (Adapted from Cytiva, 2020; © Copyright Cytiva – Reproduced with permission of the owner).

vacuum freeze-drying, the application of vacuum and condensation each contributes to around 25% of the total energy consumption (Ratti, 2001). Hence, the main target of a manufacturer or researcher working on freeze-drying process development would be to minimize the primary drying time (Patel, Doen, & Pikal 2010). A practical approach to reducing the primary drying time is to ensure a minimal amount of unfrozen water in the frozen product. Most of the water must have been converted into ice during the preceding freezing step. Here originates the advantage of the spray-freezing operation. The number of droplets that remain unfrozen after freezing (N) at temperature (T; always a negative value) is inversely related to the probability of freezing per unit time (P), which increases exponentially with decreasing temperature at a steady cooling rate (Eq. 3.1; Bigg, 1953).

$$P = \frac{1}{N}\frac{dN}{dt} = BVe^{\frac{-T}{\tau}} \tag{3.1}$$

Where B and τ are constants, V is the volume of droplets, and t is time. At a constant freezing temperature, P must be high at low temperatures, thereby reducing the number of unfrozen droplets. Thus, the concern of unfrozen droplets may not be encountered during SFD, as spray-freezing into cryogenic liquid/vapor/vapor-over-liquid deals with substantially low temperature. It has been shown that the fraction of unfrozen droplets at the end of spray-freezing a 10% coffee solution at −40°C is as low as 0.027%. However, it increases with an increase in freezing temperature and the solid concentration of the feed liquid (MacLeod et al., 2006) (Table 3.1).

Once the desired level of freezing is attained, the primary drying phase is executed by transferring the spray-frozen droplets to the pre-chilled shelves of a lyophilizer (Figure 3.2). The primary drying phase is performed at low pressure achieved using a vacuum pump that maintains the desired vacuum level in the freeze-drying chamber (Stapley, 2008). And, the product shelves are maintained at a temperature which is more than the freezing temperature but less than the critical temperatures of the food product or the formulation of biologicals to be freeze-dried. The critical temperature is characteristic of every product or formulation, below which it must be cooled and maintained for complete solidification and sublimation, respectively. Complying with

Table 3.1 The fraction of unfrozen material after spray-freezing of instant coffee solution (MacLeod et al., 2006)

Concentration (%w/w)	Freezing temperature			
	-10°C	-20°C	-30°C	-40°C
10	0.085	0.054	0.040	0.027
20	0.165	0.065	0.058	0.041
30	0.176	0.067	0.059	0.042
40	0.494	–	0.099	0.067
50	0.540	–	0.125	0.081

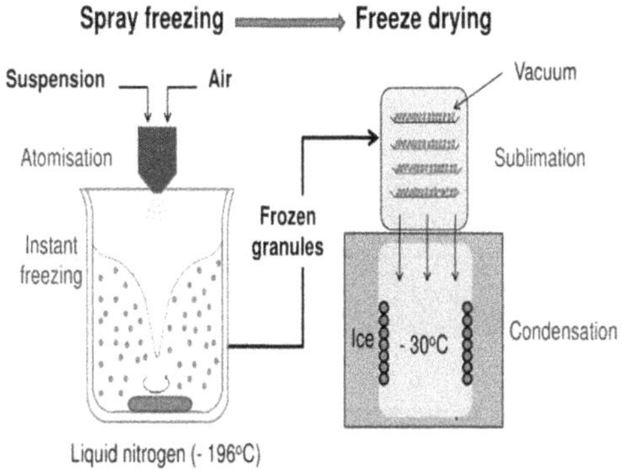

Figure 3.2 Schematic of a vacuum spray-freeze-drying process (PowderPro, 2012).

the abovementioned pre-requisite is central to preventing *'collapse,'* a major processing defect that may occur during primary drying. The three types of critical temperatures associated with the primary drying stage are *collapse temperature* (T_c), *eutectic temperature* (T_e), and *glass transition temperature* (T_g). Identification of these critical temperatures aids in the design of rational, safe, and effective freeze-drying cycles for different formulations (Franks, 1990; Pikal, 1990; Ross, Gaster & Ward, 2008). Instruments and analytical methods are available for an accurate estimation of these critical temperatures.

3.1.1 Critical temperatures

Collapse temperature (T_c) is the temperature at which a material softens to the point at which it would not be able to support its structure (Ross, Gaster & Ward, 2008). Precisely, *'collapse'* occurs when the viscosity of structural material drops below a level, at which it cannot support its weight against gravity (Bhandari, Datta, & Howes, 1997). The collapse phenomenon has a detrimental effect on the properties of the final freeze-dried product, leading to volatile loss during storage, poor reconstitution behavior, non-uniform moisture distribution, and extensive caking (Figure 3.3) (Levi & Karel, 1995; Knopp, Chongpraser, & Nail 1988). Often, the collapse temperature of a material tends to be lower than that applied during drying. Hence, as a thumb rule, the product temperature during primary drying is maintained 2–5°C below its T_c to avoid collapse and maintain an elegant cake structure.

Further, the dependence of T_c on the total solid content of feed solution has been established with clear evidence. Consequently, T_c of a formulation is always reported along with its solid content (Table 3.2). An increase in solid content shifts the T_c to higher values (Table 3.3), which is advantageous as it permits primary drying to be conducted at not very low drying temperatures.

Freeze dried with product temperature above Tc
Shelf = -10.0°C
Tc = -20.0°C
Product = -15.0°C
Result is a collapsed cake
Freeze dried with product temperature at Tc
Shelf = -15.0°C
Tc = -20.0°C
Product = -20.0°C
Result is a partially collapsed cake
Freeze dried with product temperature below Tc
Shelf = -20.0°C
Tc = -20.0°C
Product = -25.0°C
Result is an excellent cake

Figure 3.3 Illustration of the effect of shelf temperature on product collapse during primary drying stage of freeze-drying (Ross, Gaster & Ward, 2008; Image Courtesy: Biopharma Process Sytems Ltd., United Kingdom).

In addition, T_c of any feed material increases with increase in its molecular weight and a decrease in moisture content (Oetjen, 1999). The molecular weight of a formulation can be increased by adding bulking materials such as dextran, fructose, glucose, glycine, maltose, and polyethylene glycol. Thus, the phenomenon of collapse can be defined as *'time, temperature, and moisture-dependent viscous flow that results in loss of structure, reduction in pore size, and volumetric shrinkage of dried food products'* (Levi & Karel, 1995). Hence, a product development scientist needs to understand the collapse behavior of a formulation. The knowledge of collapse phenomenon will aid in the logical optimization of freeze-drying formulations by tuning the concentration of total solids or by adding high molecular weight excipients.

Glass transition temperature (T_g) is the temperature at which the frozen material changes from a glassy (brittle) to rubbery (flexible or soft) state (Abbas, Lasekan, & Khalil 2010; Ross, Gaster & Ward, 2008). When the temperature of frozen material (T) exceeds T_g ($T>T_g$), the viscosity of amorphous matrix drops to cause collapse. This decrease is a function of (T-T_g), which can be explained by the William-Landal-Ferry (WLF) relationship (Eq. 3.2) (Bhandari & Howes, 1999).

Table 3.2 Collapse temperatures of selected foods

Material	Total solids (%)	T_c (°C)	Reference
Coffee extract	25	-20.0	Fellows (2002)
Orange juice	23	-24.0	Kudra & Strumillo (1998)
Grapefruit juice	16	-30.5	Kudra & Strumillo (1998)
Sweetened concord grape juice	16	-33.5	Kudra & Strumillo (1998)
Prune extract	20	-35.0	Kudra & Strumillo (1998)
Lemon juice	9	-36.5	Kudra & Strumillo (1998)
Apple juice	22	-41.5	Fellows (2002); Bellows & King (1972); Fennema (1996); Kudra & Strumillo (1998)
Pineapple juice	10	-41.5	Kudra & Strumillo (1998)
Concord grape juice	16	-46.0	Kudra & Strumillo (1998)

Table 3.3 Influence of solid content and concentration of bulking agents on the T_c of selected formulations

Material	Total solid content (%)	Weight fraction of glycine (%)	T_c (°C)	Reference
Sucrose	7	0	-33	Kasraian, Spitznagel, Juneau, & Yim (1998)
		14	-33	
		29	-33	
		43	-33	
		57	-20/-15	
		71	-10	
		86	-10	
		100	-7	
Sucrose	5	80	-16	Passot et al. (2005)
Sucrose (1%) + PEG (1%)	6	67	-15	
Sucrose (1%) + Tween 80 (0.02%)	5	80	-18	
Maltose	5	80	-15	
Maltodextrin (DE 6)	5	80	-10	
Polyvinylpyrrolidone (PVP) (25)	5	80	-15	
PVP (25) (1%)+ polyethylene glycol (PEG) (1%)	6	67	-18	
PVP (25) (1%) + Tween 80 (0.02%)	5	80	-15	
Maltose (1%) Active pharmaceutical ingredient (2%)	5	40	-20/-10 Complete at -7°C	Ma et al. (2001)

$$log\frac{\mu}{\mu_g} = \frac{17.4(T - T_g)}{51.6 + (T - T_g)} \tag{3.2}$$

Where μ is the viscosity, μ_g is the viscosity at T_g, and T is the temperature. Similar to collapse temperature, a product must be freeze-dried at a temperature below its glass transition temperature. Unlike collapse, glass transition is a physical change that is independent of the concentration of substances in the solution (Meister & Gieseler, 2008).

In studies pertaining to spray-freeze-drying of amorphous materials, two different types of glass transition temperatures have been considered: (i) glass transition temperature of the maximally freeze-concentrated solution (T_g'), which is relevant in the frozen solution state; (ii) glass transition temperature of the drying solid phase (T_g) that holds good after the primary drying begins (Straller & Lee, 2017). Thus, to decide on the shelf temperature during the primary drying so as to avoid collapse, it is more relevant to consider T_g' than T_g. Usually, the values of collapse temperature are higher than the T_g' values by 1–3°C. Drying at a shelf temperature above T_c may result in the macrocollapse of the freeze-dried product. Provided the protein stability is preserved, drying above T_g', but below T_c has been practiced for highly concentrated protein formulations, which permit higher product temperatures and lead to more rapid drying and shorter drying times without the loss of cake structure (Colandene et al., 2007).

Unstable amorphous products that tend to form a metastable glass with incomplete crystallization during freezing are subjected to an optional thermal treatment process before primary drying, which is termed as *'annealing.'* During annealing, the frozen samples are held at a temperature between the melting point and T_g for a definite period. Annealing promotes the growth of larger ice crystals to further freeze concentrate the solute phase. As already explained, an increase in solute concentration shifts T_c to the higher side and thus permits primary drying to be conducted at a higher temperature. This serves as an alternative strategy to reduce the duration and energy cost of primary drying. Another advantage of annealing-induced larger ice crystals is the reduction in specific surface area and internal pore area of the final product, which reduces its hygroscopicity. For instance, during the spray-freeze-drying of trehalose, annealing for 2 hours at a shelf temperature of 0°C reduced the moisture uptake of the final product from about 4.5% to 3.5%, when exposed to a relative humidity of 33% for 1 hour (Sonner, Maa, & Lee, 2002).

From the above discussion, it is apparent that the cost of freeze-drying will increase drastically with feed materials whose frozen solutions have a low critical temperature (T_c, T_e, and T_g) and hence require a low drying temperature (Ishwarya, Anandharamakrishnan & Stapley, 2015). Besides, when conventional freeze-drying is employed after the spray freezing step of SFD, the energy-intensive operation necessitated by the vacuum requirement and batch mode operation leads to high fixed and operating costs (Matteo, Donsi, & Ferrari, 2003). Apart from T_c and T_g, T_e or eutectic temperature is the critical temperature above which a crystalline solute material melts and prevents any structure formation after the removal of solvent (Ross, Gaster & Ward, 2008).

3.1.1.1 Methods to determine the critical temperature of formulations and pressure to be set during the primary drying phase

The preceding sections emphasized the importance of operating the primary drying stage below the critical temperatures of the formulations. Indeed, the first step in optimizing the freeze-drying cycle for a product is to conduct a thermal characterization study to identify the critical temperatures. A product development scientist should have a thorough knowledge of the T_g', T_c and T_e, to design a robust and product-specific freeze-drying cycle that can be completed in the shortest duration (Schwegman & Neiblas, 2015). Different methods are available to determine the critical temperatures of products, of which the most commonly used, are differential scanning calorimetry (DSC) and freeze-drying microscopy (FDM). Generally, DSC is used to calculate T_g', and FDM is used to determine T_c (Horn & Friess, 2018). Case-studies on the determination of critical temperatures using each of these techniques are presented in the forthcoming sections.

3.1.1.1.1 Differential Scanning Calorimetry (DSC)

DSC is the conventionally used approach to determine the maximum temperature that a product can withstand during any drying procedure. It is an experimental technique for measuring the energy (heat) absorbed or released by a sample as it is heated or cooled at a constant temperature rate (Ford & Timmins, 1989). During the DSC analysis, a very small amount of the sample is freeze/thawed in small aluminum crucibles (Horn & Friess, 2018), to mimic the freeze-drying cycle. And, in a DSC thermogram, T_g' is the temperature at which the heat capacity of a freeze-concentrated formulation shows a noticeable change (Pansare & Patel, 2016). An example of a typical DSC thermogram is shown in Figure 3.4, in which T_g is identified as the temperature at which there is an apparent shift in the baseline.

Anandharamakrishnan (2008) adopted the DSC approach to determine the T_g' of a whey protein isolate (WPI) solution (40% w/v concentration). 15-20 mg of the WPI solution was weighed to 0.1 mg accuracy and placed in a pre-weighed aluminium pan, which was then hermetically sealed. Subsequently, the sample pan and reference pan (empty) were placed in the furnace of the DSC instrument and subjected to a heating/cooling cycle, controlled by a predefined temperature program. The sample was scanned from 20°C to −80°C at a cooling rate of 10°C/min. After the sample attained −80°C, it was again heated to −40°C at 10°C/min and held at that temperature (isothermal state) for 15 minutes. Then, the sample was recooled to −80°C at 5°C/min and scanned from −80°C to 20°C at 10°C/min. Based on the resultant DSC thermogram (Figure 3.5), the temperature of −25.8°C (corresponding to the evident shift in the baseline) was considered as the T_g' of WPI. Thus, in this study, it was ascertained that the freeze-drying of whey proteins should be operated at a temperature below −25.8°C to prevent structural collapse (Anandharamakrishnan, 2008).

3.1.1.1.2 Freeze-drying Microscopy (FDM)

FDM provides a visual indication of T_c and T_e. The freeze-drying microscope is essentially a 'micro freeze dryer' that imitates the freeze-drying process on a small scale by freezing and drying small volumes of formulation (~2 μL) under the microscope (Meister & Gieseler, 2009; Bosch, 2014). An FDM system (Figure 3.6[a]) includes a compound microscope that is placed on a vacuum-tight temperature-controlled stage, which acts as a micro-freeze-dryer. The sample behavior during the mimicked freeze-drying cycle is visualized, and the images are recorded at defined time intervals using a digital camera. The sample stage chamber (Figure 3.6[b & c]) under

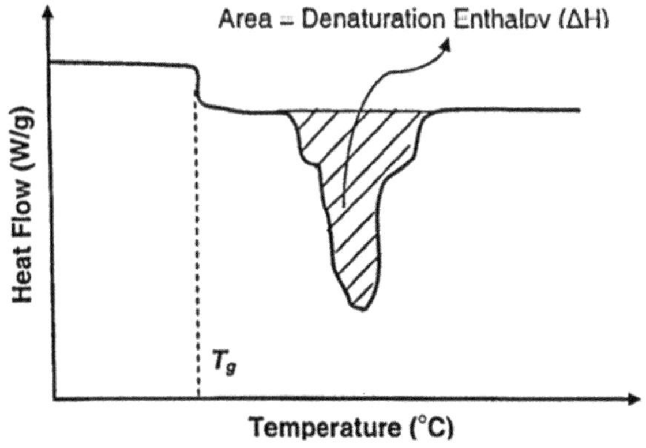

Figure 3.4 An example of DSC thermogram (Modified from Anandharamakrishnan, 2008).

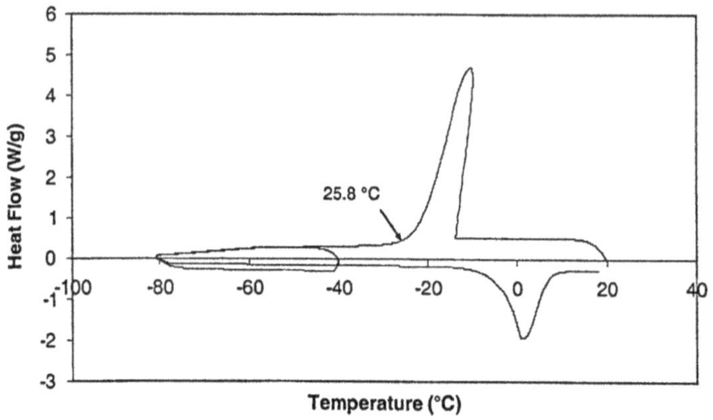

Figure 3.5 Thermal behavior of whey protein isolates solution (Anandharamakrishnan, 2008).

the microscope functions like a miniature freeze-dryer. It is cooled by liquid nitrogen supplied from a dewar through a pump, and a controller regulates the temperature of the sample stage. A vacuum pump assists in the drying of the sample (Hajare, More, Walekar & Hajare, 2012). To determine the critical temperatures, FDM induces collapse in the sample while being dried and records the images of its structural behavior and the time-point at which the collapse occurs. In an FDM image, T_c is the temperature that marks the onset of a visible collapse or full collapse (Meister & Gieseler, 2009; Bosch, 2014).

The sampling format in an FDM is shown in Figure 3.7. In a study by Horn & Friess (2018), T_c of different sugar and protein formulations was determined by FDM. 2 μL of sample along with a spacer was placed on the sample holder and covered by a glass slide. The precision-cut spacers maintain a constant thickness of the sample layer. The samples were cooled down to −50°C at the rate of 1°C/min. When the sample reached the holding temperature and began to freeze, the vacuum pump was switched ON to apply a pressure of 0.1 mbar. Then, the samples were heated to −40°C or −30°C at 5°C/min. The sample was maintained under the above conditions to achieve the required thickness of sublimation front. As mentioned above, T_c was detected as the onset of collapse in the subsequent drying step at 1°C/min to 0°C by capturing the images at an interval of 2s. The values of T_c obtained from FDM are depicted in Figure 3.8.

During FDM visualization, the sublimation front moves through the frozen sample, and the temperature-time profile is recorded in real-time (Figure 3.9[a]). The freeze-drying characteristics can be studied by increasing or decreasing the sample temperature. The T_c of the formulation can be ascertained by examining the freeze-dried structure behind the interface (Figure 3.9[b]). The temperature may be altered (reheating or recooling) to have a clearer visualization of the collapse phenomenon (Figure 3.9[c & d]). From the FDM images, it can be seen that the sample loses its structural integrity at $T>T_c$ (Figure 3.9[b]). And, the structure is regained as the sample is recooled ($T<T_c$) (Figure 3.9[c]) and collapse again on reheating (Figure 3.9[d]).

Earlier, FDM was used only for the determination of collapse temperatures. But, the present-day equipment can also identify phenomena such as melting, crystallization, skin/crust formation, effects of annealing on ice crystal growth, and solute structure (SP Scientific Inc., 2019). Apart from DSC and FDM, the other modern-day techniques for the determination of critical temperatures are thermomechanical analysis (TMA) and dynamic mechanical analysis (DMA), atomic force microscopy (AFM) including cryo-AFM for the analysis of biological samples,

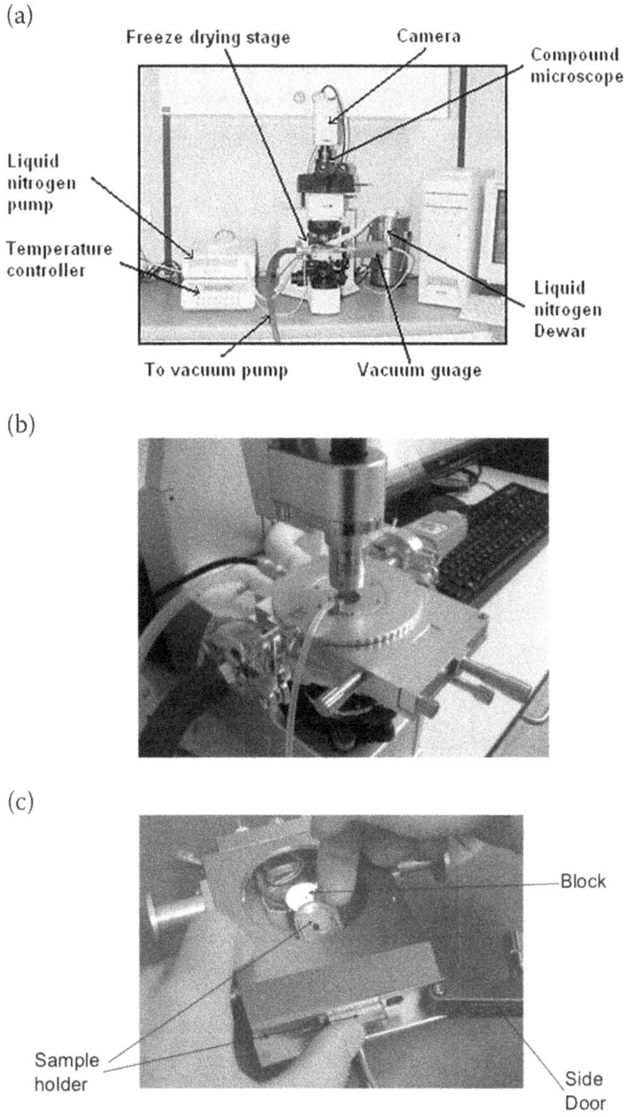

Figure 3.6 A freeze-drying microscope: (a) Components; (b) Sample stage; (c) A closer look of the sample stage (Biopharma Technology Limited, 2021; Image Courtesy: Biopharma Process Sytems Ltd., United Kingdom).

dielectric spectroscopy (DS) and electrical impedance or resistance (ER) analysis (Ward & Matejtschuk, 2010).

After identifying the critical temperature of the formulation and the shelf temperature that is typically colder than the former, the next step is to decide on the vacuum level. A scientific approach to determine the appropriate vacuum level for freeze-drying is to follow the vapor pressure of ice table (Table 1.1; Chapter 1). As a thumb rule, the system pressure is set at 20–30% of the vapor pressure of ice at the target product temperature. Sublimation would take place when the set level of system pressure (or applied vacuum) is lower than the vapor pressure of ice corresponding to the product temperature at that time-point (Barley, 2021).

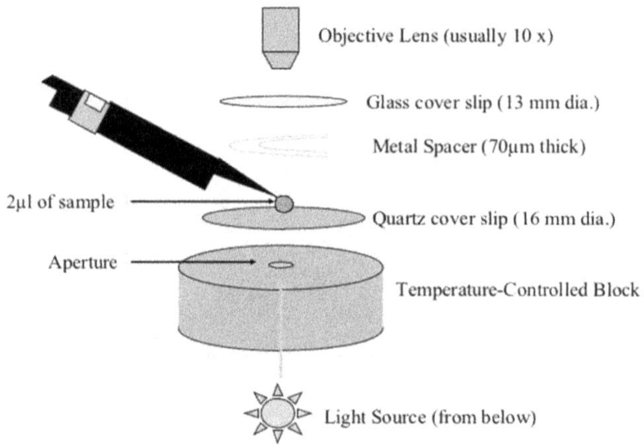

Figure 3.7 Sampling format in a FDM (Biopharma Technology Limited, 2021; Image Courtesy: Biopharma Process Sytems Ltd., United Kingdom).

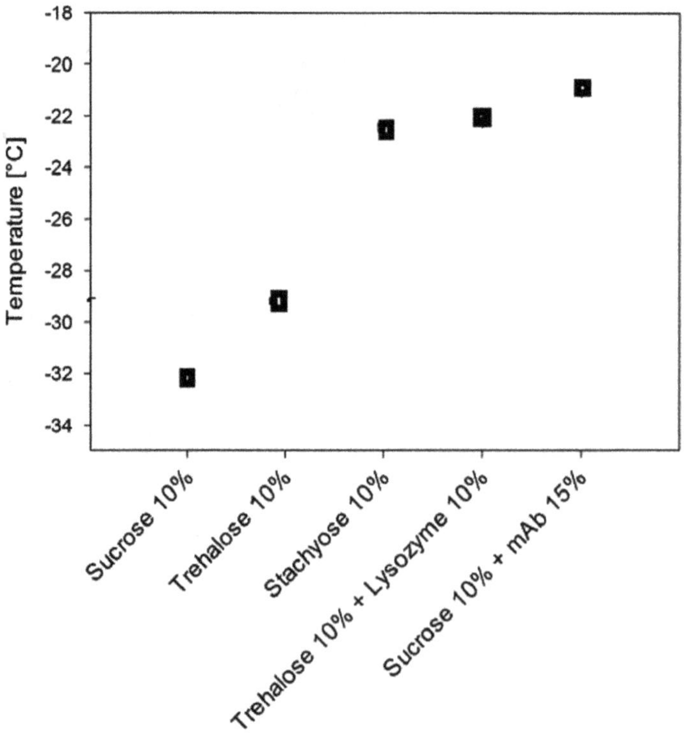

Figure 3.8 T_c values of different sugar and protein formulations determined by FDM (Modified from Horn & Friess, 2018).

At the end of the primary drying phase, 70-80% of the ice is sublimed off, and the residual water remains in the amorphous matrix. The completion of primary drying is detected by a sharp decrease in the dew point temperature (T_{dew}). T_{dew} is the temperature at which the equilibrium vapor pressure

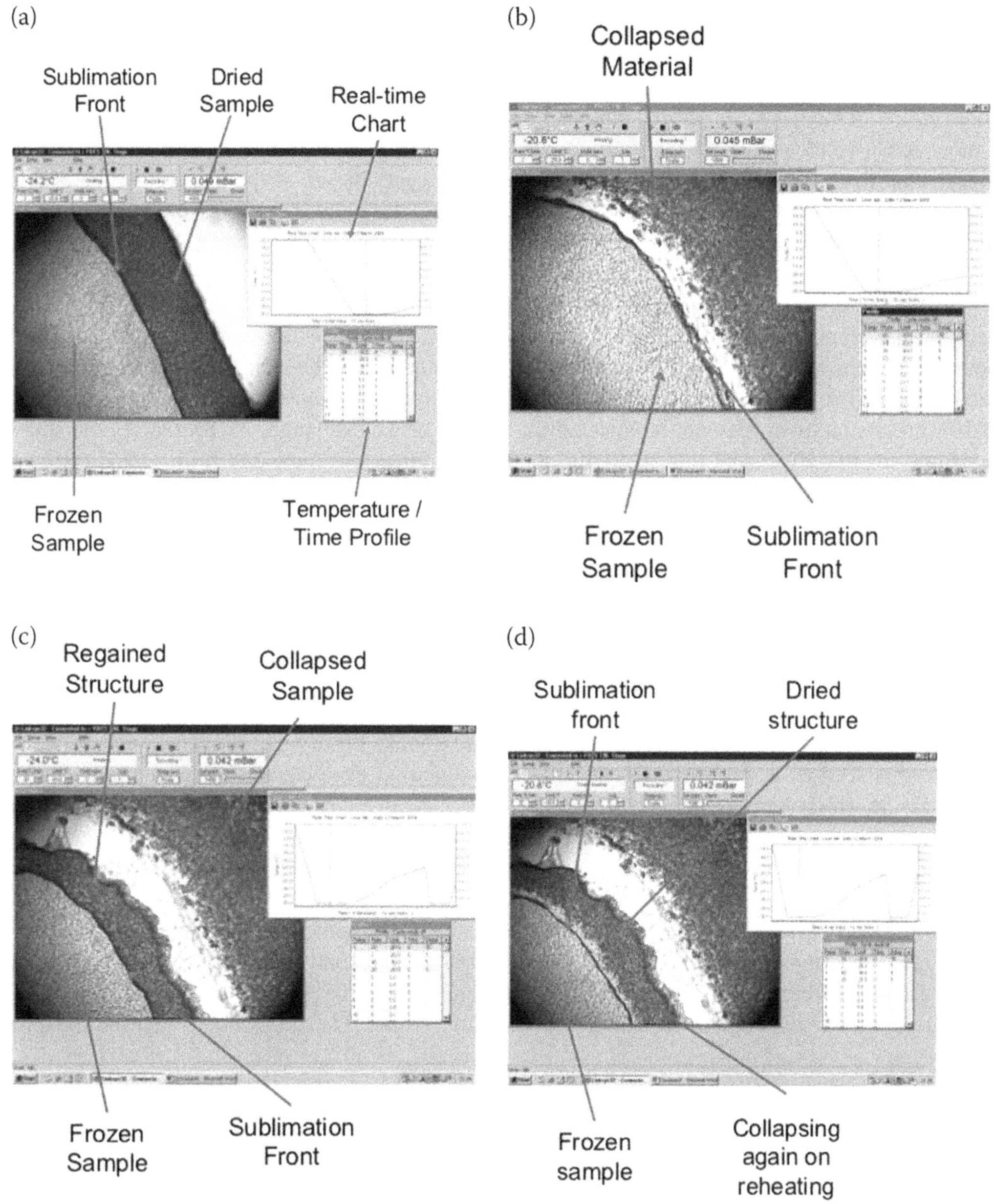

Figure 3.9 Visualization of the *collapse* phenomenon in FDM imaging (Biopharma Technology Limited, 2021; Image Courtesy: Biopharma Process Sytems Ltd., United Kingdom).

of ice is equal to the partial pressure of water. It has also been defined as the temperature at which the air is saturated with water vapor such that some of the water vapor starts condensing into liquid water. An electronic moisture sensor is used to measure the dew point. In this device, the principle of measurement is based on monitoring the changes in the capacitance of a thin film of aluminum oxide due to the adsorption of water at a given partial pressure. The change in capacitance is further converted into a voltage output which is previously calibrated to read the dew point or the partial pressure of water (Roy & Pikal, 1989). In the plot of primary drying time versus dew point, the onset of the drop in dew point or Pirani pressure indicates the completion of sublimation.

Alternatively, the endpoint of primary drying can also be detected by tracking the thermal conductivity of the gas in the drying chamber using a Pirani gauge (Nail & Johnson, 1992). This is a more sensitive technique, as the thermal conductivity of water vapor is approximately 1.6-fold higher than that of nitrogen. During primary drying, essentially all of the gas in the chamber is water vapor. The point of time during primary drying at which the Pirani pressure begins to drop implies the change in chamber gas composition from mostly water vapor to nitrogen. This is an indication that the sublimation is complete (Patel, Doen & Pikal, 2010).

Secondary drying is a comparatively short phase of desorption or evaporative drying that is performed at temperatures higher than primary drying to remove the unfrozen, residual water bound to the amorphous phase. The temperature transition during the switch-over from primary to secondary drying should be done gradually to avoid product collapse (Abdul-Fattah & Truong, 2010). Pressure may or may not be lowered during this phase (McHugh, 2018). During the spray-freeze-drying of coffee solutions, while the shelf temperature was increased from −10°C (primary drying) to +10°C (secondary drying), the corresponding system pressure was reduced from 107 Pa to 40 Pa (Ishwarya & Anandharamakrishnan, 2015).

3.1.2 Construction and operation of a vacuum spray-freeze-drying apparatus

When operated in batch mode, the spray-frozen particles are collected and transferred to the shelves of a conventional lyophilizer, wherein the primary and secondary drying phases are conducted, and the final product is obtained. The details on the construction and operation of a conventional freeze-dryer are available in standard textbooks on drying, and hence not discussed here. In this section, a brief description will be provided on an integrated vacuum spray-freeze-drying rig, in which spray-freezing and vacuum freeze-drying have been conducted as a single-step continuous process in the same system. Such a vacuum spray-freeze-drying apparatus (Zheng, 2017) is shown in Figure 3.10. The components of the vacuum-spray-freeze-dryer are listed in Table 3.4.

During the operation of a vacuum spray-freeze-dryer, the refrigerant temperature in the first refrigerating jacket is maintained at −60°C, and that in the second refrigerating jacket is held at −20°C for spray-freezing and freeze-drying, respectively. First, the pressure of the sealed container is reduced to a level of 300 Pa, using the vacuum system. Then, the liquid distribution pipe

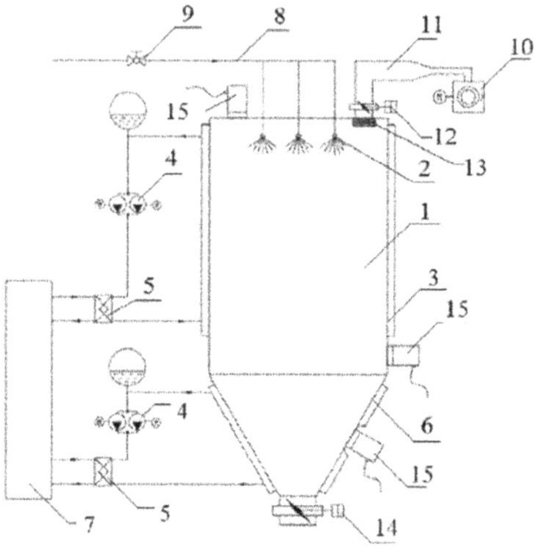

Figure 3.10 Construction of a vacuum spray-freeze-dryer (Zheng, 2017).

Table 3.4 Components of a vacuum spray-freeze-dryer and their functions

Part No.	Component	Description/Function
1	Sealed container	• Cylindroconical in shape. • Accommodates the material during the vacuum spray-freeze-drying process. • Can be thermally and cold-conducted.
2	Atomizing nozzle	• Atomizes and disperses the liquid to be freeze-dried into liquid drops and sprays the liquid drops into the container (1). • The diameter of the nozzle is 1.0 mm
3	Condensing device	• The first cooling jacket provided on the outer side of the upper portion of the container (1). • Contains refrigerant (silicone oil) inside. • Condenses a portion of the vaporized droplets.
4	Fluid pump	• Connected to the refrigerant circulation pipeline.
5	Heat exchanger	• Cools the refrigerant.
6	Refrigerating jacket (or) temperature adjusting sleeve	• Filled with a refrigerant. • Prevents the spray-frozen particles from melting. • Provides the heat of sublimation to the frozen granules during the drying process.
7	Compressor	• Functions as a refrigeration system.
8	Liquid distribution pipe	• Supplies the liquid feed to the atomizing nozzle.
9	Regulating valve	• Regulates the flow through the liquid distribution pipe.
10	Vacuum pump	• Creates a vacuum environment in the container (1) during the vacuum spray-freeze granulation stage. • Causes some of the components of the liquid drops be vaporized and absorb heat to cool and freeze the liquid drops into granules. • Performs continuous vacuum pumping during the drying process to maintain the vacuum environment
11	Pipe	• The vacuum pump (10) is connected to the connection port of the container (1) through a pipe (11), and the pipe (11) is provided with a valve (12).
12	Valve	
13	Filter	• Disposed on the connection port of the container.
14	Discharge valve	• For discharging the frozen particles. • The discharge valve can be connected to a drying device to dry the frozen particles.
15	Vibrator	• The particles are shaken to the discharge valve and remain in the loose vibrator, which is driven to vibrate by a motor.

supplies the feed liquid to the atomizing nozzle at a spray pressure of 0.4 MPa. Subsequently, the feed solution is atomized, dispersed into droplets, and sprayed into the container. The water in the droplet vaporizes and absorbs heat under the vacuum environment to lower the temperature of the droplet, as the vacuum system continues to evacuate. The first cooling jacket condenses the vaporized water on the inner wall of the container to maintain the container. The degree of vacuum in the container and the droplet temperature continues to drop with time until drying is complete. The above events are indicative of *'wet-bulb'* depression during the sublimative cooling of the primary drying phase. *'Wet-bulb'* depression (ΔT_{wb}) is the difference in temperature between the gas (T_g) and the particles (T_p) ($\Delta T_{wb} = T_g - T_p$). The dried particles are shaken by the vibration of the vibrator to enter the discharge valve through which they are discharged. The VSFD process is typically completed within a few seconds. It is effectively regulated by a control system comprising a programmable logic controller and a computer (Zheng, 2017). The vacuum spray-freeze-dryer, as described above, has been deemed suitable for the quick freeze-drying of liquid juice and medicinal extracts (Laboao, 2021).

3.1.3 Applications, advantages, and limitations of vacuum spray-freeze-drying

The applications of vacuum spray-freeze-drying in the production of wide-ranging food and bioproducts are charted in Table 3.5. From the findings of various studies presented in Table 3.5, it is evident that VSFD leads to reduced drying time compared to conventional freeze-drying. Also, it results in products with controlled particle size and size distribution (owing to the atomization step), and porous microstructure. Consequently, the products exhibit good reconstitution behavior and better stability than those obtained from conventional freeze-drying and spray drying.

Table 3.5 Applications of vacuum spray-freeze-drying process

Process	Product	Focus & Findings	Researchers
SFV/L + VSFD	Salbutamol sulfate	• Used thermal ink-jet nozzle for atomization for the production of Produced inhalable porous particles of size less than 5 μm.	Mueannoom, Srisongphan, Taylor & Hauschild (2012)
SFV/L + VSFD	Inhaler formulation of drug-loaded lipid-polymer hybrid nanoparticles	• Compared spray drying and spray freeze drying for dry powder inhaler formulation. • Nano-aggregates produced by SFD had superior aerosolization efficiency and fine particle fraction with lower mass median aerodynamic diameter.	Wang et al. (2012)
SFV/L + VSFD	Thermally-sensitive polymeric substances	• Production of dry-powder aggregates of thermally sensitive polymeric nanoparticles. • Examined morphology, production yield, flowability, and aqueous reconstitution of SFD particles with the effect of atomization rate, feed concentration, and feed rate.	Kho & Hadinoto (2011)
SFV/L + VSFD	Protein	• A systematic characterization of the SFD trehalose/mannitol/dextran particulate composition. • Produced microencapsulated protein particles using the SFD technique for sustained release.	Rochelle & Lee (2007)
SFV/L + VSFD	Active pharmaceutical ingredient	• Showed that the processing of salmeterol xinafoate (SX) by SFD technique could be a constructive approach to the production of various forms of the drug. • Drastic changes in the physical characteristics of microparticles could be achieved by changing the composition of the bulking agent (trehalose and cyclodextrin).	Rahmati et al. (2013)
SFL + VSFD	Protein (Bovine Serum Albumin, BSA)	• Investigated the effect of atomization conditions and formulation variables on particle size and stability of excipient free and zinc complexed BSA. • A significant atomization parameter was the mass flow ratio (mass of atomization N2 relative to that of liquid feed).	Costantino et al. (2000 and 2002)

Table 3.5 *(Continued)* Applications of vacuum spray-freeze-drying process

Process	Product	Focus & Findings	Researchers
		• Protein with excipients showed greater stability.	
SFL + VSFD	Bovine Serum Albumin	• Compared microparticles produced by SFV/L and SFL process. • Intense atomization and ultra-rapid freezing reduced the degree of denaturation and aggregation of the nanostructured protein.	Yu et al. (2004)
SFL + VSFD	Carbamazepine and Danazol	• SFL process enhanced the dissolution rates of poorly water-soluble compounds.	Rogers et al. (2002)
SFL + VSFD	Danazol, Hydroxypropyl -b-cyclodextrin	• Produced microencapsulated free-flowing powder consisting of API. • Exhibited better dissolution than conventional freeze-dried powder.	Rogers, Hu, et al. (2002)
SFL + VSFD	Danazol, carbamazepine	• Produced nanostructured porous microparticles with enhanced wetting and dissolution rates.	Hu et al. (2002)
SFL + VSFD	Carbamazepine (CBZ)	• Enhanced the SFL technique for the preparation of nanoparticles of CBZ, by using acetonitrile system.	Hu, Johnston, & Williams (2003)
SFL + VSFD	Trypsinogen	• Developed the model of dispersion and solidification of aqueous droplets in the turbulent flow field. • The developed model enabled the estimation of the temperature profile of the continuous and dispersed aqueous phases as well as the final size of the frozen droplets.	Henczka, Baldyga, & Shekunov (2006)
SFL + VSFD	Albuterol sulfate	• Investigated processing parameters to prepare polyethylene glycol micro-particles for drug delivery.	Barron et al. (2003)
SFL + VSFD	Inhaled drug-Kanamycin	• Physical and antibiotic properties of kanamycin powders obtained by spray freeze drying (SFD) were compared with those of raw kanamycin. • SFD procedure was optimized for the use of Kanamycin drug.	Her et al. (2010)
SFL + VSFD	Microencapsulated *Lactobacillus paracasei*	• Produced micro-capsules of *Lactobacillus paracasei* with high viability. • High viability was obtained (>60%).	Semyonov et al. (2010)
SFL + VSFD	Protein	• Produced uniform encapsulation of stable protein nanoparticles. • High loadings (10-15%) of proteins in microspheres with low burst release and low protein aggregation. All three of these goals were met in this study.	Leach et al. (2005)
SFL + VSFD; SFV/L + VSFD	Lysozyme	• Studied the influence of protein aggregation and biological activity of the enzyme. • Reduced protein aggregation, less loss of enzyme activity, and increased particle surface area were achieved for the SFL process.	Yu, Johnston, & Williams (2006)
SFL + VSFD;	Lysozyme		

(Continued)

Table 3.5 ***(Continued)*** **Applications of vacuum spray-freeze-drying process**

Process	Product	Focus & Findings	Researchers
SFV/L + VSFD		• Increased cooling rate produced a high specific surface area of the spray freeze-dried powder. • Among the two cryogens, the cooling rate for SFL into liquid nitrogen was three times slower than isopentane (i-C5) due to the Leidenfrost effect.	Engstrom, Simpson, Lai, et al. (2007)
SFV/L + VSFD	Docosahexaenoic acid (DHA)	• Investigated the plausibility of using SFD as a microencapsulation technique for bioactives that are susceptible to oxidation. • SFD resulted in microencapsulated DHA with lower oxidation (13%) compared to freeze-drying (31%) and spray-drying (33%). • Storage study indicated that SFD microencapsulated powders showed lower peroxidation than freeze-dried and spray-dried encapsulates. • The porous surface morphology of the spray-freeze-dried product resulted in good rehydration behavior. • The total processing time of SFD was 2.75-fold lesser (4 h) than conventional freeze-drying (11 h).	Karthik & Anandharamakrishnan (2013)
SFV/L + VSFD	*Lactobacillus plantarum*	• Characteristics of SFD microencapsulated *Lactobacillus plantarum* powder was compared with the spray-dried (SD) and freeze-dried (FD) counterparts. • SFD and FD processed microencapsulated powders showed 20% higher cell viability than the SD samples. • In simulated gastrointestinal conditions, the SFD and FD cells showed high tolerance (up to 4 h) than SD samples and unencapsulated cells in acidic and pepsin conditions. • The morphology of SFD samples showed spherical shaped particles numerous fine pores on the surface, which resulted in good rehydration behavior of the powdered product. • The drying time of the SFD process was 5 h compared to 8 h of the FD process.	Dolly et al. (2011)
SFV/L + VSFD	Soluble coffee	• The suitability of the SFD technique for soluble coffee processing was evaluated. • SFD and FD coffee powders exhibited a comparable aroma profile, as indicated by the electronic nose analysis. • SFD resulted in higher volatile retention (93%) than FD (77%) and SD (57%). • SFD coffee showed instantaneous solubility due to its highly porous morphology. • SFD coffee depicted monomodal particle size distribution.	Ishwarya & Anandharamakrishnan (2015)

Table 3.5 ***(Continued)*** **Applications of vacuum spray-freeze-drying process**

Process	Product	Focus & Findings	Researchers
		• SFD resulted in higher free and tapped bulk densities of the product compared to SD and FD. • SFD coffee exhibited free-flowing characteristics. • SFD resulted in a 30% reduction in drying time compared to FD.	
SFV/L + VSFD	Vanillin	• Vanillin was microencapsulated for the first time by SFD. • SFD-microencapsulated vanillin with whey protein isolate as wall material showed spherical shape with numerous fine pores on the surface, which in turn led to good rehydration ability. • Spray-freeze-dried vanillin microcapsules exhibited better thermal stability than spray-dried and freeze-dried counterparts. • The drying period was reduced from 16 h in conventional freeze-drying to 4 h in SFD.	Hundre, Karthik & Anandharamakrishnan (2015)
SFV/L + VSFD	*Lactobacillus plantarum*	• Microcapsules containing *Lactobacillus plantarum* (MTCC 5422) were produced by the SFD technique with different wall materials. • SFD microcapsules were spherical in shape and exhibited 'good' flowability and lower hygroscopicity. • Microencapsulation by the SFD method did not affect the cell viability, as indicated by good encapsulation efficiency (87.92–94.86%). • The drying time was reduced from 20 h in conventional freeze-drying to 8 h in SFD.	Rajam & Anandharamakrishnan (2015)

However, the main disadvantage of VSFD is its high fixed and operational costs, which is mainly due to its vacuum operation in batch mode (Matteo, Donsi & Ferrari, 2003). As discussed in Chapter 1, a key contributor to the operational costs is the energy cost, which is high in vacuum freeze-drying. Thus, to reduce the manufacturing and energy costs, atmospheric spray-freeze-drying was developed, which is based on Meryman's finding that vacuum is not an indispensable requirement for freeze-drying.

3.2 ATMOSPHERIC SPRAY-FREEZE-DRYING (ASFD) & ATMOSPHERIC FLUIDIZED BED SPRAY-FREEZE-DRYING (AFBSFD)

3.2.1 Theory of atmospheric freeze-drying (AFD)

The advent of atmospheric spray-freeze-drying (ASFD) and atmospheric fluidized bed spray-freeze-drying (AFBSFD) processes in the early 1990s is based on Meryman's theory (1959) of atmospheric freeze-drying (AFD) which stated that *'the diffusion of water vapor from the drying boundary through the dried shell is facilitated primarily by the vapor pressure gradient rather than*

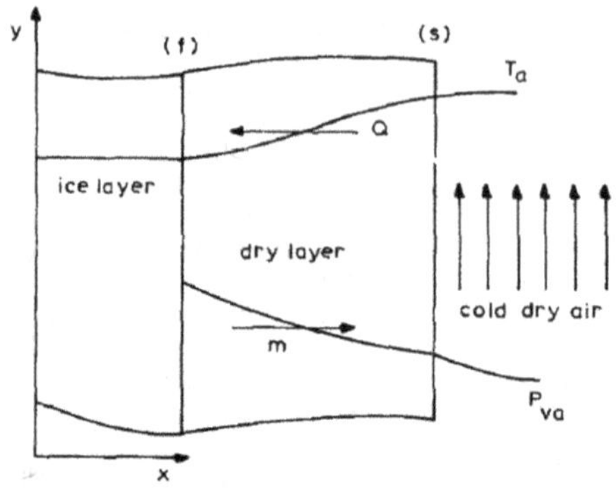

(*f*: interface; *s*: surface; *Q*: heat transfer; *m*: mass transfer; *Ta*: temperature gradient; *Pva*: partial pressure gradient)

Figure 3.11 Schematic diagram of the principle of atmospheric freeze-drying (Mumenthaler & Leuenberger, 1991).

by the absolute pressure in the system. Hence, it is possible to freeze dry at atmospheric pressure.' Schematic representation of the principle of atmospheric freeze-drying is depicted in Figure 3.11. The entire AFD process operates at positive gauge pressures. A cold desiccated gas such as air, nitrogen, or helium is used as the water removal and heat transfer medium to accomplish the sublimation of frozen material (Mumenthaler & Leuenberger, 1991; Wang, Finlay, Peppler & Sweeney, 2006). Driven by the temperature gradient (T_a) between the product surface and cold gas, the heat of sublimation (Q) is initially transported from the surface (s) to the ice-vapor interface (f). With time, the sublimation front moves toward the centerline ($y, x = 0$). Alongside the heat transfer, mass transport of water vapor also occurs through the porous dry layer to reach the surface (s). The partial pressure gradient (P_{va}) between the ice-vapor interface and the surrounding cold gas medium serves as the driving force for mass transfer. Thus, in AFD, drying occurs mainly by convection rather than diffusion (Heldman & Hohner, 1974; Mumenthaler & Leuenberger, 1991).

3.2.1.1 Advent and theory of atmospheric spray-freeze-drying (ASFD)

Atmospheric freeze-drying has a longer residence time due to the internal resistances to mass transfer (Claussen, Ustad, Strømmen, & Walde 2007). During AFD, the freeze-drying rates of small particles do not present any improvement over vacuum freeze-drying (Dunoyer & Larousse, 1961; Woodward, 1963). But, a reduction of particle size and an increase in surface mass transfer coefficient were found to increase the atmospheric freeze-drying rates compared to vacuum freeze-drying. And, the time to dry a sample to any given dimensionless moisture content was directly proportional to the square of the sample size (Heldman & Hohner, 1974). Thus, AFD was integrated with the SFD process, in which atomization led to reduction of sample dimensions. This led to the conception of the atmospheric spray-freeze-drying (ASFD) technology.

3.2.1.2 Advent and theory of atmospheric fluidized bed spray-freeze-drying (AFBSFD)

Before trying to appreciate the theory of atmospheric fluidized bed spray-freeze-drying, it would be useful to understand the mechanism of heat and mass transfer during a freeze-drying process. During conventional freeze-drying in a shelf-type freeze-dryer, the latent heat of sublimation is provided by conductive or radiative heating (Figure 3.12). However, the uniform

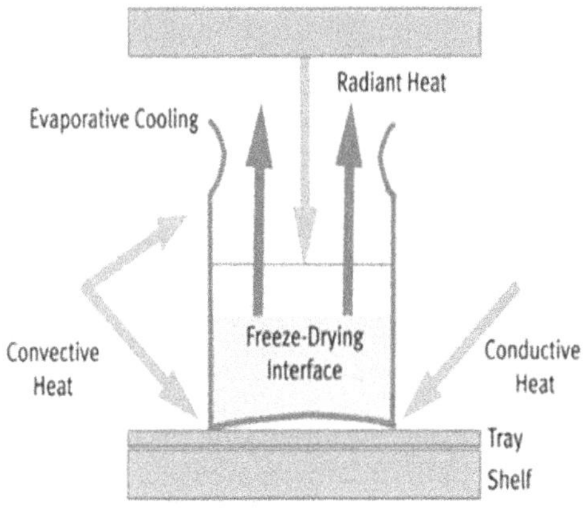

Figure 3.12 Heat transfer in a conventional freeze-dryer (Barley, 2021).

application of latent heat throughout the product cannot be guaranteed in this case. Thus, a drying method that is capable of uniform application of heat can leverage the small sample dimensions and reduce drying time without the risk of particle melting and collapse (Ishwarya, Anandharamakrishnan & Stapley, 2015). Fluidized bed drying is one such method that is known for its excellent heat and mass transfer due to good contact between particles and the drying medium. The phenomenon of fluidization occurs when a fluid (gas or liquid) flows upward through a bed of particles with sufficient velocity to support the weight of the particles without carrying them away in the fluidized stream (Geldart, 1986). Further, compared to lyophilization, the diffusion distance from the ice core surface to the external convective flow is shortened in fluidized bed freeze-drying. Due to the strong convective flow, the drying gas can easily enter and flow around or through the partially porous particles, depending on whether the frozen particles are moving and rotating or stationary (packed bed), respectively (Figure 3.13). Integrating fluidized bed drying with the ASFD process was found to be an effective approach to reduce the drying time through convection of the drying gas (Malecki et al., 1970). Thus, contacting the spray-frozen particles with cold, dry gas in a fluidized bed was proposed, and this led to the advent of atmospheric fluidized bed spray-freeze-drying (AFBSFD) (Leuenberger, Plitzko, & Puchkov, 2006; Wang et al., 2006).

3.2.2 Construction and operation of an atmospheric spray-freeze-dryer and atmospheric spray-fluidized-bed freeze-dryer

The five main components of an atmospheric spray-freeze-dryer dryer (Figure 3.14) include (Ly et al., 2019):

1. ***An aluminum top lid*** with openings to connectors for the twin-fluid nozzle, liquid nitrogen supply tube, and a copper tubing that supplies the drying gas, which has passed through a heat exchanger that is filled with liquid nitrogen to chill it.
2. ***A twin-fluid nozzle*** connected to the top lid, which is supplied with gas and liquid from a compressed air line and a peristaltic pump, respectively. This nozzle atomizes the aqueous feed solution downward into a co-current flow of a cold, dry inert gas.
3. ***An aluminium outer chamber:*** The drying gas enters through the top lid and flows into this chamber, where both spraying and drying occurs. Within this chamber, drying is carried out by the forced convection of cold, dry gas all through the frozen material until the target moisture content is attained.
4. ***A porous metal cylinder*** present within the outer chamber, which creates a gas barrier to prevent the atomized particles from adhering to the inner side of the porous cylinder. It directs the flow of

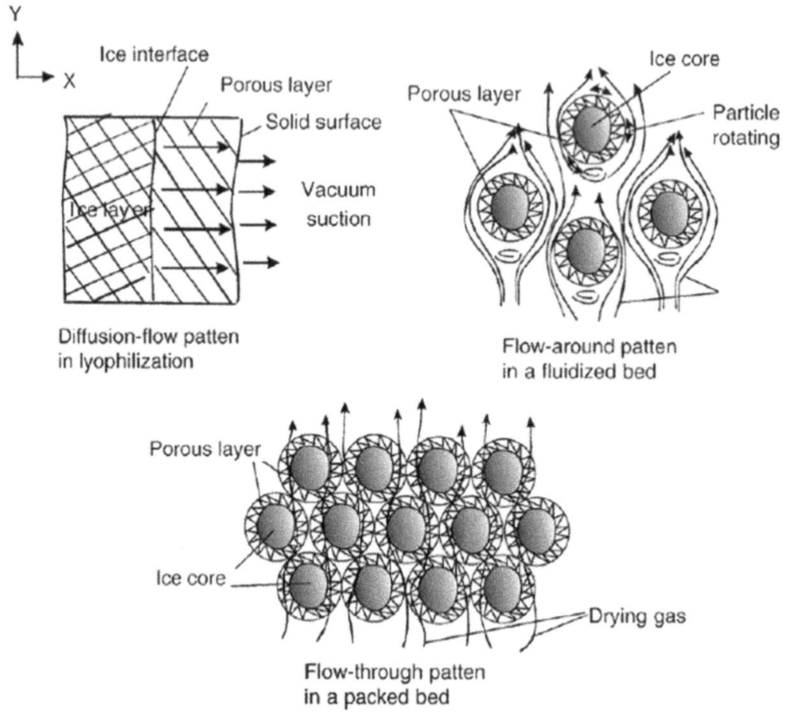

Figure 3.13 Mass transfer: conventional freeze-drying versus fluidized bed drying (Wang et al., 2006).

particles towards the filter disk and facilitates the uniform distribution of temperature, pressure, and gas flow within the core of the chamber.

5. ***A porous filter disk*** made of chemical and corrosion-resistant Stainless Steel (316L) having a pore size of 20 μm. This disk is attached to the bottom of the ASFD chamber via an aluminium piece; its function is to collect the freeze-dried powder. With time, a uniformly thick layer of powder accumulates on the filter.

Apart from the above components, a feedback system is present at the bottom of the ASFD chamber near the powder collection area, which comprises a Type-T thermocouple that controls the heating rope. By sensing the temperature near the powder, the heating rope warms the cold drying gas to the preferred operating temperature. Similarly, a thermocouple with a temperature logger placed at the exit of the ASFD chamber measures the temperature of the outlet gas. Besides, the humidity of the outlet gas is measured with a humidity probe connected to a hand-held indicator (Ly et al., 2019).

The first design of an atmospheric fluidized bed spray-freeze-dryer (Figure 3.15) was proposed by Leuenberger (1987), wherein spray freezing was carried out by *'top-spraying.'* Top-spraying involves atomizing the feed solutions onto a fluidized bed of pulverized dry ice. Subsequently, the frozen powder was dried by an upward flow of desiccated fluidizing air. But, a bottleneck in this operation was the gas velocity, which was much higher than the terminal velocity of frozen particles. This led to the entrainment or carrying-away of the frozen particles to the filter system. Thus, the drying process was completed at the surface of the filter system via a thin powder layer. But, this work demonstrated that the primary drying of frozen particles could be completed within a few hours, which is much faster than conventional freeze-drying.

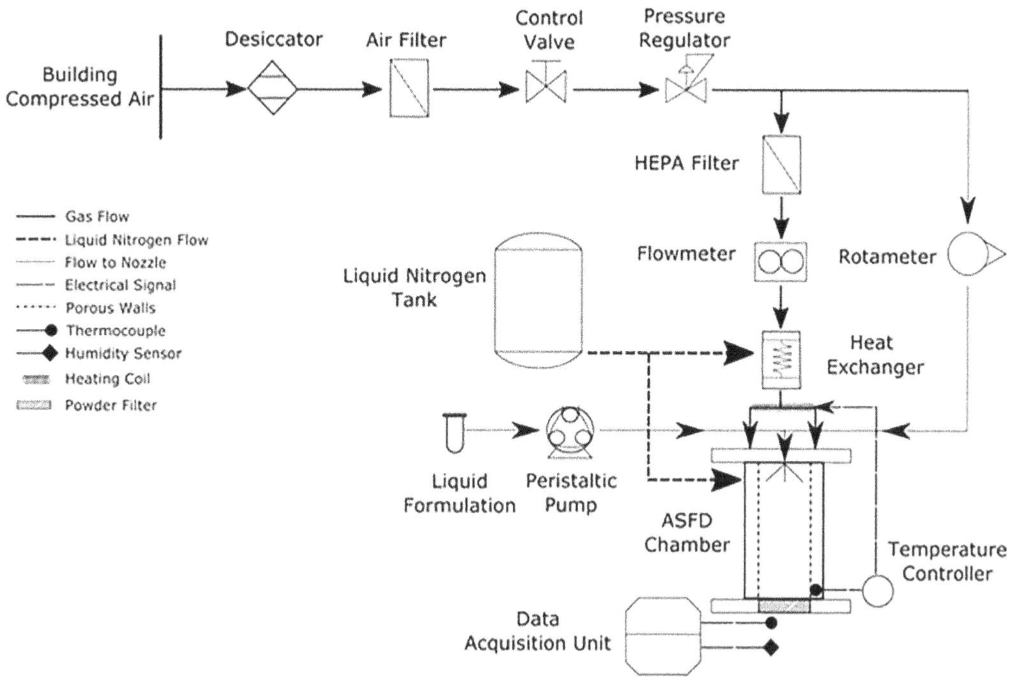

Figure 3.14 Schematic of the modern-day ASFD apparatus (Ly et al., 2019).

Subsequently, Mumenthaler & Leuenberger (1991) came up with an improved design of the atmospheric fluidized bed spray-freeze-dryer (Figure 3.16). In their set-up, the cold air stream always circulated in a closed-loop and was constantly replenished by condensing the humidity on the refrigerated surface of cooling systems. The cold air was passed either through the drying chamber or the bypass system that facilitated the withdrawal of samples at defined time intervals for moisture analysis without causing the loss of conditioned cold air. The flow rate of circulating gas was set to the required value by changing the position of different air flaps. And, its temperature was raised below or above the critical temperature of the material using an internal heater. Nevertheless, Mumenthaler could not get a stable fluidized bed due to the small size of droplets. Consequently, the product was dried as a layer at the filter, which had to be shaken during the process to avoid compaction (Mumenthaler & Leuenberger, 1991).

Later, Leuenberger (2002) developed an integrated AFBSFD system to freeze-dry liquid pharmaceutical products. Leuenberger's apparatus (Figure 3.17[a]) included a product vessel and an expansion chamber, which were confined on the top by a shakable filter and on the bottom by a fine sieve. This chamber was capable of operating the fluidized bed in a closed-loop system. The residual air and/or the residual vapor of the solvent (water/organic solvent) were used at reasonably high gas velocities to maintain the fluidized bed of the powder. The gas was heated up by a heater. The pressure gradient (ΔP) to maintain the fluidized bed was attained using a molecular pump. The solvent vapor was partly removed from the closed-loop system by the solvent recovery unit. This unit consisted of two condensers (Figure 3.17[a]) that resulted in recovery rates ranging between 90-95%. The recovered solvents showed a high purity due to the efficient filter system that prevented particulate contamination.

In the above system, a two-fluid pneumatic nozzle was used in the spray-freezing operation to top-spray the aqueous feed solution against a stream of cold air (–60°C). The size range of the

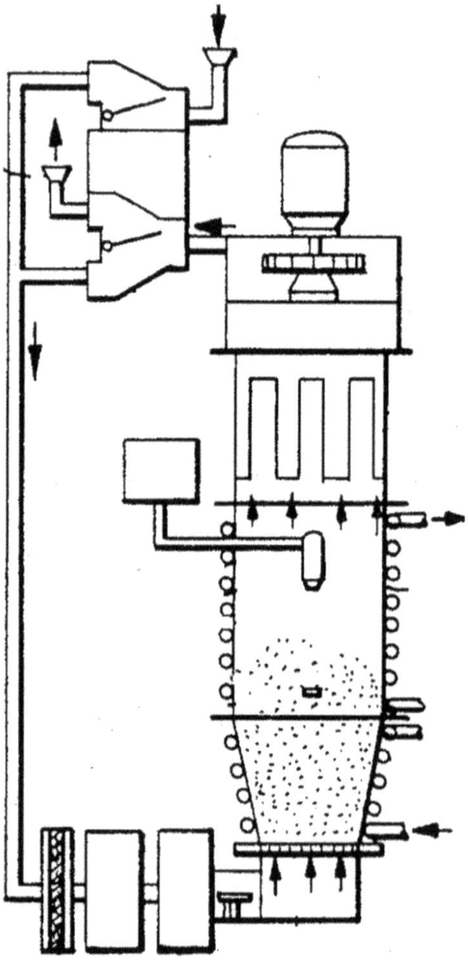

Figure 3.15 The first design of AFBSFD equipment (Leuenberger, 1987).

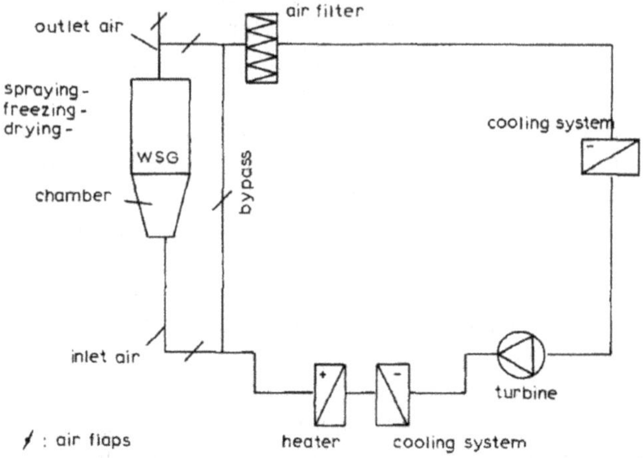

Figure 3.16 Schematic diagram of an early AFBSFD apparatus (Glatt, 1986).

(a)

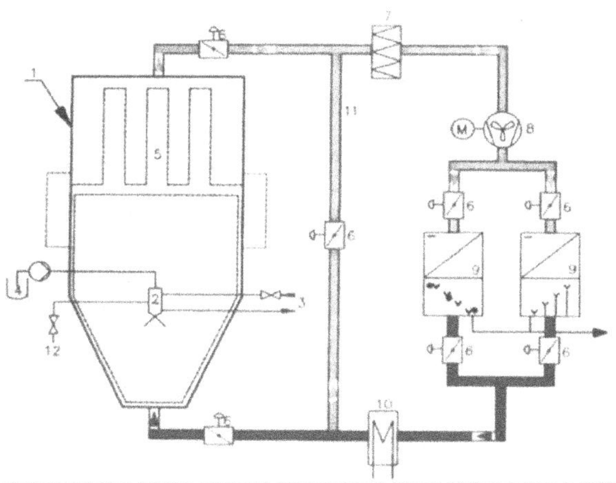

1. Spray tower; 2. Spray nozzle; 3. Heating device; 4. Spray solution; 5. Filter system; 6. Flap; 7. Airfilter; 8. Fan; 9. Refrigerator and condensers; 10. Heating system; 11. Bypass pipe; 12. Spray air for spray nozzle

(b)

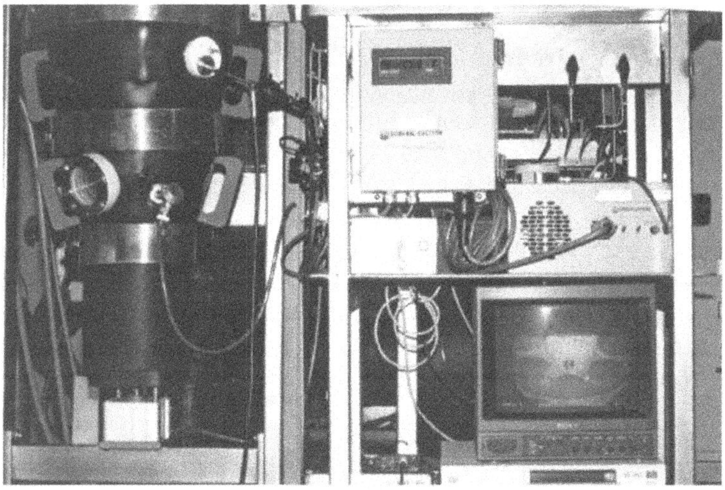

Figure 3.17 (a) Schematic of an atmospheric spray-fluidized-bed freeze-dryer with counter-current flow; (b) Photograph of an AFBSFD prototype apparatus (Leuenberger, 2002).

droplets emanating from the nozzle was 10–30 μm. Hot saltwater was circulated to prevent the freezing of aqueous feed within the nozzle (No. 3, Figure 3.17[a]). The droplets were instantaneously frozen in a counter-current stream of air at a temperature of –60°C, thus preventing the phase separation between ice of pure water and the frozen eutectic liquid within the droplet. Then, the frozen droplets were dried in the cold desiccated air stream by sublimation. A counter-current flow of drying medium and feed liquid was employed. A heater (No. 10, Figure 3.17[a]) was connected to maintain the temperature of cold air below the critical temperatures of the frozen solution. A filter (No. 5, Figure 3.17[a]) was fitted at the bottom to hold the fine powder back in the drying chamber. And, the water vapor was removed by the circulating air in the cooling systems (No. 9, Figure 3.17[a]), where the humidity condensed on the refrigerated surfaces. The equipment had appropriate measurement and control systems for the temperatures of inlet air, outlet air, drying chamber, and cooling systems, airflow rate, and dew-point temperatures to monitor the freeze-drying process continuously. A bypass connecting tube (No. 11, Figure 3.17[a]) allowed the drying chamber to be opened without heating up the conditioned and cold drying air that was circulated. But, the above system was operated in batch mode, and the counter-current operation led to the elutriation of particles that were caught by the outlet gas-filter. This led to low powder collection efficiency. A pilot plant prototype of the above system was developed by the process engineering laboratory of GlattAGin Pratteln, Switzerland (Figure 3.17[b]).

To overcome the particle elutriation problem observed in the Leuenberger's apparatus, Wang et al. (2006) studied the feasibility of conducting AFBSFD in a co-current flow configuration between the frozen particles and drying medium (nitrogen gas passed through a copper coil partially immersed in liquid nitrogen). In this process, the spray freezing of liquid droplets and conveying of frozen particles to the collection unit were performed in one step to improve the powder collection efficiency. In Wang et al.'s design (Figure 3.18), the drying chamber was essentially a cylinder having diameter and height of 5.9 cm and 34 cm, respectively. A high aspect ratio (height-to-diameter ratio) of the freeze-drying chamber or long chamber height is of significance as it influences the freezing operation. It dictates the length of time for which the droplets would be exposed to the freezing gas before deposition. In this context, a taller chamber guarantees uniform freezing. Reducing the chamber height to half of its original length led to incomplete freezing of the feed droplets even at an increased gas pressure at the cylinder and reduced chamber temperature. The chamber consisted of four parts: a top lid, a nozzle system, an air jacket, and a powder collection unit. The top lid was conical in shape to decrease spray re-circulation. The nozzle system included a two-fluid nozzle driven by a peristaltic pump and compressed nitrogen gas. A novel aspect in Wang et al.'s design was the air jacket, which consisted of an external solid wall and an internal porous wall. It facilitated the passage of cooling gas, thus preventing the adhesion of frozen particles on the porous wall. A filter disk held between two flanges and sealed with an O-ring trapped the deposited frozen particles and retained the drying powder (Wang et al., 2006).

As an additional step to prevent particle elutriation, a suitable gas velocity was identified by varying the gas flow rate from low to high velocities. The velocity at which a strong enough gas cushion was built upon the surface of the porous walls was chosen to prevent particle adhesion. This corresponded to a value of 0.11 ms^{-1} when the gas flow rate and spray-pressure were set at 0.6 ft^3/min (SCFM) and 3.0 psi, respectively. Under the above conditions, spray-freezing, freeze-

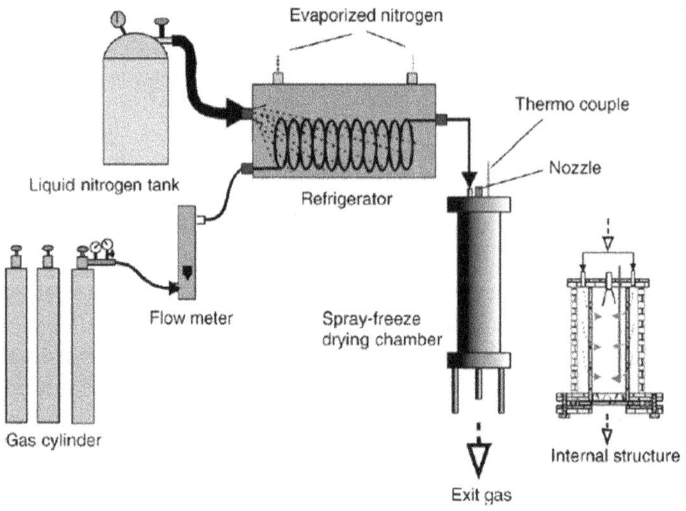

Figure 3.18 Schematic of an atmospheric spray-fluidized-bed freeze-dryer with a co-current flow (Wang et al., 2006).

drying, uniform powder deposition, and collection were successfully finished in one step for all the tested feed formulations.

3.2.3 Applications and advantages of ASFD and ASFBFD

The applications of ASFD and ASFBFD in the production of a variety of foods and biologicals are compiled in Table 3.6. Apart from the elimination of the vacuum requirement, ASFD demonstrates several other advantages. The benefits of ASFD have mainly been realized in the pharmaceutical sector. Unlike a cake that results from vacuum freeze-drying, the end-product from ASFD is a fine and light powder (Figure 3.19). Thus, the need for milling to reduce the size of drug particles can be avoided with the ASFD process (Sebastião, Robinson, and Alexeenko 2017). The production of free-flowing powder with homogeneous and ultrafine particle size, larger surface area, porous structure, and good solubility is advantageous in the pharmaceutical industry. The combination of spray-freezing-into-liquid (SFL) and AFD have resulted in high product yields, good drug stability and uniform molecular dispersion of drugs in the excipient matrices, promoted by the cryogenic temperature and fast freezing, which minimizes the time for phase separation and results in a more homogeneous dispersion of solutes throughout the frozen droplets (Sebastião, Robinson, & Alexeenko 2017). The possibility of developing an industrial scale ASFD process was explored by Rogers et al. (2003), for the bulk production of micronized SFL powders with aqueous dissolution. Collectively, ASFD is perceived as an economical alternative for VSFD in terms of reducing energy consumption whilst maintaining a high product quality (Claussen, Ustad, Strommen, & Walde 2007). It has been recommended as a fast-rate drying technology that is highly appropriate for the drying of thermo-sensitive biologicals (Sebastião, Robinson, & Alexeenko 2017).

The advantages of AFBSFD can be listed as follows (Mumenthaler & Leuenberger, 1991):

- ***True freeze-drying***: The product is maintained at a temperature below its critical temperature throughout the process;
- ***Constant drying conditions***: The drying medium circulates in a closed loop and is constantly replenished by condensation on the refrigerated surface;
- ***Shorter drying times***: Due to improved heat and mass transfer between the cold dry gas and the frozen sample;
- ***Reduced process cost and high energy savings***;
- ***Superior and uniform product quality:*** Consistent product quality with respect to its color, particle size, flowability, consistence, and moisture content of the dry product, attributed to the consistent conditions of the drying medium, which leads to a more homogeneous drying. The local temperature and/or concentration gradients in different product layers are avoided;
- ***Improved aroma retention***: Loss of volatile flavors in food products is minimized by the lower sublimation temperatures and the absence of vacuum.

3.2.4 Limitations of ASFD and ASFBFD

Despite its various benefits, a major disadvantage of ASFD is the need to circulate a very large quantity of cold, dry gas through the product to accomplish primary drying. Primary drying must be performed below the collapse temperature of the material, which can be as low as –30°C. At such low temperatures, the ice phase exerts an extremely low vapor pressure. Hence, a large amount of cold, dry gas must be circulated to create the required vapor pressure gradient to conduct freeze-drying without vacuum requirement. Recirculation of gas and the use of heat pump have been found to overcome the abovementioned limitation of ASFD to some extent. But, irrespective of these measures, a huge quantity of gas has to be supplied to the process. Refrigeration and drying requirements for such a mass of gas would jeopardize the economy of the process. Also, to pass this

Table 3.6 Applications of ASFD and AFBSFD (Adapted from Ishwarya, Anandharamakrishnan & Stapley, 2015)

Process	Product	Focus & Findings	Researchers
AFD	Mouse kidney tissue	• Proposed the concept of freeze-drying at atmospheric pressure.	Meryman (1959)
ASFD (SFV + AFD)	Low water-soluble drugs	• Obtained free-flowing powder, porous structure with high surface area, and improved bioavailability of extremely low water-soluble drugs.	Leuenberger (2002)
ASFD (SFV + AFD)	Bacteriophage D29	• The bacteriophages were atmospheric spray-freeze-dried in solutions with varying concentrations of trehalose and mannitol. • Convective transport combined with the low temperature of the drying gas resulted in a free-flowing, porous powder. • A solution of trehalose and mannitol at a mass ratio of 7:3 and a total mass concentration of 100 mg/mL led to powder with 4.9 ± 0.1% moisture content and an acceptable titer reduction of ~0.6 logs. • Trehalose in the powder was amorphous while mannitol completely crystallized during ASFD, both of which are desirable for preserving phage viability and storage in powders.	Ly et al. (2019)
ASFD (SFV + AFD)	Biologically active substances and commercial liquid foods (milk, orange juice, and coffee extract)	• An apparatus and a technique for the spray-freezing of aqueous solutions in situ at very low temperatures (= - 90°C) and for subsequent dehydration of the resulting frozen particles in a stream of cold, desiccated air was developed. • Improved heat and mass transfer between the circulating drying medium and the frozen sample in the case of ASFD. • High and homogeneous quality of the dry product with increased retention of volatile aromatic compounds in foods. • Instead of a cake, fine and free-flowing powder was obtained with a large inner surface area and good wettability and solubility properties.	Mumenthaler & Leuenberger (1991)
AFBSFD (SFV + AFBFD	Apple juice, egg albumin	• The feasibility of AFD of frozen juice in the droplet form was studied. • The sticky nature of the product influenced the drying rate.	Malecki et al. (1970)
AFBSFD (SFV + AFBFD	Bovine serum albumin (BSA)	• Combined the spray-freezing step and fluidization conveying of frozen powder using co-current flow to convey the frozen powder to the exit filter.	Wang et al. (2006)
AFBSFD (SFV + AFBFD	Protein, dextran	• Equipment developed for spray freeze drying in a fluidized bed at normal and low pressure. • Short drying time, good control of drying parameters was achieved.	Leuenberger et al. (2006)

huge mass of gas through the bed in a reasonable time demands a very high volumetric flow rate and velocity of the gas. This leads to entrainment of particles out of the fluidized bed where they must be caught by a gas filter (Mumenthaler & Leuenberger, 1991), as mentioned above. A practical way to tackle this issue would be to apply a partial vacuum to the fluidized bed process (Leuenberger et al.,

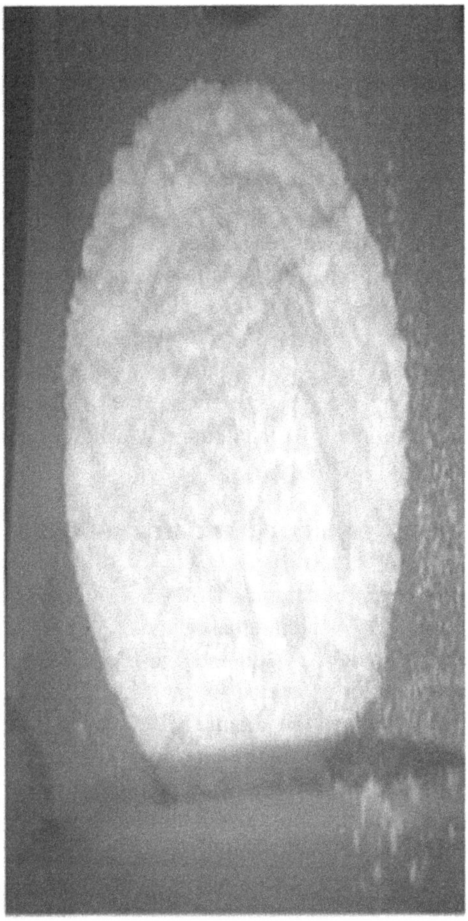

Figure 3.19 Bovine serum albumin powder produced by ASFD (Ly et al., 2019).

2006). Thus, came into existence the use of vacuum fluidized bed spray-freeze-drying process (Ishwarya, Anandharamakrishnan & Stapley, 2015). Further, the prototypes of atmospheric spray-freeze-dryers reported in the literature (Mumenthaler & Leuenberger, 1991) have been found oversized for high-value pharmaceutical products, whilst being too small for the freeze-drying of foods, owing to the uncertain values of critical temperatures for most of the food products (Leuenberger, 2002).

3.3 VACUUM FLUIDIZED BED SPRAY-FREEZE-DRYING (VFBSFD)

3.3.1 Theory

Vacuum fluidized bed freeze-drying (VFBFD) can be defined as *'a dehydration technique for the removal of ice by sublimation so that the quality of the product is very high and also it requires a smaller volume of cold gas compared to atmospheric conditions'* (Anandharamakrishnan, 2008). This process is based on the theory that after all, the drying gas is just an inert heat transfer medium that does not have any role in deciding the driving force for mass transfer. It has been observed that an *n*-fold reduction in the pressure of a system cuts-down the mass of gas required for the process by *n* times. This, in turn, reduces the gas density. The low density of gas at adequately low pressure reduces the inertial (non-viscous) drag forces on the particles and thereby prevents particle elutriation from the bed (Anandharamakrishnan et al., 2010). Yet, it is imperative to be aware that viscous forces do not drop off until extremely low pressures are reached. Hence, elutriation must always be considered as a potential processing bottleneck.

Furthermore, the water uptake capacity of air is inversely proportional to the water capacity at atmospheric pressure. Even at low pressures (applied vacuum), the total pressure of the system would be much higher than the partial pressure of water in the ice phase. As the density of air is proportional to the total pressure, drying air velocity could not be reduced. But at lower pressures, the fluidization air velocity can be set at 1.5 to 2 times higher than that at atmospheric pressure, which results in shorter drying times (Leuenberger et al., 2006). In addition, low-pressure operation increases the temperature difference between the particle and gas due to its influence on the heat transfer coefficient between the gas and the particle bed. Thus, low pressure or application of vacuum to a fluidized bed drying process is advantageous from three perspectives: (1) in reducing the mass of gas required; (2) in preventing the particle entrainment (Anandharamakrishnan et al., 2010) and (3) in reducing the drying time (Leuenberger et al., 2006).

3.3.2 Construction and operation of a vacuum spray-fluidized-bed freeze-dryer

The VSFBFD system housed at Loughborough University is composed of a spray-freezing chamber and a vacuum fluidized bed drying unit (Figure 3.20) (Anandharamakrishnan, 2008). The spray-freezing chamber is a cylindroconical-shaped vessel of height 2.01 m and diameter 0.8 m. It is fabricated of a 1 mm thick mild steel sheet, and its inner side is coated by matt-black acrylic paint to protect the chamber from corrosion. The conical portion of the chamber is equipped with nitrogen gas inlets on either side and feed connections to the hydraulic nozzle atomizer (Figure 3.21). A nozzle heater assembly encloses the hydraulic nozzle. It is an aluminium cylinder (700 mm × 100 mm), and a small electrical heater is used to heat the compressed air to maintain the temperature around the nozzle at around 16-20°C. A particle collection polystyrene box is kept beneath the chamber to collect the frozen particles at the outlet. The drying medium, i.e., nitrogen gas, is cooled by liquefied nitrogen. A gas-mixer atomizes the liquid nitrogen, which is supplied from an external dewar (Figure 3.22). The flow rate and temperature of drying gas are regulated by using a manual needle valve. The cold gas flows into the spraying chamber through the venturi and perforated copper pipes (rows of 3 mm orifices).

The main fluidization vessel (Figure 3.23) of the vacuum fluidized bed freeze-dryer has a volume of 0.022 m^3 and is made of SS 316 grade stainless steel. A gas flow inlet (0.15 m diameter) is fixed at the bottom of the vessel, and two outlets (each of diameter 0.0754 m) are present at the top on either side. The gas mixer supplies liquid nitrogen to the fluidization vessel. An annular heat exchanger controls the inlet temperature of the gas by varying the heater temperature and the compressed air flow rate. The gas is extracted, and a partial vacuum is created by a rotary-vane oil-sealed vacuum pump having a capacity of 100 m^3/hr. The material of construction of the fluidization vessel tube (height: 16 cm and diameter: 5 cm) is polycarbonate that enables visualization. The top of the polycarbonate cylinder has a fine mesh that allows the flow of low-temperature gas through the particle bed and also prevents the elutriation of particles from the bed. A 2 mm thick porous polyethylene sheet functions as the distributor plate. To measure the sample mass, the fluidization unit is suspended from a spring balance that is fixed to a hand-operated wheel (Figure 3.24). During operation, due to a high-pressure drop across the fluidized bed, the gas tends to escape from the sides of the tube. This can be prevented by pressing down the hand-operated wheel on the fluidization unit.

The VFBSFD process is carried out in two stages: spray-freezing followed by drying. Firstly, spray-freezing is carried out by atomizing the pressurized liquid feed (6 barg) through the heated hydraulic nozzle into a pre-cooled spray chamber. Before commencing the spray-freezing

Figure 3.20 Vacuum fluidized bed spray-freeze-dryer unit (Anandharamakrishnan, 2008).

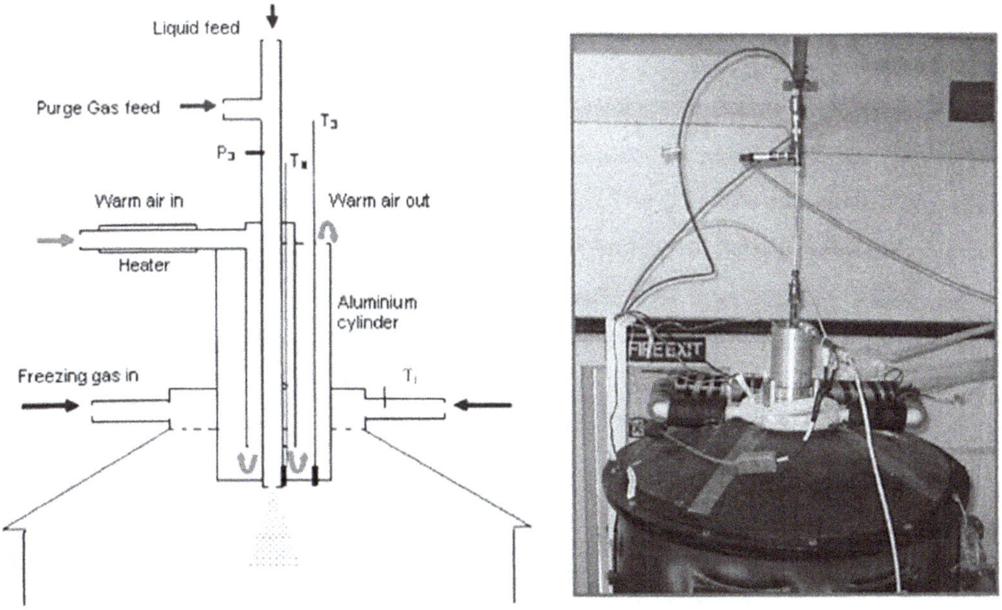

Figure 3.21 Feed connections to the nozzle (Anandharamakrishnan, 2008).

operation, the chamber is purged with the cold nitrogen gas until the spray chamber wall temperature, and gas temperature reach –85°C. At the end of the spray-freezing operation, the frozen particles are collected from the outlet of the chamber in a polystyrene cup placed inside a polystyrene box.

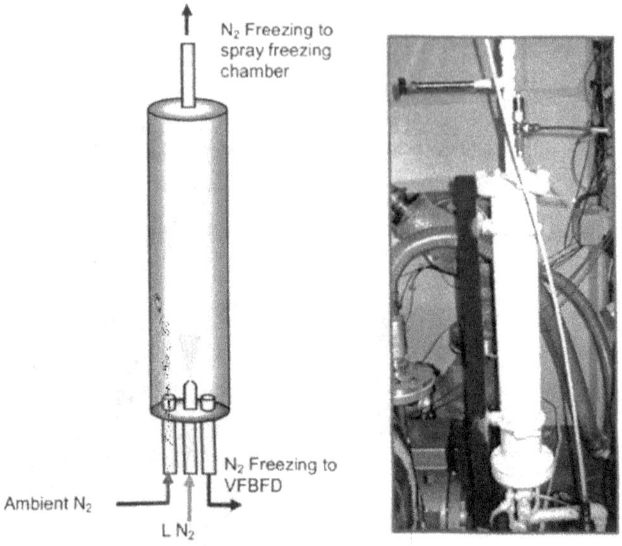

Figure 3.22 Gas mixer (Anandharamakrishnan, 2008).

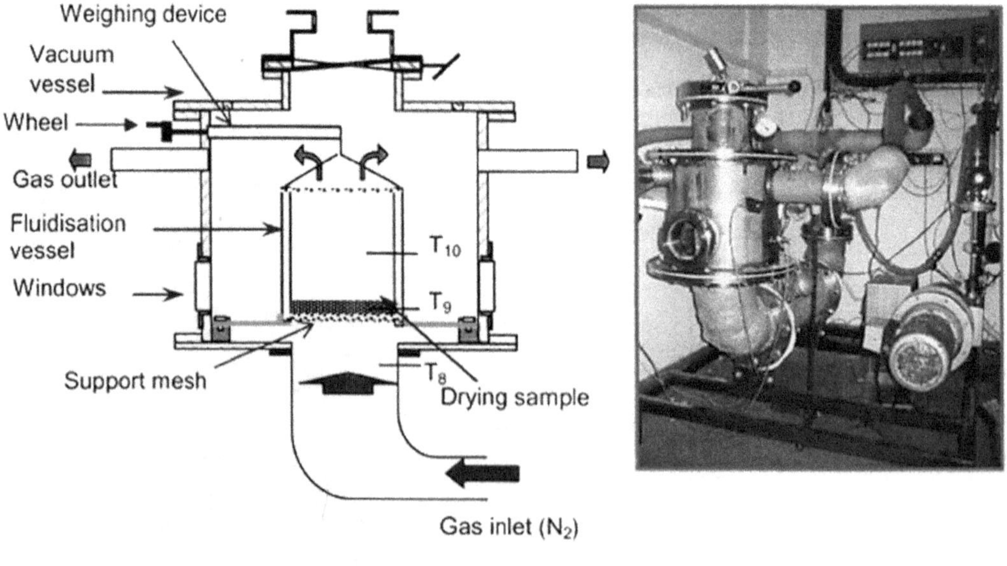

Figure 3.23 Schematic diagram and photograph of the vacuum fluidized bed freeze dryer (Anandharamakrishnan, 2008).

Then, the frozen particles are loaded into the VFBFD equipment that has previously been cooled for 20 minutes using the flow of gas from the liquid nitrogen dewar through the gas mixer. A constant inlet gas temperature is the main control variable of the VFBSFD process, which is adjusted to be below the collapse temperature of the frozen feed (Anandharamakrishnan, 2008).

Figure 3.24 Photograph of fluidization vessel (Anandharamakrishnan, 2008).

3.3.3 Applications and advantages of VFBSFD

The applications of VFBSFD are compiled in Table 3.7. From the published studies, it is evident that spray-freezing combined with VFBFD has the potential to produce pharmaceutical and high-value-added food products at a shorter time and reduced process cost compared to the currently available commercial vacuum freeze-drying processes (Anandharamakrishnan, 2008).

Table 3.7 Applications of VFBSFD process

Process	Product	Focus & Findings	Researchers
SFV + VFBFD (VFBSFD)	Proteins	• Compared the spray-freeze-drying process in a fluidized bed at normal and low pressure. • Three different pressure values were chosen: 150, 300, and 1000 mbara. • A 3-fold reduction in drying time was achieved by operating at pressures lower than at atmospheric pressure. • The shorter drying time (approximately 200 min) was attributed to a higher fluidization velocity of the particles.	Leuenberger et al. (2006)
SFV + VFBFD (VFBSFD)	Coffee, maltodextrin	• The objective of the study was to obtain better volatile retention during drying. • For coffee and maltodextrin, it took 1.75-2 h for drying instead of 24 h in a conventional freeze dryer. • Volatiles retention was comparable to that of conventional freeze-drying in both coffee and maltodextrin. • With maltodextrin, it was clear that the influence of the spray freezing step was found to be as important to volatile retention as the freeze-drying step. • Approximately half of the volatile loss in the spray freeze-dried samples was attributed to the spray freezing step.	Khwanpruk, Anandharamakrishnan, Rielly & Stapley (2008)
SFV + VFBFD (VFBSFD)	Whey protein isolate	• The drying was complete in 60, 65, and 100 min at the inlet gas temperatures of −10°C, −15°C, and −30°C, respectively, to result in a product of 8% moisture content.	Anandharamakrishnan et al. (2010)

CONCLUSIONS

Having a comprehensive understanding of the different modes of freeze-drying adopted during the SFD process would aid in developing an effective protocol for conducting spray-freeze-drying trials. From the information presented in this chapter, it is apparent that the principle, instrumentation, and operation of various freeze-drying techniques have continuously evolved across the years. A series of innovations have been built into this process to make it more precisely-controlled, automated, economical, and advantageous in terms of obtaining high-quality products. A greater understanding on the critical temperatures of the formulation and the rationale of defining the process parameters such as system pressure and gas flow rate facilitates a process scientist to design and conduct a spray-freeze-drying trial in a more scientific way. The scope for future research exists in developing economically viable, industrial-scale spray-freeze-dryers for the bulk manufacturing of food products and active pharmaceutical ingredients.

REFERENCES

Abbas, K. A., Lasekan, O., & Khalil, S. K. (2010). The significance of glass transition temperature in processing of selected Fried Food Products: A Review. *Modern Applied Science*, 4(5), 3–21.

Abdul-Fattah, A. M., & Truong, V. L. (2010). Drying process methods for biopharmaceutical products: an overview. In F. Jameel, & S. Hershenson (Eds.), *Formulation and process development strategies for manufacturing Biopharmaceuticals*. John Wiley & Sons, Inc., Hoboken, New Jersey, pp. 705–737.

Anandharamakrishnan, C., Rielly, C. D., & Stapley, A. G. F. (2010). Spray-freeze-drying of whey proteins at sub-atmospheric pressures. *Dairy Science and Technology*, 90, 321–334.

Anandharamakrishnan, C. (2008). *Experimental and computational fluid dynamics studies on spray-freeze-drying and spray-drying of proteins*. Ph.D. thesis. Loughborough University, UK.

Barley, J. (2021). *Basic Principles of Freeze Drying*. Available from: https://www.spscientific.com/freeze-drying-lyophilization-basics/ (Accessed 9 September 2021).

Barron, M. K., Young, T. J., Johnston, K. P., & Williams, R. O. (2003). Investigation of processing parameters of spray freezing into liquid to prepare polyethylene glycol polymeric particles for drug delivery. *AAPS Pharm Sci Tech*, 4(2), 1–13, Article 12.

Bellows, R. J., & King, C. J. (1972). Freeze-drying of aqueous solutions: Maximum allowable operating temperature. *Cryobiology*, 9(6), 559–561.

Bhandari, B. R., & Howes, T. (1999). Implication of glass transition for the drying and stability of dried foods, *Journal of Food Engineering*, 40, 71–79.

Bhandari, B. R., Datta, N., & Howes, T. (1997). Problems associated with spray drying of sugar-rich foods. *Drying Technology*, 15(2), 671–684.

Bigg, E. K. (1953). *The supercooling of water. Proceedings of the Physical Society. Section B*, 66, 688–694.

Biopharma Technology Limited (2021). Available from: https://biopharma.co.uk/intelligent-freeze-drying/intelligent-freeze-drying/lyostat-microscope/; http://www.biopharma.co.uk/intelligent-freeze-drying/products/lyotherm2-dta-impedance/ (Accessed 9 September 2021).

Bosch T. (2014). *Aggressive Freeze-Drying – a Fast and Suitable Method to Stabilize Biopharmaceuticals*. Ludwig-Maximilians Universität München.

Chow, K. T., Zhu, K., Tan, R. B. H., & Heng, P. W. S. (2008). Investigation of electrostatic behaviour of a lactose carrier for dry powder inhalers. *Pharmaceutical Research*, 25, 2822–2834.

Claussen, I. C., Ustad, T. S., Strømmen, I., & Walde, P. M. (2007). Atmospheric freeze drying e a review. *Drying Technology*, 25, 957–967.

Colandene, J. D., Maldonado, L. M., Creagh, A. T., Vrettos, J. S., Goad, K. G., & Spitznagel, T. M. (2007). Lyophilization cycle development for a High-concentration monoclonal Antibody formulation lacking a crystalline Bulking Agent. *Journal of Pharmaceutical Sciences*, 96(6), 1598–1608.

Costantino, H. R., Firouzabadian, L., Hogeland, K., Wu, C. C., Beganski, C., Carrasquillo, K. G., Cordova, M., Griebenow, K., Zale, S. E., & Trace, (2000). Protein spray-freeze drying. Effect of atomisation conditions on particle size and stability. *Pharmaceutical Research*, 17(11), 1374–1383.

Costantino, H. R., Firouzabadian, L., Wu, C., Hogeland, K. C., Carrasquillo, K. G., Cordova, M., et al. (2002). Protein Spray freeze drying. 2. Effect of formulation variables on particle size and stability. *Journal of Pharmaceutical Sciences*, 91, 388–395.

Cytiva (2020). The benefits of lyophilization in assay kit development - Maximizing the integrity and shelf life of reagents. Available from: https://www.cytivalifesciences.com/en/us/solutions/genomics/knowledge-center/advantages-of-lyophilization (Accessed 9 September 2021).

Dolly, P., Anishaparvin, A., Joseph, G. S., & Anandharamakrishnan, C. (2011). Microencapsulation of Lactobacillus plantarum (mtcc 5422) by spray-freeze-drying method and evaluation of survival in simulated gastrointestinal conditions. *Journal of Microencapsulation*, 28(6), 568–574.

Dunoyer, J. M., & Larousse, J. (1961). *Experiences nouvellessur la lyophilisation*. Trans. Eighth National vacuum Symposium (pp. 1059–1062).

Engstrom, J. D., Simpson, D. T., Lai, E. S., Williams, R. O., & Johnston, K. P. (2007). Morphology of protein particles produced by spray freezing of concentrated solutions. *European Journal of Pharmaceutics and Biopharmaceutics*, 65, 149–162.

Fellows, P. J. (2002). Freeze drying and freeze concentration. In: *Food Processing Technology - Principles and Practice*. Woodhead Publishing, Cambridge, pp. 401–414.

Fennema, O. (1996). Water and ice. In: O. Fennema (ed.) *Food Chemistry*, 3rd Edition. Marcel Dekker, New York, pp. 18–94.

Fonseca, F., Passot, S., Cunin, O., & Marin, M. (2004). Collapse temperature of freeze-dried *Lactobacillus bulgaricus* suspensions and protective media. *Biotechnology Progress*, 20, 229–238.

Ford, J. L., &Timmins, P. (1989). Pharmaceutical Thermal Analysis: Techniques and Applications. In Rubinstein, M. (Ed.). *Ellis Horwood Books in Biological Sciences, Series in Pharmaceutical Technology*. John Wiley & Sons, New York.

Franks, F. (1990). Freeze drying: from empiricism to predictability. *Cryo-letters*, 11, 93–110.

Geldart, D. (1986). *Gas fluidization technology*. JohnWiley, Chichester.

Glatt, A. G. (1986). CH-4133 Pratteln/Switzerland, U.S. Patent 4608,764.

Hajare, A. A., More, H. N., Walekar, P. S., & Hajare, D. A. (2012). Optimization of Freeze Drying Cycle Protocol Using Real Time Microscopy and Integrated Differential Thermal Analysis-Electrical Impedance. *Research Journal of Pharmacy and Technology*, 5(7), 985–991.

Heldman, D. R., & Hohner, G. A. (1974). An analysis of atmospheric freeze drying. *Journal of Food Science*, 39(1), 147–155.

Henczka, M., Baldyga, J., & Shekunov, B. Y. (2006). Modeling of spray-freezing with compressed Carbon dioxide. *Chemical Engineering Science*, 61, 2880–2887.

Her, J.-Y., Song, C.-S., Lee, S. J., & Lee, K.-G. (2010). Preparation of kanamycin powder by an optimized spray freeze-drying method. *Powder Technology*, 199, 159–164.

Horn, J., & Friess, W. (2018). Detection of collapse and crystallization of saccharide, protein, and mannitol formulations by optical fibers in lyophilization. *Frontiers in Chemistry*, 6.

Hu, J., Rogers, T. L., Brown, J., Young, T., Johnston, K. P., & Williams, R. O. (2002). Improvement of dissolution rates of poorly water soluble APIs using novel spray freezing into liquid technology. *Pharmaceutical Research*, 19(9), 1278–1284.

Hu, J., Johnston, K. P., & Williams, R. O. (2003). Spray freezing into liquid (SFL) particle engineering technology to enhance dissolution of poorly water soluble drugs: organic solvent versus organic/aqueous co-solvent systems. *European Journal of Pharmaceutical Sciences*, 20, 295–303.

Hundre, S. Y., Karthik, P., & Anandharamakrishnan, C. (2015). Effect of whey protein isolate and β-cyclodextrin wall systems on stability of microencapsulated vanillin by spray–freeze drying method. *Food Chemistry*, 174, 16–24.

Ishwarya, S. P., & Anandharamakrishnan, C. (2015). Spray-Freeze-Drying approach for soluble coffee processing and its effect on quality characteristics. *Journal of Food Engineering*, 149, 171–180.

Ishwarya, S. P., Anandharamakrishnan, C., & Stapley, A. G. F. (2015). Spray-freeze-drying: A novel process for the drying of foods and bioproducts. *Trends in Food Science & Technology*, 41, 161–181.

Karthik, P., & Anandharamakrishnan, C. (2013). Microencapsulation of docosahexaenoic acid by spray-freeze-drying method and comparison of its stability with spray-drying and freeze-drying methods. *Food and Bioprocess Technology*, 6, 2780–2790.

Kasraian, K., Spitznagel, T. M., Juneau, J. A., & Yim, K. (1998). Characterization of THE Sucrose/glycine/water system by differential scanning calorimetry and freeze-drying microscopy. *Pharmaceutical Development and Technology*, 3(2), 233–239.

Kho, K., & Hadinoto, K. (2011). Optimizing aerosolization efficiency of dry powder aggregates of thermally sensitive polymeric nanoparticles produced by spray-freeze-drying. *Powder Technology*, 214, 169–176.

Khwanpruk, K., Anandharamakrishnan, C., Rielly, C. D., & Stapley, A. G. F., (2008). Volatiles retention during the sub-atmospheric spray freeze drying of coffee and maltodextrin. In: Proceedings of the *16th International Drying Symposium (IDS2008)*, 9–12 November 2008, Hyderabad, India, pp. 1066–1072.

Knopp, S. A., Chongpraser, S., & Nail, S. L. (1988). The relationship between the TMDSC curve of frozen sucrose solutions and collapse during freeze-drying. *Journal of Thermal Analysis and Calorimetry*, 54, 659–672.

Kudra, T., & Strumillo, C. (1998). *Thermal Processing of Biomaterials*. Gordon and Breach Science Publishers, Amsterdam.

Laboao (2021). *Vacuum spray freeze drying machine*. Available from: https://www.laboao.com/products/spray-dryer/vacuum-spray-freeze-drying-machine (Accessed 9 September 2021).

Leach, W. T., Simpson, D. T., Val, T. N., Anuta, E. C., Yu, Z., Williams, R. O., & Johnston, K. P. (2005). Uniform encapsulation of stable protein nanoparticles produced by spray freezing for the reduction of burst release. *Journal of Pharmaceutical Sciences*, 94(1), 56–69.

Leuenberger, H. (1987). *Process of drying a particulate material and apparatus for implementing the process*. US Patent 4,608,764.

Leuenberger, H. (2002). Spray Freeze drying e the process of choice for low water soluble drugs? *Journal of Nanoparticle Research*, 4, 111–119.

Leuenberger, H., Plitzko, M., & Puchkov, M. (2006). Spray freeze drying in a fluidized bed at normal and low pressure. *Drying Technology*, 24, 711–719.

Levi, G., & Karel, M. (1995). Volumetric shrinkage (collapse) in freeze-dried carbohydrates above their glass transition temperature. *Food Research International*, 28(2), 145–151.

Ly, A., Carrigy, N. B., Wang, H., Harrison, M., Sauvageau, D., Martin, A. R., Vehring, R., & Finlay, W. H. (2019). Atmospheric spray freeze drying of sugar solution with Phage D29. *Frontiers in Microbiology*, 10.

Ma, X., Wang, W., Bouffard, R., & MacKenzie, A. (2001). Characterization of murine monoclonal antibody to tumor necrosis factor (TNF-MAb) formulation for freeze-drying cycle development. *Pharmaceutical Research*, 18(2), 196–202.

MacLeod, C. S., McKittrick, J. A., Hindmarsh, J. P., Johns, M. L., & Wilson, D. I. (2006). Fundamentals of spray freezing of instant coffee. *Journal of Food Engineering*, 74, 451–461.

Malecki, G. J., Shinde, P., Morgan, A. I., & Farkas, D. F. (1970). Atmospheric fluidized bed freeze drying. *Food Technology*, 24, 601–603.

Matteo, Di. P., Donsi, G., & Ferrari, G. (2003). The role of heat and mass transfer phenomena in atmospheric freeze drying of foods in a fluidized bed. *Journal of Food Engineering*, 59, 267–275.

McHugh, T. (2018). Freeze-Drying Fundamentals. *IFT Food Technology Magazine*. Available from https://www.ift.org/news-and-publications/food-technology-magazine/issues/2018/february/columns/processing-freeze-drying-foods (Accessed 6 September 2021).

Meister, E., & Gieseler, H. (2008). A significant comparison between collapse and glass transition temperatures. Available from: https://www.europeanpharmaceuticalreview.com/article/1479/a-significant-comparison-between-collapse-and-glass-transition-temperatures/ (Accessed 9 September 2021).

Meister E., & Gieseler H. (2009). Freeze-dry microscopy of protein/sugar mixtures: drying behavior, interpretation of collapse temperatures and a comparison to corresponding glass transition data. *Journal of Pharmaceutical Sciences*, 98, 3072–3087.

Meryman, H. T. (1959). Sublimation freeze-drying without vacuum. *Science*, 130(3376), 628–629.

Mueannoom, W., Srisongphan, A., Taylor, K. M. G., & Hauschild, S. (2012). Thermal ink-jet spray freeze drying for preparation of excipient free salbutamol sulphate for inhalation. *European Journal of Pharmaceutics and Biopharmaceutics*, 80, 149–155.

Mumenthaler, M., & Leuenberger, H. (1991). Atmospheric spray freeze-drying: a suitable alternative in freeze drying technology. *International Journal of Pharmaceutics*, 72, 97–110.

Nail, S. L., & Johnson, W. (1992). Methodology for in-process determination of residual water in freeze-dried products. *Dev Biol Stand.*, 74, 137–151; dicussion 150-1. PMID: 1592164.

Nasirpour, A., Landillon, V., Cuq, B., Scher, J., Banon, S., & Desobry, S. (2007). Lactose crystallization delay in model infant foods made With Lactose, β-Lactoglobulin, and starch. *Journal of Dairy Science*, 90(8), 3620–3626.

Oetjen, G. W. (1999). *Freeze drying*. Wiley-VCH, German.

Pansare, S. K., & Patel, S. M. (2016). Practical considerations for determination of glass transition temperature of a maximally freeze concentrated solution. *AAPS PharmSciTech*, 17(4), 805–819.

Pardo, J. M., Suess, F., & Niranjan, K. (2002). An investigation into the relationship between freezing rate and mean ice crystal size for coffee extracts. *Food and Bioproducts Processing*, 80, 176–182.

Passot, S., Fonseca, F., Alarconlorca, M., Rolland, D., & Marin, M. (2005). Physical characterisation of formulations for the development of two stable freeze-dried proteins during both dried and liquid storage. *European Journal of Pharmaceutics and Biopharmaceutics*, 60(3), 335–348.

Patel, S. M., Doen, T., & Pikal, M. J. (2010). Determination of end point of primary drying in freeze-drying process control. *AAPS PharmSciTech*, 11(1), 73–84.

Pikal, M. J. (1990). The collapse temperature in freeze drying: Dependence on measurement methodology and rate of water removal from the glassy phase. *International journal of Pharmaceutics*, 62, 165–186.

PowderPro (2012). *Spray-freezing and freeze drying*. Available from: https://powderpro.se/2012/05/29/spray-freezing-and-freeze-drying-2/ (Accessed 9 September 2021).

Rahmati, M. R., Vatanara, A., Parsian, A. R., Gilani, K., Khosravi, K. M., Darabi, M., et al. (2013). Effect of formulation ingredients on the physical characteristics of salmeterol xinafoate microparticles tailored by spray freeze drying. *Advanced Powder Technology*, 24(1), 36–42.

Rajam, R., & Anandharamakrishnan, C. (2015). Spray freeze drying method for microencapsulation of lactobacillus plantarum. *Journal of Food Engineering*, 166, 95–103.

Ratti, C. (2001). Hot air and freeze-drying of high-value foods: A review. *Journal of Food Engineering*, 49(4), 311–319.

Rochelle, C., & Lee, G. (2007). Dextran or hydroxyethyl starch in spray freeze dried trehalose/mannitol microparticles intended as ballistic particulate carriers for proteins. *Journal of Pharmaceutical Sciences*, 96(9), 2296–2309.

Rogers, T. L., Hu, J., Yu, Z., Johnston, K. P., & Williams, R. O., (2002). A novel particle engineering technology: spray-freezing into liquid. *International Journal of Pharmaceutics*, 242, 93–100.

Rogers, T. L., Nelsen, A. C., Sarkari, M., Young, T. J., Johnston, K. P., & Williams, R. O. (2003). Enhanced aqueous dissolution of a poorly water soluble drug by novel particle engineering technology: spray-freezing into liquid with atmospheric freeze-drying. *Pharmaceutical Research*, 20, 485–493.

Ross, C., Gaster, T., & Ward, K. (2008). The Importance of critical temperatures in the freeze drying of pharmaceutical products. Available from: http://www.biopharma.co.uk/wp-content/uploads/2010/07/importance_critical_temps.pdf (Accessed 9 September 2021).

Roy, M. L., & Pikal, M. J. (1989). Process Control in Freeze Drying: Determination of the End Point of Sublimation Drying by an Electronic Moisture Sensor. *Journal of Parenteral Science and Technology*, 43, 60–66.

Scoffin, K. (2014). The Importance of Critical Temperatures in Freeze-drying. *International Pharmaceutical Industry*, Spring 2014, 6(1), 96–100.

Schwegman, J. J., & Neiblas, R. (2015). Thermal characterization as part of an empirical process for developing optimized formulations and lyophilization cycles. Available from: https://www.mccrone.com/mm/thermal-characterization-part-empirical-process-developing-optimized-formulations-lyophilization-cycles/ (Accessed 9 September 2021).

Sebastião, B. I., Robinson, T. D., & Alexeenko, A. (2017). Atmospheric spray freeze-drying: Numerical modeling and comparison with experimental measurements. *Journal of Pharmaceutical Sciences*, 106(1), 183–192.

Semyonov, D., Ramon,O., Kaplun, Z., Levin-Brener, L., Gurevich, N., & Shimoni, E. (2010). Microencapsulation of *Lactobacillus paracasei* by spray freeze drying. *Food Research International*, 43, 193–202.

Sonner, C., Maa, Y.-F., & Lee, G. (2002). Spray Freeze Drying for protein powder preparation. *Journal of Pharmaceutical Sciences*, 91(10), 2122–2139.

SP Scientific Inc. (2019). Freeze-Drying Microscope Generates Critical Formulation Specific Data. Available from: https://www.labbulletin.com/articles/20190508_1/print (Accessed 9 September 2021).

Stapley, A. (2008). Freeze drying. In Evans, J. A. (Ed.). *Frozen Food Science and Technology*. Blackwell Publishing, Oxford, pp. 248–275.

Straller, G., & Lee, G. (2017). Shrinkage of spray-freeze-dried microparticles of pure protein for ballistic injection by manipulation of freeze-drying cycle. International *Journal of Pharmaceutics*, 532(1), 444–449.

Thomas, M. E. C., Scher, T. J., & Desorby, S. (2004). Lactose/β-Lactoglobulin interaction during storage of model whey powders. *Journal of Dairy Science*, 87, 1158–1166.

Wang, Y., Kho, K., Cheow, W. S., & Hadinoto, K. (2012). A comparison between spray drying and spray-freeze drying for drypowder inhaler formulation of drug-loaded lipid polymer hybrid nanoparticles. International Journal of Pharmaceutics, 424, 98–106.

Wang, Z. L., Finlay, W. H., Peppler, M. S., & Sweeney, L. G. (2006). Powder formation by atmospheric spray-freeze-drying. *Powder Technology*, 170, 45–52.

Ward, K., & Matejtschuk, P. (2010). The use of microscopy, thermal analysis and impedence measurements to establish critical formulation parameters for freeze-drying cycle development. In: *Freeze-Drying-Lyophilization of Pharmaceutical and Biological Products*, Louis Rey and Joan C. May, Informa Healthcare.

Woodward, H. T. (1963). Freeze drying without vacuum. *Food Engineering*, 35, 96–97.

Yang, G., Gilstrap, K., Zhang, A., Xu, L. X., & He, X. (2010). Collapse temperature of solutions important for lyopreservation of living cells at ambient temperature. *Biotechnology and Bioengineering*, 106, 247–259.

Yu, Z., Garcia, A. S., Johnston, K. P., & Williams, R. O. (2004). Spray freezing into liquid nitrogen for highly stable protein nanostructured microparticles. *European Journal of Pharmaceutics and Biopharmaceutics*, 58, 529–537.

Yu, Z., Johnston, K. P., & Williams, R. O. (2006). Spray freezing into liquid versus spray-freeze drying: Influence of atomisation on protein aggregation and biological activity. *European Journal of Pharmaceutical Sciences*, 27, 9–18.

Zheng, X. (2017). *Vacuum spray-freeze-drying apparatus and method.* WO2017084163.

CHAPTER 4

Morphology of Spray-Freeze-Dried Products

Morphology is the term derived from the Greek word, *morph,* meaning, *shape* or *form.* As the name implies, morphology is the study of *shapes* or *forms.* Particle morphology is an important attribute of spray-freeze-dried products. The major competitive advantage of spray-freeze-drying over other liquid drying techniques is its ability to produce particles of unique microstructure. Intact morphology is an indication of successful particle drying, which is possible only when the product bypasses structural collapse during the freeze-drying stage. The surface morphology of a spray-freeze-dried (SFD) product revealed by scanning electron microscopy can indicate the occurrence or absence of collapse. The solubility, flowability and stickiness of dried powders have a bearing on their surface morphology (Anandharamakrishnan et al., 2010). Further, the dissolution property, aerosolization efficiency and site-targeted delivery of drugs for pulmonary and ballistic delivery are also dependent on the morphology of spray-freeze-dried particles.

Generally, it is considered that a properly conducted spray-freezing step ensures no further changes to the particle microstructure during the subsequent freeze-drying stage (Anandharamakrishnan et al., 2010). Nevertheless, feed composition (Leuenberger et al., 2006; MacLeod et al., 2006) and freeze-drying temperature (Anandharamakrishnan, 2008; Anandharamakrishnan et al., 2010) exert influence on the collapse phenomenon and microstructure development in spray-freeze-dried products. So far, studies on SFD have revealed different morphological patterns, each of which have a defined correlation with the process conditions and the intended functionality of the final product. Hence, it is imperative to be familiar with the aspects described above to design customized spray-freeze-dried products for food and pharmaceutical applications.

4.1 IMAGING THE MORPHOLOGY OF SPRAY-FREEZE-DRIED PARTICLES

The most commonly employed technique for visualizing the morphology of spray-freeze-dried particles is the Scanning Electron Microscopy (SEM) and its advanced variants such as FE-SEM (field emission SEM), Cryo-SEM and E-SEM (Environmental SEM). The principle of working and instrumentation of SEM have been explained in standard references and hence not included in the purview of this book. But, an overview of the information that can be provided by the scanning electron microscope, sample preparation and visualization conditions for the SEM imaging of SFD particle morphology would be presented in this section.

The scanning electron microscope is a well-known instrument to visualize and understand the intricacies of a material's microstructure. It provides information on the topographies, morphology, compositional differences, crystal structures and orientation of a material (Swapp, *n.d.*). As the name suggests, a scanning electron microscope uses electrons instead of light to form an image. An electron gun generates the electrons that are accelerated to moderately high energy and focused upon the sample using electromagnetic fields. Interaction between these electrons and the atoms in

DOI: 10.1201/9781003019312-4

the specimen produces a signal. Each spot on the sample interacts with the electron beam and emits secondary electrons from its surface, which are counted by a detector that directs the signals to an amplifier (JEOL, *n.d.*). The final micrograph thus obtained is collated from all the electrons emitted from each spot on the sample.

Scanning electron microscopy has the edge over light microscopy, as it is capable of focusing a larger section of the sample at once, owing to its greater depth of field. The greater focal depth of SEM enables visualization of pores with clarity. Also, high-resolution images can be obtained from SEM, which facilitates the examination of closely-spaced features at high magnification. Sample preparation for imaging by SEM is relatively easy as its only requirement is that the sample must be conductive. The sample can be rendered conductive by coating it with a very thin layer of gold (Anandharamakrishnan, 2008), gold/palladium alloy or platinum. The powder is gently spread on the surface of a double-sided adhesive carbon conducting tape attached to a metal stub or specimen holder made of aluminium. The specimen holder with the sample must be immediately transferred to a sputter-coating device to minimize its exposure time to atmospheric moisture. In some cases, the sample stages are kept inside septum capped vials and purged with nitrogen, before transferring to the sputtering device. Then, the particles are coated with an ultra-thin layer of gold in an argon or nitrogen atmosphere, under a high vacuum at room temperature. Generally, the thickness of the coating is in the order of nanometers and the duration of coating step ranges between a few seconds up to 5 min. To prevent the hydration of coated samples upon exposure to atmospheric humidity and reduce oxidation of the metal coating, the prepared samples are stored in a vacuum desiccator (CMRF, *n.d.*).

The coated samples are immediately transferred to the scanning electron microscope and the samples are imaged under defined conditions of accelerator voltage (1.5–25.0 kV) and vacuum levels typically in the order of mTorr. Accelerator voltage is a critical parameter to obtain perfect images of the porous structure of spray-freeze-dried particles. It is the voltage at which the electrons are accelerated down the electron column (Hafner, 2007). It is preferable to operate the SEM at lower accelerating voltages, to obtain images of the finer surface structure of spray-freeze-dried particles. Hence, in most of the SFD studies, an accelerating voltage of 1.5–2.5 kV is usually employed (Figure 4.1[a & b]). On the other hand, a higher accelerating voltage gives better image resolution (Nessler, *n.d.*). Thus, in studies involving the application of SFD for nanoparticle production, high accelerator voltage in the range of 20–25 kV has been used (Figure 4.1[c]). However, it must not be increased beyond a certain level as it may lead to specimen damage and images that lack details on the surface morphology of specimens (JEOL, *n.d.*). Usually, the micrographs are recorded in the magnification range of 500X–10000X or even at higher magnifications up to 60000X, until the required structural details are visible. If the objective of SEM imaging is to visualize the internal microstructure of SFD particles, then cross-section of the sample is obtained by slicing the microspheres with a razor to expose the particle's interior.

4.2 MORPHOLOGICAL PATTERNS OF SPRAY-FREEZE-DRIED PARTICLES

4.2.1 Spherical particles with porous surface morphology

Commonly, particles in a spray-freeze-dried (SFD) product exhibit spherical shape with rough and porous surface morphology (Figure 4.2). The porous microstructure originates from the spray-freezing step, during which the fine feed droplets emanating from the atomizer facilitate rapid freezing and formation of fine ice crystals embedded in a solid matrix of the freeze-concentrated solute. During the primary drying phase of freeze-drying, if the product is held below its collapse temperature (T_g' or T_e of the maximally freeze concentrated solute), then the subsequent sublimation of ice leaves behind voids, thereby resulting in porous structure (Al-Hakim & Stapley,

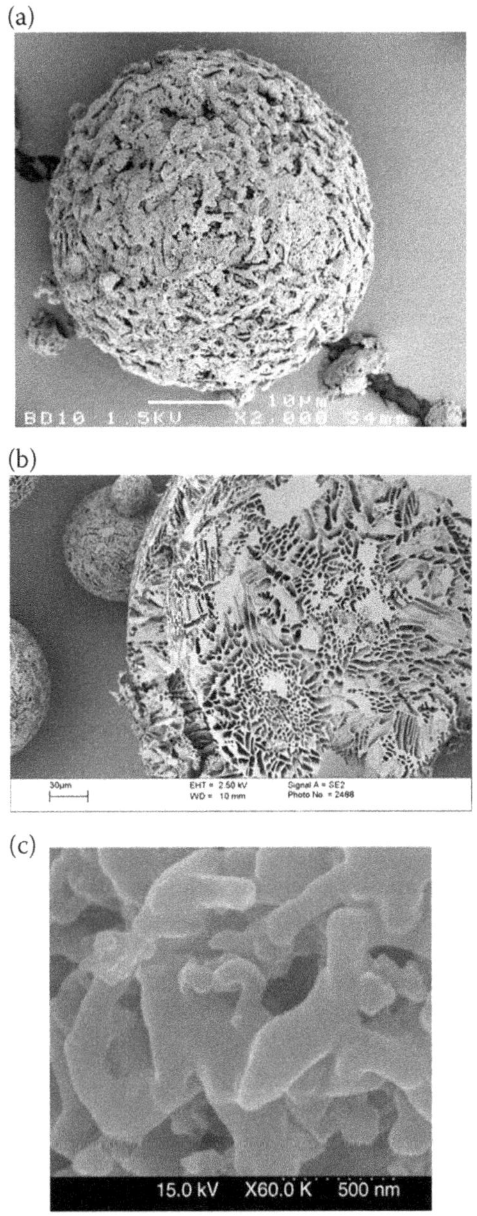

Figure 4.1 SEM images of spray-freeze-dried particles acquired at (a) low accelerator voltage (1.5 kV) at low magnification (2000X), depicting the highly porous surface of an inhalable solid dispersion powder (van Drooge et al., 2005); (b) low accelerator voltage (2.5 kV), showing the cross-section of a spray-freeze-dried maltodextrin particle (Khwanpruk et al., 2008); (c) high accelerator voltage (15.0 kV) and high magnification (60000X), for high-resolution imaging of protein nanoparticles (Yu, Johnston, & Williams 2006).

2004). The porous surface of SFD particles promotes solubility by facilitating the capillary imbibition of water on rehydration (Saguy et al., 2005). Especially, in poorly water-soluble drugs, the porous microstructure is of utmost relevance to improve their *in vivo* performance after oral administration (Dressman & Reppas, 2000). Spray-freeze-dried active pharmaceutical ingredients (APIs) with low-density and high specific surface area have demonstrated improved solubility and bioavailability (CJPH, 2016). Moreover, low-density porous particles (<0.4 g cm^{-3}) lead to the enhanced aerosol performance of drugs for pulmonary delivery (Amorij et al., 2007; Chew & Chan, 2001; Edwards, 1997; Maa et al., 1999; Mohri et al., 2010; Saluja et al., 2010; Sweeney et al., 2005; Zijlstra et al., 2007).

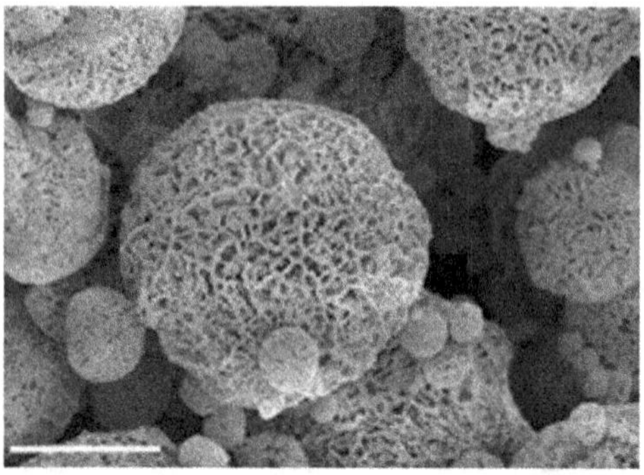

Figure 4.2 Scanning Electron micrograph of spray-freeze-dried particles with a rough and porous surface: inulin stabilized influenza subunit vaccine powder (Modified from Amorij et al., 2007).

4.2.2 Spherical particles with smooth surface morphology enclosing a porous internal microstructure

Some studies have revealed spherical SFD particles with a smooth surface layer encasing a porous and *'sponge-like'* internal microstructure (Al-Hakim & Stapley, 2004; Hindmarsh et al., 2007; Windhab, 1999). Formation of solute surface layers may result when the outer layer of the droplet is subjected to rapid cooling relative to other spots and then plunged into the glassy state without allowing much time for ice crystallization to occur (Windhab, 1999). Hindmarsh et al. (2007) opposed the above theory as large temperature gradients within the droplet are not possible. They proposed that water may evaporate from the surface before freezing begins. Consequently, the water concentration in the surface layer would drop below the level at which ice crystallization can occur. Alternatively, the smooth surface layer may form as a result of the preferential distribution of solutes at the surface soon after atomization. Hindmarsh et al. (2007) also ascertained that the thickness of the solute surface layer decreased with an increase in the freezing rate. The tendency of SFD powders to form a skin or layer that partially covers the pores on their outer surface is favorable toward improving the mechanical strength of the particles during handling (Ali & Lamprecht, 2014). Moreover, it exerts a profound influence on the powder characteristics such as stickiness and flow (Al-Hakim & Stapley, 2004; Anandharamakrishnan et al., 2010).

In some cases, the covering skin layer does not appear. One reason may be the use of a two-fluid nozzle for atomization, which introduces relatively warm air into the spray chamber, which mixes in a disordered manner with the cold chamber gas. As a result, the discrete atomized droplets may come into contact with gas pockets at different temperatures, leading to variation in cooling rates between droplets. The difference in cooling rates could affect the nucleation rates that are highly temperature-dependent and eventually lead to different surface structures (Al-Hakim & Stapley, 2004).

4.2.3 Irregular-shaped, wrinkled surface morphology

Contrarily, in the case of APIs (ex. pure proteins like insulin) intended for use in needle-free ballistic powder delivery, irregularly shaped wrinkled particles (Figure 4.3[a-d]) with reduced porosity are preferred (Straller & Lee, 2017). Hence, during the SFD of such formulations, structural collapse is intentionally induced in a controlled manner. Collapsed particles have a very low surface area and thereby form a dense and mechanical robust cake, which is the key requirement for ballistic injection of vaccine powders (Gill & Prausnitz, 2007; Matriano et al., 2002; Quan et al., 2010). Shriveled and partially-collapsed microparticles of pure proteins (BSA and BCA) were produced by two approaches: (i) inclusion of trehalose in the feed formulation (Figure 4.3[a]); (ii) manipulating the freeze-drying cycle of SFD, wherein, the primary drying phase was conducted at sub-zero shelf temperatures (−12°C to −8°C) for an extended duration of

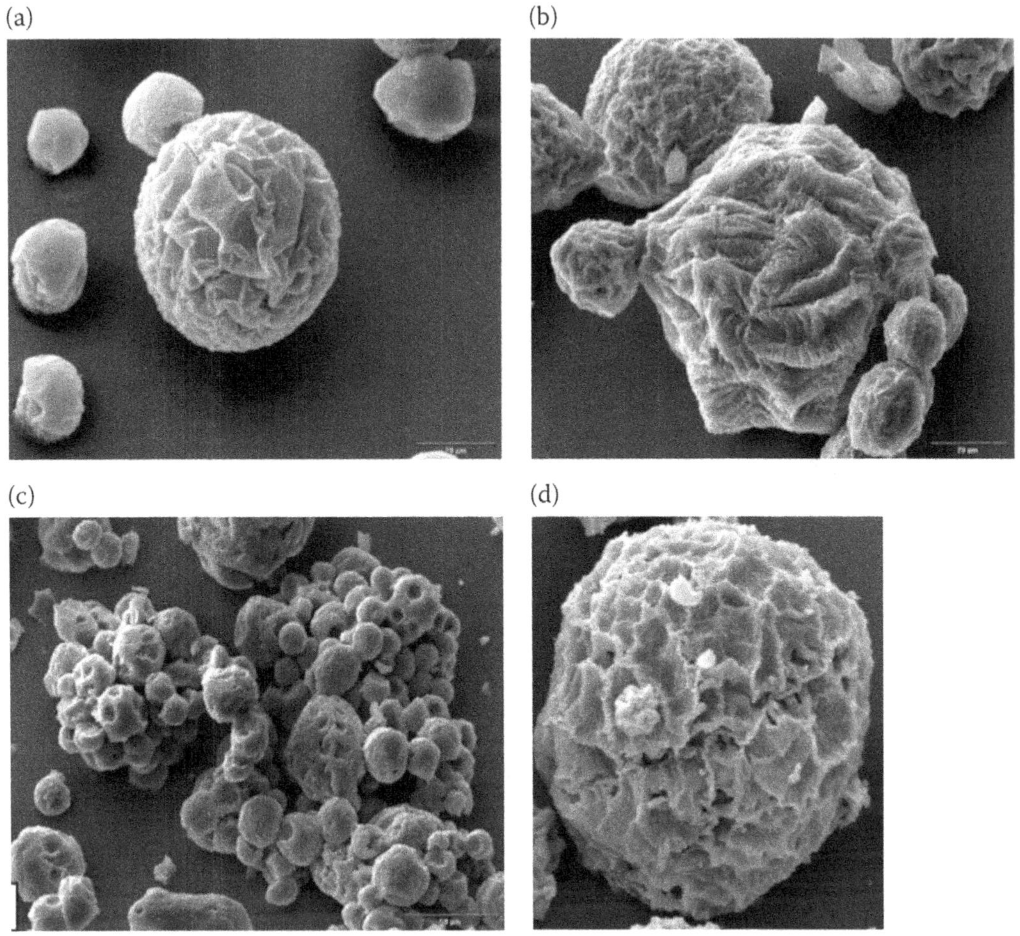

Figure 4.3 SEM images of the collapsed and wrinkled spray-freeze-dried particles of (a) BSA/trehalose mixture ratio: 1:3; (b) pure BSA made using an extended duration of primary drying for 2745 min at shelf temperature of −8°C; (c) pure BSA made using an extended duration of primary drying for 2745 min at shelf temperature of −5°C; (d) pure BCA made using an extended duration of primary drying for 2745 min at shelf temperature of −8°C (Straller & Lee, 2017).

up to 2745 min (Figure 4.3[b-d]). These approaches led to collapsed particles owing to the plasticizing effect of disaccharides on the protein and the availability of sufficient time for the plastic flow to occur, respectively (Straller & Lee, 2017).

4.3 FACTORS INFLUENCING THE MORPHOLOGY OF SPRAY-FREEZE-DRIED PARTICLES

4.3.1 Feed solid content

Feed solid content exhibits a marked influence on the surface morphology of spray-freeze-dried particles. This can be explained with the example of soluble coffee powder. Coffee feed solutions having solid contents of 10%, 20%, 30% and 40% were frozen by SFV/L approach followed by vacuum freeze-drying. The shelf temperature during primary drying ranged from −25°C to −10°C,

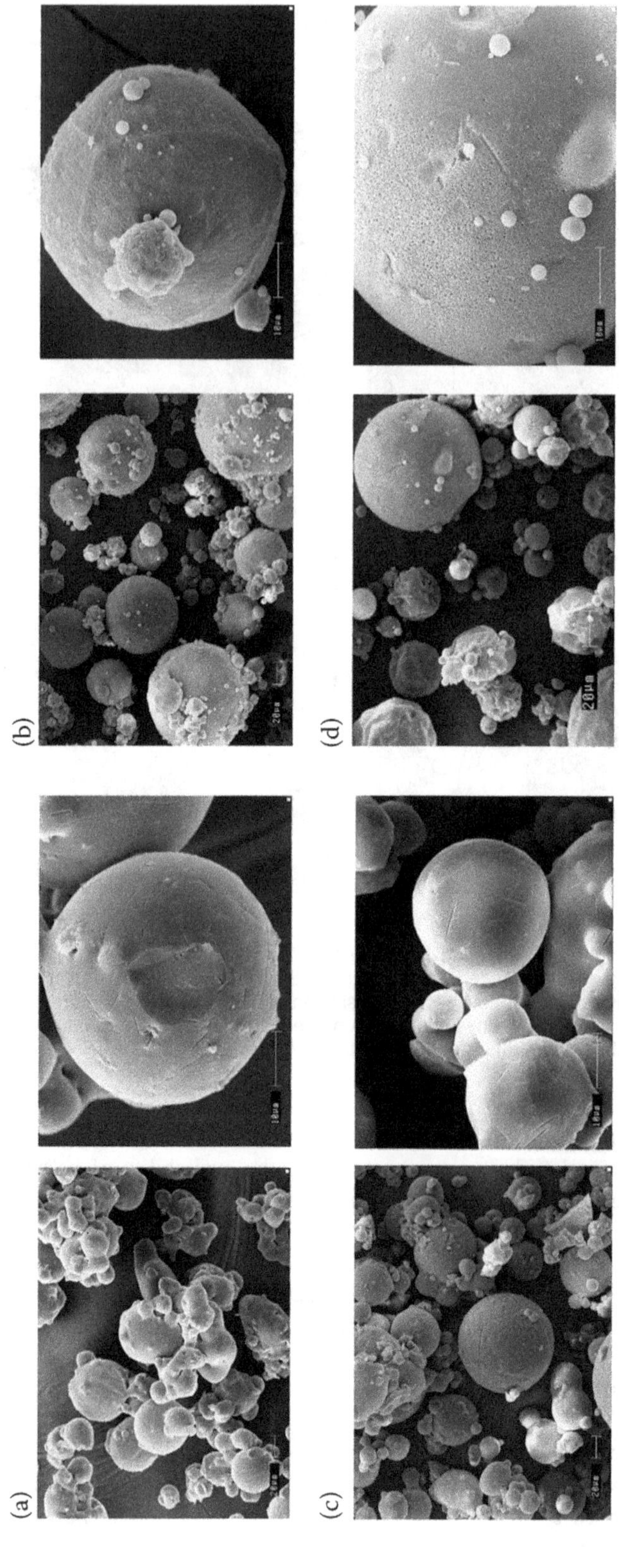

Figure 4.4 Influence of solid concentration on the surface morphology of spray-freeze-dried coffee obtained from coffee feed solution having different solid contents: (a) 10%; (b) 20%; (c) 30%; (d) 40% (Ishwarya & Anandharamakrishnan, 2015).

under an applied vacuum of 107 Pa. Under constant conditions of primary drying, the increase in feed solid content showed an improving effect on the surface morphology of spray-freeze-dried coffee particles (Figure 4.4[a-d]). From the scanning electron micrographs, it was evident that the proportion of shriveled and wrinkled particles with reduced surface porosity was more in the particles produced from feed containing 10% coffee solids (Figure 4.4[a]), thus indicating collapse. However, the proportion of collapsed particles was low at higher feed solid concentrations (Figure 4.4[b-d]) (Ishwarya & Anandharamakrishnan, 2015; unpublished data). This is justified as the collapse temperature (T_c) of a 25% coffee extract is −20°C (Bellows & King, 1972; Fellows, 2002; Kudra & Strumillo, 1998). The higher solid content of feed might have elevated the T_c, thus permitting the primary drying to proceed without collapse at a relatively higher drying temperature as that used in the above study (−25 to −10°C).

4.3.2 Feed composition

Similar to solid content, the feed composition also plays a major role in determining the morphology of SFD products. Vanillin, a widely used flavoring agent in bakery, beverages and ice-creams, was encapsulated by spray-freeze-drying using two types of wall materials: whey protein isolate (WPI) and β-cyclodextrin (β-cyd). Vanillin+β-cyd and Vanillin+WPI mixtures were atomized using a twin fluid nozzle, into cold vapor over liquid nitrogen. The frozen particles were dried by conventional freeze-drying at −24°C for 4 hours. The WPI-vanillin microcapsules showed spherical shape with numerous fine pores (Figure 4.5[a & b]). But, the β-cyd-vanillin microcapsules underwent structural collapse during freeze-drying and hence did not result in porous spherical particles (Figure 4.5[c & d]) (Hundre, Karthik & Anandharamakrishnan, 2015).

Likewise, the difference in feed formulation exerted a significant influence on the surface morphology of spray-freeze-dried probiotics (*Lactobacillus plantarum*). The variations were attributed to differences in the film-forming properties of substances that were used as carrier

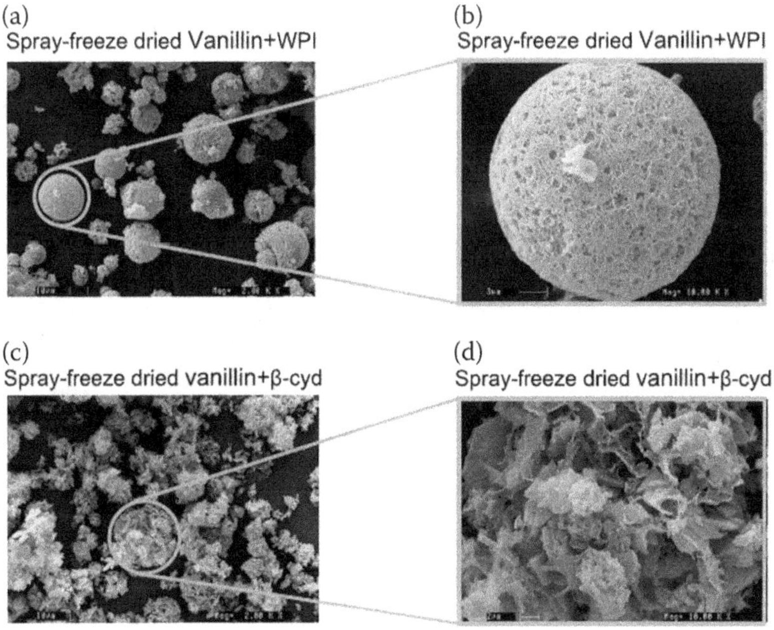

Figure 4.5 SEM images of vanillin microencapsulated by spray-freeze-drying (Hundre et al., 2015).

materials (Rajam & Anandharamakrishnan, 2015). Four types of wall material were used: (1) WPI + sodium alginate (SA); (2) WPI + fructooligosaccharide (FOS); (3) denatured WPI (DWPI) + SA; (4) DWPI + FOS. The spray-freeze-dried WPI + SA microcapsules showed a smooth outer surface, owing to the characteristic skin-forming property of whey protein (Sheu & Rosenberg, 1998; Anandharamakrishnan et al., 2007). SEM images of the cross-section of these microcapsules (Figure 4.6[a]) showed that they are solid spheres with a very fine and porous internal structure, with small air bubbles entrapped within the microstructure. The presence of air bubbles was due to atomization, as mentioned in the previous section.

On the other hand, spray-freeze-dried WPI + FOS microcapsules exhibited a slightly rough surface. And, few smaller particles were found clinging to the larger particles (Figure 4.6[b]). Spray-freeze-dried microcapsules of DWPI + SA (Figure 4.6[c]) and DWPI + FOS (Figure 4.6[d]) exhibited rough surface with protrusions, attributed to the presence of denatured whey protein isolate in the wall material formulation (Nicolai et al., 2011; Anandharamakrishnan et al., 2007). At higher magnification, the SEM images depicted the entrapment of probiotic cells within the wall material (Figure 4.6[d], inset) (Rajam & Anandharamakrishnan, 2015).

Kanamycin is an aminoglycoside antibiotic produced by strains of *Streptomyces kanamyceticus*, which is effective against several Gram-positive and Gram-negative bacteria (Puius, Stievater & Srikrishnan, 2006). It is used to treat severe bacterial infections, asthma and tuberculosis (Desai, Hancock & Finlay, 2002; Sweeney et al., 2005). Her et al. (2010) studied the

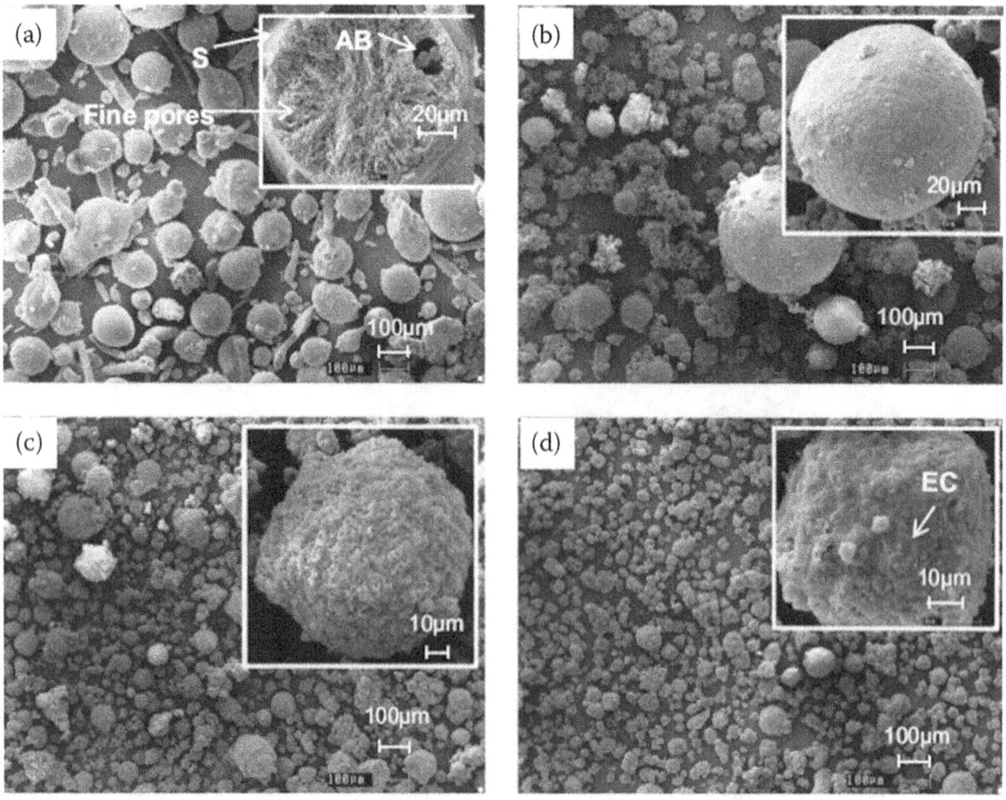

Figure 4.6 SEM images of spray-freeze-dried microcapsules containing *L. plantarum* produced with (a) WPI + SA; (b) WPI + FOS; (c) DWPI + SA; (d) DWPI + FOS wall material formulations. Inset shows the magnification at 10000X. AB - air bubble; EC - encapsulated cells; S - smooth outer skin (Rajam & Anandharamakrishnan, 2015).

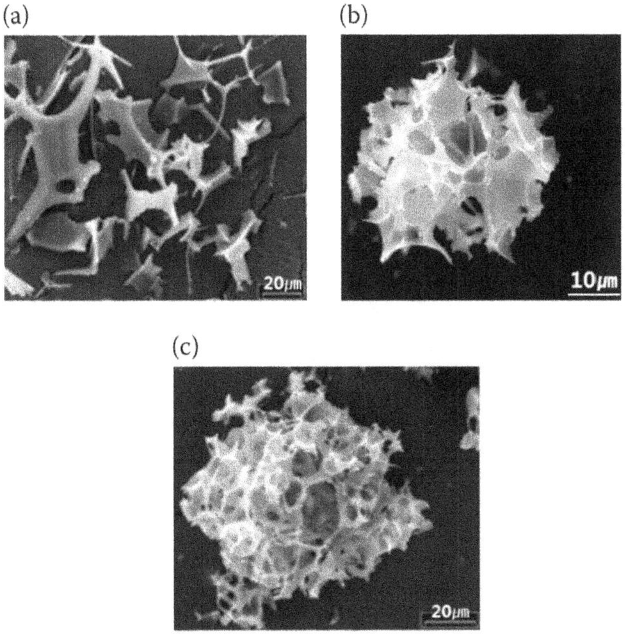

Figure 4.7 SEM images of spray freeze-dried powder at various concentrations of kanamycin: (a) 2% (w/v); (b) 10% (w/v); (c) 15% (w/v) (Her et al., 2010).

effect of kanamycin concentration (2%, 10% and 15%, w/v) on its SFD particle morphology. The SEM images revealed that the SFD powder obtained from feed containing kanamycin at 2% (w/v) did not form a spherical shape (Figure 4.7[a]). Thus, it was considered that the above concentration was not sufficient to form a spherical surface. But, SFD powder from 10% (Figure 4.8[b]) and 15% kanamycin solutions ((Figure 4.7[c])) resulted in spherical particles with a porous surface.

4.3.3 Conditions of spray-freezing

Various researchers have investigated the role of spray-freezing conditions on the morphology of SFD particles. Compared to SFV, SFL and SFV/L modes of spray-freezing exhibit rapid cooling rates and hence lead to particles with finer microstructures (Costantino et al., 2000; Engstrom, Simpson, Lai, et al., 2007; Yu et al., 2004). Due to the very rapid freezing rates, protein (BSA) nanostructured microparticles produced by SFL showed a particularly fine microstructure (Figure 4.8[a]) with a structural feature of the order of 500 nm or less (Yu et al., 2004). But, a similar study that employed the SFV/L technique (by contacting the feed spray with four liquid nitrogen sprays) did not result in a porous microstructure (Figure 4.9[b]) (Costantino et al., 2000).

Deeper insights on the effect of spray-freezing on droplet microstructure were provided by Hindmarsh et al. (2003 & 2007). These researchers studied the single-droplet freezing (in cold gas) of water and solutions containing different solutes such as sucrose, fat and whey protein concentrate. The droplet temperature was monitored by suspending it on thermocouples coated with polytetrafluoroethylene (PTFE). The authors elucidated the direct relationship between the microstructure of the solidified droplets and nucleation. In turn, the nucleation rate increases exponentially with the degree of supercooling achieved. Nucleation rate is quantified by the number of crystals appearing per unit time per unit volume. Compared to conventional freezing, considerable supercooling (20 K) is achieved during spray-freezing due to the extremely small droplet volume. A single nucleation event is sufficient

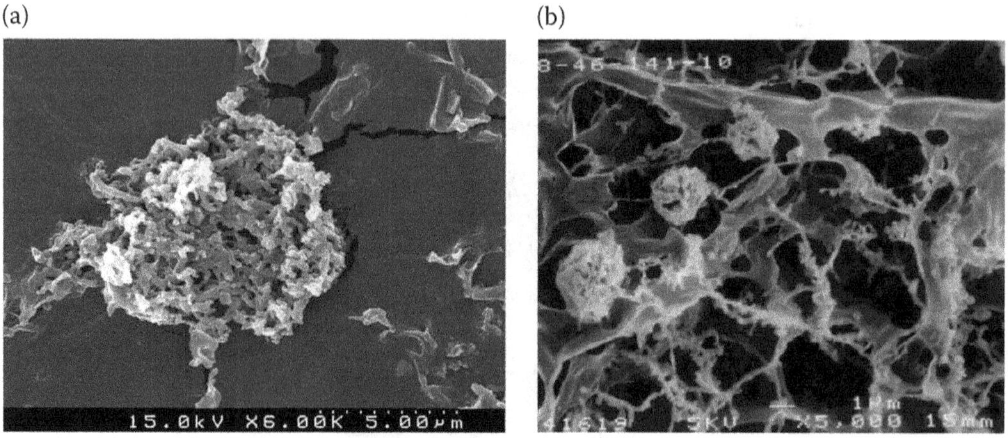

Figure 4.8 Influence of the mode of spray-freezing on SFD particle morphology: (a) SFL BSA (Yu et al., 2004); (b) SFV/L BSA (Costantino et al., 2000).

to initiate freezing throughout a sample. Hence, with small droplet volumes, unless high supercooling is reached, the nucleation probability remains very low. A droplet could remain supercooled for an indefinite duration without the occurrence of nucleation, provided the freezing gas temperature is higher than a certain nucleation temperature. If the air temperature is low enough to induce nucleation, then the nucleation temperature would exert a substantial influence on the cellular microstructure of the frozen droplet. This would be fine and homogeneous in case of rapid freezing, but coarser at slower rates. The dependence of microstructure development on the cooling rate has also been demonstrated with the

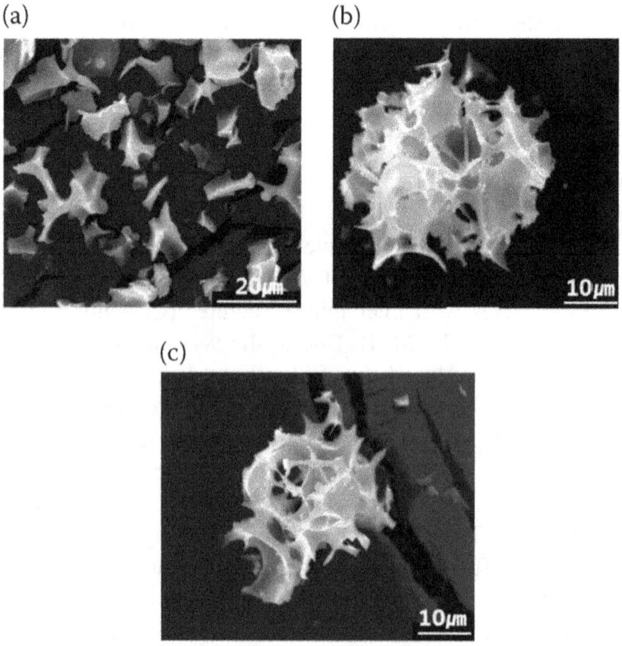

Figure 4.9 Influence of atomization conditions on the particle morphology of spray-freeze-dried kanamycin powders: (a) air pressure of 150 kPa; (b) air pressure of 100 kPa of pressure and a nozzle tip lift of 1.5 mm; (c) nozzle tip lift of 0.75 mm (Her et al., 2010).

spray-freezing of a single droplet of coffee solution (MacLeod et al., 2006). In addition to the rate of supercooling, nucleation is also driven by the rate of heat transfer from the droplet during freezing.

Atomization air pressure and nozzle tip lift were found to influence the morphology of spray-freeze-dried kanamycin powder. Two different air pressures (100 and 150 kPa) and nozzle tip lifts (1 mm and 0.75 mm) were applied. As shown in Figure 4.9(a), spherical particles did not result from the spray-freezing process carried out using 150 kPa of atomization air pressure. However, those produced at 100 kPa resulted in spherical particles (Figure 4.9[b]). The SFD kanamycin powder obtained with 0.75 mm nozzle tip had a relatively porous and spherical surface (Figure 4.9[c]) than its higher nozzle tip lift counterparts. Besides, the spraying process was unstable under the conditions of 150 kPa atomization pressure and 1.5 mm nozzle tip lift (Her et al., 2010).

4.3.4 Temperature of freeze-drying

The influence of freeze-drying temperature on the morphology of spray-freeze-dried whey protein powders was revealed in a study that employed three different inlet gas temperatures (T_i), i.e., −10°C (Figure 4.10), −15°C (Figure 4.11) and −30°C (Figure 4.12), for freeze-drying

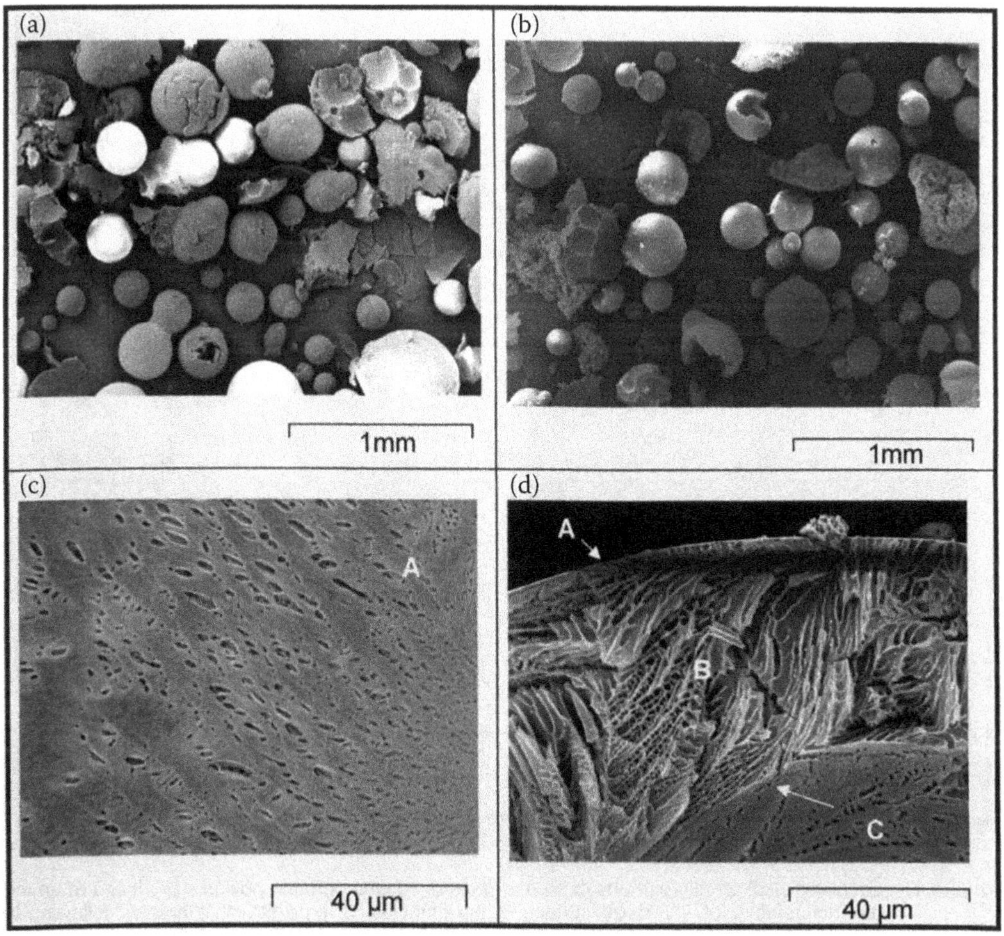

Figure 4.10 Scanning electron micrographs of spray-freeze-dried whey protein powders produced at an inlet gas temperature of −10°C (A: surface of the particles; B: pore structure created following ice sublimation, C: inside core region of the particles) (Anandharamakrishnan, 2008).

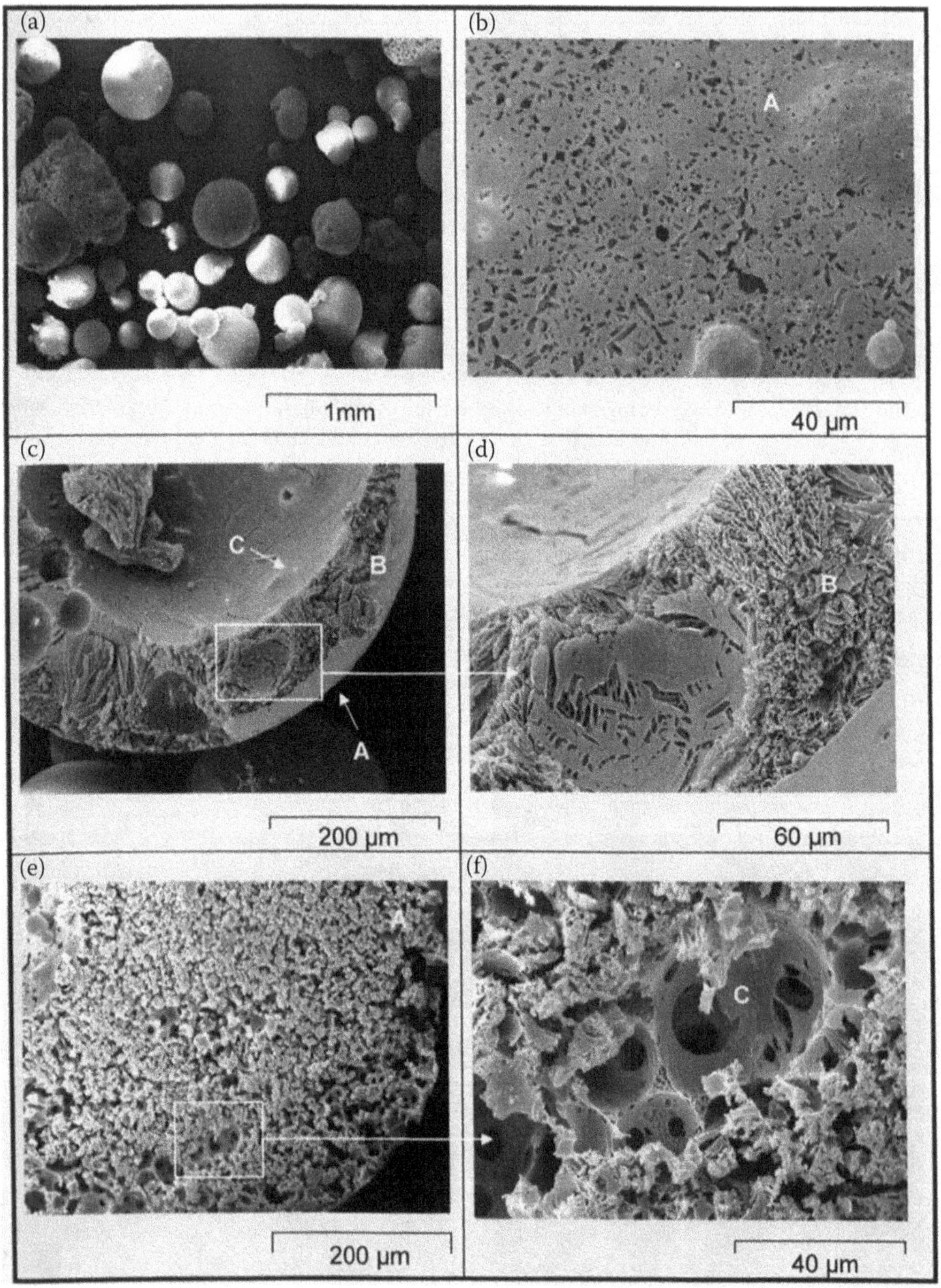

Figure 4.11 Scanning electron micrographs of spray-freeze-dried whey protein powders produced at an inlet gas temperature of −15°C (A: surface of the particles; B: pore structure created following ice sublimation, C: inside core region of the particles) (Anandharamakrishnan, 2008).

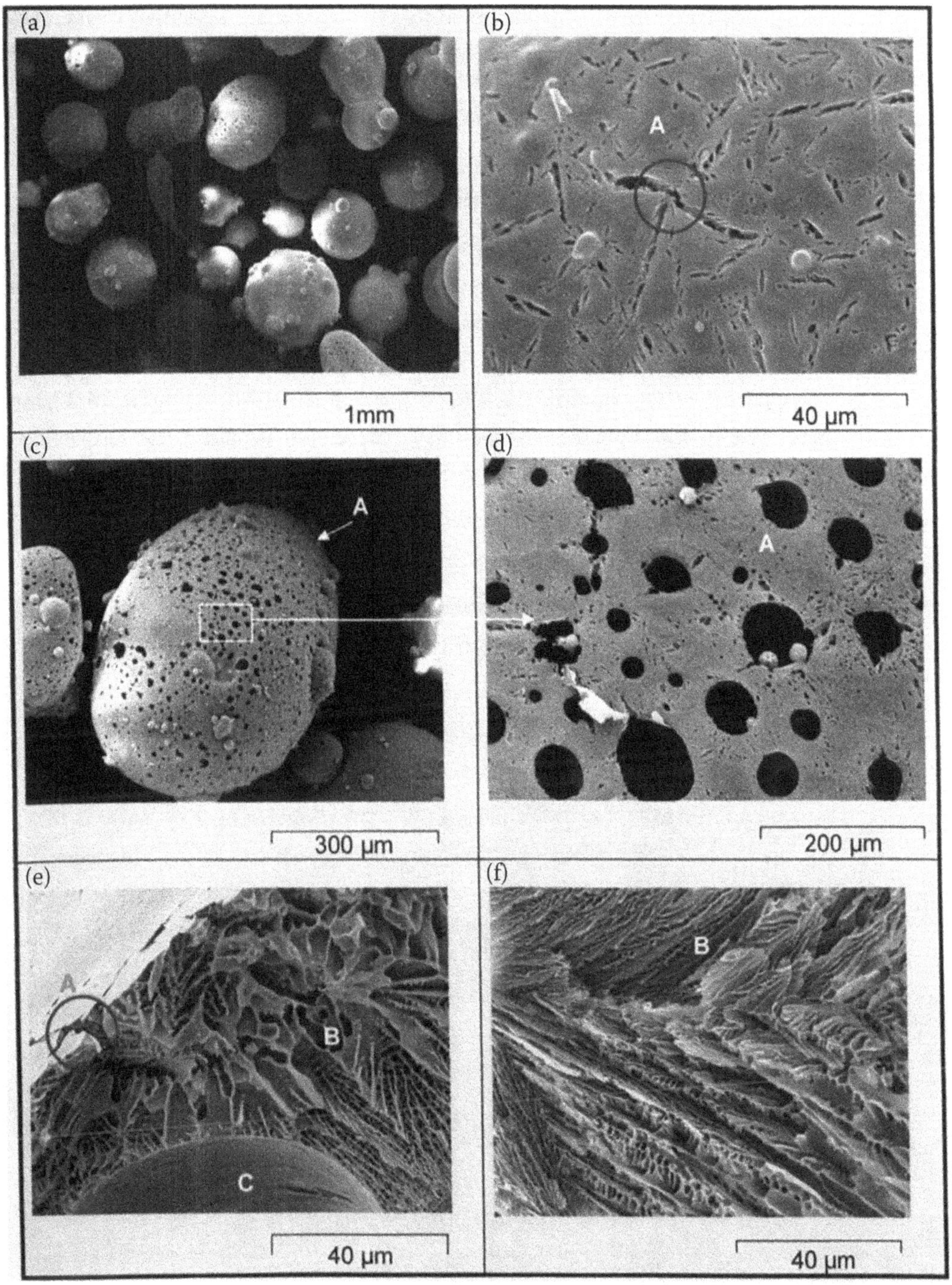

Figure 4.12 Scanning electron micrographs of spray-freeze-dried whey protein powders produced at an inlet gas temperature of −30°C (A: surface of the particles; B: pore structure created following ice sublimation, C: inside core region of the particles) (Anandharamakrishnan, 2008).

(Anandharamakrishnan, 2008). Here, it is important to note that the glass transition temperature (T_g') of frozen whey protein isolate (WPI) solution is −25.8°C. At T_i = −10°C, many broken particles with irregular shapes were observed (Figure 4.13), which was indicative of collapse. This is expected as the T_i was greater than the T_g' of frozen WPI. On the contrary, the WPI powder produced at T_i = −30°C showed an elegant porous microstructure as $T_i < T_g$'.

Irrespective of the inlet gas temperature, surfaces of the SFD-WPI particles (Figures 4.10[c], 4.11[b] and 4.12[b]) were smooth (A region) with tiny pores and an occasional surface blemish. On magnifying a spot on the sample with the surface blemish, porous structure was noticed within the particle (Figure 4.11[e-f]; C region). This is expected of a freeze-dried product, as the minute ice crystals formed during the freezing step are sublimed off during freeze-drying to result in a porous structure. Due to the rapid rates of spray-freezing, the size of both the ice crystals and surface pores were much smaller than that obtained from conventional freeze-drying. (Figures 4.10[d], 4.11[c-d] and 4.12[e-f]) shows the internal pore structure created by sublimation of ice during the primary drying period (B region). The pore routes are shown in Figure 4.12[b & e], with circular marks. Further, hollow cavities were noticed in the micrographs of all the SFD-WPI particles (Figure 4.10[d] [marked C]; 4.11 [marked C], 4.12 [marked C]), that suggested the formation of

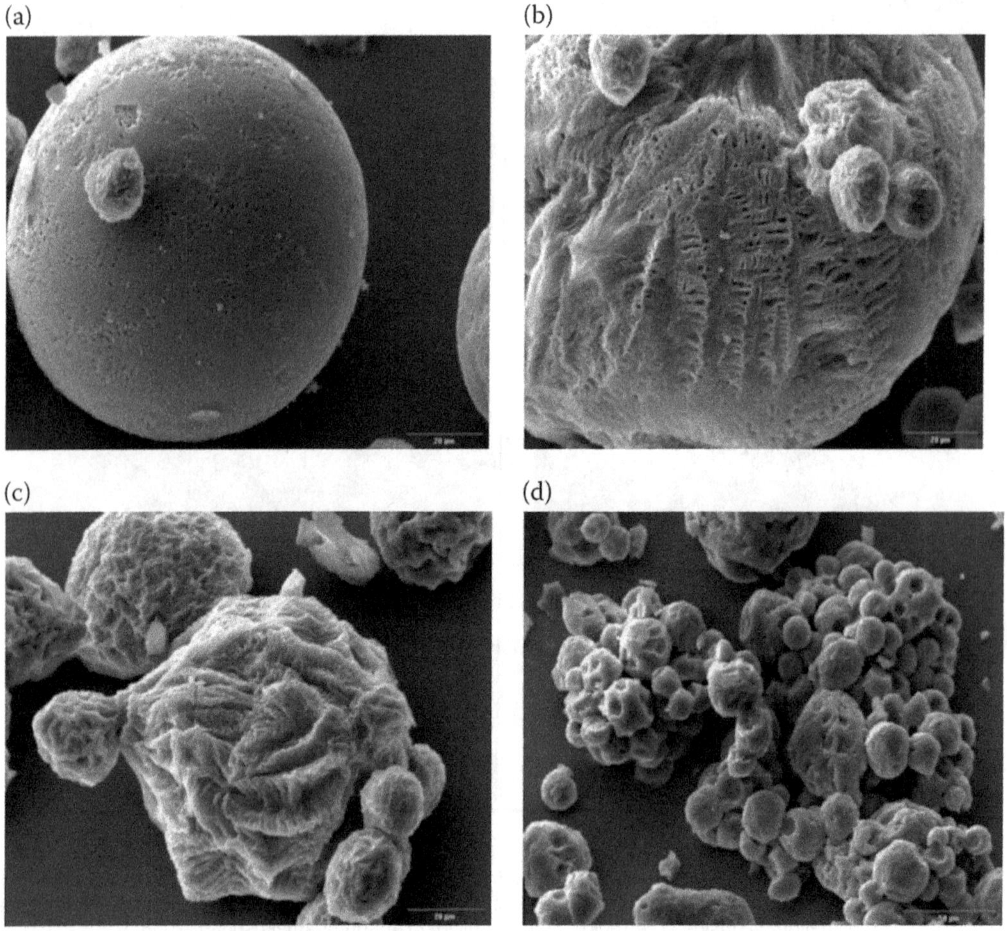

Figure 4.13 Effect of shelf temperature on the collapse of spray-freeze-dried particles of pure BSA: (a) T_{shelf} = −15°C; (b) T_{shelf} = −10°C; (c) T_{shelf} = −8°C; (d) T_{shelf} = −5°C (Straller & Lee, 2017).

gas bubbles within the particles. The bubbles were likely to originate from the dissolved gas released during atomization, wherein the feed was pressurized at 6 bar using compressed nitrogen in the feed chamber (Anandharamakrishnan, 2008).

The SFD-WPI particles produced at −30°C resembled a perforated shell wall, as depicted in (Figure 4.12[c-d]). The reason is that, during spray-freezing into vapor, whey proteins adsorb onto the ice/liquid interface. When the ice was removed during the subsequent freeze-drying stage, the voids are left behind at the center.

The influence of shelf temperature (T_{shelf}) on the collapse of vacuum-spray-freeze-dried protein (pure bovine serum albumin) powder is depicted in Figure 4.13[a-d]. When frozen at 200°C/min, the T_g' of pure BSA was found to be −11°C (Chang & Randall, 1992). At $T_{shelf} < T_g$', (−15°C), the scanning electron micrograph of the SFD particles showed larger unaffected particles (center field) except for those in the background, the top left corner that depicted some wrinkling (Figure 4.13[a]). At $T_{shelf} > T_g$' (−10°C), wrinkling of the larger particles was observed (Figure 4.13[b]). When the T_{shelf} was further increased to −8°C, serious shrinkage was observed, wherein the particles showed a highly wrinkled surface (Figure 4.13[c]). At a T_{shelf} of −5°C, shrinkage and wrinkling became too high in the SFD protein particles (Figure 4.13[d]), due to melting and aggregation.

Annealing is an optional step that is conducted before primary drying in case of unstable products. During annealing, the frozen samples are held at a temperature between the melting point and the glass transition temperature for a defined period. This leads to added freeze-concentration of the solute phase, which elevates the collapse temperature and permits the primary drying to be conducted at a higher temperature and completed within a short time. Annealing influences the product morphology as it leads to the formation of larger ice crystals, which reduces the specific surface area and the internal pore area of the final product (Ishwarya, Anandharamakrishnan & Stapley, 2015). This was demonstrated in the case of spray-freeze-dried kanamycin. Inherently, kanamycin is an amorphous product Figure 4.14[a]). In the absence of annealing, SFD kanamycin did not depict a spherical shape and smooth surface (Figure 4.14[b]), which are the key properties for inhaled drugs in addition to their aerodynamic diameter, particle size and density (Agu et al., 2001). Thus, to obtain spherical particles of spray-freeze-dried kanamycin, the annealing process was conducted at different temperatures (−1 °C, −8 °C and −15 °C) for varied durations (3 h, 5 h and 7 h), before freeze-drying. The SFD kanamycin particles obtained at all the annealing temperatures were spherical with a porous surface (Figure 4.14[c

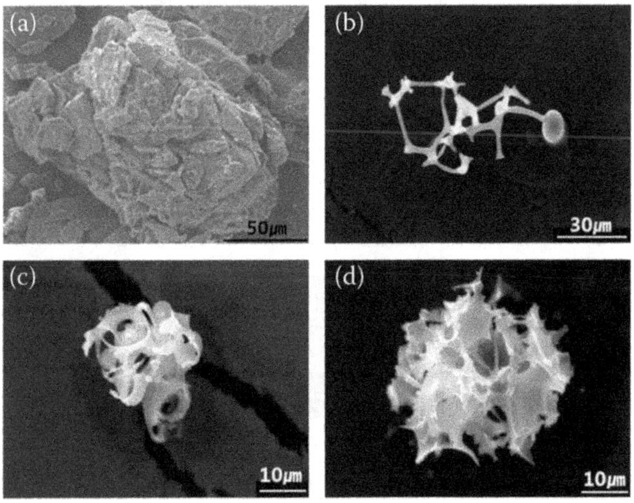

Figure 4.14 Scanning electron microscopy (SEM) image of kanamycin: (a) Kanamycin before SFD; (b) SFD powder without annealing process; (c & d) SFD powder with annealing process (Her et al., 2010).

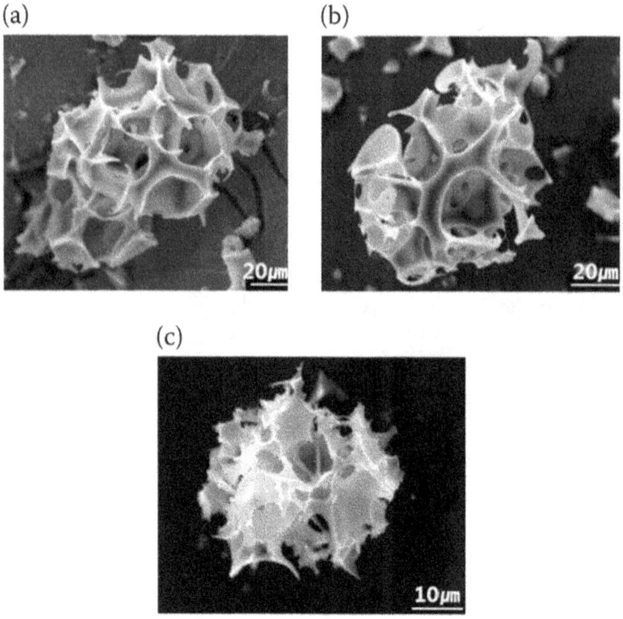

Figure 4.15 Particle morphology of spray freeze-dried kanamycin powder obtained at different annealing temperatures: (a) –1°C; (b) –8°C; (c) –15°C (Her et al., 2010).

& d]). However, annealing at –1°C (Figure 4.15[a]) and –8°C (Figure 4.15[b]) led to larger particles than those annealed at –15°C (Figure 4.15[c]).

The annealing time was a significant factor in obtaining the spherical shape and porous surface morphology of spray-freeze-dried particles. Three hours of annealing was not sufficient enough to form spherical particles of the SFD kanamycin powder (Figure 4.16[a]). But, five hours

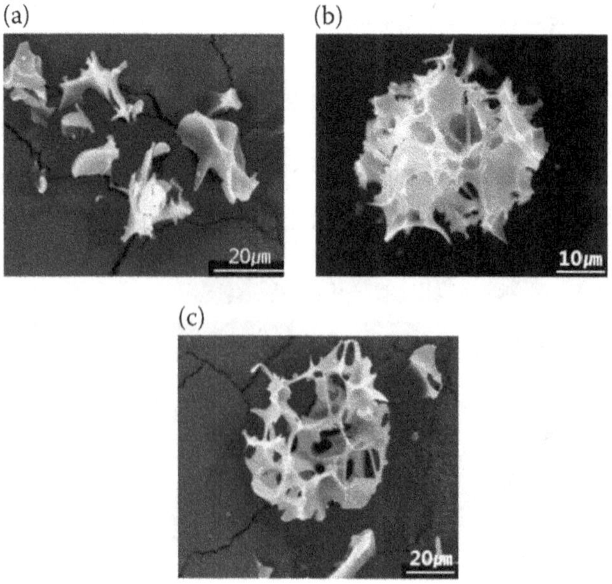

Figure 4.16 Influence of annealing time on the particle morphology of spray-freeze-dried kanamycin powders: (a) 3 h; (b) 5 h; (c) 7 h (Her et al., 2010).

(Figure 4.16[b]) and seven hours (Figure 4.16[c]) of annealing resulted in porous and spherical surface. However, it had a larger particle size than that produced after five hours of annealing. The mean particle diameters of SFD kanamycin powders with annealing times of 3 h, 5 h and 7 h were 15.1 µm, 13.5 µm and 17.5 µm, respectively. Their aerodynamic particle sizes were 3.88 µm, 3.58 µm and 4.264 µm, respectively. Thus, five hours was ascertained as the optimum annealing time for the production of SFD-kanamycin (Her et al., 2010).

CONCLUSIONS

From the above discussions, it is apparent that the spray-freeze-drying process is capable of conferring a unique microstructure upon particulates for a broad range of commercial applications in the food and pharmaceutical sectors. Different morphological patterns support varying functionalities of spray-freeze-dried particles. This chapter elaborated on the influence of feed composition and process conditions on the particle microstructure during spray-freezing and freeze-drying steps of the SFD process. A more comprehensive understanding is still required on the ways and means of tailoring the wide variety of compositions used in foods and biologicals to obtain spray-freeze-dried products with preferred morphology.

REFERENCES

Agu, R. U., Ugwoke, M. I., Armand, M., Kinget, R., & Verbeke, N. (2001). The lung as a route for systemic delivery of therapeutic proteins and peptides. *Respiratory Research*, 2, 198–209.

Al-Hakim, K., & Stapley, A. G. F. (2004). Morphology of spray-dried and spray-freeze-dried whey powders. In: Drying – Proceedings of the 14th International Drying Symposium (IDS 2004), (Vol. C pp. 1720–1726).

Ali, M. E., & Lamprecht, A. (2014). Spray freeze drying for dry powder inhalation of nanoparticles. *European Journal of Pharmaceutics and Biopharmaceutics*, 87(3), 510–517.

Amorij, J. P., Saluja, V., Petersen, A. H., Hinrichs, W. L. J., Huckriede, A., & Frijlink, H. W. (2007). Pulmonary delivery of an inulin-stabilized influenza subunit vaccine prepared by spray freeze drying induces systemic, mucosal humoral as well as cell mediated immune responses in BALB/c mice. *Vaccine*, 25, 8707–8717.

Anandharamakrishnan, C., Rielly, C. D., & Stapley, A. G. (2007). Effects of Process variables on the denaturation of whey proteins During Spray Drying. *Drying Technology*, 25(5), 799–807.

Anandharamakrishnan, C. (2008). *Experimental and computational fluid dynamics studies on spray-freeze-drying and spray-drying of proteins*. Ph.D. thesis. Loughborough University, UK.

Anandharamakrishnan, C., Rielly, C. D., & Stapley, A. G. F. (2010). Spray-freeze-drying of whey proteins at sub-atmospheric pressures. *Dairy Science and Technology*, 90, 321–334.

Bellows, R. J., & King, C. J. (1972). Freeze-drying of aqueous solutions: Maximum allowable operating temperature. *Cryobiology*, 9(6), 559–561.

Chang, B. S., & Randall, C. S. (1992). Use of subambient thermal analysis to optimize protein lyophilization. *Cryobiology*, 29(5), 632–656.

Chew, N. Y. K., & Chan, H. K. (2001). Use of solid corrugated particles to enhance powder aerosol performance. *Pharmaceutical Research*, 18, 1570–1577.

CJPH (2016). Spray freeze drying technology and its application to micronization of poorly soluble drugs [J]. *CJPH*, 47(01), 106–110.

CMRF (*n.d.*). Scanning Electron Microscopy. Available from: https://cmrf.research.uiowa.edu/scanning-electron-microscopy (Accessed on September 8, 2021), Central Microscopy Research Facility, University of Iowa.

Costantino, H. R., Firouzabadian, L., Hogeland, K., Wu, C., Beganski, C., Carrasquillo, K. G., Córdova, M., Griebenow, K., Zale, S. E., & Tracy, M. A. (2000). Protein spray-freeze drying. Effect of atomisation conditions on particle size and stability. *Pharmaceutical Research*, 17(11), 1374–1382.

Desai, T. R., Hancock, R. E. W., & Finlay, W. H. (2002). A facile method of delivery of liposomes by nebulization. *Journal of Controlled Release*, 84(1-2), 69–78.

Dressman, J. B., & Reppas, C. (2000). In vitro-in vivo correlations for lipophilic, poorly water-soluble drugs. *European Journal of Pharmaceutical Sciences*, 11.

van Drooge, D.-J., Hinrichs, W. L. J., Dickhoff, B. H. J., Elli, M. N. A., Visser, M. R., Zijlstra, G. S., & Frijlink, H. W. (2005). Spray freeze drying to produce a stable δ9-tetrahydrocannabinol containing inulin-based solid dispersion powder suitable for inhalation. *European Journal of Pharmaceutical Sciences*, 26(2), 231–240.

Edwards, D. A. (1997). Large porous particles for pulmonary drug delivery. *Science*, 276(5320), 1868–1872.

Engstrom, J. D., Simpson, D. T., Lai, E. S., Williams, R. O., & Johnston, K. P. (2007). Morphology of protein particles produced by spray freezing of concentrated solutions. *European Journal of Pharmaceutics and Biopharmaceutics*, 65, 149–162.

Fellows, P. J. (2002). Freeze drying and freeze concentration. In: *Food Processing Technology - Principles and Practice*. Woodhead Publishing, Cambridge, pp. 401–414.

Gill, H. S., & Prausnitz, M. R. (2007). Coating formulations for microneedles. *Pharmaceutical Research*, 24(7), 1369–1380.

Hafner, B. (2007). Scanning electron microscopy primer. Characterization Facility, University of Minnesota-Twin Cities, 1–29.

Her, J.-Y., Song, C.-S., Lee, S. J., & Lee, K.-G. (2010). Preparation of kanamycin powder by an optimized spray freeze-drying method. *Powder Technology*, 199, 159–164.

Hindmarsh, J. P., Russell, A. B., & Chen, X. D. (2003). Experimental and numerical analysis of the temperature transition of a suspended freezing water droplet. *International Journal of Heat and Mass Transfer*, 46, 1199–1213.

Hindmarsh, J. P., Russell, A. B., & Chen, X. D. (2007). Fundamentals of the Spray freezing of foods e microstructure of frozen droplets. *Journal of Food Engineering*, 78, 136–150.

Hundre, S. Y., Karthik, P., & Anandharamakrishnan, C. (2015). Effect of whey protein isolate and β-cyclodextrin wall systems on stability of microencapsulated vanillin by spray–freeze drying method. *Food Chemistry*, 174, 16–24.

Ishwarya, S. P., & Anandharamakrishnan, C. (2015). Spray-Freeze-Drying approach for soluble coffee processing and its effect on quality characteristics. *Journal of Food Engineering*, 149, 171–180.

Ishwarya, S. P., Anandharamakrishnan, C., & Stapley, A. G. F. (2015). Spray-freeze-drying: a novel process for the drying of foods and bioproducts. *Trends in Food Science & Technology*, 41, 161–181.

JEOL (n.d.). Available from: https://www.jeol.co.jp/en/applications/pdf/sm/844_en.pdf; https://www.jeol.co.jp/en/applications/pdf/sm/sem_atoz_all.pdf (Accessed 19 September 2021).

Khwanpruk, K., Anandharamakrishnan, C., Rielly, C. D., & Stapley, A. G. F., (2008). Volatiles retention during the sub-atmospheric spray freeze drying of coffee and maltodextrin. In: Proceedings of the *16thInternational Drying Symposium (IDS2008)*, 9–12 November 2008, Hyderabad, India, pp. 1066–1072.

Kudra, T., & Strumillo, C. (1998). *Thermal Processing of Biomaterials*. Gordon and Breach Science Publishers, Amsterdam.

Leuenberger, H., Plitzko, M., & Puchkov, M. (2006). Spray freeze drying in a fluidized bed at normal and low pressure. *Drying Technology*, 24, 711–719.

Maa, Y.-F., Nguyen, P.-A., Sweeney, T., Shire, S. J., & Hsu, C. C. (1999). Protein inhalation powders: spray drying vs spray freeze drying. *Pharmaceutical Research*, 16(2), 249–254.

MacLeod, C. S., McKittrick, J. A., Hindmarsh, J. P., Johns, M. L., & Wilson, D. I. (2006). Fundamentals of spray freezing of instant coffee. *Journal of Food Engineering*, 74, 451–461.

Matriano, J. A., Cormier, M., Johnson, J., Young, W. A., Buttery, M., Nyam, K., & Daddona, P. E. (2002). *Pharmaceutical Research*, 19(1), 63–70.

Mohri, K., Okuda, T., Mori, A., Danjo, K., & Okamoto, H. (2010). Optimized pulmonary gene transfection in mice by spray-freeze dried powder inhalation. *Journal of Controlled Release*, 144(2), 221–226.

Nicolai, T., Britten, M., & Schmitt, C. (2011). β-Lactoglobulin and WPI Aggregates: Formation, structure and applications. *Food Hydrocolloids*, 25(8), 1945–1962.

Puius, Y. A., Stievater, T. H., & Srikrishnan, T. (2006). Crystal structure, conformation, and absolute configuration of kanamycin a. *Carbohydrate Research*, 341(17), 2871–2875.

Quan, F.-S., Kim, Y.-C., Compans, R. W., Prausnitz, M. R., &Kang, S.-M. (2010). Dose sparing enabled by skin immunization with influenza virus-like particle vaccine using microneedles. *Journal of Controlled Release*, 147(3), 326–332.

Rajam, R., & Anandharamakrishnan, C. (2015). Spray freeze drying method for microencapsulation of *Lactobacillus plantarum*. *Journal of Food Engineering*, 166, 95–103.

Saguy, S. I., Marabi, A., & Wallach, R. (2005). Liquid imbibitions during rehydration of dry porous foods. *Innovative Food Science and Emerging Technologies*, 6(1), 37–43.

Saluja, V., Amorij, J.-P., Kapteyn, J. C., de Boer, A. H., Frijlink, H. W., & Hinrichs, W. L. J. (2010). A comparison between spray drying and spray freeze drying to produce an influenza subunit vaccine powder for inhalation. *Journal of Controlled Release*, 144(2), 127–133.

Sheu, T. Y., Rosenberg, M. 1998. Microstructure of microcapsules consisting of whey proteins and carbohydrates. *Journal of Food Science*, 63(3), 491–494.

Straller, G., & Lee, G. (2017). Shrinkage of spray-freeze-dried microparticles of pure protein for ballistic injection by manipulation of freeze-drying cycle. *International Journal of Pharmaceutics*, 532(1), 444–449.

Swapp, S. (*n.d.*). Scanning Electron Microscopy (SEM). Available from: https://serc.carleton.edu/research_education/geochemsheets/techniques/SEM.html (Accessed 19 September 2021).

Sweeney, L. G., Wang, Z., Loebenberg, R., Wong, J. P., Lange, C. F., & Finlay, W. H. (2005). Spray-freeze-dried liposomal ciprofloxacin powder for inhaled aerosol drug delivery. *International Journal of Pharmaceutics*, 305, 180–185.

Windhab, E. J. (1999). New developments in crystallization processing. *Journal of Thermal Analysis and Calorimetry*, 57, 171–180.

Yu, Z., Johnston, K. P., & Williams, R. O. 3rd(2006). Spray freezing into liquid versus spray-freeze drying: Influence of atomisation on protein aggregation and biological activity. *European Journal of Pharmaceutical Sciences*, 27, 9–18.

Yu, Z., Garcia, A. S., Johnston, K. P., & Williams, R. O. (2004). Spray freezing into liquid nitrogen for highly stable protein nanostructured microparticles. *European Journal of Pharmaceutics and Biopharmaceutics*, 58(3), 529–537.

Zijlstra, G. S., Rijkeboer, M., van Drooge, D. J., Sutter, M., Jiskoot, W., van de Weert, M., Hinrichs, W. L., & Frijlink, H. W. (2007). Characterization of a cyclosporine solid dispersion for inhalation. *The AAPS Journal*, 9(2) E190–E199.

CHAPTER 5

Spray-freeze-drying of dairy products

Dried milk products are an integral part of the dairy value chain. The high water content of milk (88–90%) is responsible for its perishable nature. A considerable amount of heat is required to evaporate this excess water from milk. This imposes a huge energy burden on the currently-used milk drying methods, besides causing nutrient depletion in the milk powder (Knipschildt & Andersen, 1994). Despite the above fact, spray drying has retained its monopoly in the dairy industry for more than seven decades, owing to its industry-friendly operation and commercial feasibility for bulk manufacturing. The portfolio of spray-dried dairy products includes the instant milk powder (whole and skimmed), whey protein isolate and whey protein concentrate. Alternatively, drum or roller drying and freeze-drying are also used for the drying of milk and dairy products. Infant milk formula and probiotic dairy cultures are the well-known drum-dried and freeze-dried dairy products, respectively.

It is a usual practice to homogenize the whole milk concentrate before spray drying, to avoid free fat on the powder surface. However, a small amount of free fat is inevitable (Early, 1998). During spray drying, milk solutes are redistributed in the matrix based on their diffusivities. As water is removed from the droplet surface, solutes will be drawn to the center of the droplet. Relative to smaller components like sugars, fat globules have much smaller transport velocities. Moreover, due to its rapid drying characteristics, milk fat dries faster than sugars and proteins to cover the surface of spray-dried whole milk particles (SD-WMP). Consequently, there will be a net increase in the concentration of milk fat at the particle surface due to the migration of smaller solutes towards the droplet's center (Meerdink, 1993). The presence of free fat on the particle surface reduces the wettability and increases the hydrophobicity of spray-dried milk powder (Kim, Chen, & Pearce, 2002; Millqvist-Fureby, Elofsson, & Bergenståhl 2001). In addition, the high surface-fat content causes stickiness and impair powder flowability (Hindmarsh, Russell & Chen, 2007). Also, it renders the powder susceptible to oxidation. It leads to the development of objectionable tallowiness or oxidation flavor of the milk fat during storage. Consequently, the keeping quality of spray-dried milk powder is compromised (Holm, Greenbank & Deysher, 1925; Nickerson et al., 1952; Shipstead & Tarassuk, 1953; Greenbank & Pallansch, 1962; Yazdanpanah & Langrish, 2012).

Compared to powders obtained from other drying techniques, spray-dried powders have the least porosity with closed internal pores. The solid and dense surface layer of spray-dried whole milk powders limits the water penetration during reconstitution (Rogers, Wu, Saunders, & Chen 2008), thereby impairing the instantaneous solubilization property. On rehydration, the milk powders leave behind a fatty deposition on the container, which is not savored by the consumers. The ability to rehydrate instantaneously is the key quality parameter of dairy powders. The ease of reconstitution depends on the wettability, dispersibility and solubility of powdered dairy products, which in turn are determined by their following attributes (Ding, Yu, Boiarkina, Depree, & Young,

DOI: 10.1201/9781003019312-5

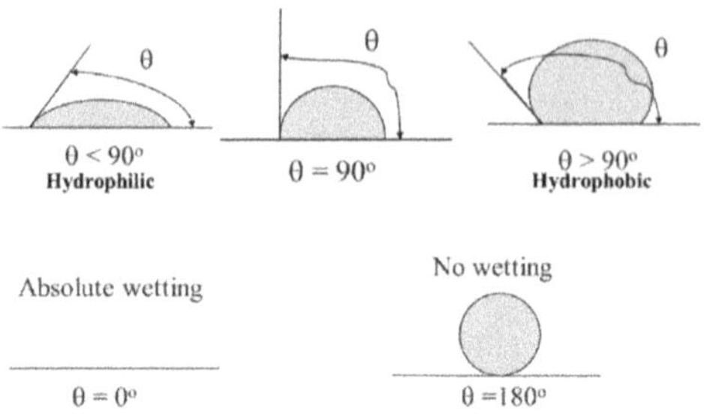

Figure 5.1 Different wetting conditions: The concept of hydrophilicity and hydrophobicity measured by the contact angle between the particle surface and water (Simpson, Hunter & Aytug, 2015).

2020; Gaiani et al., 2011; Kim, Chen & Pearce, 2002; Pathania et al., 2018; Rimpiläinen et al., 2015):

1. higher specific surface area (or) surface-to-mass ratio;
2. larger particle size;
3. porous microstructure (or) morphology;
4. pattern of distribution of fat/other solutes within the particles and,
5. hydrophilicity (or) smaller contact angle between the particle and water (Figure 5.1).

Conventionally, two approaches are employed to achieve all the above requisites that lead to dairy powders with good instant properties. First, the spray-dried dairy powder is agglomerated such that only a few fine particles are left. Agglomeration widens the pores through which the water penetrates to facilitate wetting, dispersion and dissolution. But, agglomeration adds to processing cost and time. Another approach involves the *'lecithination'* of dairy powders to achieve a small contact angle between the particle surface and water (hydrophilicity) (Walstra et al., 2005). Lecithination involves dissolving soy lecithin, a natural surfactant (or any other phospholipid), in butter oil and spraying this surfactant solution over the agglomerated milk powder, either inside or outside (in a fluidized bed) the dryer (Pisecky, 1997). Despite their good instant properties, lecithinized dairy powders have less flowability (Ilari & Loisel, 1991; Gharemann et al., 1994).

Based on the understanding obtained from the introductory chapters, we are certain that our readers would recognize the suitability of spray-freeze-drying (SFD) technique for the drying of dairy products. Since its advent, the aptness of spray-freeze-drying has been ascertained for the production of high-value products such as dairy and coffee powders, wherein rehydration impairment is a major concern. Generally, it is deemed that the spray-freeze-drying process can be adopted to dry products for which other drying techniques have proved infeasible. The merits of spray-freeze-drying often justify its higher manufacturing cost. Box 5.1 lists the specific competitive advantages of SFD over other drying techniques for the manufacturing of dairy powders. This chapter will focus on the spray-freeze-drying of different types of dairy products and their quality characteristics. Each section will present a brief prelude on the dairy product category and explain the approach and advantages of employing spray-freeze-drying for its production.

5.1 SPRAY-FREEZE-DRYING OF WHOLE MILK

Convenience is the key element of a customer-centric business. For that reason, whole milk powder (WMP) is a convenient form of liquid milk. Instant reconstitution, ease in transportation and storage are the convenience elements of WMP over liquid milk. Substantial cost savings result from the reduced storage space requirements and the non-refrigerated shipping and warehousing prospects of whole milk powder. Apart from being a rich source of protein,

BOX 5.1 COMPETITIVE EDGE OF SPRAY-FREEZE-DRYING PROCESS FOR THE DRYING OF DAIRY PRODUCTS

1. ***Higher specific surface area/surface-to-mass ratio:*** SFD is capable of producing powders with high surface-to-mass ratios of more than 80 m^2/gram of powder (Yu, Johnston & Williams, 2006).

2. ***Larger particle size:*** Unlike industrial spray drying, wherein droplets shrink as water is removed, there is no evidence of particle shrinkage during the SFD process. The possibility of obtaining larger particles from SFD is high as the water is rapidly frozen in place, and the droplets retain their shape and volume even after the water sublimes off during the freeze-drying phase. Also, the droplet is likely to expand slightly due to the expansion of ice crystals during spray-freezing. Eventually, the mean diameter of a spray-freeze-dried particle is nearly equal to the diameter of droplets emanating from the atomizer (Fig. a). Compared to conventional freeze-drying, SFD exerts precise control over the particle size of the final product due to its atomization step. The use of encapsulator nozzle to atomize whole/skim milk is known to result in a monodisperse droplet size distribution and hence a uniform particle size distribution (Rogers et al., 2008).

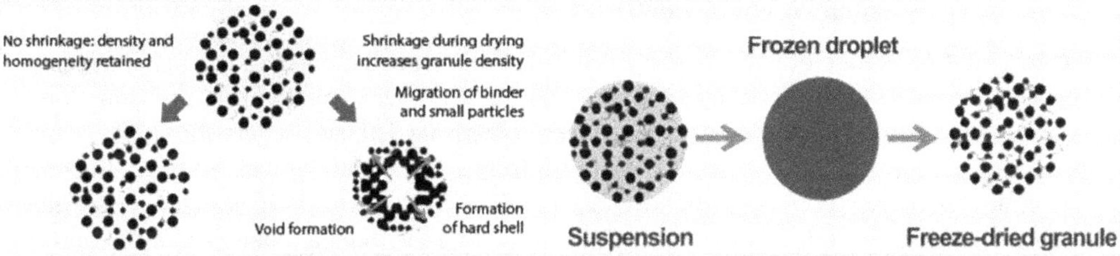

Figure a. Retention of particle size and shape: Spray-freezing Vs. Spray drying (Modified and reproduced from PowderPro®, 2020).

3. ***Porous microstructure/morphology:*** Porous microstructure is typical of any spray-freeze-dried product. Further, there exists a possibility for enhancing the porosity by reducing the total solid content of the feed. Greater the amount of solid water (ice) to be removed during the sublimation phase, more the number of pores that would be created in its place after drying. This aspect would aid in the instant solubility of spray-freeze-dried dairy powders.

4. ***Distribution of fat/other solutes within the particles:*** While slow cooling results in a dense outer layer, rapid freezing prevents the concentration of solutes on the surface (Hindmarsh, Russell & Chen, 2003 & 2007; Hindmarsh et al., 2003; Hindmarsh et al., 2005). The spray-freezing process is spontaneous enough to restrain the migration of solute components within the particle. This occurs when the feed is directly plunged into the cryogen during the spray-freezing-into-liquid (SFL) process. Under conditions of rapid freezing, there is little or no time left for the solute (fat/sugar) to be redistributed and for sufficient evaporation to occur. Consequently, the solute is concentrated at the surface (Hindmarsh et al., 2007). Notably, the solute redistribution phenomenon during spray-freezing is independent of solute concentration and solely depends on the freezing rate. Schematic representation of the above statement is presented in Fig. b, with the example of a 20% sucrose solution.

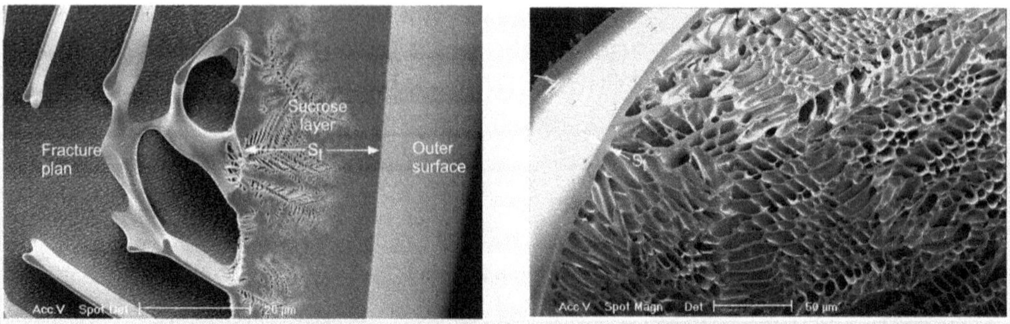

Figure b. The difference in the thickness of the surface solute layer (S_t) when a 2 mm droplet of a 20% sucrose solution was spray frozen at a low freezing rate and degree of supercooling (air velocity: 0.42 m/s; air temperature: –8°C) (image at the left) versus the same solution subjected to rapid freezing by plunging into liquid nitrogen (temperature: –192°C) with high supercooling (air velocity 0.42 m/s, air temperature –25°C) (image at the right) (Hindmarsh et al., 2007).

Moreover, the proportion of frozen particle that is finely structured is enhanced at higher freezing rates and degree of supercooling. This is a clear edge of SFD over spray drying as the spray-dried particles often have a thick, dense surface layer comprising fats and sugars that prolong the reconstitution time. Contrarily, particles that result from a frozen droplet with finer ice structure exhibit instant rehydration behavior (Rogers et al., 2008). Though liquid nitrogen is the commonly used cryogen for spray-freezing due to its ultra-low boiling point (~–196°C), much faster cooling rates can be achieved with hydrocarbons. The SFD particles of dairy powders are expected to be more hydrophilic due to the aforementioned reasons.

calcium and other nutrients, the whole milk powder also provides other functionalities. For instance, it acts as a fat replacer in the water/oil interfaces and facilitates the formulation of *'fat emulsions'*, which are intravenous dietary supplements administered to patients having deficiency in obtaining sufficient fat through diet. Whole milk powder is also used as a B2B (business-to-business) food ingredient (in place of liquid milk) by the manufacturers of ice cream, yogurt, beverages, confectionery, bakery products and infant formula. The foodservice industry also holds a significant share in the global market for whole milk powder (Persistence Market Research, 2020).

Given the nutritional benefits and market value of whole milk powder, a major bottleneck associated with this product is its quality deterioration during storage, mainly in terms of reduced instant properties and oxidation. To resolve the above apprehension, an attempt was made to produce whole milk powder by spray-freeze-drying on a laboratory scale (Rogers et al., 2008). A frequency-pulsed encapsulator nozzle (discussed in Section 2.2.4; Chapter 2; Figure 2.8) was used for atomizing the whole milk concentrate (33%, w/w), and liquid nitrogen was employed as a cryogen for the spray-freezing operation. The pulse frequency and static dispersal voltage of encapsulator nozzle were set at 1500–2000 Hz and 1.85 kV, respectively. For atomization, the whole milk feed was supplied to nozzles with two different orifice sizes (100 and 150 µm) nozzle at the rate of 2.5 mL/min. At an accurately tuned frequency of vibration, ordered and stable droplets of milk jet emanated from the encapsulator nozzle (Figure 5.2), as revealed by the built-in

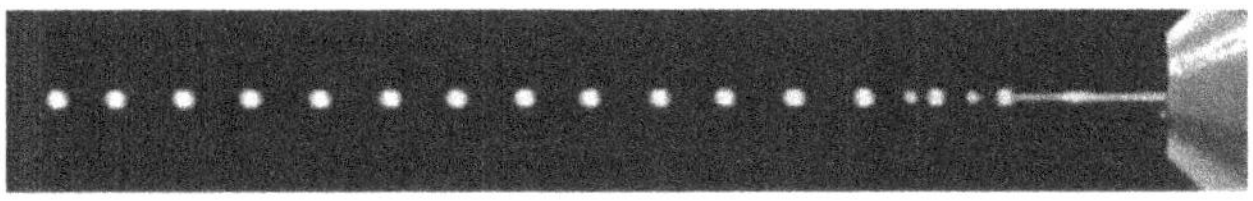

Figure 5.2 Break-up of milk jet into uniformly-sized droplets from the encapsulator nozzle (Rogers et al., 2008).

stroboscope[#]. Thus, the polydispersity of droplet/particle size distribution was avoided. The atomized droplets of homogenized whole milk were rapidly frozen in liquid nitrogen that was filled in a container placed under the stream of droplets. Subsequently, the frozen droplets were recovered and freeze-dried overnight. With respect to its microstructure, particle size and wettability, the spray-freeze-dried whole milk powder (SFD-WMP) showed notable changes relative to its industrially spray-dried counterpart (SD-WMP).

5.1.1 Microstructure of spray-freeze-dried whole milk powder

Scanning electron micrographs (SEM) of the SFD-WMP particles depicted their extremely porous surface. The particle appeared as a network structure of solids with interstitial voids (Figure 5.3[a]). A finely structured surface of SFD-WMP indicated that the SFD process did not damage the solution structure. As mentioned earlier, the highly porous surface is advantageous for enhanced wetting and solubility properties. Notably, the presence of any dense surface layer of

Figure 5.3 Differences between the microstructure of spray-freeze-dried and spray-dried whole milk particles depicted by the scanning electron micrographs: (a) surface morphology of spray-freeze-dried whole milk powder particle (Rogers et al., 2008); (b) surface morphology of spray-dried whole milk powder particle; (c) cross-section of spray-dried whole milk powder particle (Saito, 1985); (d) fractured wall of a particle of spray-dried whole milk powder showing solute surface layer (Langrish et al., 2006).

[#] An instrument that intermittently illuminates a moving object to study its motion and determine its rotary speed or vibration frequency (https://www.britannica.com/technology/tachometer).

solute was not evidenced from the scanning electron micrographs of the SFD-WMP particle, which may be attributed to the rapid freezing in liquid nitrogen (Box 5.1).

In striking contrast to the microstructure of SFD-WMP, a typical particle of whole milk powder resultant from spray drying depicted a slightly rough surface without any pores but had a crater-like scar and a tiny globule surrounded by a low brim at the junction (Figure 5.3[b]). But, the cross-section appeared porous (Figure 5.3[c]) (Rogers et al., 2008) due to the presence of tiny vacuoles and fat globules (Saito, 1985). Specifically, the presence of vacuoles within the dried particle signifies the development of surface skin. Once formed, the skin resists further removal of moisture from the partially dried particle. Then, as the particle temperature reaches the boiling point of water, water vapor of the residual moisture entrapped within the particle expands to increase the internal pressure. In response to the above, the particle expands to form vacuoles/cavities that are responsible for the formation of a dense layer of fat at the particle surface. Since the fat globules are exposed to temperatures above their melting point during spray drying, they are free to disperse and migrate to any site within the particle to create cavities within the walls (Buma, 1971a & b). These cavities may form a porous network through which the fat preferentially migrates to the surface of milk particles. As drying proceeds, the fat accumulates at the surface to form a dense solute layer (Figure 5.3[d]) (Kim, Chen & Pearce, 2003). The preferential accumulation of fat at the surface may also be ascribed to its hydrophobic nature and the overpressure within the vacuole. Also, at high drying temperatures, the skin loses its moisture to become hard and rigid, which then affects the reconstitution property (Langrish et al., 2006).

The edge of SFD process over spray drying is that the particles acquire a non-porous surface layer, as the water evaporates from the droplet either before freezing and/or solute redistribution during freezing (Hindmarsh et al., 2007), thus preventing the migration of components encountered during spray drying. Since a non-porous surface cannot support the migration of fat to the surface, the spray-freeze-dried whole milk powders are devoid of surface fat layer. Collectively, all the aforementioned factors are responsible for the reduced surface hydrophobicity and absence of stickiness in the spray-freeze-dried whole milk powders (Ishwarya, Anandharamakrishnan & Stapley, 2017; Rogers et al., 2008).

5.1.2 Particle size of spray-freeze-dried whole milk powder

The particle size distribution plot of SFD-WMP produced using the 150 μm nozzle orifice roughly fitted a lognormal distribution and showed a mode at approximately 275 μm (Figure 5.4), which is slightly larger than the nozzle diameter. The increase in particle size may be due to the coalescence of two or more neighboring droplets during or after freezing in the liquid nitrogen (Rogers et al., 2008). Conversely, possibilities exist that the atomized feed droplets agglomerated gradually and solidified as they passed through the vapor phase before contacting the surface of the cryogenic liquid and then settling into it (Kawabata et al., 2011). The mean volume diameter of spray-dried whole milk powders produced under different processing conditions ranged between

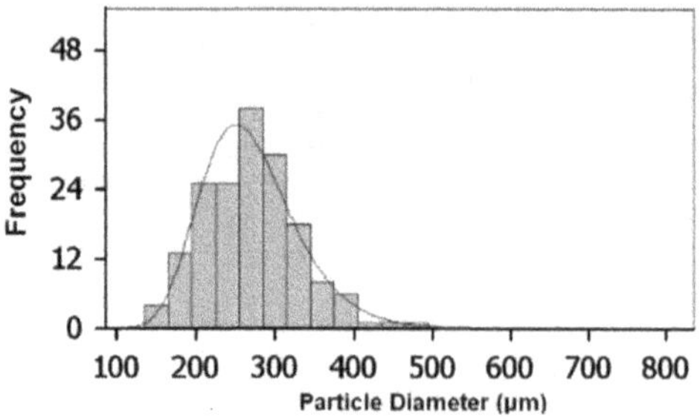

Figure 5.4 Particle size distribution plot of spray-freeze-dried whole milk powder produced using an encapsulator nozzle of orifice size, 150 μm (Rogers et al., 2008).

26–164 μm (Langrish et al., 2006; Silva & O'Mahony, 2016). Particularly, at a mean particle diameter of 164 μm, the specific surface area of spray-dried WMP was 66.1 m^2/kg (Silva & O'Mahony, 2016). Thus, due to its larger particle size, SFD can result in larger specific surface area than SD.

5.1.3 Wettability of spray-freeze-dried whole milk powder

As mentioned in Box 5.1, larger particle size and higher specific surface area of spray-freeze-dried powders are favorable towards obtaining good wettability, dispersibility and solubility. But, contrarily, the mean wetting time for spray-freeze-dried whole milk powders (145.5 s) was significantly longer than that observed for the SD-WMP (2.19 s). The SFD-WMP wetted slowly because of its low density and high porosity, due to which the same weight of powder occupied a much larger space than the other powder samples. After wetting, the particles floated on the water surface without sinking. When the wetting phenomenon was visualized under a microscope, the particles of SFD-WMP were found to disintegrate rapidly (Figure 5.5), which was attributed to its highly porous nature. As water entered the pores, it forced apart the structure to break up the particle. The larger-sized fragments resultant from the particle breakup floated on the water surface to prevent the complete contact between dry particles and water, thereby prolonging the wetting time. The researchers suggested that tailoring the spray-freeze-dried whole milk powder particles to have smaller pore size can result in denser particles that will sink upon wetting to shorten the wetting time.

The industrial whole milk powder was the fastest to wet (2.19 s) as it was lecithinized during manufacturing. Hence, it is expected that the use of surfactants at an appropriate level may improve the reconstitution behavior of SFD whole milk powders (Rogers et al., 2008). But, Tian et al. (2014) showed that lecithination did not bring about any significant change in the wetting behavior of spray-dried WMP produced from the feed at higher solid content (23% and 33%), despite using

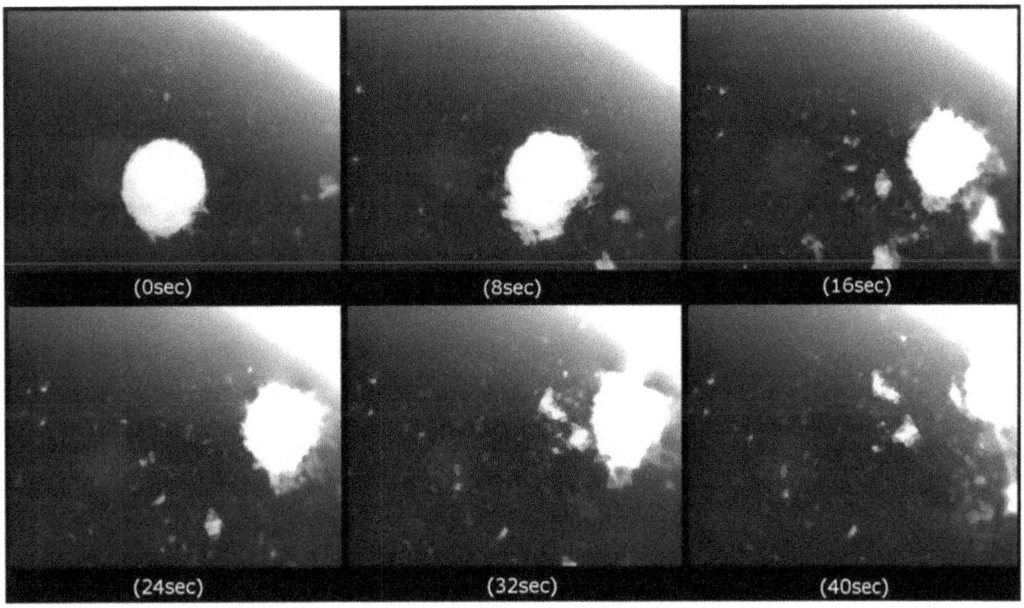

Figure 5.5 Time-lapse photography captured at a 4X magnification of a 600 μm SFD whole milk particle (150 μm nozzle; 33% reconstitution) dissolving in water (Rogers et al., 2008).

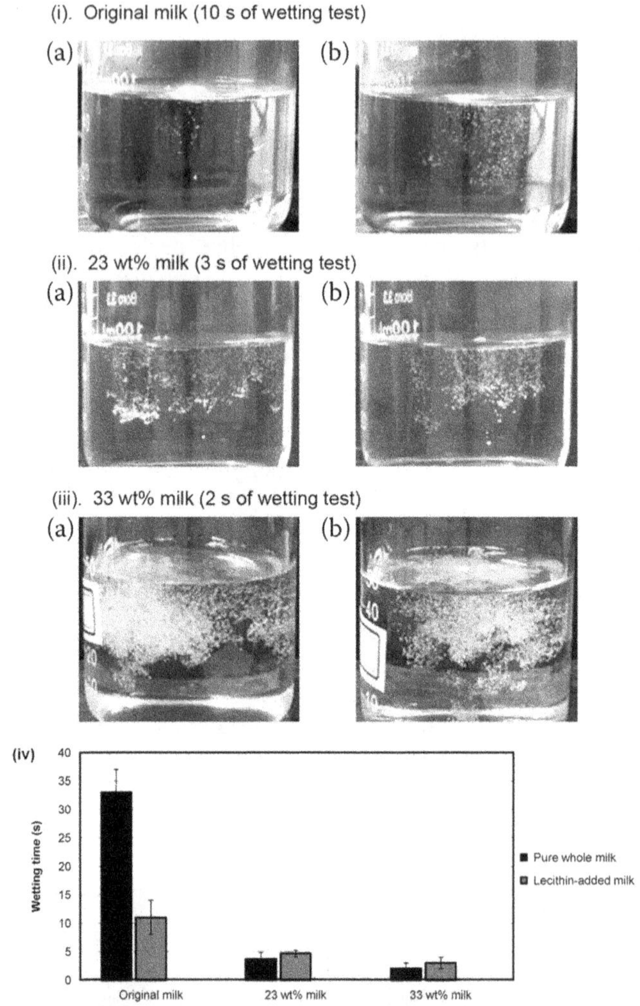

Figure 5.6 Wetting behavior of spray-dried whole milk powder: (a) pure WMP and (b) lecithinated WMP with 0.1 wt.% lecithin, produced from original milk feed and concentrated feed at 23% and 33% solid contents. Milk powder used for the wettability analysis of (i) original whole milk and (ii) 23% (w/w) milk were 0.02 g, and that used for (iii) 33% (w/w) milk was 0.05 g. The time at which each snapshot was captured is given in brackets; (iv) The differences in total wetting times of pure WMP and WMP with 0.1 wt.% lecithin addition (Modified and reproduced from Tian et al., 2014).

the same concentration of lecithin (0.1%). The pattern of wetting behavior for spray-dried whole milk powders at different initial solid concentrations is shown in Figure 5.6(i–iii). The wetting time was markedly reduced when higher feed concentrations were used for the production of SD-WMP (Figure 5.6[iv]). Similar studies are required to identify the most influential factor among feed solid content and lecithination that controls the wettability of spray-freeze-dried whole milk powders.

5.2 SPRAY-FREEZE-DRYING OF SKIM MILK

Skim milk powder (SMP) is the product obtained after removing fat and water from whole milk to contain no more than 5% moisture and 1.5% fat and a minimum milk protein content of

Table 5.1 Heat classification and applicability (*) of skim milk powder to recombined dairy products (Kneifel, 2003)

	Categories of skim-milk powder				
	Extra-low heat	**Low heat**	**Medium heat**	**Medium-high heat**	**High-heat**
Classification parameters					
Whey protein index (ADMI[a], IDF)	nd	≥6.0	5.9–4.5	4.4–1.5	≤1.4
Heat number (IDF)	nd	≤80	80.1–83.0	83.1–88.0	≥88.1
Cysteine number	26–33	34–41	42–49	50–60	≥60
Recombined products					
Pasteurized milk		*	*	*	
UHT milk			*	*	
Sterilized milk			*	*	
Evaporated milk					*[b]
Sweetened condensed milk		*	*		
Yogurt		*	*		
Cheese	*	*			
Butter					*
Ice-cream		*	*	*	*

Notes:
[a] American Dairy Products Institute (formerly American Dry Milk Institute), Chicago, Illinois, USA.
[b] Specially manufactured, 'heat-stable', high-heat powder is used.
nd: no data available.

34% (Codex Alimentarius, 2018). The milk sugars (lactose), proteins and minerals are expected to be retained in the same proportion as that present in fresh milk (ADPI, 2002). Ideally, skim milk powder must have a clean, sweet and pleasant taste, without any objectionable flavors (Bodyfelt et al., 1988). Skim-milk powder is the commonly used ingredient in recombined dairy products[*], wherein it functions as a source of milk proteins (Kneifel, 2003). For instance, SMP is added to the blend for processed cheese food or spread to promote the creaming process and improve spreadability (Gouda & Abou El-Nour, 2003). Here, SMP serves as a partial replacement for cheese solids.

Accomplishing the intended functionality of skim milk powder in recombined dairy products depends on its reconstitutability, heat stability, and viscosity properties. The raw milk composition and degree of heating to which the skim milk is subjected during powder production govern the above properties. Solubility of SMP is inversely related to the thermal exposure during powder production. Thus, it is important to determine the extent of heating applied to SMP, which can be judged by the level of protein denaturation. The proportion of undenatured whey protein is determined by calculating the *'whey protein index'*, *'heat number'* and *'cysteine number'*. Based on these indices, SMP is classified into different types, and the end-use of SMP depends on the class to which it belongs (Table 5.1) (Kneifel, 2003). From Table 5.1, it is evident that the skim milk powder belonging to the low and medium heat classes has wider applications in most of the recombined products. This suggests the suitability of a low-temperature processing technique such as spray-freeze-drying for the production of skim milk powder with versatile applications.

[*] The portfolio of recombined dairy products include pasteurized milk, UHT milk, sterilized milk, evaporated milk, sweetened condensed milk, yogurt, cheese, butter and ice-cream.

Accordingly, Rogers et al. (2008) carried out the spray-freeze-drying of skim milk on a laboratory scale for the production of SMP and compared its morphology, particle size and wettability with the industrially spray-dried equivalent. Skim milk feed solution at 33% (w/w) solid content was sonicated for 5 min for the foam to subside. Then, it was subjected to spray-freeze-drying under the same conditions specified for whole milk powder in the preceding section.

5.2.1 Morphology of spray-freeze-dried skim milk powder

The surface of SFD-SMP powder particles appeared rough and crystalline under the light microscope (Figure 5.7), probably due to the light reflected from the pores. However, under the electron microscope, the particles appeared highly porous. Similar to the SFD-WMP particles, the surface morphology of SFD-SMP particles revealed spherical-shaped particles with a highly porous and finely detailed surface, indicating the non-destructive effect of SFD process on the solution structure (Figure 5.8[a]). The SEM micrograph of a broken particle showed that the particle's interior had a very similar porous morphology as the surface (Figure 5.8[b]). This implies the rapid freezing of particles in the liquid nitrogen and the formation of a uniform ice crystal structure throughout the particle. However, some of the SFD-SMP particles were non-spherical and lacked the porous microstructure (Figure 5.8[c]). They had a smooth surface and flattened shape. It was suggested that these particles might have begun to melt when they were in the state of frozen droplets, during their transit from the nozzle tip before reaching the surface of cryogenic liquid. Alternatively, they might also have melted during the freeze-drying stage where the heat supply may have surpassed the latent heat of phase change, causing localized *'meltdowns'*. Sometimes, inadequate gold ion sputtering during the sample preparation for SEM visualization could also lead to melting away of the surface texture of particles (Rogers et al., 2008). Hence, care must be exercised during sample preparation, and a thorough interpretation of the SEM micrographs is necessary before reaching conclusions about the particle morphology of spray-freeze-dried powders.

The edge of the SFD process in obtaining a uniform microstructure of SMP particles can be ascertained by striking comparison with its spray-dried counterpart. SEM micrographs of the SMP particles produced from feed with a slightly solid content (41.2%) than that used for SFD (33%)

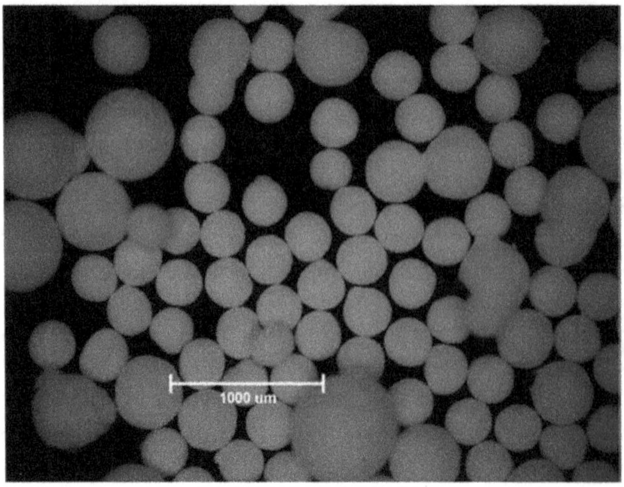

Figure 5.7 Light microscopic image of SFD skim-milk powder produced from feed having 33% solids concentration, using a 150 µm nozzle (Rogers et al., 2008).

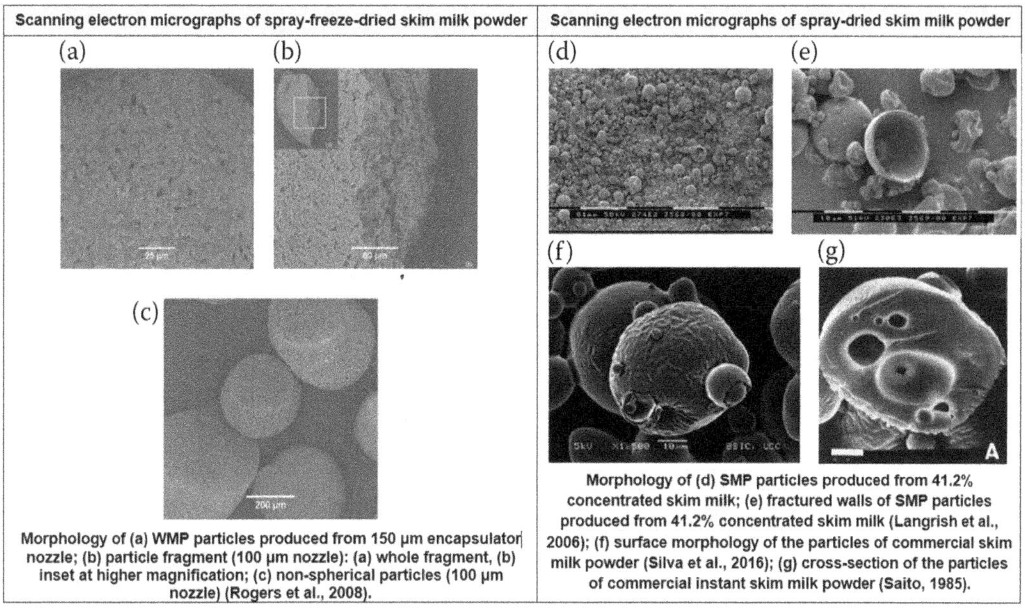

Scanning electron micrographs of spray-freeze-dried skim milk powder
Scanning electron micrographs of spray-dried skim milk powder
(a)
(b)
(c)
(d)
(e)
(f)
(g)
Morphology of (a) WMP particles produced from 150 μm encapsulator nozzle; (b) particle fragment (100 μm nozzle): (a) whole fragment, (b) inset at higher magnification; (c) non-spherical particles (100 μm nozzle) (Rogers et al., 2008).
Morphology of (d) SMP particles produced from 41.2% concentrated skim milk; (e) fractured walls of SMP particles produced from 41.2% concentrated skim milk (Langrish et al., 2006); (f) surface morphology of the particles of commercial skim milk powder (Silva et al., 2016); (g) cross-section of the particles of commercial instant skim milk powder (Saito, 1985).

Figure 5.8 Microstructure of spray-freeze-dried and spray-dried whole and skim milk powders – A comparison (Langrish et al., 2006; Rogers et al., 2008; Saito, 1985; Silva & O'Mahony, 2016).

revealed the presence of predominately smooth, spherical particles of varying size (Figure 5.8[d]). But, most of the spray-dried skim milk particles were collapsed (Figure 5.8[e]). This is because, as the particles move out of the spray chamber, their temperature and pressure reduce due to the drop in drying air temperature. The probability of particle collapse is high if the particle shell is not strong enough to withstand this pressure drop. The collapse is more prominent in the case of dilute feeds, which usually result in thinner particle shells (Langrish et al., 2006). Silva & O'Mahony (2016) also reported the relatively smaller size and smooth surface morphology of the industrially spray-dried SMP particles (Figure 5.8[f]). Even particles from the same commercial pack of spray-dried SMP did not have a similar surface microstructure. While the small particles had no dents, the larger ones showed wrinkled surfaces and dents. The cross-sections showed vacuoles and compact walls (Figure 5.8[g]) (Saito, 1985). Thus, the non-porous and non-homogeneous microstructure of spray-dried particles was affirmed, which was not observed in the spray-freeze-dried product.

5.2.2 Particle size of spray-freeze-dried skim milk powder

The particle size distribution of SFD-SMP produced from the 100 μm encapsulator nozzle followed a lognormal distribution (Figure 5.9[a]). But, the diameter of skim milk particles produced from the nozzle of larger orifice size (150 μm) showed a clear mode at 320 μm (Figure 5.9[b]), which is approximately two-fold the size of the nozzle orifice. As mentioned in the previous section on SFD-WMP, the higher modes were proposed to result due to the larger droplets formed by the coalescence of two or three droplets of size 320 μm (Rogers et al., 2008). The SFD-SMP particles produced from a feed of 33% solid concentration had a larger particle size than their spray-dried counterparts produced from a 41.5% feed (volume-weighted median diameter or d_{43} diameter: 35 ± 18 μm) (Langrish et al., 2006).

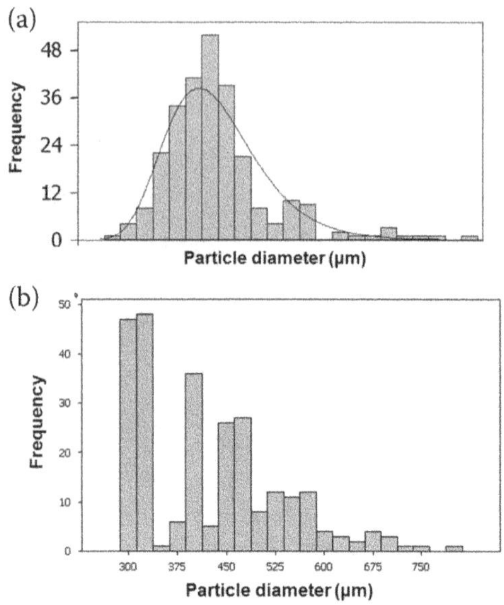

Figure 5.9 Particle size distribution plot of spray-freeze-dried skim milk powder produced using: (a) 100 μm nozzle; (b) 150 μm nozzle (Rogers et al., 2008).

5.2.3 Wettability of spray-freeze-dried skim milk powder

Unlike SFD-WMP, the advantage of the uniformly porous microstructure and larger particle size of SFD-SMP reflected in its rapid wetting and highly dispersive behaviors (wetting time: 2.31 s) compared to the spray-dried skim milk powder (wetting time: 6.20 s). Upon contacting the water surface, particles of SFD-SMP spreaded out very rapidly, probably due to the inter-particle electrostatic repulsive forces. During rehydration, the SFD-SMP powders revealed a significant level of buoyancy. The particles sank slowly under the surface of the water once wetted, after which they floated upwards once they started dissolving. The occluded air in the internal pore structure of SFD particles was considered responsible for their buoyancy. From the time-lapse microscopy images depicting the dissolution of a larger skim milk particle (600 μm in diameter) in about 1 min (Figure 5.10), it was evident that the particles fragmented into much smaller particles that radiated from the surface and then dissolved fully. The rise of particles after some time was attributed to the reduced size of the core particle, due to which the buoyancy of air that remained entrapped within the pores might have exceeded the weight of particles. The particle fragmentation is likely to be caused by the transport of water into the pore structure. Further, as they moved away from the large particle, the radiating particles tended to dissolve completely, as the water was driven across the surface, and the milk concentration reduced in the direction farther from the main particle. The whole particle disintegrated till the center dissolved eventually. Therefore, the dissolution of SFD particles is likely to be governed by individual convection rather than solute diffusion. While the former mechanism is usually observed in the case of a small clump of fine powder, the latter occurs in a single large particle (Rogers et al., 2008).

Thus, the above observations collectively demonstrated that the increased porosity of SFD-SMP leads to favorable instant properties. It is also possible that the SFD particles are more hydrophilic than their spray-dried equivalents. Notably, SFD-SMP showed rapid wettability even without lecithination or agglomeration as done in the case of spray-dried skim milk powder.

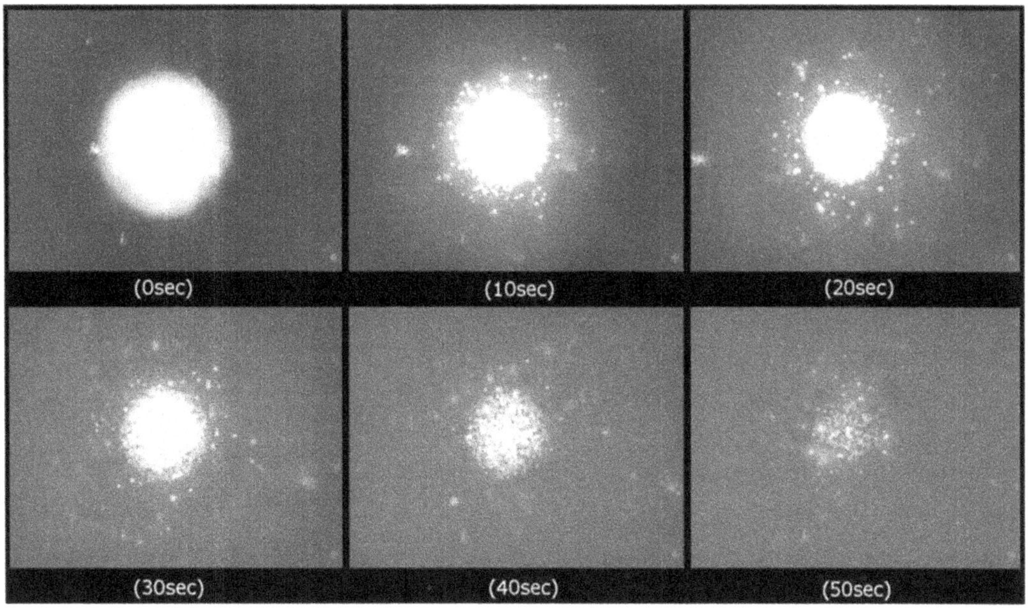

Figure 5.10 Time-lapse photography captured at a 4X magnification of a 600 μm SFD skim-milk particle (150 μm nozzle; 33% reconstitution) dissolving in water (Rogers et al., 2008).

5.3 SPRAY-FREEZE-DRYING OF WHEY PROTEIN

Whey protein is derived from whey, a by-product of the cheese manufacturing industry (Figure 5.11). It is globular in structure and composed of two major protein fractions, namely, β-lactoglobulin (51%) and α-lactalbumin (19%), besides the presence of other proteins such as immunoglobulins and serum albumin (Morr & Ha, 1993). β-lactoglobulin is a highly stable protein with gelation and emulsification properties (Schokker et al., 2000; Anema, Stockmann, & Lowe 2005). On the other hand, α-lactalbumin is the most heat-stable whey protein. Its ability to form intermolecular ionic bonds with divalent Calcium ions (Ca^{2+}) renders it resistant to thermal denaturation or unfolding (Boye et al., 1997). Notably, whey proteins have the maximum Protein Digestibility Corrected Amino Acid Score (PDCAAS) of 1.0, and hence can supply all the essential amino acids to the body.

Whey protein (WP) demonstrates a multitude of functionalities. It prevents muscle breakdown and spares glycogen during exercise. Since whey protein contains many of the components found in human breast milk, it is used as a key ingredient in infant formulas. Moreover, WP is a commonly used ingredient in formulated foods such as dairy products, baked goods, beverage mixes, sports drinks and meat products, wherein it acts as a substrate for the fortification and supplementation of micronutrients. It is also used as an encapsulant for pharmaceutical products. The broad functionalities of whey protein indicate the necessity for its cost-effectiveness and thereby a suitable drying process to convert it into its dried form. Whey protein powder is available in two different forms, i.e., whey protein isolate (WPI) and whey protein concentrate (WPC; whey proteins with lactose) (Figure 5.11). For a long time, spray drying has been the preferred process for whey protein powder production, due to its merits such as rapid drying rates, wide-ranging operating temperatures and short residence times. Nevertheless, spray drying causes a significant level of protein denaturation, especially at higher outlet temperatures (100–120°C) (Anandharamakrishnan, 2008).

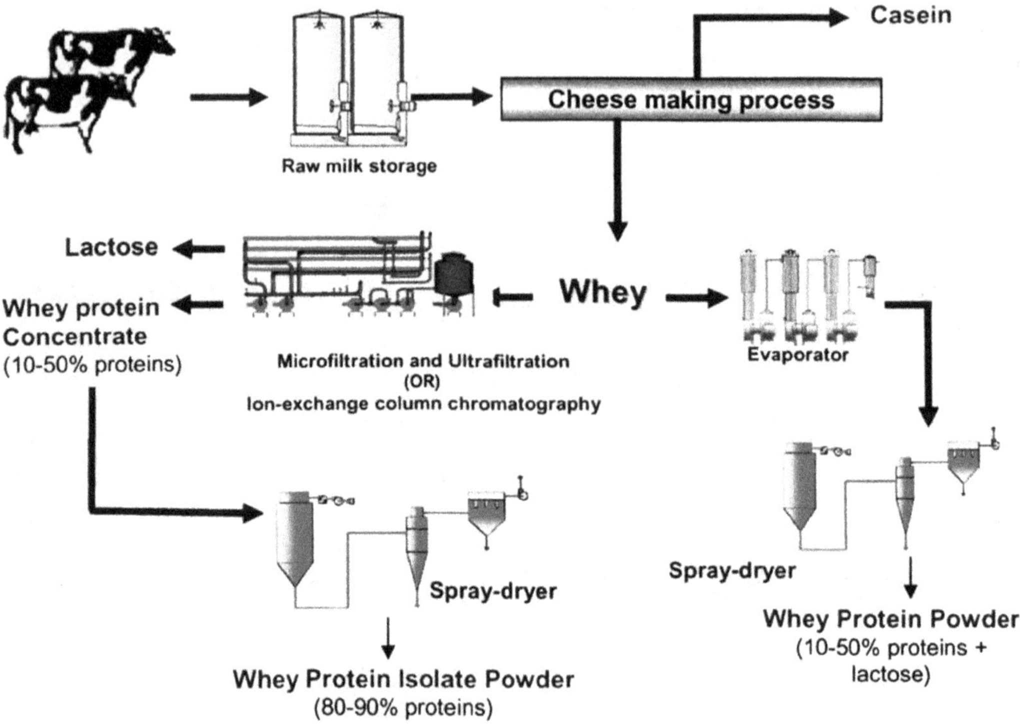

Figure 5.11 Process flow diagram of different forms of whey proteins (Anandharamakrishnan, 2008).

More than 70% of whey proteins present in the feed were found to be denatured during spray drying. The protein denaturation was relatively low at 30% with dilute feed concentration (20%, 30%) and low outlet temperature (80°C) (Figure 5.12) (Anandharamakrishnan, 2008). However, operating the spray dryer with low solid concentration of the feed and low outlet temperature increases the energy cost of the process and prolongs the drying time. Steam consumption per 1 kg of evaporated water is about 3-6 fold higher during spray drying than the corresponding consumption of a double-effect vacuum evaporator used to pre-concentrate the feed (Caric, 1994). This is of relevance as the usual production of whey powder involves concentrated whey containing 42–60% solids (Morr & Ha, 1993). All the above reasons justify the use of spray-freeze-drying for the production of whey protein powder.

Spray-freeze-drying of whey protein solutions at 20% and 30% concentrations was conducted by spray-freezing the feed into a pre-cooled spray chamber (SFV), followed by freeze-drying in a fluidized bed freeze dryer under sub-atmospheric pressure conditions (0.1 bar). The spray-chamber was previously purged with nitrogen gas from a cylinder before cooling it with liquid nitrogen from a dewar. Two different spray-chamber wall temperatures were used: –45 ± 5°C (for 30% whey solution) (Al-Hakim & Stapley, 2004) and –85 ± 2°C (for 20% whey solution) (Anandharamakrishnan, Rielly, & Stapley 2010). While the 20% WPI solution was atomized using a hydraulic nozzle, the 30% feed was atomized by a twin-fluid nozzle. As already discussed in Chapter 3, application of partial vacuum to the fluidized bed freeze-drying process alleviates the concerns associated with circulating large mass flow rates of dry gas. For instance, requirement for the mass of saturated gas would be reduced to 1/10th (~420 kg) of the amount required at 1 bar, when the process is operated at a temperature and absolute pressure of –30°C and 0.1 bar,

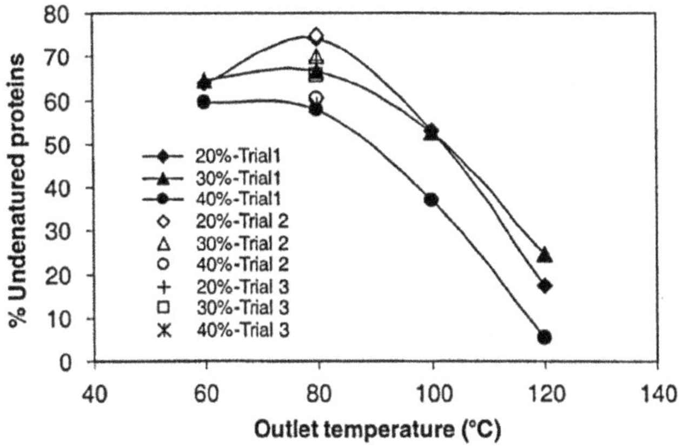

Figure 5.12 Effect of spray dryer outlet temperature and feed concentration on the denaturation of the spray-dried product as determined by DSC (Anandharamakrishnan, 2008).

respectively. As discussed earlier, though the application of vacuum adds to the capital cost and operational complexities of the manufacturing unit, it leads to a substantial reduction in the mass of dry gas required and also prevents particle elutriation from the fluidized bed. The inlet gas temperature of the fluidized bed freeze dryer was varied at three levels: −10°C, −20°C and −30°C (Anandharamakrishnan et al., 2010). Notably, whey has a collapse temperature of approximately −10°C (Al-Hakim & Stapley, 2004), which needs to be considered while selecting the freeze-drying temperature.

5.3.1 Physical properties of spray-freeze-dried whey protein powder

The moisture content, absolute density and bulk density of SFD-WPI were estimated and the values were compared with those of the spray-dried WPI powder. The bulk density of SFD-WPI (0.22 g/cm^3) was about 29% lesser than that of the spray dried particles (0.31 g/cm^3). The lower bulk density value of SFD-WPI was ascribed to their ability to retain the original dimensions of the droplet/particle soon after atomization. Similar trend was found in the absolute density values, but the margin of difference was much higher at ~67% between the spray-freeze-dried (0.41 g/cm^3) and spray-dried (1.25 g/cm^3) particles. Absolute density is the mass of any substance per unit volume of a material that does not include any free space (voids) that may exist between the particles. The lower absolute density of SFD-WPI originates from its sponge-like microstructure, which might not have the perfect interconnectivity of pores. Hence, a single surface hole may not be sufficient to cause the complete filling of the internal voids of the particle. But, the hollow shell morphology of the spray dried particles renders it susceptible to the formation of a single hole that permits the gas to fill the whole internal void space and thereby lead to high values for absolute density (Al-Hakim & Stapley, 2004).

5.3.2 Morphology of spray-freeze-dried whey protein powder

The SFD-WPI particles produced from the 20% feed solution spray-frozen at −85°C and freeze-dried in a fluidized bed under sub-atmospheric conditions were typically large, with their Sauter mean particle diameter (d_{32}) in the range of 393-489 μm. Smaller particles were found agglomerated onto the surface of a large particle (Figure 5.13[a]). The particles were spherical with their surface containing numerous fine pores and an occasional surface blemish (Figure 5.13[a, b & d]).

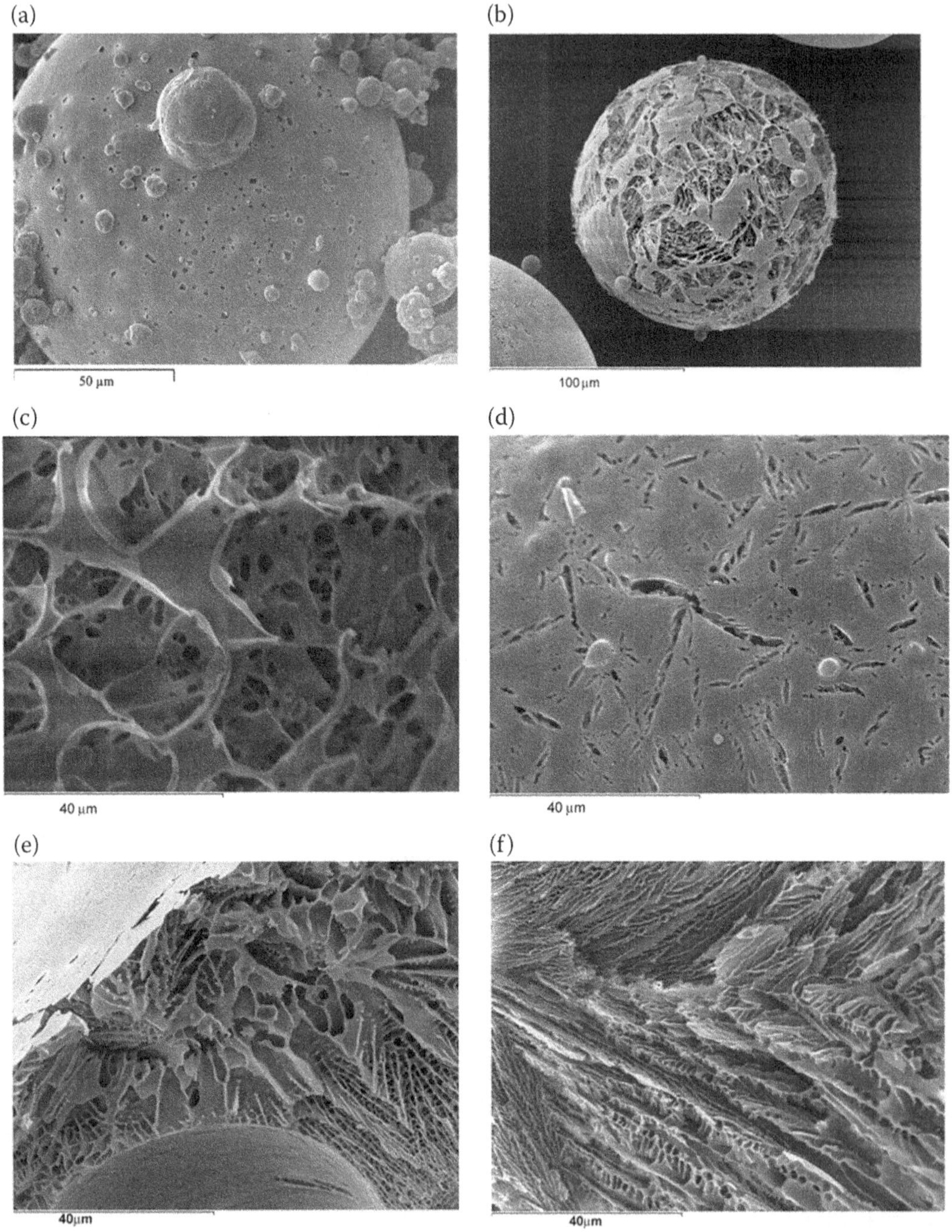

Figure 5.13 Scanning electron micrographs of spray-freeze-dried whey protein isolate powder: (a–f): Feed: 20% WPI solution; atomization by hydraulic nozzle; fluidized-bed freeze-drying under sub-atmospheric conditions (Anandharamakrishnan et al., 2010); (g, h): Feed: 30% WPI solution; atomization by twin-fluid nozzle; vacuum freeze-drying (Al-Hakim & Stapley, 2004).

Magnifying one of the surface blemishes revealed the porous structure inside the particle (Figure 5.13[e & f]), which indicated the uniformity of drying and the absence of particle collapse. Owing to the rapid freezing rate, the sizes of both the ice crystals and pores were much smaller than those observed during conventional freeze-drying of whey proteins (Anandharamakrishnan, 2008). But, SFD-WPI particles produced from a same processing batch of relatively concentrated feed solution (30%) that was spray-frozen at −45°C and vacuum freeze-dried, depicted different surface structures. The reason was not clear, but it was suggested to result from the twin-fluid nozzle atomization, which introduces relatively warm air into the chamber that might have mixed in a disordered way with the cold chamber gas in the spray cloud. As a result, different spray droplets may have contacted with different pockets of gas at varying temperatures. This could have led to the variation of cooling rates between droplets. Thus, different surface structures may have resulted from the individual droplets encountering different cooling rates, as nucleation rates are highly temperature dependent. Besides, the non-uniform surface structure was also attributed to the collisions between particles that were proposed to have occurred when some of the particles were still in their unfrozen (liquid/droplet) state (Al-Hakim & Stapley, 2004).

Different from the spray-freeze-dried WPI particles that were sponge-like with either a smooth or a rough, porous topography, the spray-dried particles were of the *skin-forming* type with hollow shells, wherein some spheres were partially deflated. The SEM micrographs of spray-dried whey protein powder showed a broader particle size distribution with a low level of particle aggregation, irrespective of the feed concentration and outlet temperatures. But, the particle shape was found to be dependent on the feed solid content and outlet temperature. While the particles were caved-in at lower outlet temperatures (60-80°C), they retained their spherical shape at higher outlet temperatures (100-120°C) at 20% and 30% feed concentrations. However, the SD-WPI particles produced from 40% feed at higher outlet temperatures were irregularly shaped (Figure 5.14) and had higher moisture content. During the falling rate period, the more concentrated feed forms a strong outer layer of dry crust much rapidly. Consequently, the droplet temperature increases

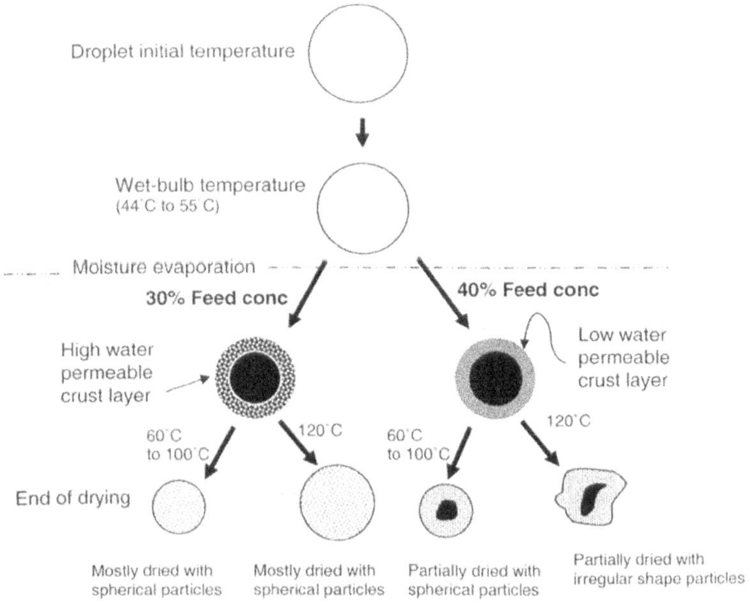

Figure 5.14 Schematic representation of crust formation during spray drying (Anandharamakrishnan, 2008).

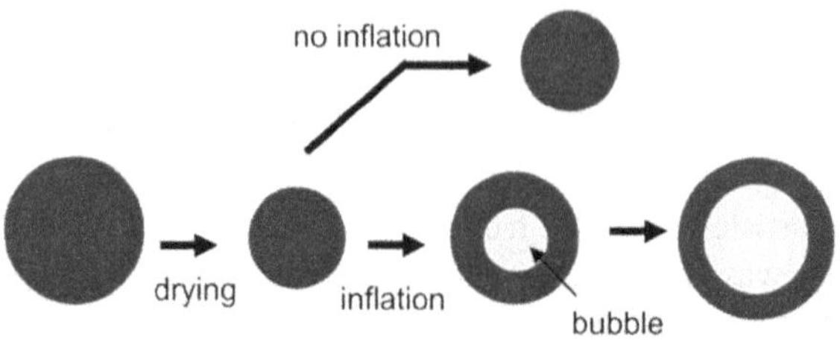

Figure 5.15 Schematic representation of the bubble inflation phenomenon during spray drying (Modified and reproduced from Etzel et al., 1996).

beyond the wet-bulb temperature to approach the dry-bulb temperature of the air. This results in a high degree of protein denaturation and aggregation (Anandharamakrishnan, 2008). Also, the impermeable crust restricts the particle expansion due to bubble inflation, which occurs when the partial pressure of moisture vapor at the droplet center exceeds the ambient pressure. Eventually, the droplet is inflated to an outer radius to result in irregularly-shaped particles (Etzel et al., 1996) (Figure 5.15). Thus, spray-freeze-dried and spray-dried whey particles showed marked differences in their morphology.

5.3.3 Solubility of spray-freeze-dried whey protein powder

The reverse-phase high-performance liquid chromatography (RP-HPLC) method was employed to estimate the soluble fractions in whey protein isolate, before and after spray-freeze-drying. The RP-HPLC chromatograms for whey protein isolate before and after spray-freeze-drying are given in Figures 5.16[a] and 5.16[b], respectively. The loss of solubility in α-lactalbumin and β-lactoglobulin was estimated from the areas of their corresponding peaks. The results showed that irrespective of the variation in the inlet gas temperature of freeze-drying, the SFD-WPI particles showed insignificant solubility loss in α-lactalbumin and 2% solubility loss in β-lactoglobulin. This result is expected as α-lactalbumin is a protein with higher stability compared to β-lactoglobulin (Anandharamakrishnan et al., 2008; Ferreira, Mendes & Ferreira, 2001) and it is the only milk protein that can bind with the Ca^{2+} ions present in the whey. This binding improves the stability of α-lactalbumin under processing conditions, especially, high temperature (Permyakov & Berliner, 2000). In fact, the dissociation of Ca^{2+} ions is one of the stages in the denaturation process of α-lactalbumin (Boye et al., 1997). Also, it has been found that the aggregation of α-lactalbumin is often not independent, but depends on the formation of aggregates with β-lactoglobulin (Schokker et al., 2000). The observation of Al-Hakim & Stapley (2004) on the solubility behavior of SFD-WPI (from 30% feed solution subjected to vacuum freeze-drying) was in consensus with the above findings. They reported complete absence of insoluble matter in the spray-freeze-dried whey powder and attributed the same to the low-temperature operation of the SFD process. Thus, SFD-WPI exhibited superior solubility irrespective of the differences in feed solid content and the modes of atomization and freeze-drying.

In contrast, the solubility of proteins in SD-WPI particles was dependent on the feed concentration and outlet temperature (Figure 5.16). The maximum solubility was detected at lower outlet temperatures (60°C and 80°C) and lower feed concentrations (20% and 30%) (Figure 5.16[a]). But, solubility reduced with an increase in outlet temperature and feed concentration. However, irrespective of the feed concentration, reduction in protein solubility was not prominent up to an outlet temperature of 80°C and 160-180°C of gas inlet temperatures. The solubility differences turned significant only when the outlet temperature exceeded 80°C. This is because the outlet temperature is not an independent parameter. It is obtained by reducing the feed

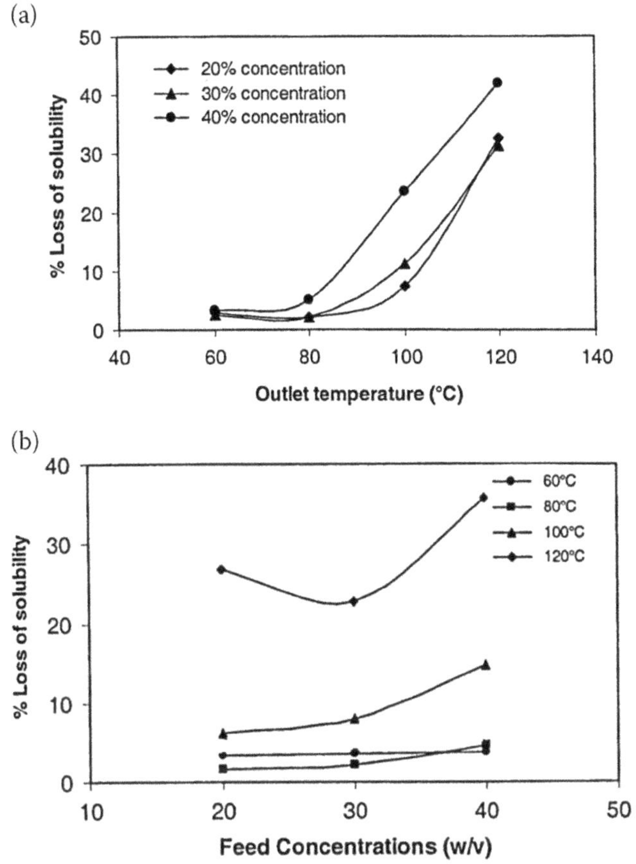

Figure 5.16 Effect of (a) spray dryer outlet temperature and (b) feed concentration on the solubility of the spray-dried WPI powder (Anandharamakrishnan, 2008).

flow rate, which leads to drier particles and higher equilibrium particle temperatures above the protein denaturation temperature (~ 75°C). The influence of feed concentration on the solubility loss of α-lactalbumin and β-lactoglobulin showed that for both the protein fractions, the loss in solubility increased by approximately 10% with the increase in feed concentration at higher outlet gas temperatures (100°C and 120°C) (Figure 5.16[b]). Specifically, SD-WPI produced from 40% feed showed the highest degree of denaturation and loss in solubility. The underlying reasons for the above could be: (i) a direct influence of feed concentration on denaturation and loss of solubility and/or (ii) an indirect effect from droplets at different initial concentrations being subjected to different temperature histories during spray drying (Anandharamakrishnan, 2008). Similarly, the solubility measurements done by Al-Hakim & Stapley (2004) showed that a small fraction (2.5%) of the spray-dried WPI powder was insoluble due to the thermally-induced denaturation of whey proteins. The insignificant loss of protein solubility after spray-freeze-drying is because of its non-involvement of heat for drying. Thus, heat-induced denaturation of proteins associated with the spray drying process is completely avoided in the SFD process (Flanders Health Blog, 2019).

CONCLUSIONS

The information presented in this chapter demonstrated the edge of spray-freeze-drying process over the conventional techniques for the drying of dairy products. The spray-freeze-dried dairy products are superior over their spray-dried counterparts with respect to particle morphology, size, wettability and solubility, all of which determine the powder reconstitution properties. In future, the dairy processing specialists must make an attempt to configure and customize the spray-freeze-drying process and equipment for the mass production of dairy powders with superior quality characteristics and nutritional profile. After all, the priorities of present-day customers are premium product quality, nutritional wellness and convenience, over and above the product cost. Therefore, it is the right time for the market positioning of spray-freeze-dried dairy products, and the product portfolio is more likely to gain high consumer acceptance in the near future.

REFERENCES

ADPI (2002). *Bulletin 916. Standards for grades of dry milk including methods of analysis.* American Dairy Products Institute, Elmhurst, IL, 14.

Al-Hakim, K., & Stapley, A. G. F. (2004). Morphology of spray-dried and spray-freeze-dried whey powders. In: *Drying - Proceedings of the 14thInternational Drying Symposium* (IDS 2004), (Vol. C pp. 1720–1726).

Anandharamakrishnan, C., Gimbun, J., Stapley, A. G. F., & Rielly, C. D. (2008). *Application of computational fluid dynamics (CFD) simulations to spray-freeze drying operations.* 16th International Drying SYMPOSIUM (IDS 2008), pp. 537–545.

Anandharamakrishnan, C., Rielly, C. D., & Stapley, A. G. F. (2010). Spray-freeze-drying of whey proteins at sub-atmospheric pressures. *Dairy Science and Technology*, 90, 321–334.

Anandharamakrishnan, C. (2008). Experimental and computational fluid dynamics studies on spray-freeze-drying and spray-drying of proteins. Ph.D. thesis. Loughborough University, UK.

Anema, S. G., Stockmann, R., & Lowe, E. K. (2005). Denaturation of β-lactoglobulin in pressure-treated skim milk, *Journal of Agricultural and Food Chemistry*, 53, 7783–7791.

Bodyfelt, F. W., Tobias, J., & Trout, G. M. (1988). Sensory evaluation of cheese (Chapter 8). In: *The Sensory Evaluation of Dairy Products*, AVINan Nostrand Reinhold, New York pp. 300–376.

Boye, J. I., Ma, C. Y., Ismail, A., Harwalkar, V. R., & Kalab, M. (1997). Molecular and microstructural studies of thermal denaturation and gelation of β-lactoglobulins A and B. *Journal of Agricultural and Food Chemistry*, 45, 1608–1618.

Buma, T. J. (1971a). Free fat in spray dried whole milk 4. Significance of free fat for other properties of practical importance. *Netherlands Milk and Dairy Journal*, 25, 88–106.

Buma, T. J. (1971b). Free fat in spray dried whole milk 5. Cohesion, Determination, influence of particle size, moisture content and free-fat content. *Netherlands Milk and Dairy Journal*, 25, 107–122.

Caric, M. (1994). *Concentrated and dried dairy products.* VCH Publishers, New York, pp. 72–125.

Codex Alimentarius (2018). Standard for milk powders and cream powder: CXS 207-1999. Available from: http://www.fao.org/fao-who-codexalimentarius/sh-proxy/en/?lnk=1&url=https%253A%252F%252Fworkspace.fao.org%252Fsites%252Fcodex%252FStandards%252FCXS%2B207-1999%252FCXS_207e.pdf (Accessed 15 August 2021)

Ding, H., Yu, W., Boiarkina, I., Depree, N., & Young, B. R. (2020). Effects of morphology on the dispersibility of instant whole milk powder. *Journal of Food Engineering*, 276, 109841.

Early, R. (1998). Milk concentrates and milk powders: In: Early, R. (Ed.), *The technology of dairy products*, 2nd Edition, Blackie Academic and Professional, London, pp. 228–300.

Etzel, M. R., Suen, S. Y., Halverson, S. L., & Budijono, S. (1996). Enzyme inactivation in a droplet forming a bubble during drying. *Journal of Food Engineering*, 27, 17–34.

Ferreira, I. M., Mendes, E., & Ferreira, M. A. (2001). HPLC/UV analysis of proteins in dairy products using a Hydrophobic Interaction Chromatographic Column. *Analytical Sciences*, 17(4), 499–501.

Flanders Health Blog (2019). Available from: https://www.flandershealth.us/therapeutic-proteins/spray-freeze-drying.html (Accessed 15 August 2021).

Gaiani, C., Boyanova, P., Hussain, R., Pazos, M. I., Karam, M. C., Burgain, J., & Scher, J. (2011) Morphological descriptors and colour as a tool to better understand rehydration properties of dairy powders. *International Dairy Journal*, 21, 462–469.

Gharemann, F., Ilari, J. L., Cantoni, P., & Boudier, J. F. (1994). Caracterisation des poudres de concentres proteiques laitiers. *Rev ENIL*, 176, 25–31.

Gouda, A., & Abou El-Nour, A. (2003). Cheeses, Processed cheese. *Encyclopedia of Food Sciences and Nutrition*, 1108–1115.

Greenbank, G. R., & Pallansch, M. J. (1962). *Proceedings of the 16thInternational Dairy Congress*, B 1002.

Hindmarsh, J. P., Russell, A. B., & Chen, X. D. (2003). Experimental and numerical analysis of the temperature transition of a suspended freezing water droplet. *International Journal of Heat and Mass Transfer*, 46, 1199–1213.

Hindmarsh, J. P., Wilson, D. I., Johns, M. L., Russell, A. B. , & Chen, X. D. (2005). NMR verification of single droplet freezing models. *AIChE Journal*, 51, 2640–2648.

Hindmarsh, J. P., Russell, A. B., & Chen, X. D. (2007). Fundamentals of the spray freezing of foods - microstructure of frozen droplets. *Journal of Food Engineering*, 78, 136–150.

Holm, G. E., Greenbank, G. R., & Deysher, E. F. (1925). The effect of homogenization, condensation and variations in the fat' content of a milk upon the keeping quality of its milk powder. *Journal of Dairy Science*, 8, 515–522.

Ilari, J. L., & Loisel, C. (1991). La ma^ıtrise de la fonctionnalite des poudres. *Process*, 1063, 39–43.

Ishwarya, S. P., Anandharamakrishnan, C., & Stapley, A. G. F. (2017). Spray-freeze-drying of dairy products. In: Anandharamakrishnan, C. (Ed.), *Handbook of Drying for Dairy Products*. John Wiley & Sons, Oxford, pp. 123–148.

Kawabata, Y., Wada, K., Nakatani, M., Yamada, S., & Onoue, S. (2011). Formulation design for poorly water-soluble drugs based on biopharmaceutics classification system: basic approaches and practical applications. *International Journal of Pharmaceutics* 420, 1–10.

Kim, E. H.-J., Chen, X. D., & Pearce, D. (2002). Surface characterization of four industrial spray-dried dairy powders in relation to chemical composition, structure and wetting property. *Colloids and Surfaces B: Biointerfaces*, 26, 197–212.

Kim, E., Chen, X. D., & Pearce, D. (2003). On the mechanisms of surface formation and the surface compositions of industrial milk powders. *Drying Technology*, 21, 265–278.

Kneifel, W. (2003). Recombined and filled milks. *Encyclopedia of Food Sciences and Nutrition*, 4921–4926.

Knipschildt, M. E., & Andersen, G. G. (1994). Drying of Milk and Milk Products. In: R. K. Robinson (Ed.), *Robinson: Modern Dairy Technology*, Springer, US, pp. 159–254.

Langrish, T. A. G., Marquez, N., & Kota, K. (2006). An investigation and quantitative assessment of particle shape in milk powders from a laboratory-scale spray dryer. *Drying Technology*, 24(12), 1619–1630.

Meerdink, G. (1993). *Drying of liquid food droplets. Enzyme Inactivation and Multicomponent Diffusion*. PhD thesis, Agricultural University Wageningen, The Netherlands.

Millqvist-Fureby, A., Elofsson, U., & Bergenståhl, B. (2001). Surface composition of spray-dried milk protein-stabilised emulsions in relation to pre-heat treatment of proteins. *Colloids and Surfaces B: Biointerfaces*, 21, 47–58.

Morr, C. V., & Ha, E. Y. W. (1993). Whey protein concentrates and isolates: processing and functional properties. *Critical Reviews in Food Science and Nutrition*, 33, 431–476.

Nickerson, T. A., Coulter, S. T., & Jenness, R. (1952). Some properties of freeze-dried milk. *Journal of Dairy Science*, 35, 77–85.

Pathania, S., Ho, Q. T., Hogan, S. A., McCarthy, N., & Tobin, J. T. (2018). Applications of hydrodynamic cavitation for instant rehydration of high protein milk powders. *Journal of Food Engineering*, 225, 18–25.

Permyakov, E. A. , & Berliner, L. J. (2000). α-lactalbumin: Structure and function. *FEBS Letters*, 473, 269–274.

Persistence Market Research (2020). Available from: https://www.persistencemarketresearch.com/market-research/whole-milk-powder-market.asp (Accessed 15 August 2021).

Pisecky, J. (1997). *Handbook of milk powder manufacture*. NiroA/S, Copenhagen.

PowderPro® (2020). Pros and cons. Available from: https://powderpro.se/background/pros-cons/ (Accessed 15 August 2021).

Rimpiläinen, V., Kaipio, J. P., Depree, N., Young, B. R., Wilson, D. I. (2015). Predicting functional properties of milk powder based on manufacturing data in an industrial-scale powder plant. *Journal of Food Engineering*, 153, 12–19.

Rogers, S., Wu, W. D., Saunders, J., & Chen, X. D. (2008). Characteristics of milk powders produced by spray freeze drying. *Drying Technology*, 26, 404–412.

Saito, Z. (1985). Particle structure in spray-dried whole milk and in instant skim milk powder as related to lactose crystallization. *Food Structure*, 4, Article 16.

Schokker, E. P., Singh, H., & Creamer, L. K. (2000). Heat-induced aggregation of β-lactoglobulin A and B with α-lactalbumin, *International Dairy Journal*, 10, 843–853.

Shipstead, H., & Tarassuk, N. P. (1953). Chemical changes in dehydrated milk during storage. *Journal of Agricultural and Food Chemistry*, 1, 613–616.

Silva, J. V. C., & O'Mahony, J. A. (2016). Flowability and wetting behaviour of milk protein ingredients as influenced by powder composition, particle size and microstructure. *International Journal of Dairy Technology*, 70, 277–286.

Simpson, J. T., Hunter, S. R., & Aytug, T. (2015). Superhydrophobic materials and coatings: A review. *Reports on Progress in Physics*, 78, 086501.

Tian, Y., Fu, N., Wu, W. D., Zhu, D., Huang, J., Yun, S., & Chen, X. D. (2014). Effects of co-spray drying of surfactants with high solids milk on milk powder wettability. *Food and Bioprocess Technology*, 7, 3121–3135.

Walstra, P., Wouters, Jan T. M., & Geurts, T. J. (2005). *Dairy Science and Technology*, 782.

Yazdanpanah, N., & Langrish, T. A. (2012). Releasing fat in whole milk powder during fluidized bed drying. *Drying Technology*, 30, 1081–1087.

Yu, Z., Johnston, K. P., & Williams, R. O. (2006). Spray freezing into liquid versus spray-freeze drying: Influence of atomization on protein aggregation and biological activity. *European Journal of Pharmaceutical Sciences*, 27, 9–18.

CHAPTER 6

Spray-freeze-drying for soluble coffee production

Coffee is one of the most important merchandise in the world. '*Soluble coffee*' is the most popular among the coffee beverages, due to its fine aroma, instant rehydration property and extended shelf-life (Burmester, Pietsch, & Eggers, 2011; Othman & Razali, 2019). Soluble or instant coffee is the dried soluble fraction of roasted and ground coffee that is sold to consumers in the form of either powder or granules. Initial stages in the manufacturing of soluble coffee are the same as that for roasted coffee that includes grading and sorting of green coffee beans, roasting, and grinding. The latter stages are those pertaining to the instantization process, involving the extraction of soluble solids followed by its concentration (evaporation or freeze-concentration) and drying (spray drying or freeze-drying), with optional pre-/post-processing steps for preserving the aromatics and agglomeration (Clarke, 2003).

Currently, the retail soluble coffee is either spray-dried or freeze-dried. Spray-drying is widely used by the instant coffee manufacturers owing to its versatile, industry-friendly and economical operation. In general, during spray-drying, the concentrated coffee extract (~50%) is atomized through a pressure nozzle fitted to the roof of a tall drying tower. The fine droplet mist of liquid coffee concentrate undergoes rapid evaporation within the spray chamber at an inlet air temperature of 148–232°C (300–450°F) and outlet air temperature of 65–121°C (150–250°F) (Ponzoni & Nutley, 1966). Finally, the dry and fine coffee powder is collected at the bottom of the tower, which has a moisture content of 2–4%, typically. Thus, the quality of coffee powder produced by spray-drying depends on the air temperature and pressure, besides the solid concentration of the extract (Huste, 1974; Ghosh & Venkatachalapathy, 2014; Koç & Kaymak, 2014). The spray-dried coffee powder may be further converted to granules by steam-agglomeration or by contacting the powder with finely atomized water to facilitate instant rehydration. Then, the agglomerates are dried on a conveyor belt with cold inlet air (Barbosa-Canovas, Ortega-Rivas, Juliano, & Yan 2005).

On the other hand, freeze-drying is employed to produce soluble coffee of premium quality. The term 'quality' includes the physical (appearance, solubility), chemical (volatile profile of aromatic compounds), and organoleptic properties (taste and aroma) of the coffee beverage (Herrera & Lambot, 2017). During freeze-drying, moisture is removed from the freeze-concentrated coffee liquor under low-temperature (<–40°C) and vacuum (30–40 Pa) (Suwelack & Kunke, 2002) to produce a flaky and porous cake. Subsequently, the cake is milled into smaller particles, which exhibits the rich aroma of roasted coffee and an instantaneous solubility. Compared to spray-dried soluble coffee, the freeze-dried product has been found to retain 17–20% more low boiling point aromatics and 75% more high boiling point volatiles (Flink, 1975; Ishwarya & Anandharamakrishnan, 2015). Not just the high level of aromatics, freeze-dried soluble coffee also has a uniformly dark color similar to that of roasted and ground coffee (Elerath, Lakes, & Esra, 1968). Thus, it is evident that freeze-drying is gentler on the product, but at the expense of energy and operational cost. Hence, the application of freeze-drying is confined to the production of finer and more expensive blends of soluble coffee (Morin et al., 2018).

DOI: 10.1201/9781003019312-6

Both spray-drying and freeze-drying approaches of soluble coffee production have their limitations. During the high-temperature spray-drying operation, the low-boiling aromatic compounds of coffee are lost (Taylor, 1983). Within the spray chamber, a portion of the spray may extend outside the downward hot airflow. This leads to the adherence of some powder to the chamber's roof and wall due to its sticky nature. The relatively low volume of hot air extends the particle residence time inside the chamber that may cause overheating of the product (Bassoli et al., 1993) and hence aroma loss. On spray-drying a 40% coffee extract, only 57% of the character-impact volatiles reminiscent of coffee aroma (ex. methyl pyrazine, 2,3-butanedione and 2-furanmethanol) are retained in the final powder (Ishwarya & Anandharamakrishnan, 2015). Likewise, energy and cost-intensiveness are the main demerits of freeze-drying. To retain the dark coffee-like color and rich aroma of roasted coffee, large bodies of the coffee extract are frozen on a continuously moving metal freezing belt. Consequently, the freezing time is longer to achieve the desired ice crystal growth. Slow-freezing of coffee extract provides sufficient time for the separation between aromatic oils and solids from the coffee extract and impairs the uniform distribution of aroma and solids. Subsequently, the coffee solids and coffee oils containing the volatile coffee aromatics rise to float on the surface of the extract during freezing (Elerath, Lakes, & Esra, 1968).

The above events lead to the formation of an impermeable film on the surface of frozen extract, which acts as a barrier for the removal of water vapor from the inner portions of the extract during the drying phase. Removing this film improves the drying rate substantially. But, it also poses drawbacks such as product wastage and loss of aromatic constituents present in the coffee extract (Elerath, Lakes, & Esra, 1968; Elerath, 1969). The primary drying phase of freeze-drying proceeds at a low heat transfer rate from the surface to the core of frozen extract. This is done to compensate for the rise in temperature of the dried coffee during sublimation and maintain it at a low temperature to avoid flavor loss (Hair & Strang, 1969). As a result, the freeze-drying time is longer (8 hours to more than 16 hours). Longer drying time promotes aroma loss as the rate at which the diffusional loss of aroma component is delayed in the frozen matrix depends on the sublimation rate and hence the total drying time (Coumans et al., 1994). Consequently, freeze-dried soluble coffee could retain only 77% of the character impact volatiles present in the coffee extract (Ishwarya & Anandharamakrishnan, 2015).

From the above discussion on the state of the art of soluble coffee manufacturing processes, the need for an improved drying technique is evident. An ideal drying technique for soluble coffee production would be the one that can freeze a bulk of aromatic coffee extract within a short time and dry it at an effective rate without losing the aroma and color. In this regard, the suitability of spray-freeze-drying for soluble coffee production is evident. True to that, the potential of spray-freeze-drying as a soluble coffee manufacturing process was unleashed by the coffee processing specialists. The association between spray-freeze-drying and soluble coffee dates back to 1968. In a patented process, Thuse et al. (1968) demonstrated the continuous production of instant coffee by directly spraying the coffee extract into a condensing vacuum freezing chamber. The spray froze during its transit and conducted the frozen particles toward the vacuum drying chamber that was equipped with a heat source. Finally, the perfectly dried particles were removed from the drying chamber.

Later, Elerath (1976) demonstrated the spray-freezing-into liquid (SFL) method to freeze the coffee extract. The above was accomplished by dispersing the coffee extract into a moderately warm immiscible liquid refrigerant maintained at a temperature of about –9.4°C (15°F) to the ice point of the extract (the temperature at which water in the extract begins to crystallize into ice). The coffee extract was pumped into the refrigerant as small droplets using a bore tube having a narrow inner diameter of 0.25 in. Alternatively, the extract and refrigerant were mixed in a venturi device in the high-velocity zone, wherein the extract entered the low-pressure zone. When the extract/refrigerant mix exited the venturi, the extract was highly dispersed as droplets in the refrigerant. The temperature of coffee extract-refrigerant mix was maintained at the critical temperature (about

–4.4 to –2.2°C) for a minimum duration of 4 minutes. The above was done to promote the formation of large ice crystals to produce a soluble coffee product that is dark and rich in color as the extract. Then, the temperature of the mix was further reduced to below –28.9°C, after which the frozen extract was separated from the refrigerant and freeze-dried (Elerath, 1976).

Subsequently, the capability of SFD as a soluble coffee production process was reestablished by several researchers. SFD was observed to yield a free-flowing coffee powder with typical quality characteristics such as good color and appearance, instant solubility and greater retention of aromatic compounds (Mumenthaler & Leuenberger, 1991; Khwanpruk et al., 2008; Ishwarya & Anandharamakrishnan, 2015). Although the superiority of spray-freeze-dried soluble coffee has been well-demonstrated, the lab-to-shop floor transformation of the process has not taken place yet. Understanding the merits of SFD process for soluble coffee production would aid in transforming it as an industrial process. This chapter would describe the process flow of soluble coffee production by spray-freeze-drying. The edge of spray-freeze-dried coffee over its spray-dried and freeze-dried counterparts will be explained in terms of its typical quality characteristics.

6.1 PROCESS FLOW FOR SOLUBLE COFFEE PRODUCTION BY SPRAY-FREEZE-DRYING

A typical process flow sheet for soluble coffee production by spray-freeze-drying is as depicted in Figure 6.1. The flow chart also indicates the process control parameters at each step. The subsequent sections would explain the different stages of soluble coffee production by the SFD process.

6.1.1 Stage-1: Atomization

After obtaining a coffee extract of desired total solid content, it is atomized into fine droplets using a hydraulic (Khwanpruk et al., 2008) or twin-fluid nozzle (Ishwarya & Anandharamakrishnan, 2015). The atomization pressure can be considered as the first process control parameter in the process (Figure 6.1). It determines the droplet size, which in turn exerts control over the temperature profile during the subsequent freezing process. Achieving a final product of defined specifications depends on the temperature profile during freezing. Smaller the droplet size, the more uniform will be the temperature field for heat transfer during the freezing step. A uniform temperature field is vital to attain uniform nucleation and formation of fine ice crystals (MacLeod et al., 2006). Khwanpruk et al. (2008) atomized a 20% coffee solution by supplying it from a pressurized feed tank at 8 barg (7.895 atm) to the hydraulic nozzle. Ishwarya & Anandharamakrishnan (2015) used a twin-fluid nozzle at a compressed air pressure of 588.39 kPa to atomize a 40% coffee feed solution.

Care must be taken to avoid high slip velocities in the spray during atomization, as it leads to minor volatile losses caused by the high droplet-gas mass transfer coefficients (Khwanpruk et al., 2008). As the atomization and freezing steps occur simultaneously, the ultra-low temperature of the cryogen may cause the feed solution to freeze within the nozzle and block its orifice. Hence, for atomizers made of stainless steel, the nozzle housings are equipped with external air heating to prevent freezing-induced blockage. In the above case, the set temperature of nozzle heater can be considered as the second process control parameter. Alternatively, the use of plastic nozzles has also been suggested (Al-Hakim et al., 2006; Ishwarya, Anandharamakrishnan, & Stapley, 2015).

6.1.2 Stage-2: Freezing

During the second stage of the SFD process, icy particles of the coffee feed droplets are formed as a result of rapid freezing upon contact with a cryogen. The commonly used cryogens for the spray-freezing of coffee extract/solution are liquid nitrogen and cold nitrogen gas. Liquid nitrogen

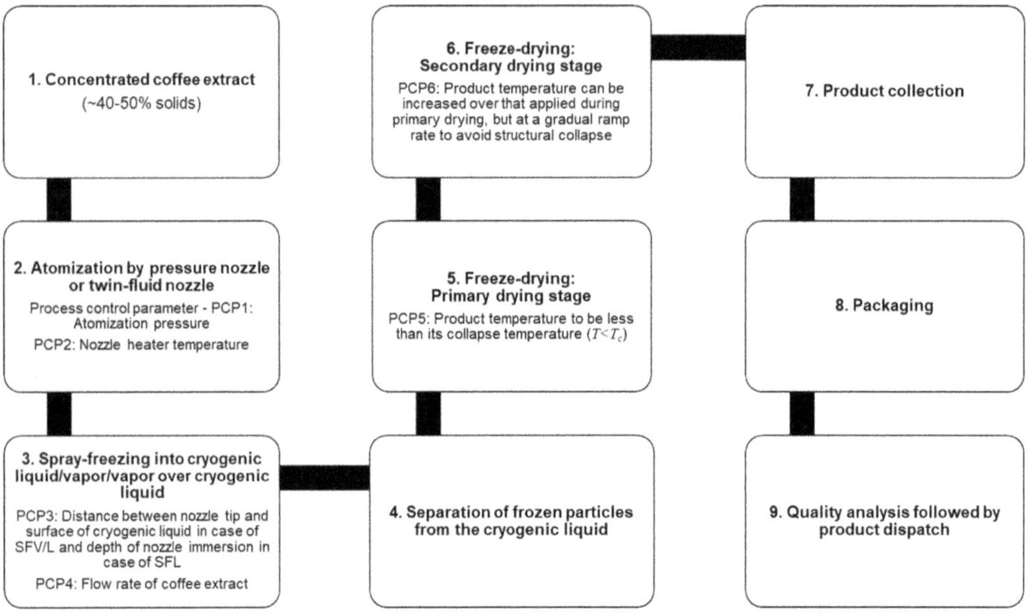

Figure 6.1 Process flow sheet for soluble coffee production by spray-freeze-drying.

is advantageous over other cryogens as it evaporates readily due to its ultra-low boiling point (−196°C). The boiling point of other cryogens such as isopentane, argon and ethanol is 27.7°C, −185.8°C and 78.37°C, respectively. The typical temperature profile during the spray-freezing of a single droplet of coffee solution is depicted in Figure 6.2. Coffee freezing occurs in a characteristic sequence comprising the four stages as described below (MacLeod et al., 2006):

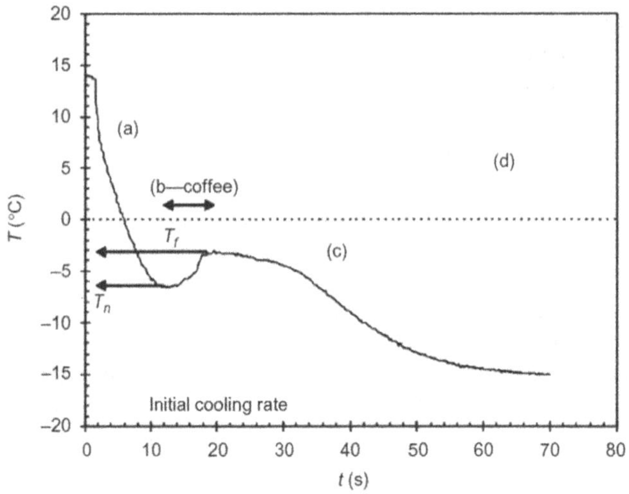

Figure 6.2 Temperature profile during the spray-freezing of coffee droplets. (Reproduced with permission from MacLeod et al., 2006).

a. ***Cooling phase:*** During this stage, the heat transfer occurs between the droplet of coffee solution and the surrounding cold, dry air by convection, radiation, and evaporation. After this phase, the less volatile materials form a thin surface layer, which reduces the mass transfer significantly.
b. ***Nucleation:*** This phase marks the beginning of ice crystal formation after the attainment of nucleation temperature (T_n). Due to the increase in solution viscosity, T_n varies inversely with the concentration of coffee extract. The dispersed coffee particles in the extract act as heterogeneous nucleation sites for ice crystal formation. Subsequently, the latent heat evolves rapidly due to *'recalescence'*, which is the phenomenon of fast crystal growth.
c. ***Freezing:*** The third phase in the spray-freezing of coffee is the heat transfer limited freezing that occurs at the equilibrium temperature (T_f). T_f decreases with an increase in solute concentration, which is known as the *'freezing point depression'*. For instance, the T_f reduced by 5 K when the concentration of the coffee extract was increased from 10% to 50% (w/w).
d. ***Cooling of the solidified droplet to ambient conditions:*** During the last stage of spray-freezing, there is still a portion of the coffee droplet which remains unsolidified due to the freeze concentration of the solution in the interstices between ice crystals.

6.1.2.1 Microstructure of spray-frozen coffee droplets

The microstructure of spray-frozen coffee droplets, especially the porous surface, is important for obtaining a soluble coffee product with instant rehydration properties. In the SEM micrographs shown in Figure 6.3(a & b), the darker regions represent pores, wherein the ice crystals were located before sublimation. The finely divided droplets emanating from the atomizer facilitate rapid freezing at the rate of about 10^6 K/s to form a large number of smaller ice crystals (Surasarang & Williams, 2016). On sublimation, these crystals leave behind pores on the surface (Costantino et al., 2000). The resultant porous microstructure of the spray-frozen droplets of coffee feed solution (Figure 6.3[b]) was relatively more homogeneous than that of the conventional freeze-dried coffee (Figure 6.3[a]). The uniformity in crystal microstructure is indicative of a homogeneous temperature field due to the low Biot number regime (Biot number is a dimensionless quantity, which is the ratio between the thermal resistances within a body and that at it surface). A smaller value for Biot number implies insignificant temperature gradients within the droplet and uniform temperature distribution through the entire droplet volume. Figure 6.3[c] depicts the presence of a thin outer skin formed as a result of the high solid content generated by the evaporation of water from the surface. This impervious layer hinders further mass transfer (MacLeod et al., 2006).

The major factors that govern the microstructure development in spray-frozen coffee are the solid content of coffee extract and freezing rate. Relative to a coffee solution at higher solid content (Figure 6.4[a]), the one at lower solid content leads to ice crystals of larger sizes and thinner walls due to reduced freeze concentration and delayed termination of crystal growth (Figure 6.4[b]). Likewise, lower freezing rate or slow-freezing results in crystals of larger average hydraulic radius (r_h) (Figure 6.4[a]) than those resultant from the droplets frozen at a faster rate (Figure 6.4[c]).

The established modes of spray-freezing the coffee extract are spray-freezing into vapor (SFV) and spray-freezing into vapor over liquid (SFV/L). In the SFV mode, the particulates are formed by spraying the concentrate through a heated hydraulic nozzle into a spray chamber that is previously purged with nitrogen gas. The nitrogen gas dehumidifies the chamber before commencing the atomization of the coffee feed (Al-Hakim et al., 2006). Then, the chamber is cooled by a co-current flow of cold nitrogen gas supplied from a liquid nitrogen dewar. By regulating the flows of liquid nitrogen and nitrogen gas to the chamber, its temperature is maintained at a definite value, say, −60±5°C, (Khwanpruk et al., 2008).

Contrastingly, in the SFV/L mode, the spray-freezing of coffee feed is accomplished in a rig that includes a twin-fluid nozzle, peristaltic pump, and a polystyrene container connected to a

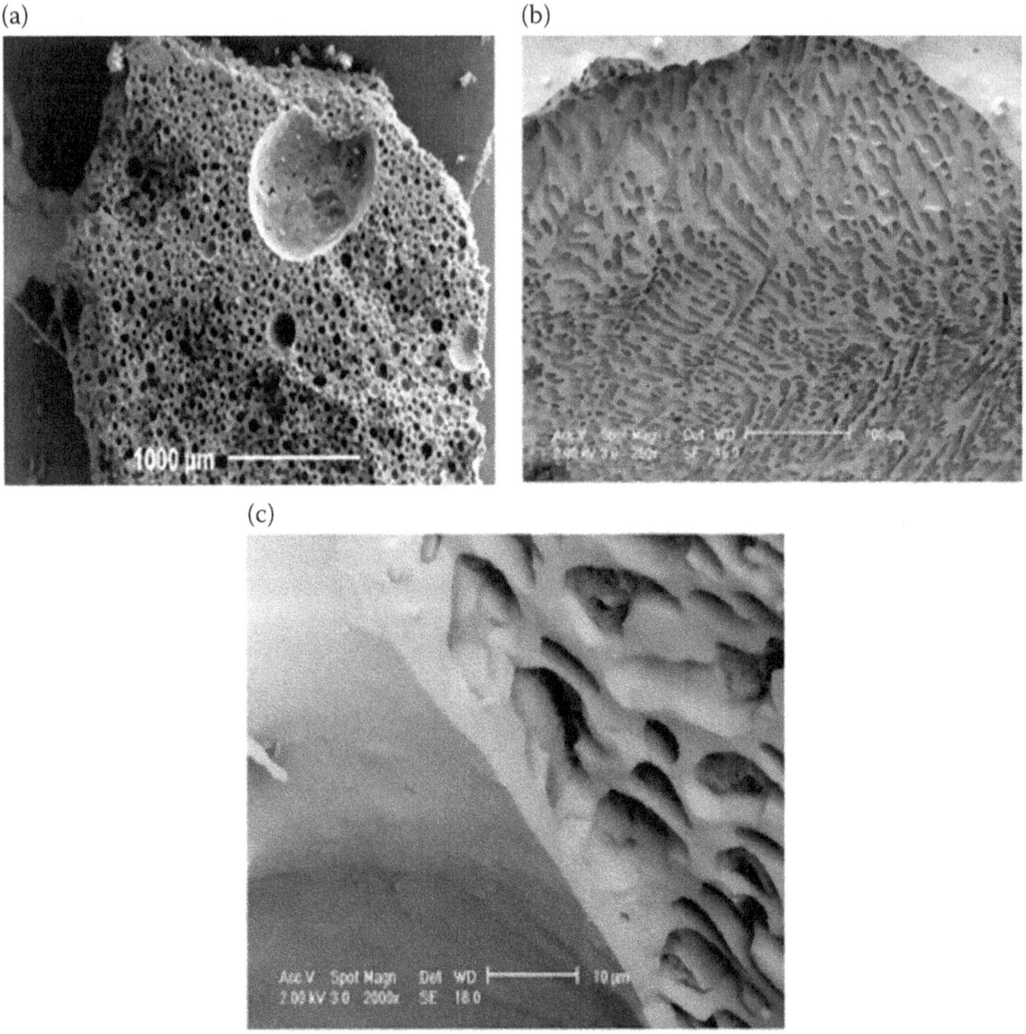

Figure 6.3 Cryo-scanning electron microscopy (SEM) images of coffee microstructure: (a) commercial freeze-dried coffee; (b) a spray-frozen droplet of coffee feed solution having a concentration of 50% (w/w); and (c) high-resolution image of the surface layer of a spray-frozen droplet of coffee feed solution having a concentration of 50% (w/w) (Reproduced with permission from MacLeod et al., 2006).

liquid nitrogen dewar. The contents (cryogen and atomized coffee droplets) in the polystyrene container are mixed using a disk blade impeller. Initially, the distance between the nozzle and the surface of liquid nitrogen in the container is optimized. This distance is significant in achieving the desired mean particle diameter of the final soluble coffee powder. When this distance is set farther apart, there are chances that the atomized feed droplets undergo slow agglomeration and solidification during their transit through the vapor phase before reaching the surface of the cryogenic liquid. This will increase the mean particle diameter of the frozen droplet and, thus, the final product (Kawabata et al., 2011). A distance of 10 cm between the nozzle and cryogen surface was used when SFV/L mode was employed to freeze a 40% coffee solution (Ishwarya & Anandharamakrishnan, 2015).

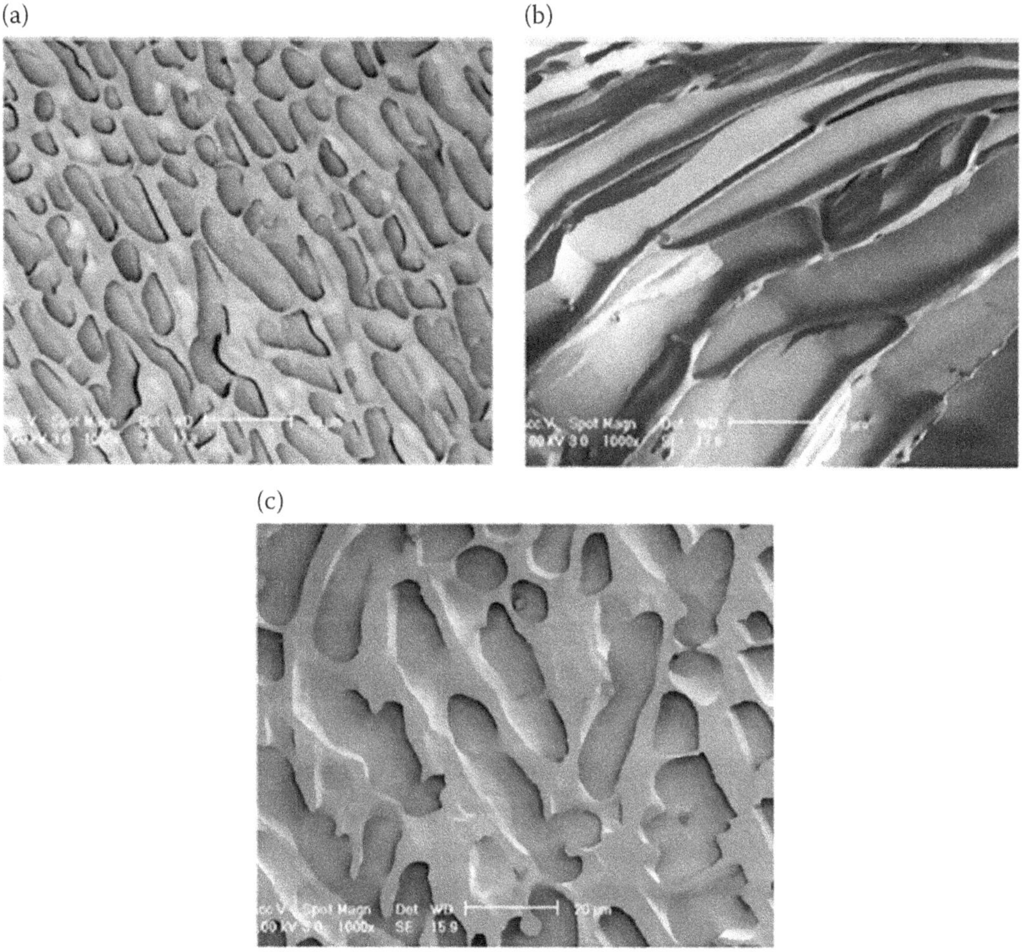

Figure 6.4 Effect of composition and cooling rate on the microstructure of spray-frozen coffee droplet: (a) 50% (w/w), rate: 0.28 m s^{-1}; (b) 10% (w/w), rate: 0.49 m s^{-1}; and (c) 50% (w/w), rate: 0.49 m s^{-1} (Reproduced with permission from MacLeod et al., 2006).

6.1.3 Stage-3: Freeze-drying

After the completion of spray-freezing, the frozen coffee particles are either sieved off from the cryogenic liquid or the latter is allowed to boil off. Then, the frozen particles are subjected to freeze-drying. Thus far, the studies pertaining to SFD processing of soluble coffee have used vacuum freeze-drying (Ishwarya & Anandharamakrishnan, 2015), atmospheric fluidized-bed freeze-drying (Mumenthaler & Leuenberger, 1991) and sub-atmospheric fluidized-bed freeze-drying (Khwanpruk et al., 2008). During vacuum freeze-drying, a product-specific temperature program for primary and secondary drying phases was devised upon considering the collapse temperature of coffee extract, which is –20°C at 25% solid concentration (Kudra & Strumillo, 1998). As explained in Chapter 3, the temperature of the primary drying stage should be less than the product collapse temperature (T_c) to prevent structural collapse. Accordingly, for a 40% coffee feed solution, the temperature range of the primary drying stage was set at –25 to –10°C under a vacuum of 107 Pa (Ishwarya & Anandharamakrishnan, 2015). With respect to soluble coffee, it is imperative to avoid structural collapse as it is associated with deteriorative reactions, volatile loss, particle shrinkage and impaired rehydration (Stapley, 2008).

The ensuing secondary drying phase was conducted at 10°C under a vacuum of 40 Pa (Ishwarya & Anandharamakrishnan, 2015). From primary to secondary drying, the temperature transition must be gradual to avoid product collapse, which may occur when the shelf temperature is higher than the glass transition temperature (T_g) of the product at that time point (Abdul-Fattah & Truong, 2010). It is important to note that the T_c increases gradually when the secondary drying starts, depending on the moisture content of the amorphous phase. The choice of temperature for secondary drying must be such that the entire sample is gradually warmed to reduce the final moisture content, but without warming too quickly to cause localized collapse. Process duration for the vacuum freeze-drying of spray-frozen coffee droplets to result in the final soluble coffee product was 30% shorter (8 hours) than that required by the conventional freeze-drying (24 hours) for soluble coffee production (Ishwarya, Anandharamakrishnan, & Stapley, 2017).

Sub-atmospheric freeze-drying is a useful approach to obtain better volatile retention in the spray-freeze-dried soluble coffee, owing to its reduced drying time compared to vacuum freeze-drying (Khwanpruk et al., 2008). In order to avoid particle collapse during sample loading, the fluidization vessel is pre-cooled to –60°C, which is well below the collapse temperature of coffee feed (–30°C at a solid concentration of 20%). Pre-cooling is accomplished by purging the fluidization vessel with a stream of cold nitrogen gas. A manual valve is installed at the inlet to the fluidized-bed apparatus for regulating the pressure within at 0.1 bara. Notably, the sub-atmospheric fluidized-bed freeze-drying process takes only 1.75 h for the production of soluble coffee, against 24 h in a conventional freeze dryer and 8 h in a vacuum spray-freeze-drying operation.

6.2 QUALITY CHARACTERISTICS OF SPRAY-FREEZE-DRIED SOLUBLE COFFEE – AN OVERVIEW

The retail price of a soluble coffee product is decided based on its compliance with the quality specifications defined by specific markets (countries). The key quality characteristics of soluble coffee can be classified under four categories:

i. ***Physical characteristics***: Moisture content, solubility, bulk density, tapped density, powder flowability indices (Carr index and Hausner ratio), and color;
ii. ***Structural characteristics***: Particle size and shape;
iii. ***Aroma profile:*** Retention of character impact aromatic compounds in soluble coffee with respect to the initial content of volatiles in the coffee extract before drying.

6.2.1 Physical characteristics

6.2.1.1 Moisture content

The moisture content of soluble coffee must be less than 5% (w/w), which is controlled by the drying conditions. Due to its hygroscopic nature, instant coffee can readily pick up moisture from the atmosphere and begin caking at about 7–8% (w/w) moisture content (Clarke, 2003). Moreover, soluble coffee is highly susceptible to cohesiveness instigated by the inter-particle liquid bridges at high moisture content, which impairs the powder flowability (Barbosa-Canovas et al., 2005). Thus, moisture content is a key quality parameter of soluble coffee that plays an important role in its handling during packaging, transportation and storage.

In the studies conducted thus far, the moisture content of spray-freeze-dried soluble coffee has been higher than the specified limit of 5%. While the moisture content was 8.7% for the SFD soluble coffee product obtained from a 40% extract (Ishwarya & Anandharamakrishnan, 2015), it was 15% in the product prepared from 20% coffee extract (Khwanpruk et al., 2008). Though the moisture content of SFD coffee was slightly less than that of the FD product (8.8%), it was higher than that of its SD counterpart (5.3%). Strikingly, in the latter study, SFD resulted in higher moisture content than freeze-drying (10.6%). The variation in moisture content values of SFD coffee may be due to differences in the initial concentration of coffee extract and the conditions applied for spray-freezing and freeze-drying. The moisture content of spray-freeze-dried soluble coffee can be further reduced by warming the drying medium or shelf during the final stages of freeze-drying (Khwanpruk et al., 2008).

6.2.1.2 Particle size and shape

Particle size is a vital parameter that influences the key quality parameters of instant coffee such as color, solubility and bulk density. Spray-freeze-dried soluble coffee showed intermediate values for particle size (i.e., mean volume diameter: 91.1 µm) and relative span factor (RSF: 2.24) between the spray-dried (50.41 µm; 1.71) and freeze-dried (636.8 µm; 3.35) coffees. Although atomization is a part of both spray-drying and spray-freeze-drying processes, the SFD coffee presented greater mean volume diameter and RSF values, at a constant feed solid content. This may be ascribed to the gradual agglomeration and solidification of the atomized feed droplets during their transit through the vapor phase before contacting the cryogenic liquid (Kawabata et al., 2011). However, the particle size distribution (Figure 6.5) of SFD coffee was monomodal, similar as that of spray-dried coffee. Contrarily, freeze-dried coffee showed a bimodal particle size distribution. Thus, the precise control of atomization over the uniformity of particle size is evident in the case of spray-dried and spray-freeze-dried coffees, which is not provided by grinding the frozen coffee slab at a low temperature during freeze-drying.

Further, roundness is a descriptor of particle shape at the mesoscale level (Merkus, 2009). It is a dimensionless number that quantifies the closeness of a particle's shape to a perfect circle (Cruz-Matías et al., 2019). Roundness is directly related to the solubility and wettability of food powders (Perea et al., 2009). Also, it is an important parameter during the packaging of soluble coffee. The protruding parts of non-spherical granules are more susceptible to break and form dust particles during packaging (Rumpler, Jacob, & Waskow, 2012). The values of roundness range between 0 and 1, and a value close to 1 signifies a perfectly circular particle shape. The roundness value for a soluble coffee particle granule can be calculated using the following formula, in which *P* and *A* are the perimeter and area of the particle, respectively (Cruz-Matías et al., 2019).

$$Roundness = \frac{4\pi A}{P^2} \tag{6.1}$$

SFD and SD coffee powders showed a roundness value of 0.71 and 0.8, respectively. But, the roundness value of freeze-dried coffee was less at 0.64 (Figure 6.6). The close-to-circle particle shape of spray-freeze-dried and spray-dried coffee particles can again be attributed to the role of atomization that promotes the formation of circular shape (Ishwarya & Anandharamakrishnan, 2015). The morphology of spray-freeze-dried soluble coffee particles has already been discussed in Chapter 4 and hence not included in this chapter.

6.2.1.3 Solubility

The solubility of instant coffee is considered 'good' when its solubilization time in freshly boiled water is less than 30 s (IS 2791, 1992). Spray-freeze-dried instant coffee was soluble within 11 s,

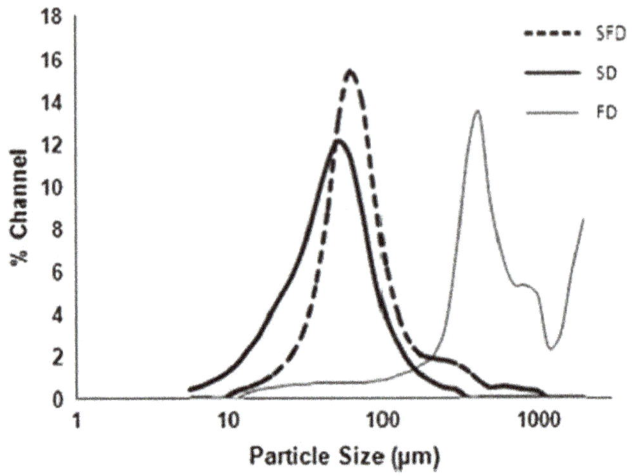

Figure 6.5 Particle size distribution – Comparison plot of spray-freeze-dried (SFD), freeze-dried (FD) and spray-dried (SD) coffee samples (Reproduced with permission from Ishwarya & Anandharamakrishnan, 2015).

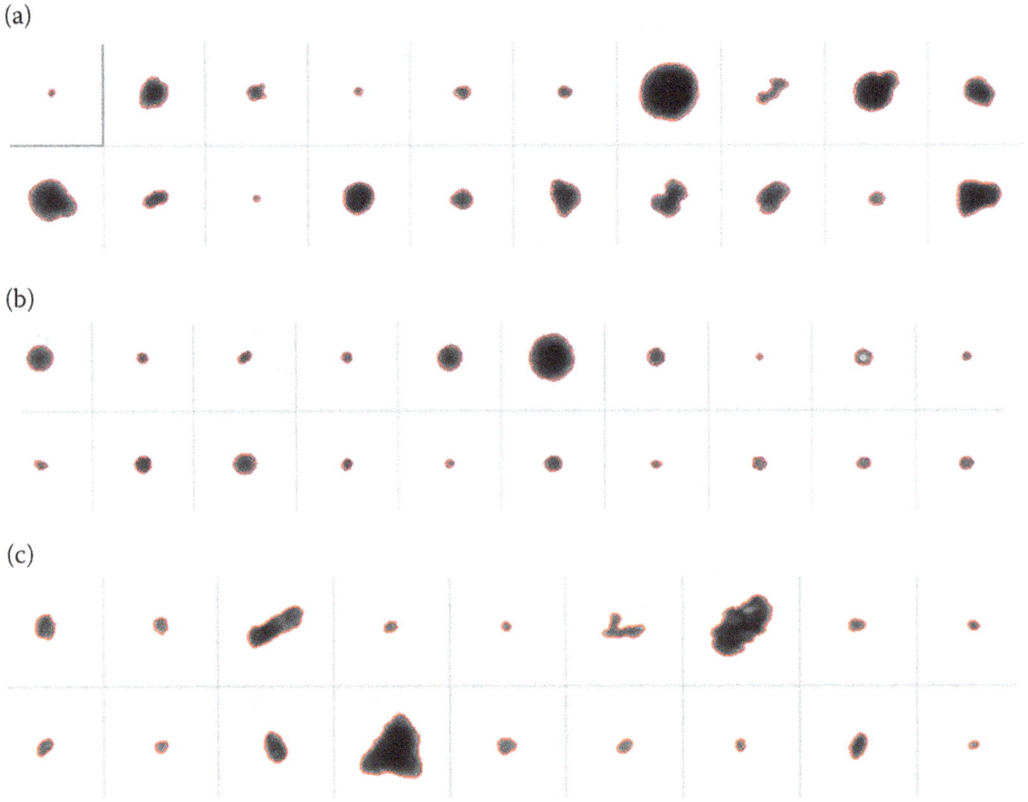

Figure 6.6 Snapshots of the particle shape of soluble coffee captured by the dynamic laser scattering based particle size analyzer: (a) spray-freeze-dried coffee; (b) spray-dried coffee; (c) freeze-dried coffee.

compared to a solubilization time of 22 s and 20 s for the freeze-dried and spray-dried coffee, respectively. Thus, the instantization effect is more prominent in the SFD soluble coffee than the SD and FD products. Solubility is a function of the particle microstructure, size and shape (Anandharamakrishnan et al., 2010; Gaiani et al., 2011). Accordingly, the porous surface microstructure of SFD soluble coffee particles (see Figure 4.4(d); Chapter 4) is responsible for its rapid reconstitution behavior. The pores on the surface act like capillary tubes that imbibe water upon rehydration (Saguy et al., 2005). Further, the reduced particle size of SFD soluble coffee compared to its freeze-dried counterpart is also a reason for its relatively high solubility. The resultant increase in specific surface area and surface energy of the particles promotes solubility (Kaptay, 2012).

6.2.1.4 Bulk density and tapped density

For any food powder, bulk density is of relevance as deviations in it would result in regulatory non-compliance, i.e., under net weight or a package that appears short of the product despite its correct net weight (Barbosa-Canovas et al., 2005). High bulk density is desirable for long-distance transportation of powdered products, as the cost for packaging and transit cost would be less (Bhandari et al., 2008). Compared to spray-drying (0.328 g/mL/0.388 g/mL) and freeze-drying (0.345 g/mL/0.361 g/mL) processes, SFD resulted in soluble coffee with higher free flow (0.612 g/mL) and tapped bulk densities (0.679 g/mL). Apart from the low-temperature operation of spray-freeze-drying process, other plausible reasons for the high density of SFD-coffee could be its higher residual moisture content, non-cohesive nature (Geldart, 1973), smaller particle size and broader particle size distribution. Smaller particle size implies reduced inter-particle voids with larger contact surface area per unit volume, thus leading to higher bulk density (Caparino et al., 2012). Besides, an inverse relationship exists between the drying temperature and the solubility and bulk density of dried powder (Hassen & Al-Kahtani, 1990).

6.2.1.5 Flowability

Flowability is an important trait of bulk powders. The term ‘flowable’ refers to the irreversible deformation of powder to cause its flow by applying external energy or force (Hadjittofis, 2018). It is one of the physical properties, the change in which can be used to judge the quality of instant coffee during storage. The flowability of instant powders is usually indicated in terms of two parameters, namely, Hausner ratio (*H*) and Carr’s index or Carr’s compressibility index (*C*). *H* is the ratio between tapped bulk density (ρ_T) and freely-settled bulk density (ρ_B) of instant coffee powder (Eq. 6.2).

$$H = \frac{\rho_T}{\rho_B} \tag{6.2}$$

Similarly, Carr’s index is calculated by the formula,

$$C = 100\frac{(\rho_T - \rho_B)}{\rho_T} \tag{6.3}$$

For free-flowing powder, the closer values of ρ_T and ρ_B result in smaller value for Carr index. Conversely, in powders with poor flowability, the delta between ρ_T and ρ_B will be high due to higher inter-particle interactions, and hence the Carr index would be larger (US Pharmacopoeia, 2006).

According to the classification of powders based on Hausner ratio and Carr index (Hayes, 1987; Turchiuli et al., 2005; Table 6.1), the flowability of SFD soluble coffee was intermediate

Table 6.1 Flow characteristics of powders (Turchiuli et al., 2005)

Carr's index	Flow character	Hausner ratio
≤10	Excellent	1.00–1.11
11–15	Good	1.12–1.18
16–20	Fair	1.19–1.25
21–25	Passable	1.26–1.34
26–31	Poor	1.35–1.45
32–37	Very poor	1.46–1.59

between spray-dried and freeze-dried coffee, with more inclination towards the former. Nevertheless, freeze-dried coffee exhibited free-flowing characteristics (Ishwarya & Anandharamakrishnan, 2015). As discussed above, the larger difference between the values of ρ_T and ρ_B (0.06 g/mL) is the reason for the reduced flowability of spray-freeze-dried coffee compared to its freeze-dried counterpart (0.016 g/mL). The difference between ρ_T and ρ_B increases with the decrease in particle size. The smaller mean particle diameter of spray-freeze-dried and spray-dried soluble coffee implies the presence of a higher amount of fines in the powder bulk. For smaller particles, the number of contact points with the neighboring particles is more, thus leading to difficulties in the rearrangement and construction of dense packing. However, when tapped, small particles in the powder bulk rolls between the particle voids and attains the densest packing condition. In a free state, large voids are formed due to the arching of particles and these voids collapse on tapping and results in a substantial difference between their free flow and tapped densities, thus increasing the values of H and C (Bodhmage, 2006). Thus, a decrease in particle size of the powder impairs powder flow (Cain, 2002). Accordingly, the flowability of soluble coffee improved with an increase in particle size in the order of SD < SFD < FD, indicated by the decrease in H value (Table 6.2). The flowability of spray-freeze-dried soluble coffee can be improved further by varying the atomization pressure to obtain an optimally larger droplet size and thereby an increased particle size.

6.2.1.6 Color

Color is a key quality parameter of the finished soluble coffee product, which determines its premium price in markets across the world (GEA, n.d.). It is the first attribute perceived by a consumer on opening the packet of an instant coffee product. While some consumers prefer dark-colored coffee, others might like a light-colored product. Particle size determines the color of instant food powders. Generally, the brightness of powder increases with the reduction in its particle size. Brightness is measured by the 'L^*' value (lightness) of the International Commission on Illumination (CIE) parameters (Sharma et al., 2013). The other two color parameters are a^* (+:

Table 6.2 Flowability of spray-freeze-dried soluble coffee vis-à-vis spray-dried and freeze-dried coffee (Adapted from Ishwarya & Anandharamakrishnan, 2015)

Product	Free bulk density (g/mL)	Tapped bulk density (g/mL)	Hausner ratio	Carr index
Spray-freeze-dried soluble coffee	0.612 ± 0.007	0.679 ± 0.008	1.11 ± 0.0001	10 ± 0.0001
Spray-dried soluble coffee	0.328 ± 0.002	0.388 ± 0.001	1.18 ± 0.009	15.5 ± 0.707
Freeze-dried soluble coffee	0.345 ± 0.006	0.361 ± 0.004	1.05 ± 0.008	4.5 ± 0.707

Table 6.3 Color parameters of soluble coffee: Spray-freeze-drying vs. freeze-drying (Adapted from Ishwarya & Anandharamakrishnan, 2015)

Product	L^*	a^*	b^*	ΔE	C
Spray-freeze-dried soluble coffee	36.95 ± 0.16	14.13 ± 0.08	28.69 ± 0.28	3.57 ± 0.324	31.98 ± 0.291
Freeze-dried soluble coffee	33.20 ± 0.60	11.2 ± 0.165	22.09 ± 0.423	4.82 ± 0.351	24.76 ± 0.439

red to −: green) and b^* (+: yellow to −: blue). SFD coffee showed the least total color difference (ΔE), compared to the color parameters of a renowned commercial brand of spray-dried soluble coffee (L^*: 35.44; a^*: 3.75; b^*: 25.47). Hence, it is evident that the SFD-coffee is closer to the color of commercial product compared to the freeze-dried coffee (Table 6.3).

The higher ΔE value of FD sample vis-à-vis commercial product can be related to its long drying time that is known to degrade product color (Contreras et al., 2008; Ozkan et al., 2005). The freezing rate determines the lightness of soluble coffee. Due to its rapid freezing rate than the conventional freeze-drying, the SFD process resulted in a brighter soluble coffee product (higher L^* value) than its FD counterpart. Generally, quick-frozen materials retain their brighter color than those frozen at a slower rate. This is because the pores originating from the sublimation of small ice crystals formed by rapid freezing scatters more light than large pores formed by slow freezing (Ceballos et al., 2012). Also, the redness, yellowness and chroma values were higher for the SFD coffee relative to the freeze-dried product (Table 6.3). Chroma (*C*) is an indicator of the intensity of product color. Generally, broader particle size distributions are responsible for the reduced color strength, as opacity or transparency of a substance is greatly influenced by the particle size (Heinrich, 2003). Thus, as mentioned in the previous section, the smaller particle size and monomodal particle size distribution of spray-freeze-dried soluble coffee compared to freeze-dried coffee are the underlying reasons for its desirable color parameters (Ishwarya & Anandharamakrishnan, 2015).

6.2.2 Aroma quality of spray-freeze-dried soluble coffee

6.2.2.1 Qualitative analysis of aroma quality by electronic nose

The aroma profile of coffee is one of the most intricate food aromas comprising more than 600–800 volatile constituents (Clarke, 2001; Schaller, Bosset & Escher, 1998). Apparently, spray-freeze-drying is expected to retain more character impact aromatics during the drying of coffee extract to result in a flavorful final product. A well-established approach to judge the aroma quality of soluble coffee is the sensory analysis or cup test by expert tasters. Nevertheless, the inferences obtained from the sensory analysis are subjective, and the correctness of quality discrimination solely depends on the skill of the tasters (Rodríguez, Durán, & Reyes 2010). Therefore, an objective technique such as the 'electronic nose' is useful in evaluating the aroma quality of instant coffee (Schaller et al., 1998). Essentially, electronic nose or e-nose is an instrument that mimics the human nose to detect and analyze gaseous mixtures to distinguish between different classes of similar odor-emitting products (Bartlett, Elliot & Barcher, 1997; Lorwongtragool, Wongchoosuk, & Kerdcharoen 2010 ; Wongchoosuk, Lutz & Kerdcharoen, 2009; Wongchoosuk et al., 2010). E-noses have been successfully used to discriminate between different coffee blends (Pardo et al., 2000; Singh, Hines & Gardner, 1996; Ulmer et al., 1997) and coffee powders produced by different drying techniques (Ishwarya & Anandharamakrishnan, 2015). The components of the e-nose instrument include a headspace sampling system, an array of chemical sensors, electronic circuitry and data analysis software (Gardner & Bartlett, 1999).

In the above context, e-nose was used for a qualitative discriminative analysis of the difference in aroma quality between the spray-freeze-dried (SFD), freeze-dried (FD) and spray-dried (SD) soluble coffee (Ishwarya & Anandharamakrishnan, 2015). Principal component analysis (PCA) - a

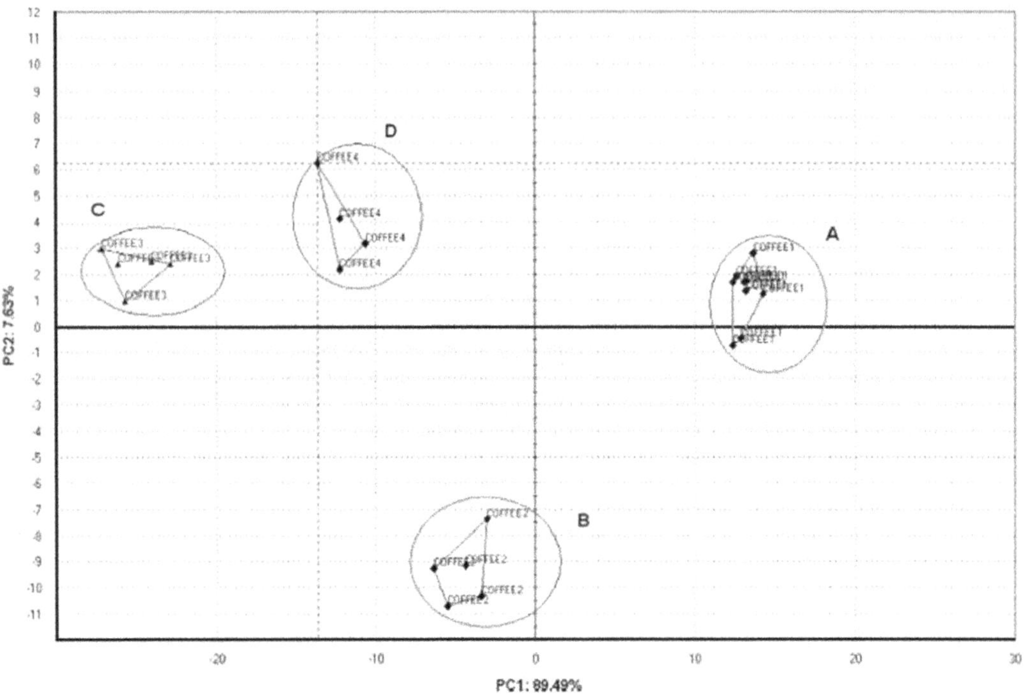

Figure 6.7 PCA plot – electronic nose analysis (A: commercial instant coffee; B: spray-dried coffee; C: freeze-dried coffee; D: spray-freeze-dried coffee) (Reproduced with permission from Ishwarya & Anandharamakrishnan, 2015).

linear method that is effective in differentiating the response of the electronic nose to simple and complex odors (Gardner, 1991) was applied to infer the data obtained from the E-nose system. Figure 6.7 illustrates the PCA two-dimensional plot generated by the e-nose system, after the headspace analysis of commercially available instant coffee powder, and the SFD, FD and SD coffee powders. From the PCA plot, it was evident that 89.49% of the variation in data was explained by PC1 and 7.63% of the variation by PC2. This discrimination implied the difference between the headspace vapors and hence the aroma of differently dried coffee products.

The response clusters of spray-freeze-dried and freeze-dried soluble coffee remained in the same quadrant, thus indicating the similarities in their odor profile. The low-temperature operation of SFD and FD processes was proposed as a reason for the similarity in the aroma profile. The difference in drying time between the SFD and FD processes might be a reason for the slight difference between the positioning of clusters of the corresponding soluble coffee products in the PCA plot. On the other hand, the spray-dried coffee formed a significantly discrete cluster relative to the other samples, which can be attributed to its high-temperature operation (Ishwarya & Anandharamakrishnan, 2015). Thus, e-nose was able to ascertain the superior aroma quality of spray-freeze-dried soluble coffee on parity with the freeze-dried coffee. The latter is well-known for its enhanced aroma properties due to the lower temperature processing.

6.2.2.2 *Quantitative profiling of aromatic volatiles in spray-freeze-dried soluble coffee by gas chromatography-mass spectroscopy (GC-MS)*

Gas chromatography-mass spectroscopy (GC–MS) is the standard method for the identification and quantification of volatile constituents in foods and beverages (Mellon, 1994). The volatile molecules eluting from the column are identified based on their fragmentation patterns, by the complementing separation and sensitivity abilities of the gas chromatography and mass spectroscopy techniques, respectively. It is possible to detect a volatile concentration of as low as 1 ppb using the GC-MS approach (Pardo & Sberveglieri, 2002).

Volatile retention in the spray-freeze-dried soluble coffee was quantified by comparing the aroma profiles of coffee extract (40%) and the final dried product, using headspace solid-phase microextraction (HS-SPME) followed by gas chromatography-mass spectrometry (GC-MS). The results were compared against the profiles of spray-dried and freeze-dried coffees. Initially, 1 g of sample was taken in a screw-capped vial with polytetrafluoroethylene (PTFE) septa and equilibrated in a thermostatic oven at 70°C for 1 hour. Then, volatiles from the coffee extract and powder samples were adsorbed onto a 60 μm polydimethylsiloxane/divinylbenzene (PDMS/DVB) Stableflex fiber. Volatile adsorption was achieved by contacting the fiber with the sample headspace for 10 min. The volatiles adsorbed by the fiber were thermally desorbed in the GC-MS injector port at 250°C for 3 min. A polyethylene glycol (PEG) polar capillary GC column was used with helium as the carrier gas. The oven temperature program is as follows:

- Holding at 40°C for 1 min;
- Heating at the rate of 3°C/min from 40°C to 150°C;
- Heating at the rate of 5°C/min from 150°C to 230°C;
- Holding at 230°C for 5 min.

The above program was formulated by considering the boiling points of the character impact volatiles formed after coffee roasting and extracted into the coffee brew. The mass spectrometer was operated at a source temperature of 180°C, and an electron energy of 70 eV. 40 to 400 was the mass range in which the mass spectra were obtained.

The mass spectra of coffee extract and dried samples thus obtained were compared with those listed in the library. Out of the twenty volatile compounds (Table 6.4) that could be identified confidently, six were chosen as marker compounds, based on their impact on the coffee aroma. The marker compounds were selected such that they are representative of the six major categories of volatile compounds in coffee: acids, alcohols, furans, ketones, pyrazine and pyridine (Fisk et al., 2012; Marin et al., 2008; Zambonin et al., 2005) and encompass both high-boiling and low-boiling volatile compounds. Accordingly, the compounds chosen were 2,3-butanedione (boiling point [bp]: 88°C), pyridine (bp: 115.4°C), acetic acid (bp: 117.9°C), methyl pyrazine (bp: 135°C), 2-furanmethanol (bp: 171°C), and maltol (bp: 284.7°C). Notably, the spray-freeze-dried soluble coffee showed the highest average retention at 93% of all the selected marker compounds (Figure 6.8). Freeze-dried and spray-dried products were found to retain 77% and 57% of the character-impact aromatics, respectively.

The loss of volatiles during spray-drying is justified due to its high-temperature operation. The boiling point of most of the character impact volatile compounds in coffee is lesser than the inlet temperature of spray-drying. This is of particular relevance during the initial stages of spray-drying, i.e., atomization and constant rate drying period, wherein the major volatile losses occur as the formation of protective dry skin, or solid crust is not complete. Besides, loss of aromatic compounds is promoted by the morphological changes that happen during the bubble inflation phenomenon, which reduces the path length of diffusion for the volatiles from within the droplet to its surface (King, 1990).

The underlying mechanisms responsible for the significantly high retention of aromatic compounds in the spray-freeze-dried soluble coffee would be explained in the subsequent sections.

6.2.2.3 Mechanism of volatile retention during spray-freeze-drying

i. ***Relative volatility***

The significantly higher retention of coffee aromatics during spray-freeze-drying relative to spray-drying and freeze-drying processes can be explained by the concept of 'relative volatility'.

Table 6.4 Volatile compounds identified from the chromatograms of soluble coffee samples (Ishwarya & Anandharamakrishnan, 2015)

S.No.	Retention time (min)	Compound	Main m/z ions observed in MS spectra
1.	2.56	2,3-Pentanedione	43, 57, 100
2.	4.414	Pyridine	79, 52
3.	4.799	Pyrazine	80, 53
4.	6.27	Methyl pyrazine	94, 67
5.	7.11	1-Hydroxy, 2-propanone	43, 44, 42
6.	8.06	2,5-Dimethyl pyrazine	108, 42
7.	8.26	1,4-Benzenediamine	108, 107, 42
8.	10.11	Methyl ethyl pyrazine	121, 122, 94
9.	10.34	2-Ethyl, 5-methyl pyrazine	121, 122, 94
10.	12.16	Acetic acid	60, 43, 45
11.	12.60	2-Furancarboxaldehyde	109, 53, 81
12.	13.71	2,3-Butanedione	43, 86
13.	14.17	1-(2-Furanyl), ethanone	95, 110, 41
14.	15.88	2-Furanmethanol acetate	98, 81, 140
15.	16.75	5-Methyl, 2-furancarboxaldehyde	109, 53
16.	18.52	Butyrolactone	42, 86, 41
17.	20.36	2-Furanmethanol	98, 97, 41
18.	26.12	3-Methyl, pyridazine-5-one	112, 41, 55
19.	30.305	Maltol	126, 94, 43
20.	31.78	1H-Pyrrole, 2-carboxaldehyde	95, 94, 66

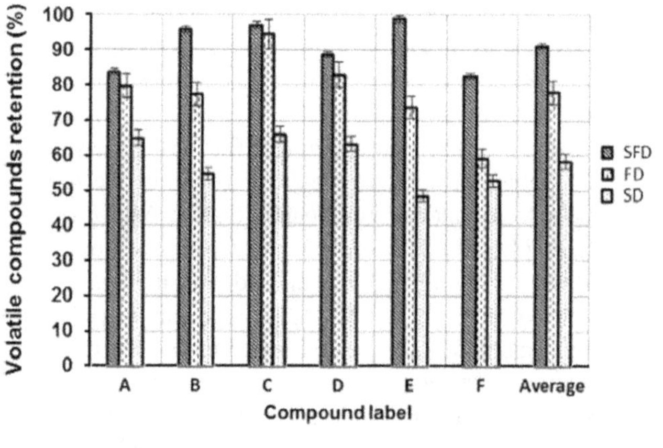

Figure 6.8 Volatile retention of selected compounds after Spray Freeze-drying (SFD), freeze-drying (FD) and spray-drying (SD) of the coffee extract (A: methyl pyrazine; B: acetic acid; C: 2,3-butanedione; D: pyridine; E: 2-furanmethanol; F: maltol) (Reproduced with permission from Ishwarya & Anandharamakrishnan, 2015).

Relative volatility is a measure of the vapor pressure of a pure compound relative to that of water at the same temperature. It controls the phase change of a volatile compound to its gaseous phase at a given temperature (Goubet et al., 1998). Higher the relative volatility, the greater would be the loss of flavor compound (Bhandari, 2005). Moreover, the relative volatility of a compound and temperature are directly related, which is governed by the Clausius-Clapeyron equation (Eq. 6.4) that gives the rate of increase in vapor pressure per unit increase in temperature.

$$\frac{dp}{dT} = \frac{L}{T\,(V_v - V_l)} \tag{6.4}$$

In the above equation, p is the saturation vapor pressure, T is the temperature, L is the latent heat of evaporation and V_v, and V_l are the specific volumes at temperature T of the vapor and liquid phases, respectively. Accordingly, the vapor pressure of any pure volatile compound will decrease non-linearly with temperature. Thus, lower the processing temperature as in SFD and FD, lesser would be the vapor pressure of volatile aromatic compounds, relative volatility and volatile loss.

For instance, the relative volatility of methyl pyrazine was found to decrease from 19.3 to 3.4 when the processing temperature was decreased from 60°C to 25°C (Clarke, 2001). Notably, methyl pyrazine was the character impact compound that was present at a relatively higher concentration in the differently-dried soluble coffee products. Considering the above rate of reduction in relative volatility at ~0.5/°C and the Clausius-Clapeyron relationship, one can appreciate the extent of the drop in relative volatility at the ultra-low temperature employed during SFD and FD. This could be the reason for the multiple-fold reduction in the relative volatility of methyl pyrazine, which led to its higher retention of 88.5% and 84.5% during SFD and FD, respectively, over 66.2% retention obtained with SD (Ishwarya & Anandharamakrishnan, 2015). The above theory is applicable to explain the higher retention of other volatiles during SFD and FD, compared to SD.

Both spray-freeze-drying and freeze-drying are low-temperature drying processes. Yet, SFD soluble coffee retained 16% more aromatic compounds than the freeze-dried coffee, which can be related to the differences in freezing temperature and drying time, as would be explained below.

ii. ***Concentration of dissolved solids in the frozen matrix***

In addition to relative volatility, another vital factor that influences volatile retention during SFD and FD processes is the percentage of dissolved solids in the freeze-concentrated matrix that encapsulates the volatiles during the freezing step. The amount of dissolved solids in the matrix increases with a reduction in the freezing temperature (Chandrasekaran, 1969). When the freezing temperature is reduced below the vitrification point of water, the solid concentration is increased further. Vitrification point is the temperature at which a substance changes to its glassy state or a non-crystalline amorphous solid form (Varshneya, 2006). The vitrification temperature of water is −146 ± 4°C, below which it exists as a brittle, glassy substance (Yannas, 1968). Thus, the temperature during the spray-freezing step of SFD corresponding to the boiling point of liquid nitrogen is −196°C, which is well below the vitrification point of water. Consequently, the concentration of dissolved solids in the frozen matrix might have been higher than that during the conventional freeze-drying process in which the freezing occurs at −40°C. Hence, the aroma components are permanently encapsulated by the dissolved solids and hence protected from losses during spray-freeze-drying.

iii. ***Drying time***

The drying time for achieving the final soluble coffee product with spray-freeze-drying was 30% less compared to freeze-drying. This is of relevance as the deceleration of the rate of aroma loss by diffusion in the frozen matrix depends on the sublimation rate and hence the total drying time (Coumans et al., 1994). Despite the above mechanisms of volatile retention, the minor volatile loss (~7%) during SFD has been ascribed to the high slip velocities in the spray during atomization, which leads to high droplet-gas mass transfer coefficients (Khwanpruk et al., 2008).

6.3 MERITS OF SPRAY-FREEZE-DRYING AS A SOLUBLE COFFEE PRODUCTION PROCESS

Compared to conventional freeze-drying, the advantages of employing spray-freeze-drying for soluble coffee production are manifold. Firstly, the handling costs are reduced, because the operation can be rendered continuous by atomizing the coffee extract into the freezing chamber that is also equipped to accomplish drying. As a result, the requirement for loading trays is eliminated. The resultant dried particles have a very fine texture, porous microstructure and rapid reconstitution properties with water. Consequently, the sensory quality of spray-freeze-dried soluble coffee is superior when compared to similar products dried by spray-drying and freeze-drying. Further, the residence time of the product (the total time during which the product is under treatment) is reduced to just a fraction of that encountered in the conventional freeze-drying process.

CONCLUSIONS

The discussions presented in this chapter ascertain the competitive edge of spray-freeze-drying over the conventional soluble coffee manufacturing processes (spray-drying and freeze-drying). Also, it is evident that the SFD process for soluble coffee production has evolved but at a slower pace due to the limited number of studies in this discipline. The spray-freeze-dried soluble coffee outweighs its conventionally processed counterparts in every aspect, i.e., superior aroma quality obtained by retaining the characteristic low-boiling aromatic compounds of coffee, porous microstructure, instantaneous solubility, good flowability, high bulk density and good packaging and transportation characteristics. Despite the research-based evidence that demonstrates the premium quality of spray-freeze-dried soluble coffee, efforts to industrialize the process have still not commenced. The monopoly of spray-drying and freeze-drying in the instant coffee manufacturing sector and the operational limitations associated with the spray-freeze-drying process (as discussed in Chapter 1) are the bottlenecks in establishing it as an industrial process for soluble coffee production. In addition, the cost per kilogram of spray-freeze-dried soluble coffee depends on the drying temperature, which in turn depends on the concentration of coffee extract and its corresponding critical temperatures (eutectic point (T_e), glass transition temperature (T_g), collapse temperature (T_c), as discussed in Chapter 3). The cost increases drastically with the increase in solid content of coffee extract, as the highly freeze-concentrated matrix will have a low critical temperature and hence would demand an ultra-low drying temperature (Ishwarya et al., 2015). Thus, addressing these challenges would be the future responsibility and scope of research for coffee manufacturing specialists. If the efforts materialize, soon we can expect to purchase a retail package of premium-quality spray-freeze-dried soluble coffee from a hypermarket and relish its rich aroma and flavor.

REFERENCES

Abdul-Fattah, A. M., Truong, V. L. (2010). Drying process methods for biopharmaceutical products: an overview. In: Jameel, F., Hershenson, S. (Eds.), *Formulation and Process Development Strategies for Manufacturing Biopharmaceuticals*. John Wiley & Sons, Inc., Hoboken, NJ, pp. 705–737.

Al-Hakim, K., Wigley, G., & Stapley, A. G. F. (2006). Phase Doppler anemometry studies of spray freezing. Trans. ChemE, Part A, *Chemical Engineering Research and Design*, 84, 1142–1151.

Anandharamakrishnan, C., Rielly, C. D., & Stapley, A. G. F. (2010). Spray-freeze-drying of whey proteins at sub-atmospheric pressures. *Dairy Science and Technology*, 90, 321–334.

Barbosa-Canovas, G. V., Ortega-Rivas, E., Juliano, P., & Yan, H. (2005). Bulk properties. In: *Food Powders: Physical Properties, Processing, and Functionality*. Kluwer Academic/Plenum Publishers, New York, pp. 55–88.

Bartlett, P. N., Elliot, T. M., & Barcher, J. W. (1997). Electronic noses and their application in the food industry. *Food Technology*, 51, 44–48.

Bassoli, D. G., Sumi, A. P., Akashi, Y., Uchida, H., De Castro, A. S., Ohtani, N., Obayashi, T., Nakayama, M., Shigekane, A., Tamura, Y., Tomita, M., Takahashi, M., & Narui, M. (1993). Instant coffee with natural aroma by spray-drying. *Technologie*, 712–718.

Bhandari, B. (2005). Spray drying – an encapsulation technique for food flavors. In: A. S. Mujumdar (Ed.), *Drying of Products of Biological Origin*. Science Publishers, Enfield, USA.

Bhandari, B. R., Patel, K. C., & Chen, X. D. (2008). Spray drying of food materials – process and product characteristics. In: Chen, X. D., Mujumdar, A. S. (Eds.), *Drying Technologies in Food Processing*. Blackwell Publishing, United Kingdom, pp. 113–157.

Bodhmage, A. (2006). *Correlation between physical properties and flowability indicators for fine powders*. PhD Thesis. Department of Chemical Engineering, University of Saskatchewan, Saskatchewan.

Burmester, K., Pietsch, A., & Eggers, R. (2011). A basic investigation on instant coffee production by vacuum belt drying. *Procedia Food Science*, 1, 1344–1352.

Cain, J. (2002). An alternative technique for determining ANSI/CEMA standard 550 flowability ratings for granular materials. *Powder Hand. Proc.*, 14, 218–220.

Caparino, A., Tang, J., Nindo, C. I., Sablani, S. S., Powers, J. R., Fellman, J. K., 2012. Effect of drying methods on the physical properties and microstructures of mango (Philippine 'Carbao' var.) powder. *Journal of Food Engineering*, 111, 135–148.

Ceballos, A. M., Giraldo, G. I., Orrego, C. E. (2012). Effect of freezing rate on quality parameters of freeze dried soursop fruit pulp. *Journal of Food Engineering*, 111, 360–365.

Chandrasekaran, S. K. (1969). *Volatiles retention during drying of food ingredients*. PhD thesis. University of California, Berkeley.

Clarke, R. J. (2001). Technology III. Instant coffee. In: Clarke, R. J., Vitzthum, O. G. (Eds.), *Coffee Recent Developments*. Blackwell Science, London, pp. 125–137 (Chapter 6).

Clarke, R. J. (2003). Coffee: Instant. In: B. Caballero, L. C. Trugo, & P. Finglas (Eds.), *Encyclopedia of Food Sciences and Nutrition*, 2nd Edition, Elsevier Science, Netherlands, pp. 1493–1498.

Contreras, C., Martín-Esparza, M. E., Chiralt, A., Martínez-Navarrete, N. 2008. Influence of microwave application on convective drying: effect on drying kinetics, and optical mechanical properties of apple and strawberry. *Journal of Food Engineering*, 88, 55–64.

Costantino, H. R., Firouzabadian, L., Hogeland, K., Wu, C. C., Beganski, C., Carrasquillo, K. G., Cordova, M., Griebenow, K., Zale, S. E., & Tracy, M. A. (2000). Protein spray freeze drying. Effect of atomization conditions on particle size and stability. *Pharmaceutical Research*, 17, 1374–1383.

Coumans, J. W., Piet, J. A. M., Kerkhof, & Bruin, S. (1994). Theoretical and practical aspects of aroma retention in spray drying and freeze drying. *Drying Technology*, 12, 99–149.

Cruz-Matías, I., Ayala, D., Hiller, D., Gutsch, S., Zacharias, M., Estradé, S., & Peiró, F. (2019). Sphericity and roundness computation for particles using the extreme vertices model. *Journal of Computational Science*, 30, 28–40.

Elerath, B. E. (1969). Method of freeze-drying coffee. United States Patent US3443962A.

Elerath, B. E. (1976). *Process for freezing coffee extract prior to lyophilization*. United States Patent US3961424.

Elerath, B. E., Lakes, M. N. J., & Esra, P. (1968). Process for continuously freezing coffee extract. United States Patent US3373042A.

Fisk, I. D., Kettle, A., Hofmeister, S., Virdie, A., & Kenny, J. S. (2012). Discrimination of roast and ground coffee aroma. *Flavour*, 1–14.

Flink, J. (1975). Freeze Drying and Adv. In: S. A. Goldblith et al. (Eds.), *Food Tech*. Academic Press.

Gaiani, C., Boyanova, P., Hussain, R., Murrieta Pazos, I., Karam, M. C., Burgain, J., & Scher, J. (2011). Morphological descriptors and colour as a tool to better understand rehydration properties of dairy powders. *International Dairy Journal*, 21, 462–469.

Gardner J., & Bartlett, P. (1999). *Electronic Noses*. Oxford University Press, Oxford, U.K..

Gardner, J. W. (1991). Detection of vapours and odours from a multisensory array using pattern recognition Part I. Principal component and colour analysis. *Sensors and Actuators B: Chemical*, 4, 109–115.

GEA (*n.d.*). Instant Coffee, GEA Niro process technologies for the instant coffee industry. Available from: https://www.gea.com/en/binaries/GEA%20Process%20Technology%20for%20Instant%20Coffee_tcm11–23912.pdf (Accessed 14 August 2021), GEA Process Engineering A/S, Denmark.

Geldart, D. 1973. Types of gas fluidization. *Powder Technology*, 7, 285–292.

Ghosh, P. & Venkatachalapathy, N. (2014). Processing and drying of coffee – A review. *International Journal of Engineering Research & Technology*, 3, 784–794.

Goubet, I., Le Quere, J. L., & Voilley, A. J. (1998). Retention of aroma compounds by carbohydrates: influence of their physiochemical characteristics and of their physical state – a review. *Journal of Agricultural and Food Chemistry*, 46, 1981–1990.

Hadjittofis, E. (2018). Interfacial Phenomena in Pharmaceutical Process Development. *Ph.D. Thesis*, Submitted to Imperial College London South Kensington Campus SW7 2AZ London, United Kingdom, October 2018.

Hair, E. R. & Strang, D. A. (1969). *Method of freeze drying coffee extracts*. US 3,486,907.

Hassen, A., & Al-Kahtani, B. H. H. (1990). Spray drying of roselle (*Hibiscus sabdariffa* L.) extract. *Journal of Food Science*, 55, 1073–1076.

Hayes, G. D. (1987). Food Engineering Data Handbook. John Wiley & Sons, New York, USA.

Heinrich, Z. (2003). *Colour chemistry: syntheses, properties and applications of organic dyes and pigments* (Third revised edition). VHCA, Verlag, Helvetica ChimicaActa AG, Zurich (Switzerland) and Wiley-VCH GmbH & Co. KGaA, Weinheim (Federal Republic of Germany).

Herrera, J. C., & Lambot, C. (2017). The coffee tree-genetic diversity and origin. In: Folmer, B. (Ed.), *The Craft and Science of Coffee*. Academic Press, Elsevier Inc., London, pp. 1–14.

Huste, A. (1974). Process for Producing High Quality Spray Dried Coffee. United States Patent US3798342A.

IS 2791 (1992) (Reaffirmed 2009). Soluble coffee powder [FAD 6: Stimulant Foods] (Third revision). Bureau of Indian Standards, New Delhi.

Ishwarya, S. P., & Anandharamakrishnan, C. (2015). Spray-Freeze-Drying approach for soluble coffee processing and its effect on quality characteristics. *Journal of Food Engineering*, 149, 171–180.

Ishwarya, S. P., Anandharamakrishnan, C., & Stapley, A. G. F. (2015). Spray-freeze-drying: a novel process for the drying of foods and bioproducts. *Trends in Food Science & Technology*, 41, 161–181.

Ishwarya, S. P., Anandharamakrishnan, C., & Stapley, A. G. F. (2017). Spray-freeze-drying of dairy products. In: Anandharamakrishnan, C. (Ed.), *Handbook of Drying for Dairy Products*. John Wiley & Sons, Oxford, pp. 123–148.

Kaptay, G. (2012). On the size and shape dependence of the solubility of nanoparticles in solutions. Int. J. Pharm., 430, 253–257.

Kawabata, Y., Wada, K., Nakatani, M., Yamada, S., & Onoue, S. (2011). Formulation design for poorly water-soluble drugs based on biopharmaceutics classification system: basic approaches and practical applications. *International Journal of Pharmaceutics*, 420, 1–10.

Khwanpruk, K., Anandharamakrishnan, C., Rielly, C. D., & Stapley, A. G. F. (2008). Volatiles retention during the sub-atmospheric spray freeze drying of coffee and maltodextrin. In: Proceedings of the *16th International Drying Symposium (IDS2008)*, 9–12 November 2008, Hyderabad, India, pp. 1066–1072.

King, C. J. (1990). Spray drying food liquids and the retention of volatiles. *Chemical Engineering Progress*, 86, 33–39.

Koç, B., & Kaymak, F.-E. (2014). The effect of spray drying processing conditions on physical properties of spray dried maltodextrin. *Baltic Conference on Food Science and Technology*, 13, 243–247.

Kudra, T., & Strumillo, C. (Eds.) (1998). *Thermal Processing of Biomaterials*. Gordon and Breach Science Publishers, Amsterdam.

Lorwongtragool, P., Wongchoosuk, C., & Kerdcharoen, T. (2010). Portable artificial nose system for assessing air quality in swine buildings. In: *Proceedings of ECTI-CON2010 7th International Conference of Electrical Engineering/Electronics, Computer*, Telecommunications and Information Technology Association, pp. 532–535, ISBN 978-1-4244–5606-2, Chaing Mai, Thailand, May 19–21, 2010.

MacLeod, C. S., McKittrick, J. A., Hindmarsh, J. P., Johns, M. L., & Wilson, D. I. (2006). Fundamentals of spray freezing of instant coffee. *Journal of Food Engineering*, 74, 451–461.

Marin, K., Pozrl, T., Zlatic, E., & Plestenjak, A. (2008). A new aroma index to determine the aroma quality of roasted and ground coffee during storage. *Food Technology and Biotechnology*, 46, 442–447.

Mellon, F. (1994). Mass spectroscopy. In: R. Wilson (Ed.), *Spectroscopic techniques for food analysis*, VCH.

Merkus H. G. (2009). Particle Size, Size Distributions and Shape. In: *Particle Size Measurements. Particle Technology Series*, Volume 17. Springer, Dordrecht.

Morin, J.-F., Jamin, E., Guyader, S., & Thomas, F. (2018). Coffee (Chapter 17). In: J.-F. Morin & Lees, M. (Eds.), *Food integrity handbook: a guide to food authenticity issues and analytical solutions*, EU-funded FoodIntegrity project, pp. 295–314.

Mumenthaler, M., & Leuenberger, H. (1991). Atmospheric spray freeze-drying: a suitable alternative in freeze drying technology. *International Journal of Pharmaceutics*, 72, 97–110.

Othman, N. T. A., & Razali, M. E. F. M. (2019). Drying of instant coffee in a spray dryer. *Jurnal Kejuruteraan*, 31, 295–301.

Ozkan, A. I., Akbudak, B., & Akbudak, N. (2005). Microwave drying characteristics of spinach. *Journal of Food Engineering*, 78, 577–583.

Pardo, M., Niederjaufner, G., Benussi, G., Comini, E., Faglia, G., Sberveglieri, G., Holmberg, M., & Lundstrom, I. (2000). Data preprocessing enhances the classification of different brands of espresso coffee with an electronic nose. *Sensors and Actuators B: Chemical*, 69, 2000.

Pardo, M., & Sberveglieri, G. (2002). Coffee analysis with an electronic nose. *IEEE Transactions on Instrumentation and Measurement*, 51, 1334–1339.

Perea, M. J., Arzate, I., Terres, E., Alamilla, L., Calderon, G., & Guttierrez, G. (2009). Morphological characterization of powder milk and their relationship with rehydration properties. In: Proceedings of the *5th CIGR Section IV International Symposium on Food Processing, Monitoring Technology in Bioprocesses and Food Quality Management*, 31st August–2nd September, 2009, Potsdam, Germany.

Ponzoni, G. B., & Nutley, N. J. (1966). Soluble Coffee Process. US Patent 3261689A.

Rodríguez, J., Durán, C., & Reyes, A. (2010). Electronic nose for quality control of Colombian coffee through the detection of defects in "Cup Tests". *Sensors (Basel, Switzerland)*, 10, 36–46.

Rumpler, K., Jacob, M., Waskow, M. (2012). Method for production of enzyme granules and enzyme granules produced thus. United States Patent US20070093403A1.

Saguy, S. I., Marabi, A., & Wallach, R. (2005). Liquid imbibitions during rehydration of dry porous foods. *Innovative Food Science and Emerging Technologies*, 6, 37–43.

Schaller, E., Bosset, J. O., & Escher, F. (1998). 'Electronic noses' and their application to food. *LWT*, 31, 305–316.

Sharma, M., Kadam, D. M., Chadha, S., Wilson, R. A., Gupta, R. K., 2013. Influence of particle size on physical and sensory attributes of mango pulp powder. *International Agrophysics*, 27, 323–328.

Singh, S., Hines, E., & Gardner, J. (1996). Fuzzy neural computing of coffee and tainted-water data from an electronic nose. *Sens. Actuators B*, 30, 185–190.

Stapley, A. (2008). Freeze drying. In: Evans, J. (Ed.), *Frozen Food Science and Technology*. Blackwell Publishing, Oxford, pp. 248–275.

Surasarang, S. H., & Williams III, R. O. (2016). Pharmaceutical cryogenic technologies. In: Williams, R. O. III, Watts, A. B., Miller, D. A. (Eds.), *Formulating Poorly Water Soluble Drugs*. American Association of Pharmaceutical Scientists, Springer, Switzerland, pp. 527–608.

Suwelack, W. & Kunke, D. (2002). *Process for freeze drying coffee extract*. US 6,428,833 B1.

Taylor, A. H., 1983. Encapsulation system and their applications in the flavor industry. *Food Flavour Ind.*, 768–772.

Thuse, E., Ginnette, L. F., & Derby, R. (1968). *Spray freeze drying system*. United States Patent US3362835.

Turchiuli, C., Fuchs, M., Bohin, M., Cuvelier, M. E., Ordonnaud, C., & Peyrat-Maillard, M. N. (2005). Oil encapsulation by spray drying and fluidized bed agglomeration. *Innovative Food Science and Emerging Technologies*, 6, 29–35.

Ulmer, H., Mitrovics, J., Noetzel, G., Weimar, U., & Gopel, W. (20006). Odours and flavours identified with hybrid modular sensor systems. *Sensors and Actuators B*, 43, 24–33.

US Pharmacopoeia (2006). US Pharmacopoeia XXIX In: *Proceedings of the US Pharmacopoeial Convention*, Physical test <1174> Powder flow, p. 3017.

Varshneya, A. K. (2006). *Fundamentals of inorganic glasses*, 2nd Edition, Society of Glass Technology, Sheffield.

Wongchoosuk, C., Lutz, M., & Kerdcharoen, T. (2009). Detection and classification of human body odor using an electronic nose. *Sensors*, 9, 7234–7249.

Wongchoosuk, C., Wisitsoraat, A., Tuantranont, A., & Kerdcharoen, T. (2010). Portable electronic nose based on carbon nanotube-SnO_2 gas sensors: feature extraction techniques and its application for detection of methanol contamination in whiskeys. *Sensors and Actuators B*, 147, 392–399.

Yannas, I. (1968). Vitrification temperature of water. *Science*, 160, 298–299.

Zambonin, C. G., Balest, L., De Benedetto, G. E., & Palmisano, F. (2005). Solid-phase microextraction–gas chromatography mass spectrometry and multivariate analysis for the characterization of roasted coffees. *Talanta*, 66, 261–265.

CHAPTER 7

Spray-freeze-drying for the Encapsulation of Food Ingredients and Biologicals

The scope of spray-freeze-drying is beyond just drying. Encapsulation is an auxiliary application of spray-freeze-drying, which gained momentum in the food and biopharmaceutical sectors in the recent years. Spray-freeze-drying (SFD) is an apt addition to the list of existing encapsulation techniques such as spray drying, freeze-drying, emulsification, extrusion and fluidized bed coating. High-temperature operation and prolonged processing times are the critical bottlenecks that affect the encapsulation efficiency obtainable from spray drying and freeze-drying, respectively. The main purpose of encapsulation is to protect sensitive bioactive ingredients from the adverse processing and environmental conditions. Hence, SFD's low-temperature operation, high cooling rates and unique microstructure of products justify it as an effective encapsulation process (Anandharamakrishnan & Ishwarya, 2015). Studies have established the proficiency of SFD process for the encapsulation of various active ingredients that fall under the categories of food ingredients and supplements, nutraceuticals and drugs (Figure 7.1). This chapter intends to introduce to the readers, the facet of spray-freeze-drying as an encapsulation technique. The aspects emphasized in this chapter are the principle of encapsulation by spray-freeze-drying and the factors influencing encapsulation efficiency, core stability and release from the polymeric matrix of SFD-encapsulates. The given elements are explained with pertinent case-studies from the literature.

7.1 A PRELUDE TO ENCAPSULATION

Encapsulation is a process to entrap an active compound within a stable, protective substance to produce encapsulates of varied size and functional properties. While the encased constituent is termed as the *core*, the carrier material is known as the *wall*. The core can either be a food component (flavors, specialty lipids, vitamins and nutraceuticals) or an active pharmaceutical ingredient (API). Likewise, the wall material can belong to any of the following category of biological macromolecules: carbohydrates (starch, cellulose derivatives and gums), proteins (whey protein, casein, gelatin, gluten and soy protein isolate) or lipids (fatty acids, glycerol, waxes and phospholipids). The final product is referred to as the *encapsulate*. Depending on their size, the encapsulates can be classified as macrocapsules (>5000 μm), microcapsules (0.2–5000 μm) and nanocapsules (2000 Å – 0.2 μm) (Anandharamakrishnan & Ishwarya, 2015). Generally, smaller size and larger surface area of encapsulates are desirable for the improved dissolution, sustained release and enhanced intestinal absorption of the encapsulated bioactive substance.

DOI: 10.1201/9781003019312-7

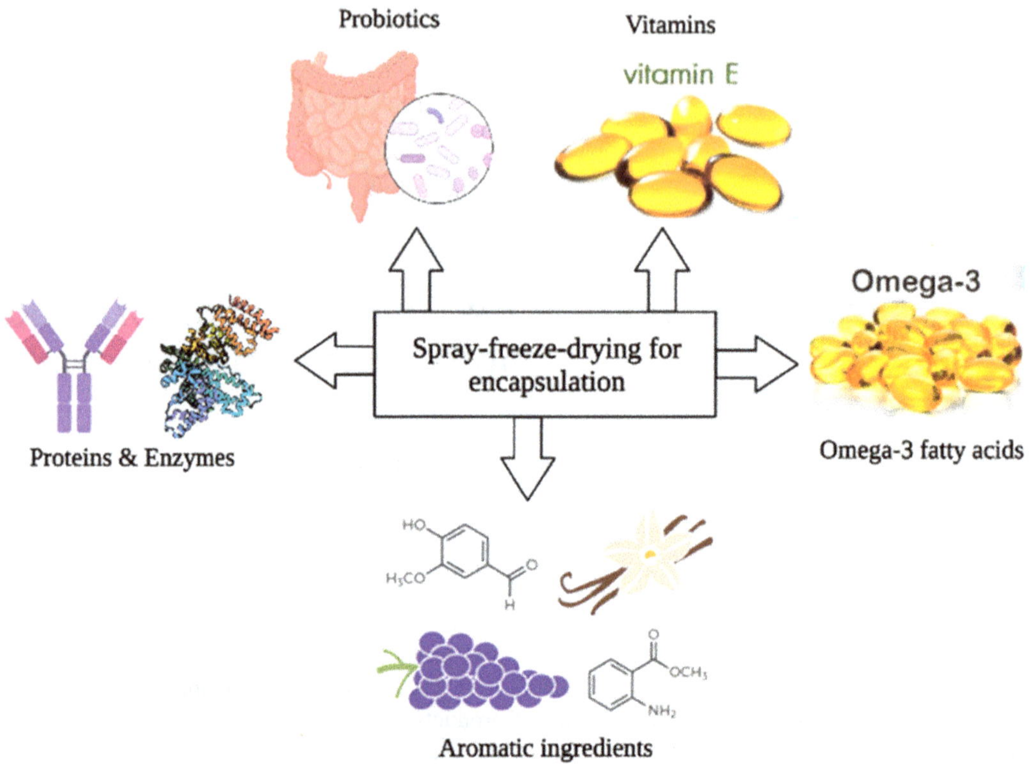

Figure 7.1 Spray-freeze-drying for encapsulation: The avenues of application (inset figures: reproduced with permission from https://biorender.com/; Sepidarkish et al., 2019; Brunning, 2021).

7.1.1 Classification of encapsulation techniques

Based on the method by which core is entrapped within the wall material, an encapsulation technique can be categorized as mechanical/physical or chemical (Table 7.1). In other words, encapsulation can be achieved by a top-down or bottom-up approach (Ezhilarasi et al., 2013) (Figure 7.2). A top-down approach involves precise tools to achieve size reduction and structuring of encapsulates. Conversely, in the bottom-up approach, encapsulates are formed by self-assembly and self-organization of molecules, influenced by factors such as pH, temperature, concentration and ionic strength (Augustin & Sanguansri, 2009; Sanguansri & Augustin, 2006; Mishra et al., 2010).

7.2 PRINCIPLE OF ENCAPSULATION BY SPRAY-FREEZE-DRYING

Spray-freeze-drying is a mechanical or top-down approach of encapsulation. The atomization step of SFD brings about the size reduction of an emulsion or suspension into finely divided droplets. The emulsion or suspension contains the core compound emulsified or dispersed in a wall material solution. This results in a large air-liquid interfacial area, which promotes rapid freezing on contact with a cryogenic liquid at ultra-low temperature. The solidification and structuring of encapsulates is achieved during the subsequent freezing and freeze-drying steps. During freezing, the solvent (which is often water in food products) is solidified into tiny ice crystals and the core component is concentrated in the wall matrix. The size of ice crystals is related to the freezing rate,

Table 7.1 Classification of encapsulation techniques (Anandharamakrishnan & Ishwarya, 2015)

Mechanical/Physical Processes	Chemical Processes
Spray drying	Simple coacervation
Freeze-drying	Complex coacervation
Spray-freeze-drying	Ionotropic gelation
Spray cooling	Interfacial polymerization
Spray chilling	Solvent evaporation
Centrifugal suspension preparation	Liposome entrapment
Co-crystallization	Inclusion complexation
Emulsification	Solvent exchange method
Fluidized bed coating	
Centrifugal extrusion	
Spinning disk	
Pressure extrusion	
Hot-melt extrusion	
Electrospraying or electrohydrodynamic technique	

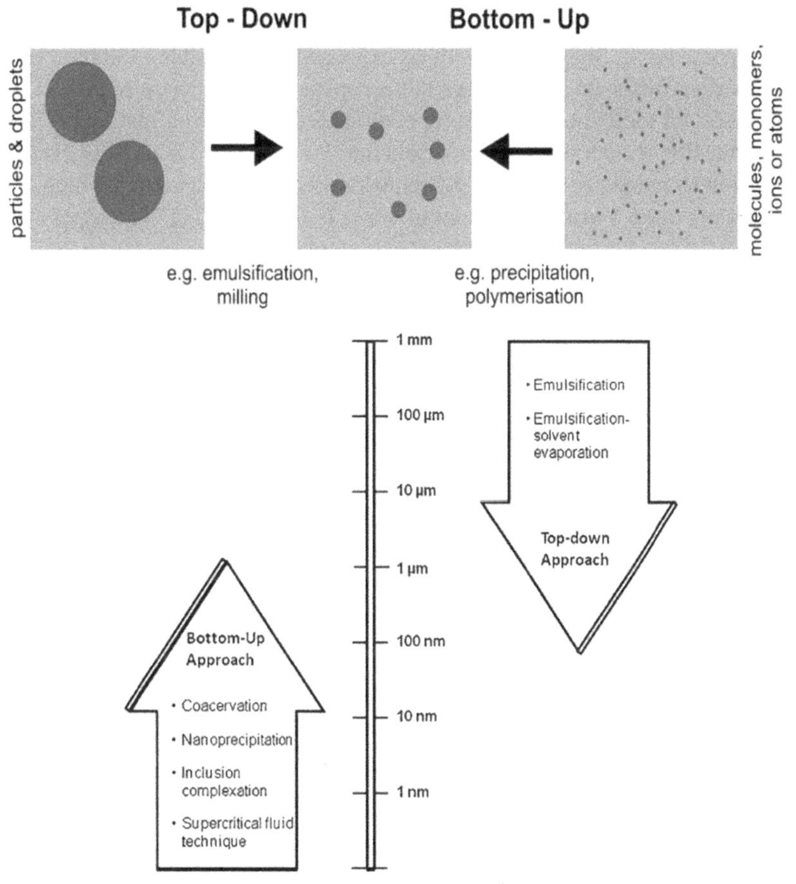

Figure 7.2 Schematics of the top-down and bottom-up approaches of encapsulation and the size-scale of the resultant encapsulates (Ezhilarasi et al., 2013; Velikov & Pelan, 2008).

with rapid freezing resulting in fine crystals. The dimensions of ice crystals is crucial in determining the final particle size of the spray-freeze-dried encapsulates. Thus, during freezing, the active core is permanently encapsulated by the dissolved solids in the freeze-concentrated wall matrix and hence protected from the processing conditions (Coumans et al., 1994). Then, in the course of freeze-drying, the polymeric wall material dries rapidly and encloses the core, owing to its relatively rapid drying rate than the latter. Further, the porous microstructure of encapsulates is created by the sublimation of ice crystals during the primary stage of the freeze-drying step (Anandharamakrishnan & Ishwarya, 2015) (Figure 7.3).

7.3 FACTORS INFLUENCING ENCAPSULATION BY SPRAY-FREEZE-DRYING

There are several factors that influence the process of encapsulation by spray-freeze-drying at each of its stage (Figure 7.4). During atomization, feed formulation is an influential factor in overcoming the tendency of droplets to aggregate owing to their higher surface energy. Droplet aggregation leads to core instability. More particularly, when the core is a proteinaceous substance, aggregation at droplet-air interfaces leads to denaturation, which inactivates and prevents it from exerting its intended function at the target site. Inclusion of freezing adjuvants in the feed formulation aids in preventing particle agglomeration and denaturation. Selection of appropriate wall materials and preparing an emulsion of the feed constituents (core, wall, emulsifier and excipient) rather than a simple dispersion before atomization also lead to improvement of core stability. The influence of feed composition on encapsulation efficiency during SFD process will be explained with suitable examples in later sections.

Another important factor that determines the encapsulation efficiency of core ingredient is the proportion of dissolved solids in the freeze-concentrated matrix of wall material. This increases in a drop with the freezing temperature (Chandrasekaran, 1969). The edge of SFD over freeze-drying as an encapsulation technique is the ultra-low temperature of spray-freezing, which is less than the vitrification point of water. Vitrification point of water (–146 ± 4°C) is the temperature at which it

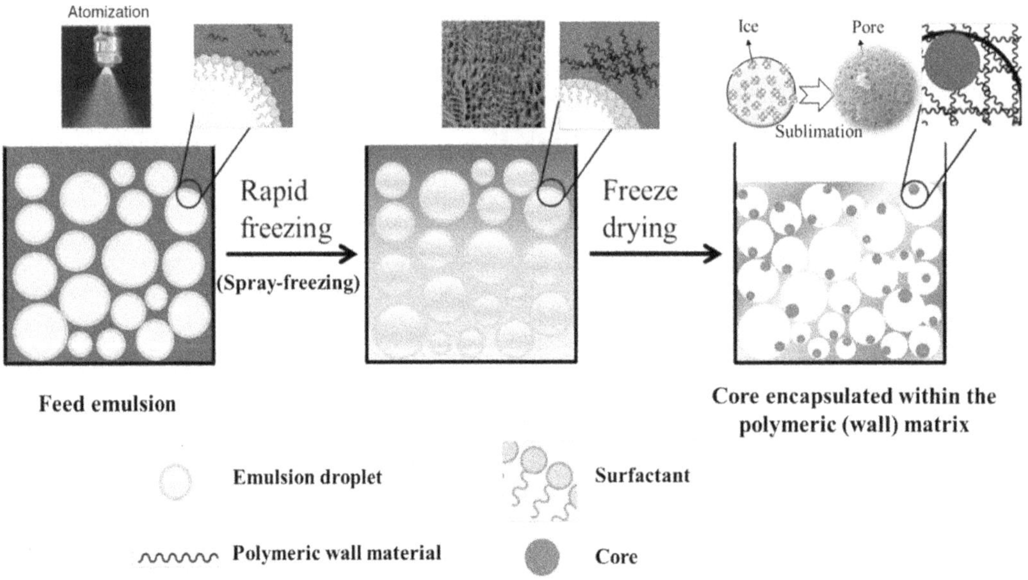

Figure 7.3 Principle of encapsulation by spray-freeze-drying (modified from Hundre et al., 2015; Ishwarya et al., 2017; Wais et al., 2016).

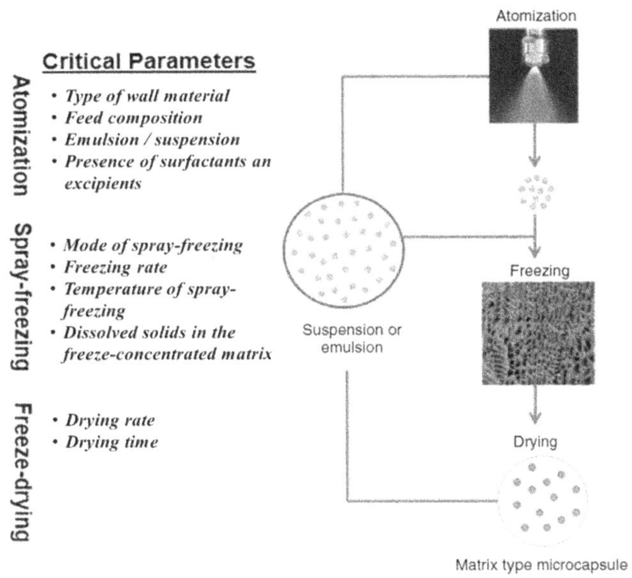

Figure 7.4 Factors influencing encapsulation by spray-freeze-drying (modified from Anandharamakrishnan & Ishwarya, 2015).

prevails in the form of a brittle glassy substance (Yannas, 1968). Freezing temperature below the vitrification point of water promotes increase in the concentration of dissolved solids. In most SFD trials, the temperature of spray-freezing corresponds to that of the cryogen, i.e., liquid nitrogen, the boiling point of which is −196°C. This is well below the vitrification point of water to facilitate the increase in concentration of dissolved solids after freezing. The synergistic effect of higher solid concentration and lower sublimation temperature can possibly lead to the early onset of selective diffusion phenomenon (Thijssen & Rulkens, 1968) and ultimately greater encapsulation efficiency.

7.4 SELECTION OF WALL MATERIALS FOR ENCAPSULATION BY SPRAY-FREEZE-DRYING

The precise choice of wall material is central to any encapsulation technique and SFD is no exception to this thumb rule. The nature of carrier substance determines the stability of feed emulsion or dispersion before drying, and controls the flowability and mechanical stability of encapsulates after drying (Anandharamakrishnan & Ishwarya, 2015). Following a systematic approach to choose the appropriate wall material for a particular core ingredient will lead to considerable saving of time, energy, efforts and material, all of which would eventually improve the process economics. The five major properties that govern the wall material selection for encapsulation by SFD are its *solubility, emulsifiability, film-forming ability, viscosity* and *collapse temperature*, explained as follows.

7.4.1 Solubility

Encapsulation by spray-freeze-drying often deals with aqueous feed formulations including oil-in-water emulsions. Hence, the primary criterion for wall material selection is its solubility in water (Gouin, 2004), which is also vital in deciding the trigger mechanism for core release

Table 7.2 Commonly used wall materials for encapsulation by spray-freeze-drying

Core	Wall Material	Reference
Docosahexaenoic acid (DHA)	Whey protein isolate (WPI)	Karthik & Anandharamakrishnan, 2013
Vanillin	WPI	Hundre et al., 2015
Lactobacillus plantarum (Probiotics)	WPI	Dolly et al., 2011
Lactobacillus plantarum	• WPI + Sodium alginate (SA) • WPI + Fructooligosaccharide (FOS) • Denatured WPI (DWPI) + SA • DWPI + FOS	Rajam & Anandharamakrishnan, 2015
Bromelain (Aerosol preparation)	Maltodextrin	Lavanya et al., 2020
Small interfering RNA (siRNA)	Mannitol	Liang et al., 2018

(Anandharamakrishnan & Ishwarya, 2015). For instance, maltodextrin is a preferred wall material for drying-based encapsulation techniques, mainly for its high solubility in water (Table 7.2).

7.4.2 Emulsifiability

Emulsifiability of a wall material is the maximum amount of core that it can hold per unit weight to form a stable emulsion. This property of the wall material becomes crucial when the core is a hydrophobic substance such as essential oil or specialty lipid, wherein the wall functions to stabilize the core before encapsulating it upon drying (Hogan et al., 2001). Besides, an emulsifiable wall material promotes the attainment of smaller droplet size during the emulsification step, which enhances the emulsion stability, encapsulation efficiency and eventually the structural stability of encapsulates. In the given context, the superior emulsifying ability of whey protein (concentrate or isolate) qualifies it as an ideal choice of wall material for the SFD encapsulation of hydrophobic cores such as lipids, fat-soluble vitamins and flavorings (Anandharamakrishnan & Ishwarya, 2015). Accordingly, whey protein isolate has been used as wall material for the SFD encapsulation of docosahexaenoic acid (DHA) (Karthik & Anandharamakrishnan, 2013) and vanillin (Hundre, Karthik & Anandharamakrishnan, 2015) (Table 7.2).

7.4.3 Film-forming ability

Film-forming involves the quick formation of a thin and dense layer of wall material around the atomized droplets containing the core, as moisture is removed during the freeze-drying step of SFD. This is relevant, as one of the main functions of a wall material is to form a protective barrier around the core against oxygen diffusion during the storage of encapsulates. A wall material with ideal film-forming ability is the one which shows a drying curve with rapidly declining drying rate with reduction in moisture content. Higher the drying rate of a wall material to form a denser protective membrane around the core, better will be the encapsulation efficiency and stability of the encapsulated core. Also, wall materials which dry out fast offer better protection to volatile core compounds during the initial phase of drying. Among the different wall materials used for SFD-based encapsulation, whey protein isolate shows smooth, skin-forming behavior owing to its major constituent, i.e., β-lactoglobulin that exhibits excellent emulsifying properties (Jouenne & Crouzet, 2000). Other wall materials with superior film-forming ability include gum Arabic, high-amylose starch, chitosan and pullulan (Anandharamakrishnan & Ishwarya, 2015).

7.4.4 Viscosity

A wall material which presents low viscosity at high solid concentration of its aqueous suspension is favorable in terms of ease of atomization and crust formation to protect the core. Moreover, viscosity of feed emulsion or suspension is directly related to the droplet size after atomization and eventually the particle size of encapsulate. Thus, higher viscosity would reduce the specific surface area of encapsulates and thereby affect the delivery of its intended functionality. This is especially relevant in the encapsulation of poorly water soluble drugs and bioactives with limited bioavailability, where larger specific surface area facilitates increased intestinal absorption and controlled release in the intended target site of action. Further, low viscosity facilitates Newtonian flow behavior of feed and mitigates the need for a pumping device, so that encapsulation by spray-freeze-drying can be conveniently operated as a continuous process without the need for differential torques. Maltodextrin is known for its ability to form a low-viscosity solution at high solid content. Thus, it is suitable for the encapsulation of hydrophilic core substances (Anandharamakrishnan & Ishwarya, 2015). When used in combination with other wall materials that exhibit pseudoplastic flow behavior, maltodextrin can reduce the viscosity though it is not possible to modify the non-Newtonian rheology (Prata et al., 2013).

7.5 PERTINENT CASE-STUDIES ON ENCAPSULATION BY SPRAY-FREEZE-DRYING

7.5.1 Encapsulation of food ingredients

7.5.1.1 Docosahexaenoic acid

Docosahexaenoic acid (DHA; 22:6n–3) is an omega-3 polyunsaturated fatty acid (Figure 7.5), which is known for its health benefits towards the normal functioning of cardiovascular and immune systems (Gutiérrez, Svahn, & Johansson, 2019; Yamagata, 2017). It is abundantly found in the brain and eye, besides being a vital structural component of heart tissue. Certain algae (ex. genus *Schizochytrium*) and fatty fishes such as salmon, herring, mackerel and tuna are rich sources of DHA (U.S. Agricultural Research Service Food Data Central, *n.d.*). Nevertheless, the intake of DHA is often too low to leverage its functionalities, mainly due to its oxidative instability and negative effect on the palatability of foods. Encapsulation serves as a one-stop solution to mitigate the oxidation and sensorial impact of DHA on the product in which it is incorporated. But, employing a high-temperature-based encapsulation technique such as spray drying could lead to oxidation of DHA during its early stages, when the crust formation is yet to begin. In this background, employing SFD for the encapsulation of DHA is advantageous.

Accordingly, an oil-in-water microemulsion containing algal oil (DHA: 38%) (core), WPI (wall material) and Tween-40 (emulsifier) was spray-freeze-dried using the SFV/L mode of atomization/spray-freezing (liquid nitrogen as cryogen), followed by vacuum freeze-drying (–24°C). The SFD process yielded spherical micron-scale particles with numerous fine pores (Figure 7.6) and an encapsulation efficiency (%EE) of 71%. Though the %EE as obtained earlier was less than that resultant from spray drying (83%) and freeze-drying (73%), the oxidative stability of SFD-DHA microcapsules was

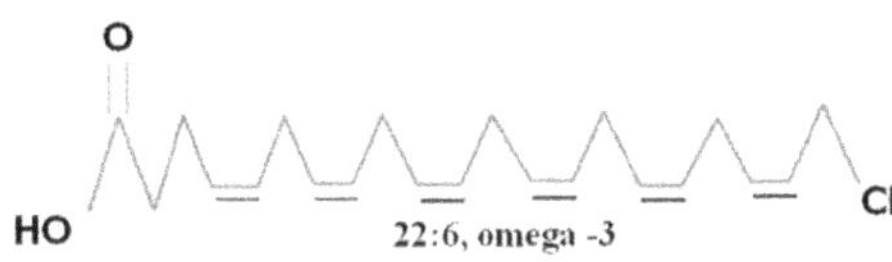

Figure 7.5 Molecular structure of docosahexaenoic acid (DHA) (Valenzuela & Valenzuela, 2013).

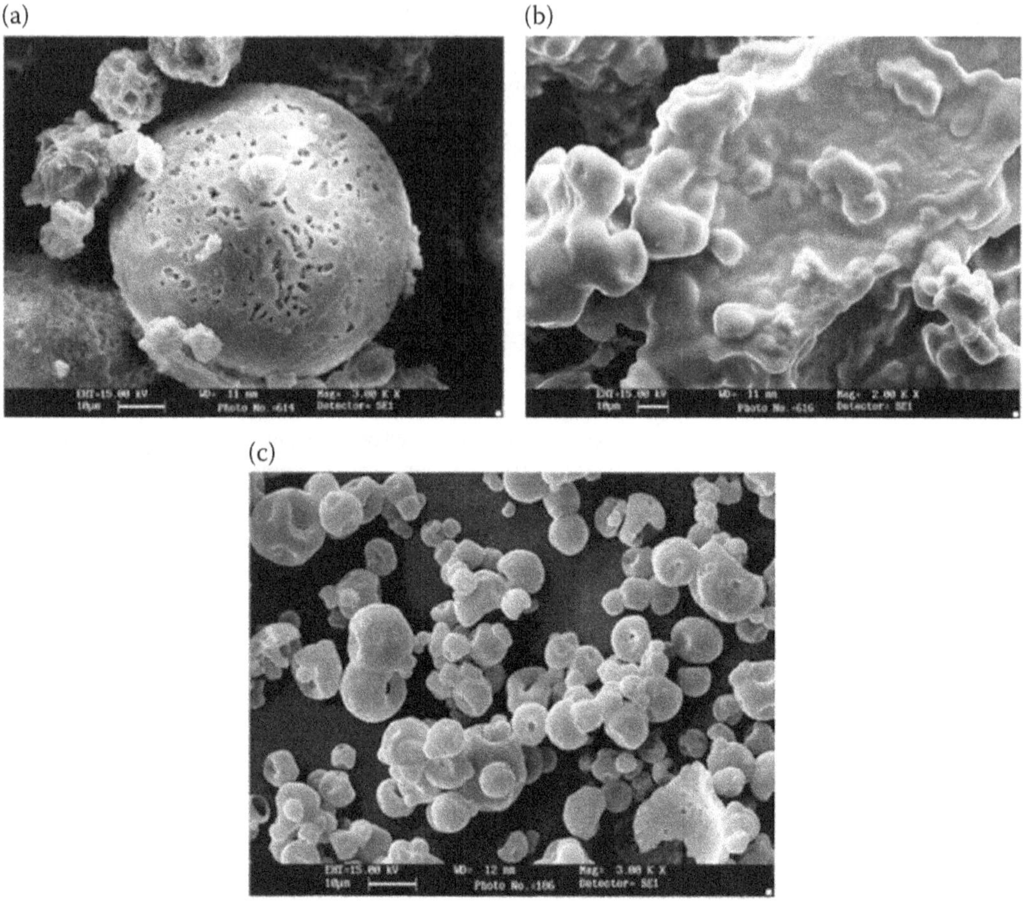

Figure 7.6 Morphology of DHA microencapsulates (a) spray-freeze-dried; (b) freeze-dried; (c) spray-dried (Karthik & Anandharamakrishnan, 2013).

substantially high. The percentage oxidation of SFD-DHA microcapsules was merely 13%, compared to 33% and 31% oxidation observed in the case of spray-dried (SD) and freeze-dried (FD) products, respectively. Irrespective of the drying method, the peroxide value of DHA microencapsulates showed a progressive increase during a 36-day storage period. However, the degree and rate of peroxidation was lower for the SFD microencapsulates than the unencapsulated oil and the spray-dried and freeze-dried counterparts (Figure 7.7). The lower operating temperature and shorter drying time (4 hours of SFD vis-à-vis 9 hours of FD) were considered responsible for the higher oxidative stability of SFD-DHA microencapsulates than their SD and FD equivalents. In turn, the shorter drying time of SFD process than freeze-drying resulted from the reduced dimension of feed liquid caused by its atomization step. This is because drying times vary approximately with the square of sample thickness (Anandharamakrishnan et al., 2010). Further, the oxidative stability of DHA microcapsules was almost two-fold higher under refrigerated storage (4 ± 1°C) than when stored at room temperature (29 ± 2°C) (Figure 7.7).

An additional advantageous feature of SFD-DHA microcapsules was their excellent rehydration behavior relative to the spray-dried (without pores) and freeze-dried encapsulates. The time for complete dissolution of DHA microcapsules in water (24 ± 2°C) reduced in the order of SD > FD > SFD with the values at 89–102 s, 36–48 s and 30–50 s, respectively. The porous structure was deemed to be responsible for the superior rehydration behavior of SFD-DHA microcapsules.

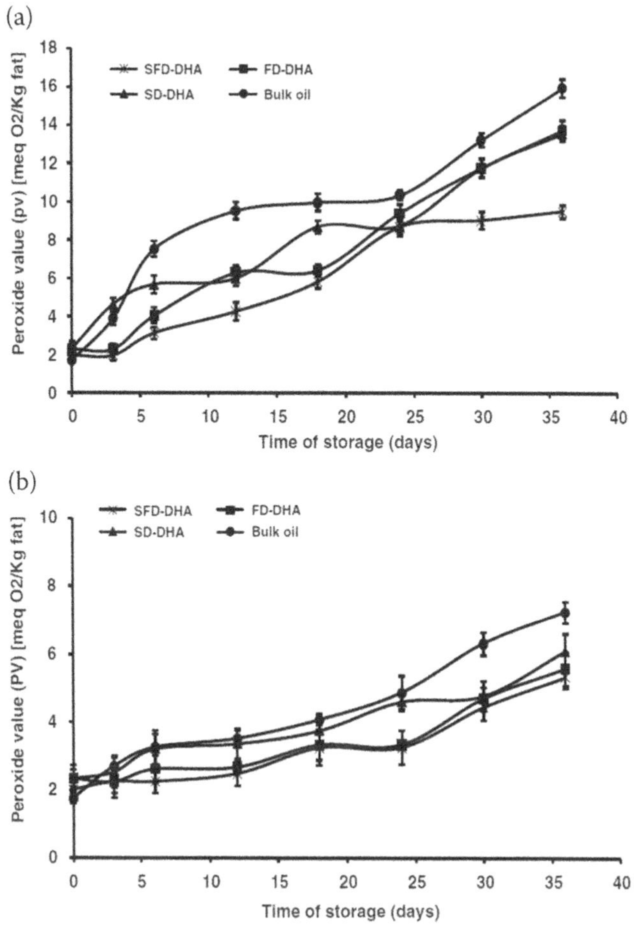

Figure 7.7 Oxidative stability of DHA microcapsules and unencapsulated oil stored under: (a) room temperature (29 ± 2°C); (b) refrigeration temperature (4 ± 1°C) (Karthik & Anandharamakrishnan, 2013).

At the same time, it was also stated as the reason for the lower encapsulation efficiency of SFD encapsulates compared to the non-porous spray dried microcapsules (Figure 7.6) (Karthik & Anandharamakrishnan, 2013). Thus, SFD proved to be an effective process for the encapsulation of algal DHA oil and its conversion into shelf-stable microcapsules in dry, powdered form.

In addition to algal oil, fish oil is an excellent source of omega-3 polyunsaturated fatty acids such as DHA and eicosapentaenoic acid (EPA, C20:5 ω-3). Short shelf-life is the major limitation associated with the further processing of fish oil. Thus, similar to DHA algal oil, spray-freeze-drying was employed for the encapsulation of fish oil to obtain high-quality powdered encapsulates than spray drying and freeze-drying. Porous and shelf-stable microcapsules of fish oil with improved digestibility was obtained at a high yield of 95%. The oxidative stability of SFD-fish oil microcapsules was assessed using propanal as marker, which is one of the main volatile compounds formed during the oxidative decomposition of omega-3 fatty acids (Ramakrishnan et al., 2014). Results showed that the propanal content of SFD microcapsules was lower than that of SD and FD samples (Figure 7.8). Propanal content is related to the surface oil content and the values of moisture content, both of which were lower for the SFD microcapsules and higher for the FD and SD samples. Consequently, SFD resulted in fish oil microcapsules with superior oxidative stability. The absence of cracks in the microstructure

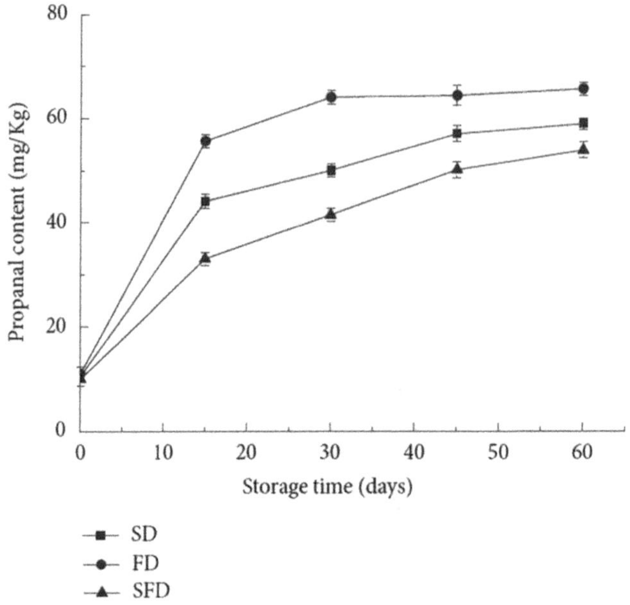

Figure 7.8 Relationship between storage time and propanal content of fish oil microcapsules produced by different drying methods (Pang et al., 2017).

of spray-freeze-dried fish oil microcapsules (Figure 7.9) was considered responsible for preventing the deterioration and oxidation of fish oil (Pang et al., 2017) during storage (Anwar et al., 2010).

7.5.1.2 Vanillin

Vanillin – chemically a 4-hydroxy-3-methoxybenzaldehyde or vanillic aldehyde (Figure 7.10) is a flavoring agent, which occurs naturally (ex. in vanilla beans) as well as produced synthetically

Figure 7.9 Morphology of spray-freeze-dried (5000x) fish oil microcapsule (Pang et al., 2017).

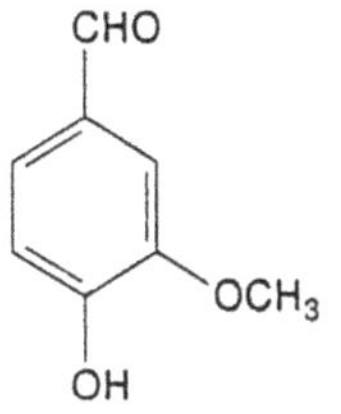

Figure 7.10 Structure of vanillin (Walton, Mayer & Narbad, 2003).

(Labuda, 2009). It has wide-ranging applications in the food, pharmaceutical and perfumery sectors. Vanillin is the most important volatile compound present in cured vanilla beans, which imparts sweet and creamy notes to enhance the vanilla flavor in food products (Ranadive, 2006). However, the natural sources contain vanillin only at a low concentration, ranging from 1.2–2.5%. Consequently, vanillin from natural sources like vanilla beans constitutes merely 1% of its global consumption, while the rest is synthesized from lignin and eugenol (Gallage & Møller, 2015; Harshvardhan, Suri, & Goswami, 2017). This makes vanillin an expensive flavoring ingredient. Currently, vanilla is traded as an ethanolic extract at a concentration of 35% (v/v). But, these alcoholic extracts suffer from disadvantages such as prolonged extraction times and presence of organic residues at high concentration. Besides, high volume of ethanolic extract is required to acquire the desired flavoring intensity in the end-products. On the other hand, vanillin in dry crystalline form is lost due to sublimation at temperature above 40°C. The loss increases progressively with further increase in temperature (Frenkel and Havkin-Frenkel, 2006) (Figure 7.11). All the aforementioned lacunae regarding its stability against processing and environment conditions justify vanillin as a suitable candidate for encapsulation by spray-freeze-drying.

Vanillin was encapsulated by spray-freeze-drying using different wall material compositions: β-cyclodextrin (β-CD), WPI and a combination of β-CD and WPI. The quality and stability of the resultant vanillin encapsulates were compared against those obtained from spray drying and freeze-

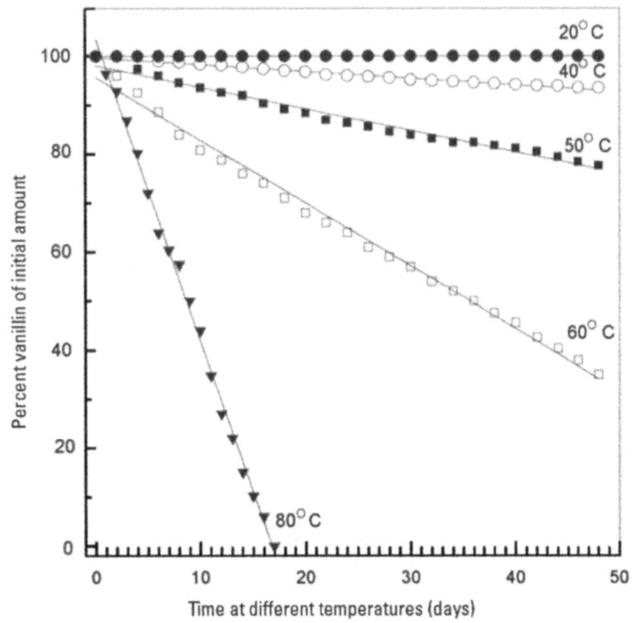

Figure 7.11 Change in the amount of dry crystalline vanillin (500 mg) as percentage of initial amount at different temperatures (Frenkel & Havkin-Frenkel, 2006).

drying. Moisture content of the SFD [vanillin + WPI] and SD [vanillin + WPI] microencapsulates were comparable (6.63% and 6.40%, respectively). Similarly, the moisture content of SFD [vanillin + β-CD] and SD [vanillin + β-CD] microcapsules were closer and also presented the lowest values (4.15% and 3.56%, respectively) amongst the other wall material compositions. Irrespective of the type of wall material used, freeze-dried vanillin encapsulates showed the highest moisture content (6.70–6.99%). The lower moisture content of SFD vanillin encapsulates than the FD counterpart was attributed to the size reduction of feed material by the atomization step. This is justified as the final moisture content of a dried product is controlled by the diffusion of water, which is speeded up by particle/droplet size reduction and the consequent increase in surface area (Heldman & Hohner, 1974; Malecki, Shinde, Morgan, & Farkas, 1970).

Similar to the SFD-DHA microencapsulates, the SFD-WPI vanillin microcapsules also showed spherical shape with numerous fine pores on the surface and superior rehydration ability. Whereas, the SD-WPI and FD-WPI vanillin encapsulates depicted non-porous–spherical and porous–flaky microstructures, respectively (Figure 7.12). However, sphericity and porosity of particles were not

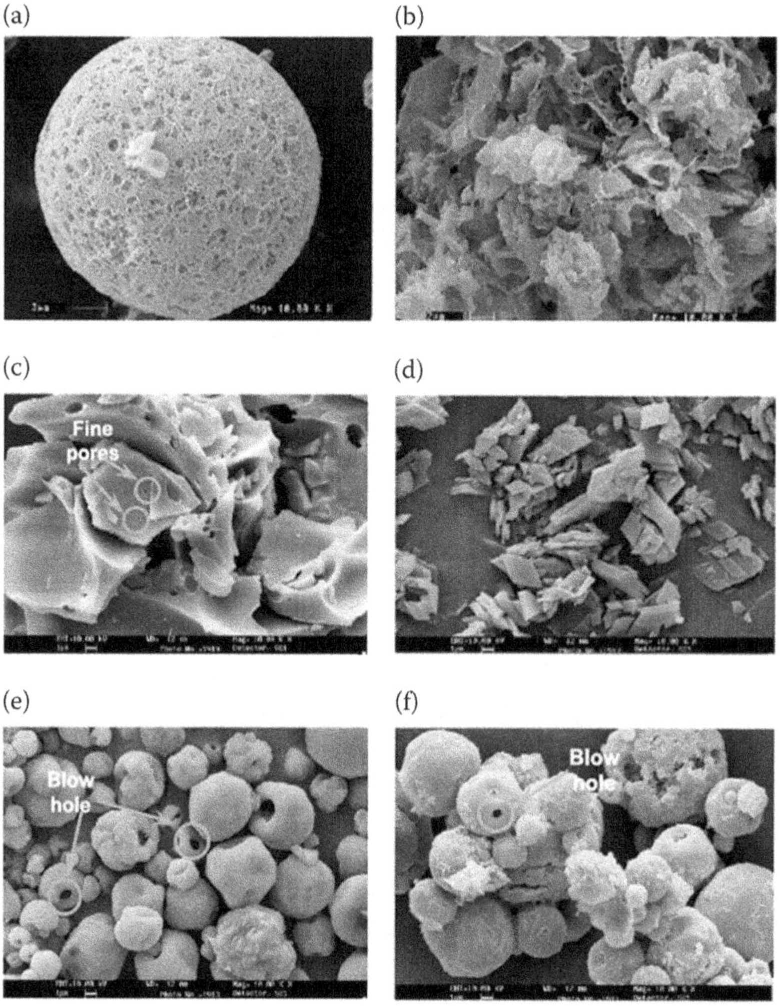

Figure 7.12 Morphology of vanillin microencapsulates: (a) SFD [Vanillin + WPI]; (b) SFD [Vanillin + β-CD]; (c) FD [Vanillin + WPI]; (d) FD [Vanillin + β-CD]; (e) SD [Vanillin + WPI]; (f) SD [Vanillin + β-CD] (Hundre et al., 2015).

obtained with β-CD as wall material, as it underwent structural collapse during spray-freeze-drying. Thus, WPI was found to be a suitable wall material than β-CD for the encapsulation of vanillin by SFD. This finding was further confirmed by the lower mean particle diameter of SFD-WPI vanillin encapsulate (24.8 μm) compared to that of the encapsulate with β-CD as wall material (165.4 μm). Similar results were observed with the FD vanillin encapsulates, and the same was ascribed to the agglomeration of ice crystals during freezing stage. In addition to lower mean particle diameter, the SFD-WPI vanillin encapsulate showed the lowest span value (0.931), compared to other formulations (1.340–2.953). Span is a measure of the homogeneity of particle size distribution. It is a three point specification computed from the D10, D50 and D90 values, which are the diameters at 10%, 50% and 90% cumulative volume, respectively. It is a statistically robust representation of the complete particle size distribution for most particulate materials (Sun et al., 2010). Larger the span value, wider is the particle size distribution. Thus, the SFD-WPI vanillin encapsulate had a relatively uniform particle size distribution. However, β-CD did not result in a homogeneous particle size distribution of the SFD-vanillin encapsulates (span: 2.953), due to the same reasons as mentioned earlier for sphericity and porosity.

The encapsulation efficiency of SFD-WPI vanillin microcapsules was lesser (~70%) than that of the SD-vanillin encapsulates (~85%). The smooth and non-porous surface of spray-dried WPI + vanillin encapsulates (Figure 7.12[e]) was stated as the reason for its high encapsulation efficiency. Nevertheless, the SFD encapsulates with WPI as wall material showed the highest loading efficiency of vanillin (39.22%), compared to FD-WPI (34.30%), SD-WPI (31.55%), SD-β-CD (31.76%), FD-β-CD (35.31%), SFD-β-CD (37.71%) and SD-WPI+β-CD (32.94%). While encapsulation efficiency considers the ratio of difference between total oil and surface oil to the total oil, loading efficiency is the weight of encapsulated oil per unit weight of the encapsulate. The superior skin forming behavior of WPI without any pores or cracks was stated as responsible for the lower surface oil content and effective retention of core component.

In addition, spray-freeze-dried vanillin + WPI encapsulate showed better thermal stability than the spray dried and freeze-dried microencapsulates. From the thermogram (relationship between weight loss due to release of vanillin and increase in temperature) (Figure 7.13), it can be seen that the SFD and FD encapsulates were capable of delaying the release of vanillin, compared to the SD samples. The same was evident from their lower degree of weight loss. The thermogram of {vanillin + β-CD} encapsulates showed a steep decline in weight and that of {vanillin + WPI} depicted a gradual reduction. The former indicates faster degradation of sample, and the latter represents gradual, constant and continuous release of vanillin. Thus, vanillin encapsulates produced by SFD with WPI as wall material showed better thermal stability and controlled release pattern compared to those produced by SD and FD (Hundre, Karthik & Anandharamakrishnan, 2015).

7.5.1.3 Vitamin E

Vitamin E (α-tocopherol) is a fat-soluble vitamin with antioxidant properties. It is a vital nutrient for the proper functioning of immune system, vision and bone health (Reddy & Jialal, 2020). But, the hydrophobic nature (poor aqueous solubility) of Vitamin E and its instability under processing and storage conditions challenges its incorporation in food products (Gonçalves, Estevinho, & Rocha, 2016; Tomas & Jafari, 2018). Besides, its poor water solubility limits the absorption of vitamin E in gastrointestinal tract and thereby reduces its oral bioavailability. According to the Food and Drug Administration (FDA), bioavailability is defined as the *'rate and extent that a drug or active pharmaceutical compound is absorbed and becomes available at the*

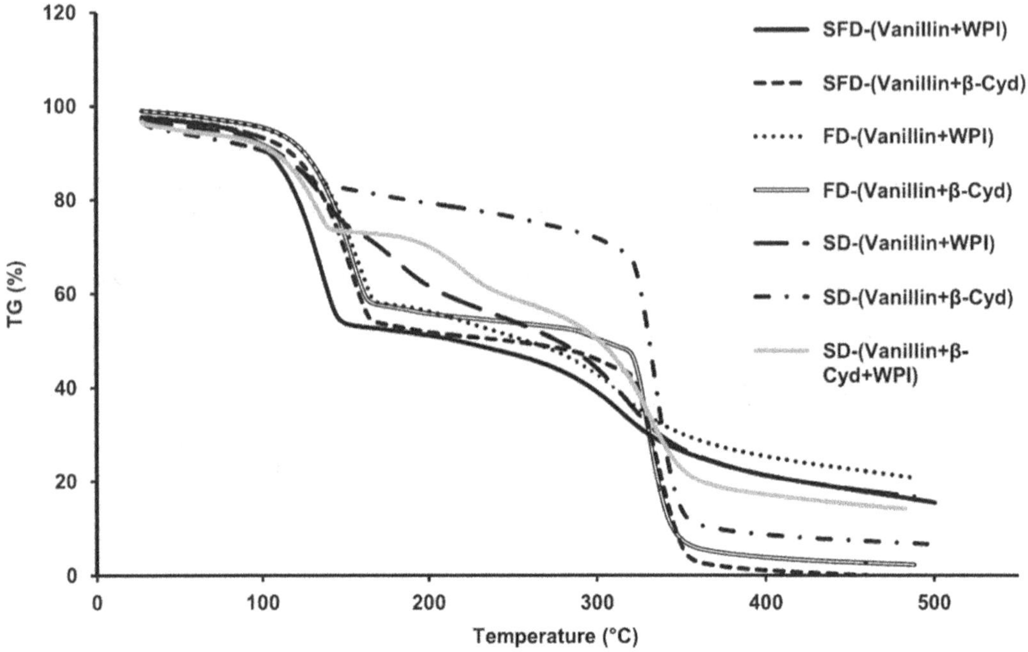

Figure 7.13 Thermograms of vanillin encapsulates (Hundre et al., 2015).

site of action' (Ting et al., 2014). Bioavailability (*F*) of a bioactive compound is given by the following equation:

$$F = F_a \times F_g \times F_h \tag{7.1}$$

Where F_a is the fraction of bioactive compound available for intestinal absorption, F_g is the fraction that permeates the intestinal epithelium without being metabolized at gut wall and F_h is the fraction that escapes the first pass hepatic metabolism and reaches the systemic circulation. Encapsulation of vitamin E into hydrophilic carrier matrices is a well-established approach to protect it against adverse conditions. But, obtaining stable encapsulates of vitamin E in powder form involves hitches such as reducing its vulnerability to chemical degradation at high temperatures, oxygen levels and light exposures (Loewen, Chan, & Li-Chan, 2018; Moeller et al., 2018). Thus, spray-freeze-drying is viewed as a technique that overcomes the given complexities.

Vitamin E microcapsules were prepared using the SFD approach. Initially, a nanoemulsion of vitamin E was prepared by microfluidizing a coarse oil-in-water (O/W) emulsion containing 10% of lipid phase and 90% of water phase. The lipid phase was a mixture of vitamin E (2%, w/w) and sunflower oil (8%, w/w). The aqueous phase contained 0.1% (w/w) of saponin (surfactant). The freshly prepared nanoemulsion was mixed with whey protein isolate (wall material) at the ratio of 1:3, by high shear homogenization. Then, the nanoemulsion was spray-freeze-dried using the SFV/L mode of spray-freezing followed by vacuum freeze-drying to produce vitamin E microcapsules.

In addition to the physical and structural properties and encapsulation efficiency, this study evaluated the dissolution characteristics and *in vivo* oral bioavailability of the SFD-Vitamin E encapsulates. As observed during the encapsulation of DHA and vanillin, SFD-Vitamin E encapsulates also showed higher encapsulation efficiency, porous internal structure and a fair flow behavior* as ascertained by its Hausner ratio (1.19) and Carr index (16), relative to the spray-dried and freeze-dried microcapsules. These findings gain relevance as improving the dissolution rate of a bioactive compound in the gastrointestinal (GI) tract is one of the important strategies to improve its oral bioavailability. Spray freeze-dried and freeze-dried microcapsules of Vitamin E showed a higher dissolution rate than the spray-dried microcapsules. Porous structures formed during the sublimation of ice crystals resulted in the good dissolution behavior of freeze-dried and spray-freeze-dried microcapsules (Figure 7.14[a]) (Parthasarathi & Anandharamakrishnan, 2016). Impaired dissolution of spray-dried microcapsules was attributed to the thermal denaturation of whey protein during spray drying. Solubility of the two major constituents of WPI, namely, α-lactalbumin and β-lactoglobulin reduce by up to 40% due to heat-induced protein aggregation at high temperatures (100–120°C) (Anandharamakrishnan 2008).

The *in vivo* bioavailability was evaluated by orally administering the vitamin E microcapsules to male Wistar rats at a dosage of 40 mg/kg body weight. Results of the *in vivo* analysis (Figure 7.14[b]) clearly established the improved absorption of vitamin E from the spray freeze-dried microcapsules. The pharmacokinetic parameters such as maximum plasma concentration (C_{max}), time of attaining C_{max} (T_{max}) and area under the curve (AUC) were high for the SFD-Vitamin E encapsulates than the SD and FD microcapsules. Rats administered with spray-freeze-dried Vitamin E microcapsules achieved the C_{max} of 9.449 μg/mL after 3 hours (T_{max}). But, spray-dried and freeze-dried microcapsules achieved C_{max} of 7.348 and 7.693 μg/mL, respectively, only after 4 hours (T_{max}) (Figure 7.14[b]). Similarly, the AUC increased in the order of SD < FD < SFD microcapsules, with the corresponding values at 109.84, 104.38 and 124.46 μg/(mL × h). Thus, the pharmacokinetic parameters explained the significant improvement in the oral bioavailability of vitamin E, delivered by the spray-freeze-dried microcapsules over the spray-dried and freeze-dried microcapsules (Parthasarathi & Anandharamakrishnan, 2016). The porous microstructure of SFD microcapsules enhances the dissolution and absorption rate of the encapsulated bioactive in the GI tract. Moreover, the larger specific surface area provided by the typical internal structure of SFD microcapsules is unique, which is not obtainable from the spray drying and freeze-drying processes. This study is a sound technical evidence that established spray-freeze-drying based encapsulation as a promising approach to enhance the oral bioavailability of lipophilic bioactives such as Vitamin E.

7.5.1.4 Bromelain

Bromelain is a bioactive compound with proven therapeutic actions against asthma, platelet aggregation, sinusitis and bronchitis (Cabral, Said & Oliveira, 2009; Secor et al., 2005). However, oral administration of bromelain is not effective due to its poor absorption. When administered to human volunteers at a dosage of 4 g/day, only 10 μg/mL was released in the plasma (Hale et al., 2002). But, the effective concentration of bromelain required to deliver its intended action, i.e., to remove bromelain-sensitive cells from the entire blood is 500 μg/mL (Castell et al., 1997). While increasing the dosage is one approach, using the pulmonary route of administration is a more effective one, as it can reduce the dosage requirement by multiple-fold (Morozov et al., 2014). The key traits of a therapeutic product intended for pulmonary delivery are its large, porous, spherical and low-density particles to facilitate deposition in the alveolar region of lungs. Other characteristics of importance include particle size, moisture content and hygroscopicity. A particle

* Carr's index of up to 10% signifies excellent flowability, between 10 and 15% indicates *good* flowability; between 16 and 20% implies*poor* flowability; between 32 and 37% specifies very poor flow and values greater than 38% point to extremely poor flowability. On the other hand, Hausner ratio less than 1.25 indicates free flowing powder and that above 1.25 designates poor flowability (USP 30-NF 25, 2007).

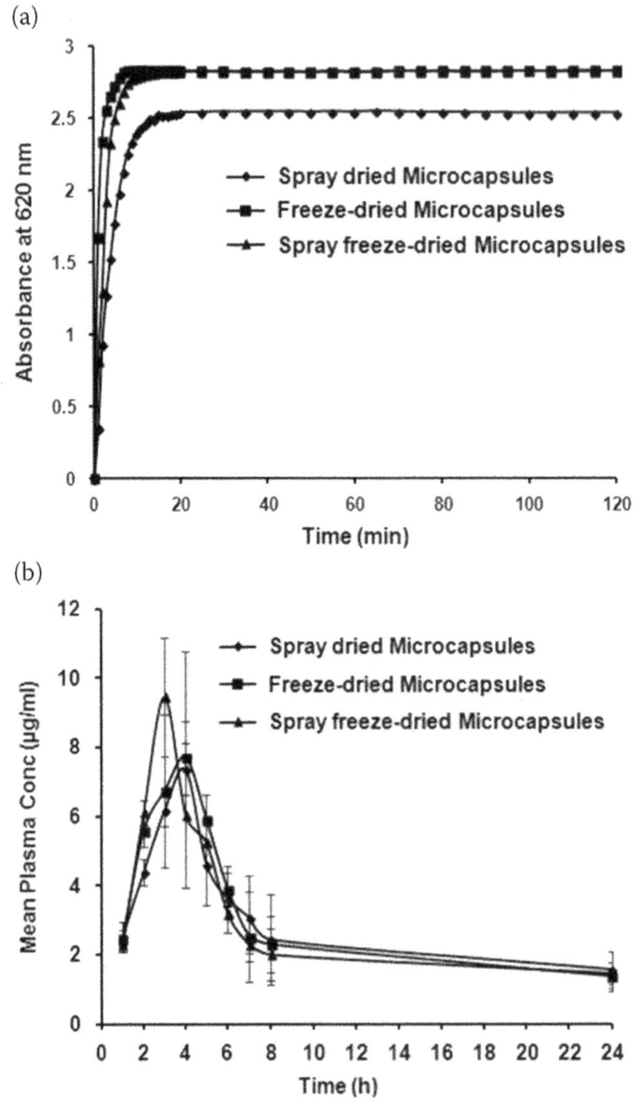

Figure 7.14 Spray-freeze-dried Vitamin E microcapsules vis-à-vis spray-dried and freeze-dried encapsulates: (a) Dissolution profile; (b) Plasma concentration-time profiles of vitamin E in rats following oral administration of 40 mg/kg equivalent dose of vitamin E from the microcapsules (Parthasarathi & Anandharamakrishnan, 2016).

mass density of less than 0.4 g/mL, particle size in the range of 10–20 µm and mass median aerodynamic diameter (MMAD) in the range of 1–5 µm are considered favorable for the optimal performance of aerosol particles (Agu et al., 2001; Edwards et al., 1997). Spray-freeze-drying is a fitting technique for the making of aerosols as it is capable of producing large, spherical, porous and low-density particles with good aerosolization efficiency and storage stability (Ali & Lamprecht, 2014; Yu et al., 2016). The porous nature of SFD particles reduces the density and renders it suitable for inhalation (Lavanya et al., 2020).

Combining SFD process with the encapsulation approach using suitable wall materials is a viable approach to produce stable bromelain aerosols for pulmonary supplementation (Lavanya et al., 2020). Maltodextrin was used as the wall material at three different core-to-wall ratios: 1:10,

1:25 and 1:50. The choice of maltodextrin was based on its role as cryoprotectant (Ali & Lamprecht, 2014) and excipient to increase the emitted dosage of the active molecule (Crowder & Hickey, 2006; Smyth & Hickey, 2005). Chitosan was added in small concentration to enhance the absorption of active molecule by overcoming the rigidity between the para-cellular junctions (Vllasaliu et al., 2012). The solution containing bromelain, maltodextrin and chitosan (in acetic acid) was stirred for 30 min at 800 rpm and homogenized at 18000 rpm for 10 min using a high speed homogenizer. The given solution was atomized by spray-freezing-into-vapor over liquid nitrogen (−196°C) method (SFV/L) using a twin-fluid nozzle, followed by vacuum freeze-drying at a temperature of −80°C to +30°C and pressure of 100 Pa, through the primary and secondary stages.

The resultant bromelain aerosol had a moisture content and water activity in the range of 3.58–3.91% and 0.54–0.56, respectively. Low moisture content of dry powder aerosols is important in preventing particle agglomeration and emission failure from the inhaler (Lavanya et al., 2020). As expected, the spray-freeze-dried bromelain aerosol particles were spherical and porous (Figure 7.15) with particle size in the range of 4.71 – 5.56 μm. Moreover, the SFD-bromelain aerosols showed low density (0.30–0.38 g/mL) and good to excellent flowability (Hausner ratio: 1.07–1.13; Carr index: 7.14–11.56).

Further, the activity of bromelain in the SFD aerosol was evident from the enzyme activity (as % to the initial enzyme activity) after encapsulation, which was in the range of 77.8–94.5% for different core-to-wall ratios. The maximum enzyme activity was obtained at the highest concentration of maltodextrin (1:50), owing to its role as a cryoprotectant that protected the bromelain against freezing and drying induced shocks. Conversely, the highest anti-inflammatory activity of bromelain aerosols (65%) was obtained at the lowest core-to-wall ratio (1:10). The *in vitro* analysis showed that the release of bromelain from the spray-freeze-dried aerosol was slow and sustained for 12 h, during which 46–55% of the core was released. Regarding the aerodynamic properties of the SFD-bromelain aerosol, the calculated mass median aerodynamic diameter (MMAD) was in the range of 2.97 to 3.33 μm. These values were within the desirable range of 1–5 μm suitable for deep-lung deposition, as mentioned earlier. The SFD formulation showed 95% aerosolization from the total loaded dose with an emitted dosage of 97.02% and fine particle fraction (FPF) of 84.28%. About 70% of FPF is adequate to deliver particles to the alveolar region. A combination of low MMAD and high FPF of aerosols is suitable for deep-lung deposition (Patil-Gadhe et al., 2014). Thus, spray-freeze-drying based encapsulation is a suitable approach to produce aerosols with desired characteristics for pulmonary supplementation of food bioactives with optimal activity, stability and absorption (Lavanya et al., 2020).

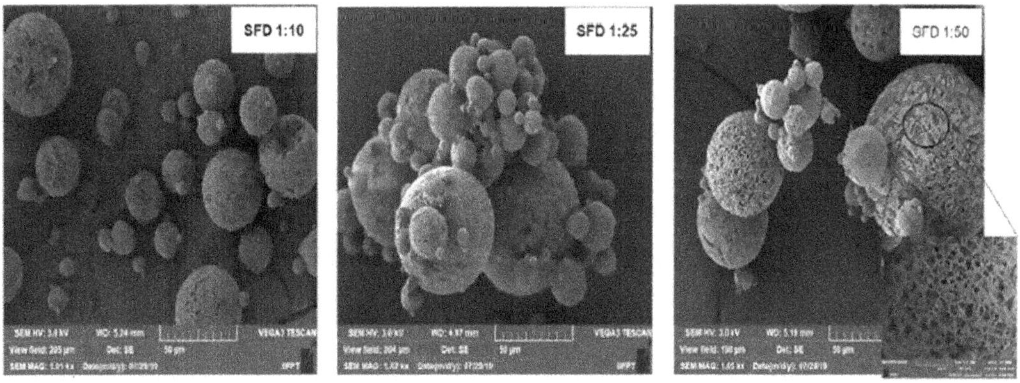

Figure 7.15 Scanning electron micrographs of spray-freeze-dried bromelain aerosols at different core-to-wall ratios (Lavanya et al., 2020).

7.5.1.5 Probiotics

Probiotics are *'live microorganisms that, when administered in adequate amounts, confer a health benefit on the host'* (FAO/WHO Food and Agricultural Organization of the United Nations and World Health Organization, 2001 & 2002). The most common probiotic organisms are gram-positive bacteria, belonging to the genera of *Lactobacillus* and *Bifidobacterium. Escherichia coli, Sporolactobacillus inulinus*, other bacterial species of *Propionibacteria, Enterobacteria* and *Bacillus* genera, and yeasts such as *Saccharomyces cerevisiae (boulardii)* have also been accepted for use as probiotics (Mercenier et al., 2003; Seth & Maulik, 2011). Besides having a positive effect on the general well-being of host, probiotics are beneficial in alleviating the clinical symptoms or conditions of Crohn's disease, diarrhea, gastroenteritis, inflammatory bowel disease, irritable bowel syndrome and lactose intolerance (Kumar & Salminen, 2016). The benefits of probiotics are delivered to the consumers in the form of a dietary supplement, food ingredient or a drug. With respect to foods, dairy products are the predominant delivery vehicles for probiotics. However, the claim as probiotic foods is valid only when the concentration of live probiotic bacteria is more than 10^7 CFU per gram of the product at the time of consumption (FAO/WHO Food and Agricultural Organization of the United Nations and World Health Organization, 2001). The given regulation mandates the encapsulation of probiotics to protect them from the adverse conditions of food processing. Though spray drying is a renowned process for probiotic encapsulation, thermal injury to the cells caused by the high-temperature operation of spray drying is unavoidable. Due to its low temperature operation and uniquely structured products, SFD is considered advantageous over spray drying in preserving the viability and stability of probiotic cells during processing and transit through the upper regions of gastrointestinal tract, before reaching their target site of action, i.e., the large intestine.

Lactobacillus plantarum was encapsulated and converted into powder-form by spray-freeze-drying, using whey protein isolate as wall material at two different core-to-wall ratios (1:1 and 1:1.5). The spray-freeze-dried probiotic microcapsules were obtained after the two-stage operation of SFV/L (cold vapor over liquid nitrogen) followed by conventional freeze-drying for 5 hours. The characteristics of SFD-probiotic encapsulates (moisture content, morphology, cell viability, storage stability and survival of encapsulated bacteria in simulated gastric conditions) were compared with those of the spray-dried (SD) and freeze-dried (FD) microcapsules. The moisture content of probiotic microcapsules increased in the order of SD (2.81–3.23%) < SFD (3.48–3.59%) < FD (3.61–5.6%). During SFD and FD, the moisture content was low at higher wall material (WPI) content (1:1.5). This is because, the whey proteins adsorb to the ice/liquid interface during the freezing stage. Consequently, during the ensuing stage of freeze-drying, when the ice crystals were removed, voids were retained on the surface. Therefore, increase in the proportion of WPI would have enhanced the pore formation and sublimation of ice, eventually resulting in a product with lower moisture content. This phenomenon was just the opposite of SD, wherein, higher WPI content promotes strong crust formation, which impedes diffusion of water vapor and thereby moisture removal (Dolly et al. 2011).

The viability of probiotics in the spray-freeze-dried microcapsules was 20% higher than that in the spray-dried encapsulates. It is important that the probiotic cells reach the large intestine in a viable state and in good numbers such that they are beneficial to the host. Hence, they must survive the acidic conditions in the stomach and resist the action of bile in the small intestine (Anal & Singh, 2007; Michida et al., 2006). The unencapsulated free cells showed a steep decline in viability (zero viability in 4 h) on exposure to acid, pepsin and bile; whereas, the encapsulated cells showed tolerance to bile for up to 2 hours, with a relative viability of 1–1.2 at the end of 4 hours of exposure. On exposure to acidic conditions, freeze-dried and spray-freeze-dried probiotics showed maximum survival at the core-to-wall ratio of 1:1.5. The higher wall material content and low temperature operation of SFD and FD were deemed responsible for the protective effect upon live cells. Furthermore, when stored at 4°C for 40 days, the SFD probiotic microcapsules showed merely 3–5% of survival loss compared to a 10% loss observed in spray-dried capsules. The slight survival loss of SFD-probiotic cells was attributed to the atomization stress on live cells during the spray-freezing stage. Overall, the synergistic effect of low-temperature operation of SFD and high wall material content led to a greater survival rate and better storage stability of the encapsulated probiotic cells (Dolly et al. 2011).

The type of wall material was found to have a major influence on the characteristics of spray-freeze-dried probiotic encapsulates. Different wall material formulations such as (i) whey protein isolate (WPI) + sodium alginate (SA); (ii) WPI + fructooligosaccharide (FOS); (iii) denatured WPI (DWPI) + SA and (iv) DWPI + FOS, were used for the encapsulation of *Lactobacillus plantarum* (MTCC 5422) by SFD. Among the given combination of wall materials, DWPI + FOS resulted in probiotic microcapsules with higher encapsulation efficiency (94.86%) and better storage stability during 60 days (specific rate of viability loss, $k_T = 0.0026$ day^{-1}), due to the following reasons (Rajam & Anandharamakrishnan, 2015). The other formulations showed encapsulation efficiency and k_T in the range of 87.92% to ~90% and 0.009 to 0.0155 day^{-1}, respectively.

- ***Higher moisture content:*** The DWPI + FOS probiotic microcapsules had the highest moisture content (6.32%) than the other formulations (5.57–5.95%). Higher moisture content of encapsulates is of relevance to cell viability after drying and subsequent storage (Zayed & Roos, 2004). In its denatured state, whey protein exhibits higher water holding capacity. As a result, moisture is held tightly within the particle structure, thus resisting its removal during drying (Pérez-Gago et al., 1999).
- ***Ability of FOS to penetrate the cell membrane:*** Due to its tendency to penetrate cell membrane, FOS can partially replace the water molecules in the cells. Consequently, the cells are protected from the deleterious effects during freezing and freeze-drying (Schwab et al., 2007). Moreover, the presence of carbohydrates in the wall material formulation increases the glass transition temperature, which enables the viable cells to attain the glassy phase without nucleating intracellular ice (Fowler & Toner, 2006).
- ***Poor water vapor permeability of DWPI and cell membrane stabilization by FOS*** effectively protects the cells during storage (Pérez-Gago et al., 1999; Perez-gago & Krochta, 2001; Schwab et al., 2007).
- ***Prebiotic/bifidogenic effect of FOS*** improves the survival of microencapsulated probiotics during storage (Chen et al., 2005).
- ***Rough surface morphology:*** Particle microstructure with rough surface and protrusions was ascribed to the denatured whey protein isolate in the wall material formulation. At higher magnification, the scanning electron micrographs of DWPI + FOS microcapsules showed entrapment of the probiotic cells by the wall matrix (Figure 7.16[d]). The effective entrapment of cells by this wall material composition can be due to the three dimensional gel network formed by DWPI as thermal denaturation imparts gelation property upon WPI. During thermal denaturation, initially the globular protein partially unfolds to expose the free sulfhydryl groups, which rapidly interchange with existing disulphide bonds to produce new inter and intramolecular disulphide bonds that lead to protein aggregation and eventually a three-dimensional network (Anandharamakrishnan et al., 2007; Nicolai et al., 2011).

All the other spray-freeze-dried microcapsules (Figure 7.16[a-c]) were almost spherical-shaped with uniform size, with the differences seen only in the surface morphology. The protective effect of FOS is enhanced in the presence of WPI and DWPI due to their film-forming properties and the formation of a stronger gel network (Parthasarathi et al., 2013). The SFD WPI + SA microcapsules showed smooth outer surface due to the characteristic skin forming ability of whey protein and porous internal structure with tiny air bubbles entrapped inside (Anandharamakrishnan et al., 2007; Sheu & Rosenberg, 1998). Smooth surface morphology occurs when the drying rate of wall material decreases rapidly with reduction in moisture content. This aspect is important for achieving better protection and stability of the probiotic cells as a denser wall material membrane

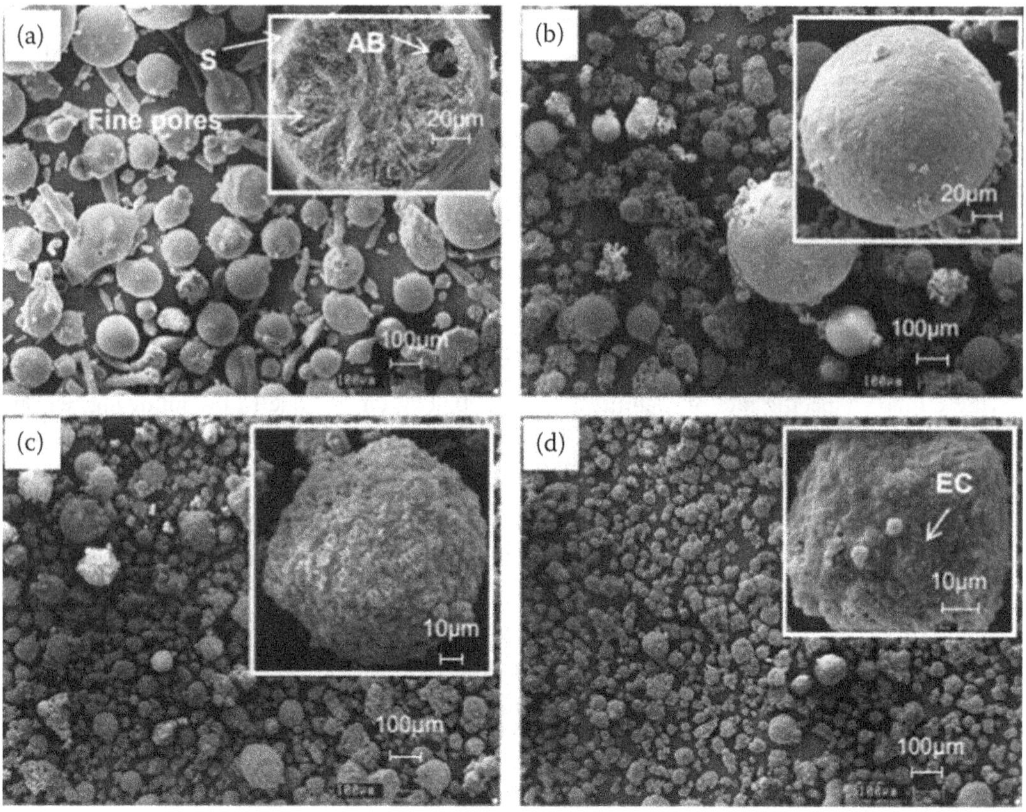

Figure 7.16 SEM images of spray freeze dried microcapsules of *L. plantarum* produced with different wall material formulations: (a) WPI + SA; (b) WPI + FOS; (c) DWPI + SA and (d) DWPI + FOS (Rajam & Anandharamakrishnan, 2015). AB, air bubble; S, smooth outer skin; EC, encapsulated cells.

is formed around the live cells when wall material with a high drying rate is used. The air bubbles are more likely to be occluded during the atomization/spray-freezing step (Anandharamakrishnan et al., 2010). On the other hand, the SFD WPI + FOS microcapsules depicted a slightly rough surface with few smaller particles attached to the larger particles. The lower glass transition temperature and hygroscopic nature of FOS may have led to the stickiness. Likewise, the SFD DWPI + SA encapsulates exhibited an extremely porous surface (Figure 7.16[c]). Thus, it is evident that the difference in wall material composition governs the morphology of spray-freeze-dried probiotic microcapsules (Rajam & Anandharamakrishnan, 2015).

In certain cases, stabilizers are added to the SFD feed formulation containing probiotic cells and wall material for an enhanced protective effect (Semyonov et al., 2010). This is relevant as the SFL method of atomization induces both thermal and osmotic stresses upon the probiotic cells. In the case of *Lactobacillus paracasei*, cell viability was reduced by the fusion between its cell membrane and denatured proteins, during encapsulation by spray-freeze-drying. But, addition of trehalose and maltodextrin as stabilizers in the feed formulation prevented viability loss during both the spray-freezing and freeze-drying stages of SFD. The protective effect of trehalose and maltodextrin is governed by *vitrification*, which is the transformation of feed material to glassy state below the glass transition line of state diagram. During conventional freezing, ample time is available for ice formation and freeze concentration such that the solution attains a nearly equilibrium state of maximal freeze concentration. Consequently, the unfrozen regions turn highly

Table 7.3 Application of spray-freeze-drying for the encapsulation of probiotics – An overview of processing conditions and product characteristics (modified from Ishwarya, Anandharamakrishnan, & Stapley, 2017)

Core	Cryogen	Nozzle Type	Feed Rate	Air-flow Rate / Air Pressure	Particle Size	Reference
Lactobacillus plantarum microencapsulates	Liquid nitrogen	Twin-fluid nozzle (0.5mm diameter)	6 mL/min	25 m^3/h	53.99 to 105.07 µm	Rajam & Anandharamakrishnan (2015)
Lactobacillus casei microencapsulates	Liquid nitrogen	Twin-fluid nozzle	5 mL/min	20 and 30 kPa	24.8 µm	Her et al. (2015)
Lactobacillus paracasei microencapsulates	Liquid nitrogen	Pneumatic nozzle	a. 0.15 mL/min b. 0.3 mL/min c. 0.8 mL/min	2.12 L/min 3.08 L/min 4.52 L/min	400 to 1800 µm	Semyonov *et al.* (2010)

viscous to vitrify at this concentration (Engstrom et al., 2007). The aforementioned phenomenon holds true during the SFD encapsulation of probiotic cells as well, due to which water is immobilized in the vitrified viscous glass. The low mobility of water prevents loss of cell viability caused by damage to cell membrane and protein unfolding. The low molecular weight of trehalose further promotes the vitrification phenomenon by decreasing the size of water crystal in the intermembrane space. Ultimately, changes in the physical state of the membrane lipids are prevented and mechanical stresses in membranes are lessened, all of which improves cell viability (Koster et al., 2000). Table 7.3 summarizes the processing parameters and particle size of probiotic encapsulates produced by SFD.

7.5.2 Biologicals encapsulated by spray-freeze-drying

Proteins are the predominantly encapsulated biologicals by spray-freeze-drying. Instability during preparation, storage and release is a major apprehension during the encapsulation of proteins within polymeric carrier matrices for controlled release systems (Wang, Chua & Wang, 2004). Particularly, when present in an emulsion medium (oil-in-water [O/W], water-in-oil [W/O] or water-in-oil-in-water [W/O/W; double emulsion]), the proteins are more likely to be denatured at the oil-water interface created during the emulsification process. Proteins adsorb to such interfaces owing to their amphiphilic nature and then unfold, aggregate or experience adverse changes that eventually lead to the loss of their biological activity (Pérez & Griebenow, 2001; Sah, 1999 [a & b]; Van de Weert et al., 2000). In many instances, medically undesirable aggregates are formed as a result of harsh encapsulation methods (Fu et al., 1999; Lu & Park, 1995; Nihant et al., 1994). Thus, SFD is perceived as an effective approach to overcome the given challenges. A non-aqueous encapsulation approach has been practiced, wherein, SFD's application is in the initial stage for the production of dehydrated protein microparticles. The aforementioned is accomplished with the addition of suitable excipients in the feed solution to circumvent protein denaturation during the atomization and freeze-drying stages of spray-freeze-drying.

Generally, using sugars as adjuvants in the feed formulation during SFD is known to cause preferential hydration and stabilization of proteins during freezing. The resultant solid microparticles from SFD are more stable and easy to be encapsulated in a polymeric carrier matrix such as polylactide-co-glycolide (PLG) in a subsequent step, by *oil-in-oil* (O/O) or *solid-in-oil-in-water (S/O/W) encapsulation* processes. Encapsulation into polymer matrix facilitates sustained release of the drug over a period of 1–3 months (Hutchinson & Furr, 1990), by maintaining the intactness of unreleased proteins and protecting them against degradation in the human body. This avoids frequent administration of therapeutic proteins by invasive routes such as subcutaneous injection (Wang, Chua & Wang, 2004). Two case-studies on the application of SFD for the encapsulation of bovine serum albumin (BSA) and Immunoglobulin G (IgG) into polymeric matrices are presented in the subsequent sections.

7.5.2.1 Bovine serum albumin

As discussed earlier, bovine serum albumin (BSA) was subjected to a non-aqueous oil-in-oil (O/O) encapsulation procedure comprising SFD followed by coacervation[$] to enclose the protein microparticles within polymeric shell. Initially, an aqueous solution of BSA without and with trehalose as an excipient (to protect BSA against denaturation during the encapsulation process) was atomized through a two-fluid nozzle in a stainless steel chamber. For excipient-free BSA feed formulation, the frozen slurry was taken in glass dishes and freeze-dried in a drying cabinet with its shelves pre-cooled to –40°C for the primary drying phase. The chamber pressure was 300 mTorr and the shelf temperature was increased to +10°C for the secondary drying stage. For feed containing trehalose (at 4 times the concentration of BSA), primary and secondary drying were carried out under the temperature/pressure conditions of –26°C/96 mTorr (for 3 days) and 20°C/10–20 mTorr (for 2 days), respectively. The given changes in freeze-drying conditions might have been included to account for the change in collapse temperature caused by the addition of trehalose.

In the second stage of encapsulation process, the spray-freeze-dried BSA microparticles were enclosed within PLG using the coacervation approach. PLG in combination with polaxamer in methylene chloride solution acted as the polymer phase. Poly(dimethylsiloxane) was used as the coacervating agent. The encapsulation of BSA microparticles into the polymeric envelope formed by coacervation was achieved by rotor-stator homogenization. The coacervating mixture containing the microspheres was hardened in heptane under constant agitation conditions at room temperature. The hardened microspheres were separated by filtration, washed twice with heptane and dried for 24 h under vacuum at room temperature, to obtain the BSA encapsulates.

The stability of BSA during different stages of the encapsulation process was evaluated by Fourier Transform Infrared (FTIR) Spectroscopy. Results showed that the SFD-BSA microparticles produced without the addition of trehalose had undergone significant structural perturbations. Spectral changes were observed in the amide I region (1700–1615 cm^{-1}). Decrease in α-helical (from 54 to 31%) and increase in β-sheet (from 8 to 22%) contents was observed. But, further changes in the protein secondary structure during the subsequent encapsulation within PLG microspheres was not very significant. Nevertheless, the trehalose excipient-containing powder resisted the changes in protein secondary structure caused by the encapsulation-induced stresses. Thus, it is evident that, in the absence of excipient, protein aggregated due to the exposed thiol and disulfide groups. But, use of trehalose as excipient prevented the formation of soluble BSA aggregates. In addition, when the spray-freeze-dried BSA-trehalose microparticles were encapsulated into the PLG or PLG/polaxamer microspheres, 88–94% of the entrapped BSA was released as monomer. This shows the protective effect of trehalose and polymers (PLG and PLG/polaxamer) on BSA, due to which the protein could retain its native structure and remain non-aggregated during the SFD, coacervation and the initial stages of *in vitro* release.

Contrarily, the presence of excipient or difference in the protein payload did not have any significant effect on the *in vitro* release duration of BSA from the polymeric microspheres. More

[$] Coacervation encompasses the preparation of a polymer-rich solution and employing it to enclose a core substance. The mechanism of encapsulation is based on the interaction between two oppositely charged polyelectrolytes in water to form the coacervate. Subsequently, the core ingredient is either deposited, suspended or emulsified in the coacervate solution, after which the polymer entraps the core. Changing the temperature or pH or adding concentrated ionic salt solution are the approaches used to induce the coacervate to encapsulate the core (Anandharamakrishnan & Ishwarya, 2015).

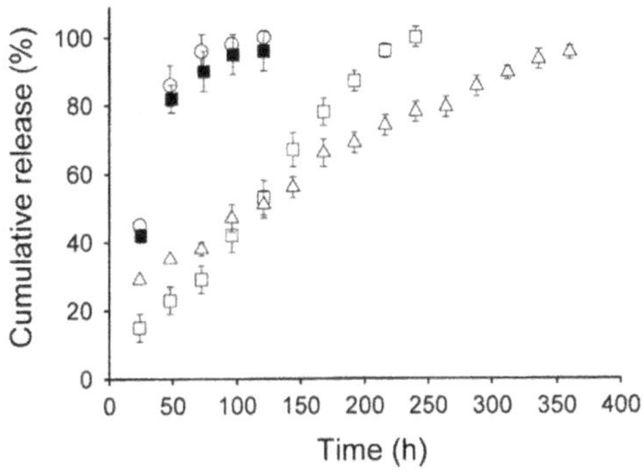

Figure 7.17 Cumulative release of (■) spray-freeze dried BSA and (○) BSA-Tre from PLG microspheres, (□) spray-freeze-dried BSA-Tre from poloxamer/PLG micro spheres and (△) spray-freeze-dried BSA-Tre from poloxamer/PLG microspheres (Carrasquillo et al., 2001).

than 80% of the protein was released within 2 days. But, employing a combination of PLG and polaxamer increased the release duration of BSA (Figure 7.17). The presence of polaxamer caused a significant increase in the microsphere diameter during the coacervation process. As a result, the burst-released protein content decreased from 45% to 15% (in the presence of polaxamer), attributed to the lesser amount of BSA powder particles on the surface of microspheres (Carrasquillo et al., 2001).

7.5.2.2 Immunoglobulin G (IgG)

Human Immunoglobulin G (IgG) is a protective protein known as antibody, which is produced by the body for defense against foreign cells such as bacteria and viruses (Wong, Wang & Wang, 2001). It is used in the treatment for congenital or acquired hypogammaglobulinemia and secondary IgG deficiencies experienced by patients with cancer, nephropathies, gastrointestinal disturbances and burns (Dwyer, 1984). IgG is also beneficial in alleviating immune disorders related to pregnancy (Kwak et al., 1996). As mentioned earlier, rather than administering through intravenous and intramuscular routes, controlled delivery systems are advantageous for the targeted delivery of IgG (Wong et al., 2001).

IgG was converted into solid particles using spray-freeze-drying. Different excipients such as mannitol, trehalose and zinc acetate dehydrate were included in the feed formulation to protect the protein against denaturation during ultrasonic atomization by SFV/L method. The particle size was controlled by regulating the frequency of ultrasonic atomizer and the feed flow rate of IgG solution through the ultrasonic nozzle. SFD of IgG solution without any excipients resulted in agglomerated particles with many pores and channels on the surface. Addition of zinc acetate as excipient was found to reduce the particle size to around 10 μm. Further, the formation of complex conjugates between IgG and zinc acetate followed by its precipitation promoted the formation of separated and smaller microparticles of IgG–zinc acetate with reduced specific surface area and prevented particle agglomeration (Costantino et al., 2000; Lam, Duenas & Cleland, 2001). Moreover, instead of becoming more concentrated after ice crystallization and being exposed to ice-water interface, the protein complex separated from the bulk water phase before any damage was caused by the subsequent freeze-drying stage of SFD process (Costantino et al., 2000; Wang, Chua, & Wang, 2004).

In addition, mannitol and trehalose were found to retain the molecular integrity and immunoactivity of IgG by more than 90%, after the SFD process. The aforementioned was verified using the size exclusion chromatography – high pressure liquid chromatography (SEC-HPLC) and Enzyme-linked Immunosorbent Assay (ELISA). Mannitol and trehalose improved the stability of IgG conformation by the mechanism of *preferential hydration*, which is based on the reduction in the chemical potential of the protein by decreasing its external surface area. The stability of the folded state might also be responsible for the protein's stability to unfolding. Conversely, the excipients may also impart stability by forming glassy matrix to decrease the flexibility and mobility of protein molecules (Tonnis et al., 2015).

Smaller particle size is a major pre-requisite for encapsulating the spray-freeze-dried IgG microparticles into PLG using the S/O/W process. The aforementioned was achieved by diluting the IgG feed solution by 5 and 10 times. But, the morphology of resultant particles showed that the size was similar irrespective of the feed dilution and their shape was no more spherical. At lower concentration, protein in the droplets could not hold the spherical shape after freeze-drying. However, feed dilution resulted in more porous and delicate structures (Figure 7.18). Subsequently, S/O/W encapsulation of IgG was carried out by dispersing 83 mg of SFD-IgG particles into 5 mL of 4.6% w/v PLG + dichloromethane (DCM) solution by ultrasonication. Then, the given mixture was dispersed into 0.5% aqueous solution of polyvinyl alcohol (PVA) by stirring until the evaporation of DCM is complete. Finally, the solid particles were collected by filtration, rinsed with deionized water and freeze-dried overnight to obtain the IgG encapsulates.

The stability of IgG in the S/O/W encapsulate was found to be higher than its unencapsulated counterpart in the form of double emulsion. Compared to the double emulsion system, S/O/W was proficient in retaining the secondary (2°) structure (α-helix) of IgG after encapsulation. The aforementioned was attributed to the protein stabilizing role of the excipients incorporated in the feed solution of IgG during the spray-freeze-drying process. Mannitol preserved the 2° structure of IgG during the S/O/W process. But, the underlying mechanism is unknown. Moreover, the release of IgG from the S/O/W microspheres showed a higher initial burst than that from a double-emulsion of IgG, which increased with increase in payload of protein in the microparticles

(a) (b) (c) (d)

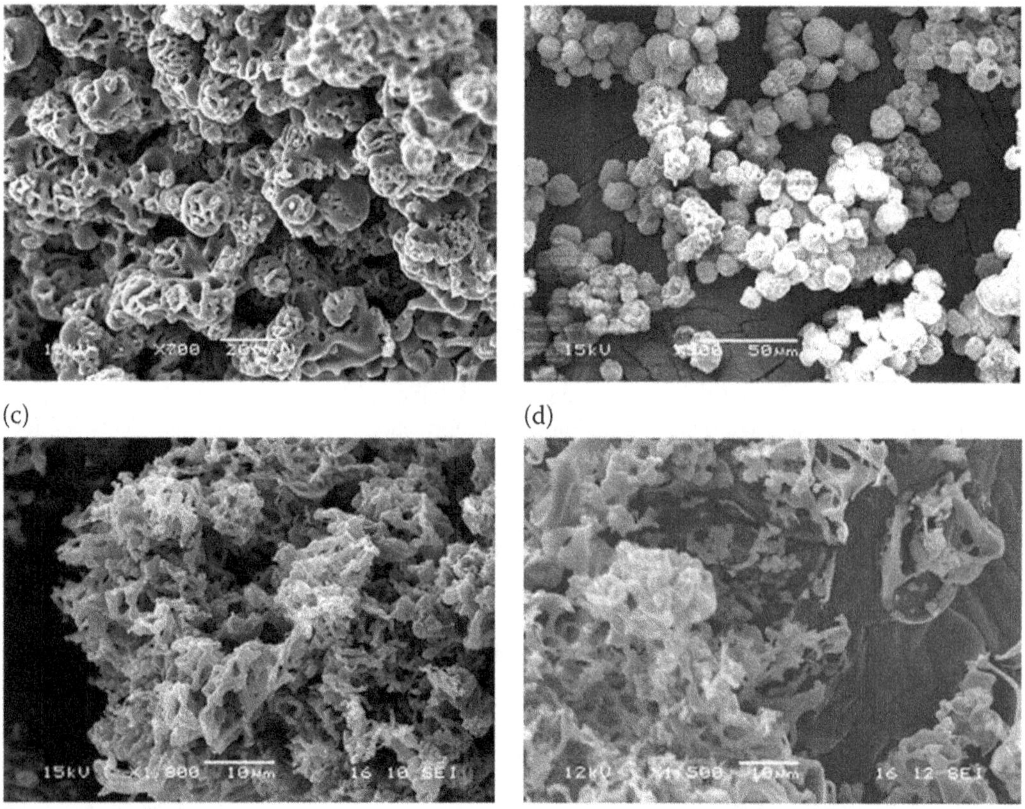

Figure 7.18 Scanning electron micrographs of spray-freeze-dried IgG particles prepared from different feed formulations: (a) Pure IgG; (b) zinc acetate:IgG = 1:1; (c) 5 times dilution with mass ratio of Zn:IgG = 1:1; (d) 10 times dilution with mass ratio of Zn:IgG = 1:13.7 (Wang et al., 2004).

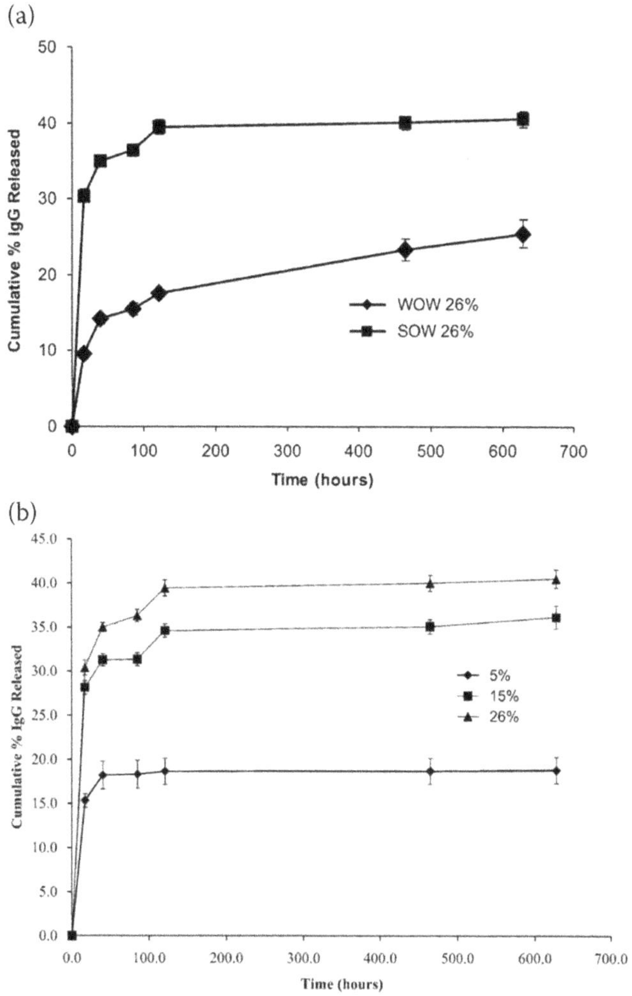

Figure 7.19 (a) Cumulative percentage of IgG released for microspheres of 26% drug-loaded PLG 50:50 microspheres fabricated using the S/O/W suspension method and the double-emulsion method for a period of 28 days; (b) Cumulative percentage of IgG released for PLG 50:50 microspheres fabricated using the S/O/W method for a period of 28 days, using various drug loadings (Wang et al., 2004).

(Figure 7.19) (Wang et al., 2004). Higher protein loading improved the diffusion gradient and enhanced its diffusion rate out of the microsphere (Gaspar et al., 1998).

Apart from using excipients and applying a two-stage non-aqueous encapsulation process comprising SFD followed by a bottom-up method, the following approaches can preserve the stability of proteins and protect them against the encapsulation-induced stresses during spray-freeze-drying:

- ***Applying the spray-freezing-into-liquid (SFL) mode of atomization instead of SFV/L:*** Rapid freezing in SFL than SFV/L reduces the time required for aggregation or diffusion of proteins to interface (water–air and water–ice), where the proteins are likely to denature. Particularly, the chances of protein denaturation at the air-particle interface are more. This was confirmed by the much less denaturation of proteins that were subjected to just SFL than that subjected to SFV/L followed by freeze-drying. During SFL, the protein feed solution is sprayed below the surface of cryogenic liquid, thus avoiding the transit of protein through the vapor gap encountered in SFV/L. This reduces the time for protein denaturation at the air–liquid interface that occurs during the SFV/L process. Further, during SFL, the atomization is intense due to the large pressure drop through the small orifice nozzle, which results in ultrafine droplets. These extremely small droplets undergo instantaneous (rapid) freezing to produce nanostructured microparticles with higher encapsulation efficiency (Yu et al., 2004). Though reduced specific surface area (SSA) improves protein stability, larger SSA is desired for the sustained release of protein. Thus, striking a trade-off between protein stability and SSA is critical while using SFD for the encapsulation of proteins (Costantino et al., 2002).

- ***Emulsification step before spray-freeze-drying:*** Adding an emulsifier to the feed formulation and homogenizing it before atomization facilitates the formation of considerably smaller and uniform droplet sizes, which on further atomization and instant freezing result in stable protein encapsulates with micron or sub-micron dimensions (Ishwarya, Anandharamakrishnan, & Stapley, 2015).

CONCLUSIONS

From the various aspects discussed in this chapter, it is apparent that encapsulation by spray-freeze-drying is governed by the choice of wall materials and feed composition. With respect to the process, the mode of atomization is influential with SFL being more advantageous than SFV or SFV/L. Nevertheless, with respect to both food and biological applications, spray-freeze-drying has established itself as an effective encapsulation process with encapsulation efficiency in the range of 80–95%. In addition, spray-freeze-drying has an edge over the conventional encapsulation techniques such as spray drying and freeze-drying, due to its ability to provide encapsulates with controlled particle size, larger specific surface area and controlled release characteristics. Potential scope for future research exists on conducting in-depth investigations on the potential factors that influence the freezing kinetics and droplet structure during spray freezing and the microstructure of spray-freeze-dried encapsulates using advanced imaging and image processing techniques. With whey protein being the commonly used wall material for encapsulation by spray-freeze-drying hitherto, novel choices of carrier substances, freezing adjuvants and excipients can be explored to enhance the encapsulation efficiency and release characteristics of SFD encapsulates. An inventive approach is the need-of-the-hour to justify the choice of spray-freeze-drying for the encapsulation of high-value food ingredients and active pharmaceutical ingredients to produce good-quality encapsulates.

REFERENCES

Agu, U. R., Ugwoke, I. M., Armand, M., Kinget, R., & Verbeke, N. (2001). *Respiratory Research*, 2(4), 198.

Ali, M. E., & Lamprecht, A. (2014). Spray freeze drying for dry powder inhalation of nanoparticles. *European Journal of Pharmaceutics and Biopharmaceutics*, 87(3), 510–517.

Anal, A. K., & Singh, H. (2007). Recent advances in microencapsulation of probiotics for industrial applications and targeted delivery. *Trends in Food Science & Technology*, 18(5), 240–251.

Anandharamakrishnan, C. & Ishwarya, S. P. (2015). *Spray Drying Techniques for Food Ingredient Encapsulation*, John Wiley & Sons, Ltd., Chichester, UK, and the Institute of Food Technologists, Chicago, IL.

Anandharamakrishnan, C. (2008). Experimental and computational fluid dynamics studies on spray-freeze-drying and spray-drying of proteins. Ph.D. thesis. Loughborough University, United Kingdom.

Anandharamakrishnan, C., Rielly, C. D., & Stapley, A. G. (2007). Effects of Process variables on the denaturation of whey proteins During Spray Drying. *Drying Technology*, 25(5), 799–807.

Anandharamakrishnan, C., Rielly, C. D., & Stapley, A. G. F. (2010). Spray-freeze-drying of whey proteins at sub-atmospheric pressures. *Dairy Science and Technology*, 90, 321–334.

Anwar, S. H., Weissbrodt, J., & Kunz, B. (2010). Microencapsulation of fish oil by spray granulation and fluid bed film coating. *Journal of Food Science*, 75(6), E359–71.

Augustin, M. A., & Sanguansri, P. (2009). Nanostructured materials in the food industry. *Advances in Food and Nutrition Research*, 58(4), 183–213.

Brunning, A. (special to C&EN) (2021). *Faking flavours with chemistry: The science of artificial flavours – in C&EN.* Available from: https://cen.acs.org/food/food-science/Periodic-Graphics-Faking-flavors-chemistry/99/i14 (Accessed 10 September 2021).

Cabral, A. C. S., Said, S., & Oliveira, W. P. (2009). Retention of the enzymatic activity and product properties during spray drying of pineapple stem extract in presence of maltodextrin. *International Journal of Food Properties*, 12(3), 536–548.

Carrasquillo, K. G., Stanley, A. M., Aponte-Carro, J. C., De Jésus, P., Costantino, H. R., Bosques, C. J., & Griebenow, K. (2001). Non-aqueous encapsulation of excipient-stabilized spray-freeze dried BSA into poly(lactide-co-glycolide) microspheres results in release of native protein. *Journal of Controlled Release*, 76(3), 199–208.

Castell, J. V., Friedrich, G., Kuhn, C. S., & Poppe, G. E. (1997). Intestinal absorption of undegraded proteins in men: Presence of bromelain in plasma after oral intake. *American Journal of Physiology-Gastrointestinal and Liver Physiology*, 273(1), G139–46.

Chandrasekaran, S.K. (1969). *Volatiles retention during drying of food ingredients.* PhD thesis. University of California, Berkeley.

Chen, K. N., Chen, M. J., Liu, J. R., Lin, C. W., & Chiu, H. Y. (2005). Optimization of incorporated prebiotics as coating materials for probiotic microencapsulation. *Journal of Food Science*, 70(5), M260–M266.

Costantino, H. R., Firouzabadian, L., Hogeland, K., Wu, C. C., Beganski, C., Carrasquillo, K. G., Cordova, M., Griebenow, K., Zale, S. E., & Trace (2000). Protein spray-freeze drying. Effect of atomisation conditions on particle size and stability. *Pharmaceutical Research*, 17(11), 1374–1383.

Costantino, H. R., Firouzabadian, L., Wu, C., Hogeland, K. C., Carrasquillo, K. G., Cordova, M., et al. (2002). Protein Spray freeze drying. 2. Effect of formulation variables on particle size and stability. *Journal of Pharmaceutical Sciences*, 91, 388–395.

Coumans, J. W., Piet, J. A. M., & Kerkhof Bruin, S. (1994). Theoretical and practical aspects of aroma retention in spray drying and freeze drying. *Drying Technology*, 12 (1 and 2), 99–149.

Crowder, T., & Hickey, A. (2006). Powder specific active dispersion for generation of pharmaceutical aerosols. *International Journal of Pharmaceutics*, 327(1-2), 65–72.

Dolly, P., Anishaparvin, A., Joseph, G. S., & Anandharamakrishnan, C. (2011). Microencapsulation of Lactobacillus plantarum (mtcc 5422) by spray-freeze-drying method and evaluation of survival in simulated gastrointestinal conditions. *Journal of Microencapsulation*, 28(6), 568–574.

Dwyer, J. M. (1984). Thirty years of supplying the missing link. History of gamma globulin therapy for immunodeficient states. *The American Journal of Medicine*, 76, 46–52.

Edwards, D. A., Hanes, J., Caponetti, G., Hrkach, J., Ben-Jebria, A., Eskew, M. L., Mintzes, J., Deaver, D., Lotan, N., & Langer, R. (1997). Large porous particles for pulmonary drug delivery. *Science*, 276(5320), 1868–1872.

Engstrom, J. D., Simpson, D. T., Lai, E. S., Williams, R. O., & Johnston, K. P. (2007). Morphology of protein particles produced by spray freezing of concentrated solutions. *European Journal of Pharmaceutics and Biopharmaceutics*, 65, 149–162.

Ezhilarasi, P. N., Karthik, P., Chhanwal, N., & Anandharamakrishnan, C. (2013). Nanoencapsulation techniques for food bioactive components: A Review. *Food and Bioprocess Technology*, 6(3), 628–647.

FAO/WHO (Food and Agricultural Organization of the United Nations and World Health Organization) (2001). *Health and nutritional properties of probiotics in food including powder milk with live lactic acid bacteria.* World Health Organization [online].

FAO/WHO (Food and Agricultural Organization of the United Nations and World Health Organization) (2002). *Joint FAO/WHO working group report on drafting guidelines for the evaluation of probiotics in food.* Food and Agricultural Organization of the United Nations [online].

Fowler, A., & Toner, M. (2006). Cryo-injury and biopreservation. *Annals of the New York Academy of Sciences*, 1066(1), 119–135.

Frenkel, C., & Havkin-Frenkel, D. (2006). The Physics and Chemistry of Vanillin. *Perfumer & Flavorist*, 31, 28–36.

Fu, K., Griebenow, K., Hsieh, L., Klibanov, A. M., & Robert Langer. (1999). FTIR characterization of the secondary structure of proteins encapsulated within PLGA microspheres. *Journal of Controlled Release*, 58(3), 357–366.

Gallage, N. J., & Møller, B. L. (2015). Vanillin–bioconversion and bioengineering of the most popular plant flavor and its de novo biosynthesis in the vanilla orchid. *Molecular Plant*, 8, 40–57.

Gaspar, M. M., Blanco, D., Cruz, M. E. M., Alonso, M. J., 1998. Formulation of l-asparaginase-loaded poly (lactide-co-glycolide) nanoparticles: influence of polymer properties on enzyme loading, activity and in vitro release. *Journal of Controlled Release*, 52, 53–62.

Gonçalves, A., Estevinho, B. N., & Rocha, F. (2016). Microencapsulation of vitamin A: A review. *Trends in Food Science & Technology*, 51, 76–87.

Gouin, S. (2004). Microencapsulation. *Trends in Food Science & Technology*, 15(7-8), 330–347.

Gutiérrez, S., Svahn, S. L., & Johansson, M. E. (2019). Effects of Omega-3 Fatty Acids on Immune Cells. *International Journal of Molecular Sciences*, 20(20), 5028.

Hale, L. P., Greer, P. K., & Sempowski, G. D. (2002). Bromelain treatment alters leukocyte expression of cell surface molecules involved in cellular adhesion and activation. *Clinical Immunology*, 104(2), 183–190.

Harshvardhan, K., Suri, M., & Goswami, A. (2017). T. Biological approach for the production of vanillin from lignocellulosic biomass (*Bambusa tulda*). *Journal of Cleaner Production*, 149, 485–490.

Heldman, D. R., & Hohner, G. A. (1974). An analysis of atmospheric freeze drying. *Journal of Food Science*, 39(1), 147–155.

Her, J.-Y., Kim, M. S., & Lee, K.-G. (2015). Preparation of probiotic powder by the spray freeze-drying method. *Journal of Food Engineering*, 150, 70–74.

Hogan, S. A., McNamee, B. F., O'Riordan, E. D. & O'Sullivan, M. (2001). Microencapsulating properties of whey protein concentrate 75. *Food Engineering and Physical Properties*, 66, 675–680.

Hundre, S. Y., Karthik, P., & Anandharamakrishnan, C. (2015). Effect of whey protein isolate and β-cyclodextrin wall systems on stability of microencapsulated vanillin by spray–freeze drying method. *Food Chemistry*, 174, 16–24.

Hutchinson, F. G., & Furr, B. J. A. (1990). Biodegradable polymer systems for the sustained release of polypeptides. *Journal of Controlled Release*, 13(2-3), 279–294.

Ishwarya, S. P., Anandharamakrishnan, C., & Stapley, A. G. F. (2015). Spray-freeze-drying: a novel process for the drying of foods and bioproducts. *Trends in Food Science & Technology*, 41, 161–181.

Ishwarya, S. P., Anandharamakrishnan, C., & Stapley, A. G. F. (2017). Spray-freeze-drying of dairy products. In: Anandharamakrishnan, C. (Ed.), *Handbook of Drying for Dairy Products*. John Wiley & Sons, Oxford, pp. 123–148.

Jouenne, E., & Crouzet, J. (2000). Effect of pH on Retention of Aroma Compounds by β-Lactoglobulin. *Journal of Agricultural and Food Chemistry*, 48(4), 1273–1277.

Karthik, P., & Anandharamakrishnan, C. (2013). Microencapsulation of docosahexaenoic acid by spray-freeze-drying method and comparison of its stability with spray-drying and freeze-drying methods. *Food and Bioprocess Technology*, 6, 2780–2790.

Koç, M., Koç, B., Yilmazer, M. S., Ertekin, F. K., Susyal, G., & Bağdatlıoğlu, N. (2011). Physicochemical characterization of whole egg powder microencapsulated by spray drying. *Drying Technology*, 29(7), 780–788.

Koster, K. L., Lei, Y. P., Anderson, M., Martin, S., & Bryant, G. (2000). Effects of vitrified and nonvitrified sugars on phosphatidylcholine fluid-to-gel phase transitions. *Biophysical Journal*, 78(4), 1932–1946.

Kumar, H., & Salminen, S. (2016). Probiotics. In B. Caballero, P.M. Finglas & F. Toldra (Eds.), *Encyclopedia of Food and Health*. Academic Press, Oxford, pp. 510–515.

Kwak, J. Y. H., Kwak, F. M. Y., Ainbinder, S. W., Ruiz, A. M., & Beer, A. E. (1996). Elevated peripheral blood natural killer cells are effectively downregulated by immunoglobulin G infusion in women with recurrent spontaneous abortions. American *Journal of Reproductive Immunology*, 35(4), 363–369.

Labuda, I. (2009). Flavor compounds. In M. Schaechter (Ed.), *Encyclopedia of Microbiology (Third Edition)*. Elsevier/Academic Press, Germany pp. 305–320.

Lam, X. M., Duenas, E. T, & Cleland, J. L. (2001). Encapsulation and stabilization of nerve growth factor into poly(lactic-co-glycolic) acid microspheres. *Journal of Pharmaceutical Sciences*, 90(9), 1356–1365.

Lavanya, M. N., Preethi, R., Moses, J. A., & Anandharamakrishnan, C. (2020). Production of bromelain aerosols using spray-freeze-drying technique for pulmonary supplementation. *Drying Technology*, 39(3), 358–370.

Liang, W., Chan, A. Y. L., Chow, M. Y. T., Lo, F. F. K., Qiu, Y., Kwok, P. C. L., & Lam, J. K. W. (2018). Spray freeze drying of small nucleic acids as inhaled powder for pulmonary delivery. *Asian Journal of Pharmaceutical Sciences*, 13(2), 163–172.

Loewen, A., Chan, B., & Li-Chan, E. C. Y. (2018). Optimization of vitamins A and D3 loading in re-assembled casein micelles and effect of loading on stability of vitamin D3 during storage. *Food Chemistry*, 240, 472–481.

Lu, W., & Park, T. G. (1995). In vitro release profiles of Eristostatin from biodegradable polymeric microspheres: Protein aggregation problem. *Biotechnology Progress*, 11(2), 224–227.

Malecki, G. J., Shinde, P., Morgan, A. I., & Farkas, D. F. (1970). Atmospheric fluidized bed freeze drying. *Food Technology*, 24, 601–603.

Mercenier, A., Pavan, S., & Pot, B. (2003). Probiotics as biotherapeutic agents: Present knowledge and future prospects. *Current Pharmaceutical Design*, 9(2), 175–191.

Michida, H., Tamalampudi, S., Pandiella, S. S., Webb, C., Fukuda, H., & Kondo, A. (2006). Effect of cereal extracts and cereal fiber on viability of lactobacillus plantarum under gastrointestinal tract conditions. *Biochemical Engineering Journal*, 28(1), 73–78.

Mishra, B., Patel, B. B., & Tiwari, S. (2010). Colloidal nanocarriers: a review on formulation technology, types and applications toward targeted drug delivery. *Nanomedicine: Nanotechnology, Biology and Medicine*, 6, 9–24.

Moeller, H., Martin, D., Schrader, K., Hoffmann, W., & Lorenzen, P. C. (2018). Spray- or freeze-drying of casein micelles loaded with Vitamin D2: Studies on storage stability and in vitro digestibility. *LWT*, 97, 87–93.

Morozov, V. N., Kanev, I. L., Mikheev, A. Y., Shlyapnikova, E. A., Shlyapnikov, Y. M., Nikitin, M. P., Nikitin, P. I., Nwabueze, A. O., & van Hoek, M. L. (2014). Generation and delivery of nanoaerosols from biological and biologically active substances. *Journal of Aerosol Science*, 69, 48–61.

Nicolai, T., Britten, M., & Schmitt, C. (2011). β-lactoglobulin and WPI aggregates: Formation, structure and applications. *Food Hydrocolloids*, 25(8), 1945–1962.

Nihant, N., Stassen, S., Grandfils, C., Jerome, R., Teyssie, P., & Goffinet, G. (1994). Microencapsulation by Coacervation of Poly(Lactide-Co-Glycolide) 3. Characterization of the Final Microspheres. *Polymer International*, 34, 289–299.

Pang, Y., Duan, X., Ren, G., & Liu, W. (2017). Comparative study on different drying methods of fish oil microcapsules. *Journal of Food Quality*, 2017, 1–7.

Parthasarathi, S., & Anandharamakrishnan, C. (2016). Enhancement of oral bioavailability of vitamin E by spray-freeze drying of whey protein microcapsules. *Food and Bioproducts Processing*, 100, 469–476.

Parthasarathi, S., Ezhilarasi, P. N., Jena, B. S., & Anandharamakrishnan, C. (2013). A comparative study on conventional and microwave-assisted extraction for microencapsulation of garcinia fruit extract. *Food and Bioproducts Processing*, 91(2), 103–110.

Patil-Gadhe, A. A., Kyadarkunte, A. Y., Pereira, M., Jejurikar, G., Patole, M. S., Risbud, A. & Pokharkar, V. B. (2014). Rifapentine-proliposomes for inhalation: In vitro and in vivo toxicity. *Toxicology International*, 21(3), 275–282.

Pérez, C., & Griebenow, K. (2001). Improved activity and stability of lysozyme at the water/CH_2Cl_2 interface: Enzyme unfolding and aggregation and its prevention by polyols. *Journal of Pharmacy and Pharmacology*, 53, 1217–1226.

Perez-Gago, M., & Krochta, J. (2001). Denaturation time and temperature effects on solubility, tensile properties, and oxygen permeability of whey protein edible films. *Journal of Food Science*, 66(5), 705–710.

Pérez-Gago, M., Nadaud, P., & Krochta, J. (1999). Water vapor permeability, solubility, and tensile properties of heat-denatured versus native whey protein films. *Journal of Food Science*, 64(6), 1034–1037.

Prata, A. S., Garcia, L., Tonon, R. V., & Hubinger, M. D. (2013). Wall material selection for encapsulation by spray drying. *Journal of Colloid Science and Biotechnology*, 2(2), 86–92.

Rajam, R., Karthik, P., Parthasarathi, S., Joseph, G. S., & Anandharamakrishnan, C. (2012). Effect of whey protein – alginate wall systems on survival of microencapsulated lactobacillus plantarum in simulated gastrointestinal conditions. *Journal of Functional Foods*, 4(4), 891–898.

Rajam, R., & Anandharamakrishnan, C. (2015). Spray freeze drying method for microencapsulation of lactobacillus plantarum. *Journal of Food Engineering*, 166, 95–103.

Ramakrishnan, S., Ferrando, M., Aceña-Muñoz, L., Mestres, M., & De Lamo-Castellví, S., & Güell, C. (2014). Influence of emulsification technique and wall composition on physicochemical properties and oxidative stability of fish oil microcapsules produced by spray drying. *Food and Bioprocess Technology*, 7(7), 1959–1972.

Ranadive, A. S. (2006). Inside look: Chemistry and biochemistry of vanilla flavor. *Perfumer & Flavorist*, 31(3), 8–44.

Reddy, P., & Jialal, I. (2020). Biochemistry, Fat Soluble Vitamins. Available from: https://www.ncbi.nlm.nih.gov/books/NBK534869/ (Accessed 10 September 2021).

Sah, H. (1999a). Stabilization of proteins against methylene chloride/water-interface induced denaturation and aggregation. *Journal of Controlled Release*, 58, 143–151.

Sah, H. (1999b). Protein instability toward organic solvent/water emulsification: implications for protein microencapsulation into microspheres. *PDA Journal of Pharmaceutical Science & Technology*, 53, 3–10.

Sanguansri, P., & Augustin, M. A. (2006). Nanoscale materials development - a food industry perspective. *Trends in Food Science & Technology*, 17(10), 547–556.

Schwab, C., Vogel, R., & Gänzle, M. G. (2007). Influence of oligosaccharides on the viability and membrane properties of *Lactobacillus reuteri* TMW1. 106 during freeze-drying. *Cryobiology*, 55(2), 108–114.

Secor, E. R., Carson, W. F., Cloutier, M. M., Guernsey, L. A., Schramm, C. M., Wu, C. A., & Thrall, R. S. (2005). Bromelain exerts anti-inflammatory effects in an ovalbumin-induced murine model of allergic airway disease. *Cellular Immunology*, 237(1), 68–75.

Semyonov, D., Ramon, O., Kaplun, Z., Levin-Brener, L., Gurevich, N., & Shimoni, E. (2010). Microencapsulation of *Lactobacillus paracasei* by spray freeze drying. *Food Research International*, 43(1), 193–202.

Sepidarkish, M., Morvaridzadeh, M., Akbari-Fakhrabadi, M., Almasi-Hashiani, A., Rezaeinejad, M., & Heshmati, J. (2019). Effect of omega-3 fatty acid plus vitamin E co-supplementation on lipid profile: A systematic review and meta-analysis. *Diabetes & Metabolic Syndrome: Clinical Research & Reviews*, 13(2), 1649–1656.

Seth, S. D., & Maulik, M. (2011). Probiotics a pharmacologist's perspective. In: Balakrish Nair, G., Yoshifumi, T. (Eds.), *Probiotic Foods in Health and Disease*, CRC Press, Taylor & Francis Group, Boca Raton, New York, Oxon, Science Publishers, Enfield, New Hampshire, USA, pp. 41–47.

Sheu, T. Y., & Rosenberg, M. (1998). Microstructure of microcapsules consisting of whey proteins and carbohydrates. *Journal of Food Science*, 63(3), 491–494.

Smyth, H. D., & Hickey, A. J. (2005). Carriers in drug powder delivery. *American Journal of Drug Delivery*, 3(2), 117–132.

Sun, Z., Ya, N., Adams, R. C., Fang, F. S., 2010. Particle size specifications for solid oral dosage forms: a regulatory perspective. *American Pharmaceutical Review*, 13(4), 70–73.

Thijssen, H. A. C., & Rulkens, W. H. (1968). Retention of aromas in drying food liquids. *De Ingenieur*, JRG 80 (47), Ch45–Ch56.

Ting, Y., Jiang, Y., Ho, C. T., & Huang, Q. (2014). Common delivery systems for enhancing in vivo bioavailability and biological efficacy of nutraceuticals. *Journal of Functional Foods*, 7, 112–128.

Tomas, M., & Jafari, S. M. (2018). *Influence of Food Processing Operations on Vitamins*. Elsevier, London, UK.

Tonnis, W. F., Mensink, M. A., de Jager, A., van der Voort Maarschalk, K., Frijlink, H. W., & Hinrichs, W. L. (2015). Size and molecular flexibility of sugars determine the storage stability of freeze-dried proteins. *Molecular Pharmaceutics*, 12(3), 684–694.

U.S. Agricultural Research Service Food Data Central (*n.d.*). Available from: https://fdc.nal.usda.gov/ (Accessed on 10 September 2021).

USP (2007). <1174> Powder flow. USP30 NF 25.

Valenzuela, A., & Valenzuela, R. (2013). Omega-3 docosahexaenoic Acid (DHA) and mood disorders: Why and how to provide Supplementation? Mood Disorders. *Intech Open*, 10.5772/53322.

Van de Weert, M., Hennink, W. E., and Jiskoot, W. (2000). Protein instability in poly(lactic-coglycolic acid) microparticles. *Pharmaceutical Research*, 17, 1159–1167.

Velikov, K. P., & Pelan, E. (2008). Colloidal delivery systems for micronutrients and nutraceuticals. *Soft Matter*, 4(10), 1964–1980.

Vllasaliu, D., Casettari, L., Fowler, R., Exposito-Harris, R., Garnett, M., Illum, L., & Stolnik, S. (2012). Absorption-promoting effects of chitosan in airway and intestinal cell lines: A comparative study. *International Journal of Pharmaceutics*, 430(1-2), 151–160.

Wais, U., Jackson, A. W., He, T., & Zhang, H. (2016). Nanoformulation and encapsulation approaches for poorly water-soluble drug nanoparticles. *Nanoscale*, 8(4), 1746–1769.

Walton, N. J., Mayer, M. J., & Narbad, A. (2003). Vanillin. *Phytochemistry*, 63(5), 505–515.

Wang, J., Chua, K. M., & Wang, C.-H. (2004). Stabilization and encapsulation of human immunoglobulin G into biodegradable microspheres. *Journal of Colloid and Interface Science*, 271(1), 92–101.

Wong, H. M., Wang, J. J., & Wang, C.-H. (2001). In vitro sustained release of human immunoglobulin G from biodegradable microspheres. *Industrial and Engineering Chemistry Research*, 40(3), 933–948.

Yamagata, K. (2017). Docosahexaenoic acid regulates vascular endothelial cell function and prevents cardiovascular disease. *Lipids in Health and Disease*, 16, 118.

Yannas, I. (1968). Vitrification temperature of water. *Science*, 160 (3825), 298–299.

Yu, Z., Garcia, A. S., Johnston, K. P., & Williams, R. O. (2004). Spray freezing into liquid nitrogen for highly stable protein nanostructured microparticles. *European Journal of Pharmaceutics and Biopharmaceutics*, 58(3), 529–537.

Yu, H., Tran, T.-T., Teo, J., & Hadinoto, K. (2016). Dry powder aerosols of curcumin-chitosan nanoparticle complex prepared by spray freeze drying and their antimicrobial efficacy against common respiratory bacterial pathogens. *Colloids and Surfaces A: Physicochemical and Engineering Aspects*, 504, 34–42.

Zayed, G., & Roos, Y. H. (2004). Influence of trehalose and moisture content on survival of *Lactobacillus salivarius* subjected to freeze-drying and storage. *Process Biochemistry*, 39(9), 1081–1086.

Web references

https://biorender.com

CHAPTER 8

Spray-freeze-dried Particles as Novel Delivery Systems for Vaccines and Active Pharmaceutical Ingredients

Development of delivery systems for drugs, vaccines and active pharmaceutical ingredients (API) is a flourishing field with continuous advancements. For any active component, the mode of delivery has a major influence on its efficacy in the human system. The delivery systems are of two types: (1) conventional delivery system (CDS) and, (2) novel delivery system (NDS). Conventional delivery systems are designed for oral or parenteral (subcutaneous, intramuscular and intravenous) administration, intended for immediate release and rapid absorption of the drug. Thus, consuming a pill or taking a jab is the common approach of CDS. But, a major apprehension with CDS is that the plasma concentration of active ingredient (AI) in any individual at a particular time can fall below the minimum effective dose or rise above the maximum safe dose, which have consequences such as toxicity and wastage of the AI. For instance, patients who are under medication for insulin-dependent diabetes mellitus (IDDM) or type-1 diabetes are vulnerable to complications resulting from wide fluctuations in blood-glucose concentrations, despite taking regular injections of insulin. These fluctuations arise mainly because of the release pattern of AI, by which its plasma concentration increases to attain a maximum (peak) and then drops (Fig. 8.1). Thus, the CD systems entail repeated administration and are suitable only for short duration applications. Moreover, active components delivered through the oral route are susceptible to the acidic pH of stomach. Their bioavailability may be reduced due to incomplete absorption or first-pass metabolism.

Apart from insulin, most of the inactivated vaccines including the subunit, split, virosome and whole inactivated virus (WIV) are administered through intramuscular or subcutaneous route. These conventional routes of vaccination suffer from various limitations. Firstly, these demand several resources such as syringes, sterile needles and the assistance of healthcare professionals. This increases the cost of vaccination and the likelihood of needle stick injuries. Fear of needles and pain reduce the preference of vaccines administered by intramuscular and subcutaneous routes (Amorij et al., 2010). Many people are vulnerable to infection even after vaccination through conventional routes. This can be overcome by eliciting a strong immune response at the entry site of the virus, which is possible with the pulmonary route of vaccine administration as it is capable of neutralizing the virus in the upper respiratory tract. Vaccine stability is yet another parameter of importance. The currently available vaccines administered through conventional route, for instance, the influenza vaccines and COVID-19 vaccines require refrigerated storage and transport. All the above factors necessitate the development of vaccines that are shelf-stable under ambient conditions and can be self-administered through a non-parenteral route, which would initiate effective immune response at a relatively low dose (Murugappan, 2014). Thus, a vaccine administered through non-conventional routes is perceived as an effective approach to simplify mass vaccination drives mandated during a pandemic situation.

DOI: 10.1201/9781003019312-8

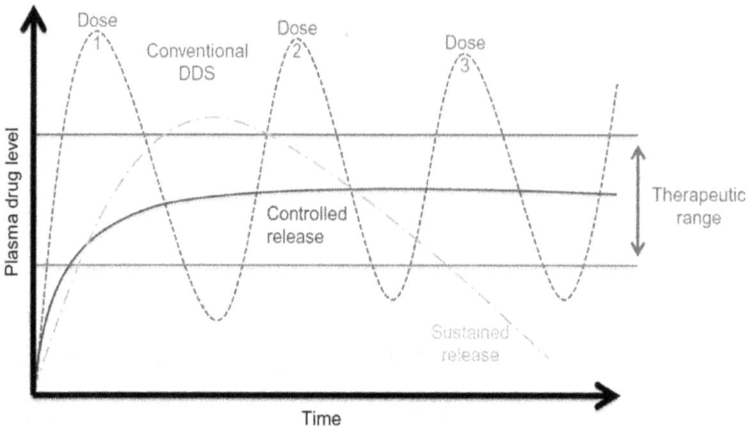

Figure 8.1 Controlled release and sustained release vis-à-vis conventional delivery systems (Abu-Thabit & Makhlouf, 2018).

In the above context, the novel delivery systems are designed to improve the therapeutic effectiveness of an active ingredient, over and above that offered by the conventional delivery systems. A novel delivery system accomplishes the above by the controlled release (CR) approach (Jain, 1995), through pulmonary or transdermal routes of administration. In contrast to the CDS, CR systems deliver an active ingredient to a specific site at a predefined rate and concentration, for a definite duration, which can be in the order of days to years without being influenced by the environmental conditions (Fig. 8.1). Consequently, a constant concentration of the AI is maintained in the blood or target tissue. The controlled release process might exhibit the following profiles (McClements, 2012):

- ***Burst release:*** quick release of most of the encapsulated compound in a short time;
- ***Sustained release:*** facilitate slow and extended release of AI for a prolonged period of time at a constant rate (Sahilhusen & Mukesh, 2014);
- ***Triggered release:*** the active compound is released in response to environmental stimuli such as pH, ionic strength, enzyme activity or temperature;
- ***Targeted release:*** promote delivery of the entrapped AI to a specific site of the gastrointestinal (GI) tract (i.e. mouth, stomach, small intestine or colon).

In addition to release pattern, the novel delivery systems are capable of mitigating the shortcomings with respect to production and properties of active ingredients, especially, in improving the stability of vaccine formulations. As per a recent market research report, the global market for novel drug delivery systems is expected to grow at a compounded annual growth rate (CAGR) of approximately 2.2% until 2024. Accordingly, the market value for NDS is estimated to reach 30300 million USD in 2024, which is 12.5% more than that attained in 2019 (26500 million USD) (Ahmad, 2021). The novel delivery systems outweigh the CDS in terms of their ability to maintain the concentration of AI in blood within the therapeutic limits, thus preventing wastage and toxicity.

With novel delivery systems, the active ingredients are mainly delivered via the pulmonary and transdermal routes, in which dry powders function as delivery vehicles. The advent of spray-freeze-drying (SFD) process is a key milestone in the successful development of stable dry powders for inhalation and transdermal delivery systems. As already discussed in Chapter 7, SFD is capable of producing powder particles with an aerodynamic particle diameter in the desirable range of 1–5 μm that can be effectively deposited in the lungs to elicit the intended therapeutic action. This is beneficial as WIV for pulmonary delivery must be formulated into powder particles with an aerodynamic particle size raging between 1–5 μm (Frijlink & de Boer, 2005; Zanen, Go, & Lammers, 1994; Westerman, Heijerman, & Frijlink, 2007), which is certainly possible by spray-freeze-drying (Amorij et al., 2007). Also, studies have demonstrated that SFD is capable of producing vaccines in a stable solid-state with superior physical and immunogenic stability. The same is achieved by incorporating the live and attenuated viruses in a suitable glassy matrix of sugars such as dextran, inulin and trehalose (Huang et al., 2004; Maa et al., 2004; Wang, 2000). This chapter is intended to present the fundamentals of novel delivery systems and the applications of spray-freeze-drying in fabricating dry powder systems suitable for administration through the pulmonary and transdermal routes.

8.1 RATIONALE AND APPLICATIONS OF SPRAY-FREEZE-DRYING FOR THE DEVELOPMENT OF PULMONARY DELIVERY SYSTEMS

8.1.1 Pulmonary delivery system – A prelude

Pulmonary delivery system can be defined as *'a non-invasive system for the systemic delivery of therapeutic macromolecules through inhalation of the formulation via mouth followed by its deposition in the lower respiratory tract (the lungs, comprising the trachea, bronchi, bronchioles and alveoli) and alveolar absorption'*. The large alveolar surface area, thin epithelial membrane lining and a high degree of vascularization[#] in the alveolar region promote the rapid absorption of an active ingredient at a low dose (Courrier, Butz, & Vandamme, 2002; Gill et al., 2007). In the lungs, gaseous exchange takes place through more than 280 billion capillaries. The blood flow rate as high as 5700 mL/min facilitates the rapid absorption of AI administered through the pulmonary route (Okamoto & Danjo, 2008). Also, the likelihood for enzymatic degradation of the AI is less during pulmonary delivery owing to the low enzyme activity in lungs (Chaurasiya & Zhao, 2021). But, these advantages are realized only when the premature clearance of dry powder particles for inhalation (DPI) from the deposition site (mucociliary clearance: phagocytosis by alveolar macrophages) is avoided. Early clearance of DPI reduces the concentration of its dosage delivered at the target site, which can be avoided by tailoring the particle morphology (size and shape) for optimal inhalation (Hamedinasab et al., 2020). Hence, it is imperative to understand the specifications of particles required for pulmonary delivery.

The particle size, more specifically, the *'aerodynamic particle diameter'* of DPI is an important parameter that governs its site of deposition in the respiratory tract and thereby the efficacy of pulmonary delivery (Fig. 8.2). The aerodynamic diameter of a particle can be defined as the diameter of a sphere having a density of 1 g cm^{-3} (*cf.* density of water), which settles in stagnant air at the same velocity as the particle of interest (Cooper & Alley, 1990). Its relationship with the geometric diameter of the particle is governed by the following equation:

$$Aerodynamic\ \ diameter = Geometric\ \ diameter \times \sqrt{\frac{\rho_P}{\rho_0 \chi}} \tag{8.1}$$

Where ρ_p and ρ_0 are the particle and unit densities, respectively and χ is the dynamic shape factor (De Boer et al., 2002). Moreover, for particles finer than 100 μm, their aerodynamic diameters are larger than the geometric diameters and the vice-versa holds true for particles coarser than 100 μm (Chen & Fryrear, 2001). In addition to the mean aerodynamic diameter of the DPI, their

[#] The term *'vascularization'* refers to the presence of a capillary network that is capable of delivering nutrients/drugs to the cells (Rouwkema, Rivron, & van Blitterswijk, 2008).

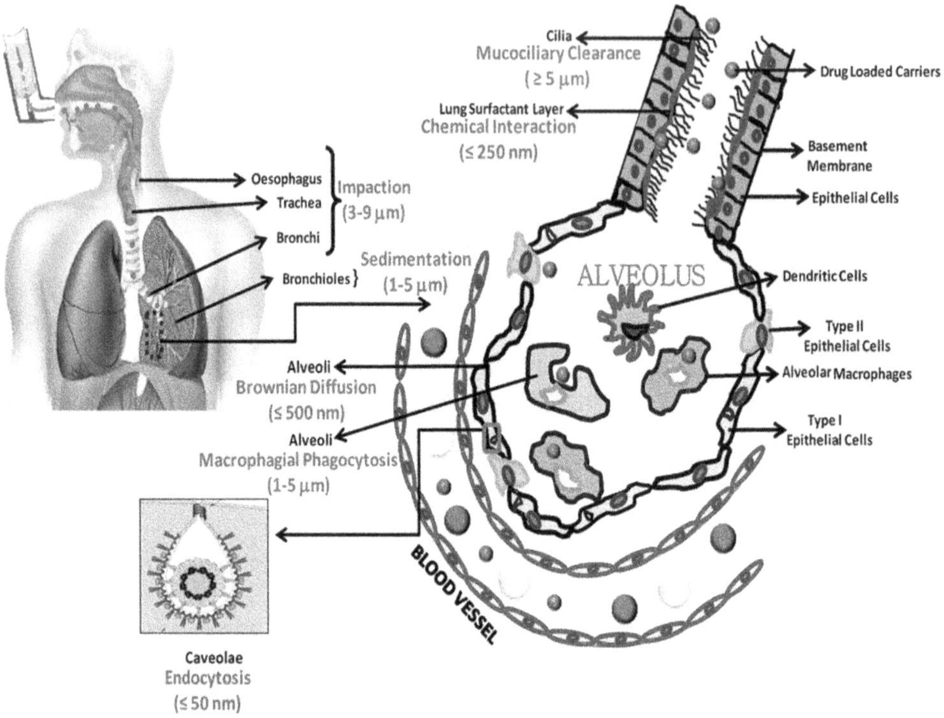

Figure 8.2 Influence of particle size of the DPI on their site of deposition in the pulmonary system, their clearance and absorption (Dhand et al., 2014).

aerodynamic particle size distribution (APSD) is also relevant, which is a virtual measurement of the diameter of a particle in airflow. APSD is also vital in controlling the site of particle deposition in the human respiratory tract (de Boer et al., 2002 & 2017).

According to an established theory, the aerodynamic diameter of dry powder inhaler particles should be in the range of 1–5 μm for an effective deep lung deposition (Bosquillon et al., 2001; Byron et al., 2010; Kanig, 1963; Shi et al., 2009). Sedimentation causes the particles of above size range to be deposited in the bronchiolar region. Particles with size larger than 5 μm are deposited in the upper respiratory tract and cleared by the mucociliary clearance mechanism (Stahlhofen et al., 1990). Brownian diffusion* causes the smaller particles (<1 μm) to be deposited in deeper alveolar regions, from where these are quickly absorbed and hence pose a risk of systemic toxicity (Dhand et al., 2014). Due to their low inertia, particles smaller than 0.5 μm are exhaled without being deposited (Heyder & Rudolf, 1984; Paranjpe & Müller-Goymann, 2014; Yang, Peters, & Williams, 2008). In addition, particles with size less than 1 μm exhibit higher inter-particle cohesiveness than those with size larger than 1 μm. Aerosol formation is hindered by high as well as low interparticle cohesiveness (Chaurasiya & Zhao, 2021). At the same time, round particles or particles with size in the range of 1.5–3 μm are more prone to mucociliary clearance (El-Sherbiny et al., 2011). Thus, for a successful deep-lung deposition, the inhaled particles must be adequately small to prevent deposition in the upper airways by sedimentation impaction and large enough to escape exhalation (Dhand et al., 2014).

* Brownian diffusion is the random wiggling motion of small airborne particles in static air, due to the continuous bombardment by neighboring gas molecules (Liu, 2010).

8.1.2 Quantifying the performance of aerosolized dry powder particles: Cascade impactor analysis

Generally, the *in vitro 'cascade impactor analysis'* is used to obtain the aerodynamic particle size distribution of aerosolized dry powder particles generated from a medical aerosol generator. An impactor utilizes the *'inertial aerodynamic forces'* to separate particles of different size classes (Fig. 8.3[a]). Inside an impactor, the particles are carried through an orifice along with a gas stream and directed towards an impaction or collection plate that is usually filled with either demineralized water or agar. The impaction plate abruptly changes the direction of particle flow into a 90° turn. Extremely slow particles cannot follow the change in flow direction and hence collide with the plate. Cascade impactors are constructed by combining individual impactors in series (Fig. 8.3[b]) that are arranged in the descending order of cut diameter. Cut diameter is the smallest aerodynamic diameter that is retained by a stage, which is more for the top stage than the one beneath. The reduction in cut diameter from one stage (top) to another (bottom) is achieved by changing the nozzle diameter as smaller nozzles create higher particle impaction velocities (Lindsley et al., 2017). Alternatively, the number of orifices and the distance of collection plate from the orifice plate can also be changed.

A typical six-stage cascade impactor is shown in Fig. 8.3(c). Here, a strong jet of air emanating from the device nozzles impacts on flat sampling surfaces. Larger particles are impacted and collected first on the stages of lower velocity, followed by the smaller particles that pass and get collected on the stages of higher velocity. In other words, particles with adequate inertia will impact on the related collection plate of specific stage, finer particles would remain entrained in the air jet and proceed to the next stage, wherein the process recurs. Consequently, each stage collects finer sized particles than its preceding stage (Lashkar, 2012). Therefore, particles collected from stages 3 to 5 (low impactor stages) with aerodynamic diameter less than or equal to 5.0 μm signify the respirable-sized drug dose. On the other hand, particles deposited in the throat and stage-1 (high impactor stages) represent the oropharyngeal deposition in the oropharynx region (Niwa, Mizutani, & Danjo, 2012; Ali & Gary, 2015). The collection plates are separable, such that the particles collected can be weighed and analyzed. After the last stage, a dry glass filter (Fig. 8.3[b]) is present, which collects the particles that are smaller than the cut diameter of particles collected in the preceding stage (Sjoholm et al., 2001). A plot between the cumulative mass of the particles and the upper limit of the particle size range corresponding to each stage is known as the *'impactor collection efficiency curve'* or the particle size distribution graph (Fig. 8.3[d]).

Other allied parameters of importance associated with the particle size and size distribution of DPI formulations are the mass median aerodynamic diameter (MMAD) and its geometric standard deviation (GSD), fine particle dose (FPD) and fine particle fraction (FPF). The definitions of these are given in Table 8.1. The size of drug particles should be respirable with MMAD ≤ 5.0 μm, so as to obtain high therapeutic efficacy. Also, a monodispersed or uniform particle size distribution indicated by low GSD is essential for deep lung deposition and systemic pulmonary delivery. But, most of the therapeutic aerosols are polydisperse with their GSD values ranging between 2 and 3. Aerosols are said to be polydisperse if their GSD ≥1.22 (ersjournals, *n.d.*). Further, the FPF should be adequately high for drug aerosolization efficiency (Banga, 2015; Peng et al., 2016).

Apart from size and shape, the other interrelated particle characteristics that influence the aerosolization, ejection from treatment devices and bioavailability of the DPI formulations are moisture content, hygroscopicity, bulk density, surface morphology and surface charge. Relative to particles with smooth surface, those with porous surface are less cohesive and more dispersible, as the irregular surface prevents inter-particle proximity (Chew & Chan, 2002). This is because, particles with irregular shape have less contact area with weak Van der Waals forces (attraction of intermolecular forces), along with a low tendency for aggregation (Zeng et al., 2000). For a successful pulmonary delivery, low hygroscopicity of particles is preferred as moisture uptake by

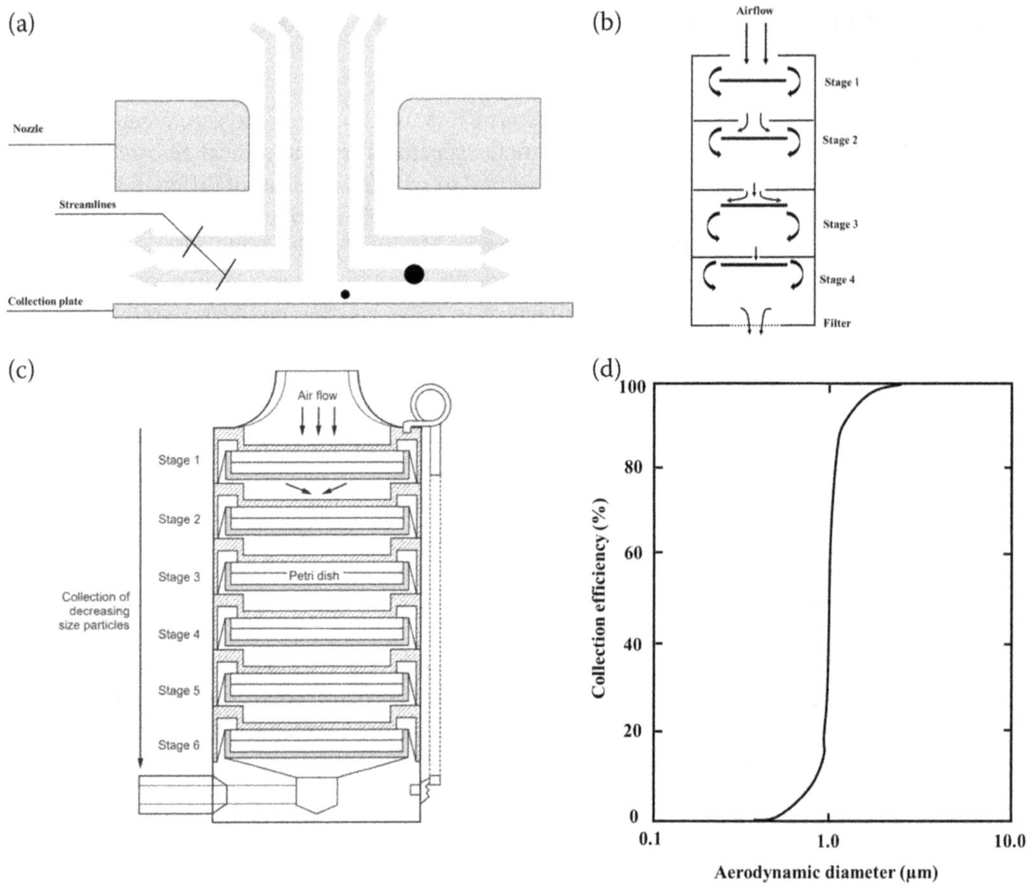

Figure 8.3 (a) Working principle of the impactor device; (b) General construction of a cascade impactor (Adapted and redrawn from Cooper & Alley, 1990); (c) Schematic representation of a 6-stage Anderson Cascade impactor (Pepper, Gerba & Brendecke, 2011); (d) impactor collection efficiency curve (Adapted and redrawn from Cooper & Alley, 1990).

Table 8.1 Lexicon of commonly used terms related to pulmonary delivery (ersjournals, *n.d.*)

Term	Definition
Mass median aerodynamic diameter (MMAD)	The diameter at which 50% (by mass) of the aerosol particles are larger and 50% are smaller. In other words, MMAD divides the particle size distribution of aerosol into two halves.
Geometric standard deviation (GSD)	A measure of the spread of particle size distribution, given by the ratio of median diameter to the diameter at ±1 SD (σ) from the median diameter.
Total emitted dose (TED)	The mass of drug discharged per actuation, which is actually available for inhalation.
Fine-particle dose (FPD)	Mass fraction of the particles with size less than 5 μm in size, with respect to total emitted dose.
Fine-particle fraction (FPF)	Ratio between fine particle dose and total emitted dose.

DPI increases the bulk density and changes the surface charge and aerodynamic size of powder particles (Zhu et al., 2007). And, particles with bulk density less than 0.4 g/cm^3 are suitable for aerosolization and deep lung deposition (Chaurasiya & Zhao, 2021).

Similarly, surface charge of DPI is also related to its size, shape and surface morphology. Smaller particles offer larger active surface area to transfer surface charge, which increases the inter-particle cohesiveness and the cohesiveness between particles and surface wall of the inhaler device and reduces the fine particle fraction (FPF). Likewise, elongated and rough particles are more likely to obtain surface charges than spherical and smooth particles (Chakrabarty et al., 2008), due to enhanced inter-particle and particle-surface contact areas (Vladykina, Deryagin, & Toporov, 1985; Guo et al., 2015). Aerosolization behavior of DPI depends on the surface charge of particles. When aerosolized in the device during inhalation, the powder becomes substantially charged, which are then transferred to the active ingredients (Mehrani, Bi, & Grace, 2007). Surface charge of particles influence the deposition brought about by cohesive attraction and are vital for the deposition in lower airways (Hickey, 2004).

Besides the advantages mentioned initially, the major merits of pulmonary delivery over conventional delivery systems are:

i. non-invasiveness;
ii. greater permeability, rapid drug uptake and improved bioavailability than intravenous dosage forms due to the larger surface area of DPI (~100 m^2) for solute transport (Laube, 2005; Patton & Byron, 2007);
iii. rapid onset of drug action;
iv. targeted or direct delivery of active ingredient to the affected area;
v. low dosage requirement to obtain the desired pharmacological action, which leads to reduced systemic side effects;
vi. the AI can escape the intestinal and hepatic first-pass metabolism;
vii. prospect of maintaining product sterility during administration (Banker & Rhodes, 2002; Cole & Mackay, 1990; Hillery et al., 2001; Sciarra & Cutie, 1991);
viii. possibility of self-administration without the assistance of medical professionals;
ix. eliminates the need for needles, syringes and thereby prevents needle-stick injury;
x. capability to induce local immune response at the entry site of respiratory viruses, besides eliciting systemic response (Cassetti, Katz, & Wood, 2006).

Thus, spray-freeze-drying is an appropriate technique for the production of DPI, as it is known for its ability to produce stable dry powders with low bulk density and porous microstructure. In general, the highly porous, less-dense and spherical particles produced by SFD demonstrate good aerosol behavior (Emami et al., 2018). Also, SFD produces relatively larger particles (geometric diameter) than spray drying, as the atomized feed droplets might gradually agglomerate and solidify during their transit through the vapor phase before reaching the surface of the cryogenic liquid (Ishwarya & Anandharamakrishnan, 2015; Kawabata et al., 2011). The aforesaid features of SFD is relevant from two perspectives. According to a theory for spherical particles, MMAD is given by the product of geometric diameter and square root of particle density (Vanbever et al., 1999). Hence, a combination of low density and large geometric diameter for DPI can result in a MMAD that falls in the range suitable for inhalation, due to the reduced particle adhesion and cohesiveness (Edwards, Hanes et al., 1997). SFD is a feasible method to produce particles with wide particle size distribution and low density to target both the lower airways and the main bronchioles (Amorij et al., 2007; Mohri et al., 2010).

Further, unlike freeze-drying, SFD particles seldom require an additional particle size reduction step. This aspect of SFD reduces processing-induced activity loss of APIs and facilitates the blending of dry powder aerosols with an additional mucoadhesive compound (Garmise, Staats, & Hickey, 2007). This feature of SFD is relevant as DPIs prepared by freeze-drying followed by

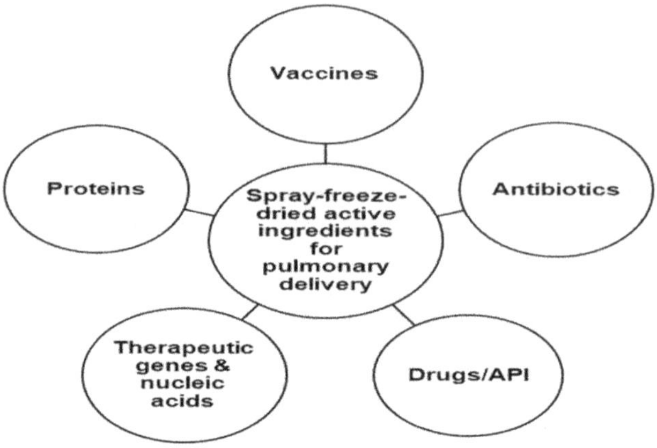

Figure 8.4 Types of active ingredients processed by spray-freeze-drying for pulmonary delivery.

milling have reduced efficacy due to the leakage of AI caused by stresses induced during the individual drying and milling operations. Consequently, substantial amount of AI are lost, which promotes the auto-adhesive properties of powder that lead to suboptimal dispersion of the AI (Desai et al., 2002). However, the instability of proteinaceous active ingredients during the atomization step limits the aerosolization behavior of SFD particles due to aggregation tendency. Loss of protein stability during SFD can be avoided by adding suitable stabilizing excipients to the feed formulation. The commonly used excipients are polyols (ex. mannitol), sugars (ex. lactose, trehalose), oligosaccharides (ex. inulin, β-cyclodextrin) and surfactants (ex. Tween 20, Tween 80) (Grasmeijer et al., 2013; Mensink et al., 2017; Depreter, Pilcer, & Amighi, 2013).

8.1.3 Spray-freeze-dried particles for pulmonary delivery: Pertinent case-studies

Hitherto, various investigators have employed spray-freeze-drying to produce dry powder inhaler formulations containing different categories of active ingredients (Fig. 8.4). This section of the chapter is aimed at providing insights to the inferences from earlier studies, with specific emphasis on the following aspects: (1) the rationale of choosing SFD for DPI production over the conventional techniques (spray drying or freeze-drying followed by milling); (2) optimal conditions of SFD that lead to particles effective in pulmonary delivery; (3) typical characteristics of SFD-DPI and (4) the competitive edge of SFD over conventional techniques for DPI production.

8.1.3.1 Vaccine powders

An inulin-stabilized influenza subunit vaccine was produced by spray-freeze-drying. The feed solution in HEPES buffered saline (HBS: 2mM Hepes; 150 mM NaCl; pH 7.4) comprised the heamagglutinin (HA; antigen) and inulin (4 kDa) at concentrations of 275 μg/ml and 5.5% (w/v), respectively. SFV/L followed by vacuum freeze-drying for 43 hours was adopted for the vaccine production. Aerosol behavior of the vaccine powder was characterized using the cascade impactor and laser diffraction analyses. Further, pulmonary immunization was tested by delivering the dry powder vaccine into the lungs of BALB/c mice by endotracheal insufflation, using a dry powder insufflator. Biochemical and antigenic properties of the vaccine powder were evaluated to determine its antigen-specific antibody responses and to verify that its conformation and antigenicity were not affected. Besides, the structural integrity of viral subunit in the SFD powder was tested using a proteolytic assay.

As expected, the SFD vaccine powder particles were large, spherical and porous. All the above characteristics are favorable for the deposition of vaccine over the large surface area of lungs. Larger the particle size, the higher their site of deposition in the airways. With respect to aerodynamic properties, the aerosolized vaccine powder had a FPF of 38% and an aerodynamic diameter of 5.3 μm with polydisperse (broad) size distribution, all of which facilitate deposition in the lower airways. Thus, the ability of SFD vaccine powder to be deposited throughout the lungs was ascertained. From polyacrylamide gel electrophoresis (PAGE), it was evident that the protein band related to the intact HA monomer (~75 kDa) was identical to that of the SFD powder. Similarly, the antigenic integrity of HA was also intact after SFD, with a potency recovery of 104%.

The reconstituted SFD vaccine powder stimulated an immunogenic response (IgG titer) that was equivalent to that elicited by the unprocessed influenza subunit vaccine in BALB/c mice. Notably, the IgG titer caused by the SFD-vaccine powder was higher than that caused by intramuscular (*i.m*) administration and liquid vaccine aerosols (*l.i*) delivered via the pulmonary route using liquid insufflators. Furthermore, *i.m* and *l.i* immunization of mice elicited only minor IgA titers in the nasal and lung lavages. Contrarily, pulmonary immunization with SFD influenza vaccine powder induced significantly higher influenza specific IgA titers in both nose and lungs in seven and eight mice, respectively (Amorij et al., 2007). This is of high relevance as induction of IgA (the main effector antibody that delivers mucosal immunity) in the respiratory tract is highly advantageous in terms of offering protection against influenza virus at its port of entry (Clements et al., 1986; Renegar & Small, 1991).

The above said capability of SFD vaccine to induce strong mucosal immune responses in the respiratory tract via the pulmonary route is an edge over the spray-dried influenza subunit vaccine powder. The latter was found to augment the serum IgG antibody levels after pulmonary immunization of rats, but did not elicit mucosal IgA antibodies in the respiratory tract (Smith et al., 2003). The higher solubility in lung fluids and broader size distribution of SFD particles could be the possible reason for the difference in the immunogenic responses. Having a particle size of 1–5 μm, the spray-dried formulation used by Smith et al. (2003) was found suitable for delivery to the lower airways. Also, the spray-dried formulation promoted phagocytic clearance from the lung by alveolar macrophages (Geiser, 2002), which led to inefficient stimulation of immune responses (Ferro et al., 1987; Gong et al., 1994; Thepen et al., 1992). Overall, the above study by Amorij et al. (2007) showed that the biochemical integrity, structural integrity and immunogenicity of the vaccine were not affected by either the SFD process or the use of inulin as a stabilizer. Thus, the feasibility of vaccination with a non-adjuvanated SFD subunit influenza vaccine powder via pulmonary administration was established.

In a later study, a comparative analysis was done to determine the best stabilizer for the preparation of SFD-influenza vaccine powder for pulmonary delivery. The stabilizers considered for the study were inulin, dextran and a mixture of dextran and trehalose. Regardless of the type of stabilizer used, all the SFD-vaccine powders showed large specific surface area (SSA) due to their porous microstructure as depicted by SEM. However, powders with inulin and dextran as excipients had a greater SSA (100–120 m^2/g) than the dextran-trehalose mixture (~75 m^2/g). The authors pointed out that the higher viscosity of inulin and dextran-containing feed solutions led to slower nucleation of ice crystals during the freezing stage, eventually leading to smaller pore sizes of final particles, than those of SFD dextran-trehalose particles.

Further, the SFD powders were reconstituted and visualized by transmission electron microscopy (TEM) to evaluate the effect of SFD process on the structural integrity of whole inactivated virus (WIV) that governs its immunogenicity. The TEM micrographs showed that the stress associated with SFD process did not influence the structural integrity and particulate nature of WIV in the reconstituted aqueous solutions of SFD-WIV containing inulin, dextran or dextran-trehalose. The images revealed the presence of WIV particles having size in the range of 100–150 nm, with evident spikes on the viral membrane (Figs. 8.5[a–c]). This further confirmed the protective ability

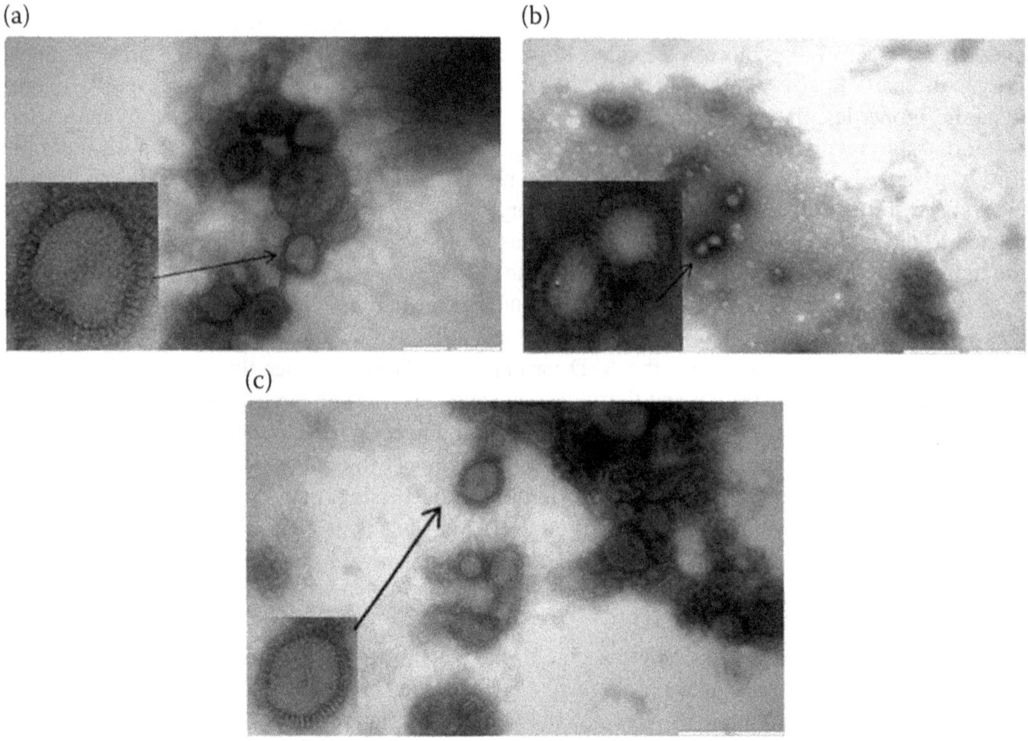

Figure 8.5 TEM images of spray-freeze-dried WIV in the presence of (a) inulin; (b) dextran; (c) dextran-trehalose, after reconstitution in water (Murugappan et al., 2013).

of the stabilizers used in this study. Not limited to the fresh powders, the stabilizers protected the immunogenicity of antigen during a 3-month storage period at temperatures up to 40°C. However, the same was not observed with unprocessed WIV, which lost its hemagglutination titer when stored at −20 and 4°C for 1 month. The storage stability of SFD-WIV was further confirmed by the unaffected protein patterns observed in the PAGE experiment (Murugappan et al., 2013).

Contrary to the above results, when dispensed through a dry powder insufflator, the stored SFD-vaccine powders induced significantly lower antibody (IgA in the nose and bronchoalveolar lavage washes of mice and serum IgG) and hemagglutination inhibition (HI) titers than the fresh samples. The SFD powders showed very poor dispersing capacities as both the primary particles and large agglomerates exited the insufflator. This hampered delivery to lungs indicates the possible changes in the physical properties of SFD vaccine powders during storage, which were not detected by laser diffraction or BET experiments that employed a RODOS dispenser, which is known to be a very powerful disperser (Jaffari et al., 2013). The above observation indicates the importance of the inhaler used during a real-time clinical situation (Murugappan et al., 2013). The authors suggested that the powder delivery and thereby the elicited immunogenicity can be improved by using inhalers such as Novolizer® or Twincer® (Fig. 8.6), which are superior powder dispersers with good de-agglomeration efficiency than the dry powder insufflators used for research purpose (de Boer et al., 2002; Fenton et al., 2003; Friebel & Steckel, 2010; Saluja et al., 2010). The technical features of Novolizer® and Twincer® that render them effective powder dispensers are presented in Table 8.2.

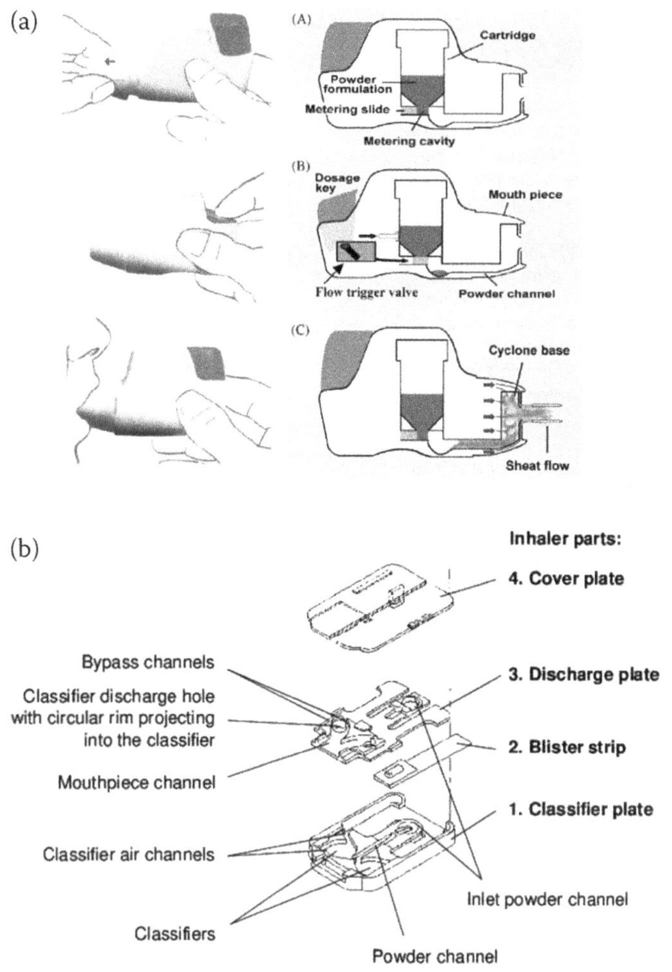

Figure 8.6 Inhalation procedure and device characteristics of (a) Novolizers® (Kohler, 2004); (b) Twincer® (de Boer et al., 2006).

8.1.3.2 Antibiotics

Ciprofloxacin is a fluoroquinolone antibiotic, which is commonly used to treat mild-to-moderate infections in the urinary and respiratory tract caused by susceptible organisms. It acts by binding to the bacterial DNA gyrase, which is the key enzyme for DNA replication. Gram-negative bacteria are more sensitive to the action of ciprofloxacin than Gram-positive bacteria (National Center for Biotechnology Information, 2021). Conventionally, liposomal formulations of ciprofloxacin were produced by lyophilization and jet milling, the limitations of which were mentioned in Section 8.1.1. Hence, spray-freeze-drying has been used as an alternative technique to prepare liposomal ciprofloxacin powder. The phospholipid-to-ciprofloxacin and lactose-to-ciprofloxacin ratios in the feed formulation were 5:1 and 17:1, respectively. Strikingly, the highly porous SFD product with large surface area showed improved MMAD (2.8 ± 1 μm) and FPF (60.6 ± 12.2%) than those manufactured by jet-milling (FPF: 45%). Moreover, the particles demonstrated high encapsulation efficiency (>70%) in pulmonary fluids such as Bovine mucin, porcine lung lavage and cystic fibrosis sputum (diluted 5X). Also, the SFD-liposomal ciprofloxacin particles exhibited a spontaneous *in vitro*

Table 8.2 Technical features of Novolizer® and Twincer® (Based on the information from de Boer et al., 2006; Kohler, 2003 & 2004)

Novolizer®	Twincer®
• Features an inspiratory flow rate threshold that elicit a multiple feedback mechanism to the patient each and every time a defined minimum inspiratory flow rate (minimum 35–50 l/min) is attained. • Confirmation of every correct inhalation by optical, acoustic and taste feedback (Fig. 8.8). • A color change from green to red in the lower window of the device and a 'click' sound indicates the exceeding of minimum inspiratory flow threshold. Following the trigger, the dose counter system is reset. • In addition, presence of lactose as drug carrier particles initiates a sweet taste in the patient's mouth to confirm the release of powder from the inhaler. • Low to medium airflow resistance and a smooth rise of pressure drop at higher flow rates during inhalation.	• The basic construction includes three plate-like parts and a blister strip for the powder formulation with the micronized drug. The plate-like parts have many projections and depressions which comprise the air flow passages and the blister chamber. • The ratio between the diameter of discharge hole and the classifier chamber and the height of the rims around the discharge holes can be varied to control the residence time of powder in the classifier and the retention efficacy of carrier or sweeper particles. • Bypass channels around the classifier discharge holes in the discharge plate are used to reduce the inhaler accumulations and to control the total inhaler resistance.

formation of liposomes in aqueous media, with greater efficiency in ionic solutions. A dry powder aerosol that is capable of lipid formation in the airway surface liquid (ASL) is advantageous as it removes the delicate liposomal particles from its production and delivery stages. Due to its improved MMAD, FPF and spontaneous dissolution ability, the ciprofloxacin released from the SFD-liposomal dry aerosol powder was found to deposit in the most distal tracheobronchial generation, at a concentration of 5 mg/L out of a dosage of 20 mg powder (Sweeney et al., 2005). The above concentration is significantly more than the minimum inhibitory concentration (MIC) of various bacteria that cause respiratory infection, such as, *Pseudomonas aeruginosa* (4 mg/l), *Streptococcus pyogenes* (1 mg/L), *Neisseria gonorrhoeae* (0.004 mg/L), *Bacillus anthracis* (1.6 mg/L) (Zhanel et al., 2002).

Similar to the above, an inhalable liposomal dry powder of clarithromycin (CLA) was prepared by ultrasonic spray-freeze-drying (USFD: ultrasonic atomization → rapid freezing → freeze-drying). Different cryoprotectants such as sucrose, mannose and trehalose were used (Ye et al., 2017). Clarithromycin is a semi-synthetic macrolide antibiotic that is used to treat a wide range of bacterial infections, including the elimination of *Helicobacter pylori* in the treatment of peptic ulcer disease. It inhibits bacterial growth by binding to the 50S ribosomal subunit and interrupting their RNA-dependent protein synthesis (National Center for Biotechnology Information, 2021). The rationale of adopting USFD is to leverage its ability to result in large porous particles with a narrow size distribution (Bi et al., 2008; Ye et al., 2017). True to the above fact, USFD-CLA particles showed spherical shape and rough porous surface morphology with micron-scale particle size, high aerosolization efficiency (>85%) and FPF (up to 50%). Moreover, the drug recovery was as high as 85%. When rehydrated, the liposomal CLA showed 80% encapsulation efficiency with narrow particle size distribution (polydispersity index, PDI < 0.4). The product showed superior storage stability after 3 months of storage at 25°C and 60% RH. Among the different cryoprotectants used, sucrose demonstrated the best protective effect during freeze-drying and mannitol exhibited good moisture protection effect owing to its crystalline nature. The latter was stated as the reason for the high stability of USFD-CLA-DPI to the high humidity storage milieu. Addition of 5% sucrose and 15% mannose (w/w) to the feed formulation imparted both moisture protective effects and aerosolization efficiency upon the liposomal dry powder prepared using USFD. Further, increasing the mannitol concentration enhanced the porosity and aerosolization efficiency of USFD-CLA-DPI (Ye et al., 2017).

In yet another study, the dry powder aerosol of kanamycin was prepared by spray-freeze-drying method. Kanamycin (2-(aminomethyl)-6-[4,6-diamino-3-[4-amino-3,5-dihydroxy-6-(hydroxymethyl) tetrahydropyran-2-yl]oxy-2-hydroxy-cyclohexoxy]-tetrahydropyran-3,4,5-triol) is an aminoglycoside bactericidal antibiotic, which is isolated from the bacterium *Streptomyces kanamyceticus*. It is useful in treating a wide variety of bacterial infections, tuberculosis and gonorrhea by administration through the oral, intravenous and intramuscular routes (National Center for Biotechnology Information, 2021; Puius, Stievater, & Srikrishnan, 2006). Raw kanamycin is an amorphous material, which renders its conversion into dry form, a challenging task. Usually, amorphous APIs are more sensitive to degradation and have a tendency to be more hygroscopic than their crystalline counterparts (Manufacturing Chemist, 2019). This impacts the aerosolization behavior of the formulation containing API. Spray drying of kanamycin has often been done in combination with excipients such as hydrophobic amino acids, predominantly leucine, to reduce moisture sorption and particle adhesion (Momin et al., 2017). Nevertheless, the use of excipients has its own drawbacks such as the need to deliver a larger mass of powder to the lung. This is a concern for kanamycin as its dose is substantial. Hence, the need emerged for an excipient-free drug-only formulation of kanamycin (Momin et al., 2017). This was not found to be possible with kanamycin-only spray-dried formulations, which showed higher moisture content (7 ± 0.4%) and lower FPF (71.1 ± 6.6%) due to moisture uptake by the formulation that resulted in particle sticking and bridging (Adhikari et al., 2020). On the other hand, freeze-drying may not be suitable for the drying of amorphous APIs such as kanamycin, owing to its prolonged drying time. Generally, rapid drying is advantageous for the drying of sensitive and unstable substances. The above challenges might have probably instigated the use of spray-freeze-drying for kanamycin production.

Annealing approach was included in the spray-freeze-drying process to obtain excipient-free kanamycin particles with spherical shape and smooth surface. Annealing for a period of 2 hours at 0°C has been found to reduce the hygroscopicity of amorphous raw materials such as trehalose, by reducing the moisture sorption from 4.5% to 3.5%, after exposing to a relative humidity of 33% for 1 hour (Sonner, Maa, & Lee, 2002). A similar approach was adopted for kanamycin as well. The optimum annealing conditions were deduced as follows:

- Annealing temperature: −10°C
- Annealing time: 5 hours
- Concentration of kanamycin: 10% (w/v)
- Pressure: 100 kPa
- Nozzle tip lift: 1 mm

The resultant kanamycin powder showed porous particles with spherical surface, as evident from the scanning electron micrograph shown in Fig. 8.7. The geometric and aerodynamic

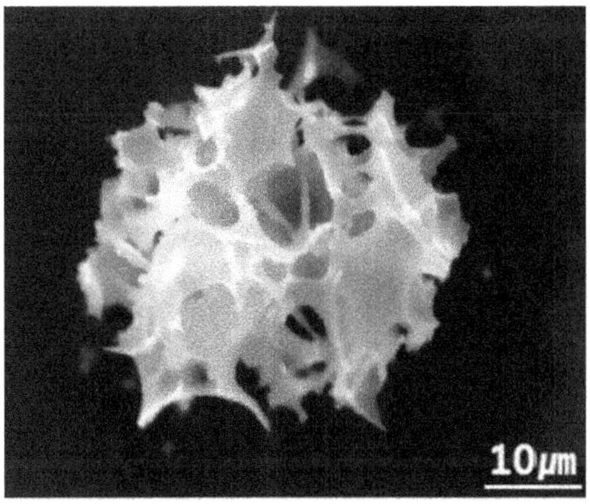

Figure 8.7 Scanning electron microscopy (SEM) image of spray-freeze-dried kanamycin powder obtained with optimum annealing conditions (Her et al., 2010).

diameters of the dry powder kanamycin were 13.5 μm and 3.584 μm, respectively, which were well within the favorable range for inhalation. The antimicrobial activity of raw and spray-freeze-dried kanamycin was tested by the standard agar disk diffusion method. Results showed that the disk diameter of raw and SFD kanamycin for *Escherichia coli* and *Staphylococcus aureus* were 21.07 mm/20.73 mm and 21 mm/20.1 mm, respectively. Thus, the antibiotic activity of kanamycin after SFD was as high as 98.4% for *E. coli* and 95.7% for *S. aureus*. Thus, the SFD process did not affect the antibiotic activity of kanamycin (Her et al., 2010).

8.1.3.3 Proteins

Preparing dry powder aerosol of protein is indeed complicated due to its sensitive nature to the processing conditions of the drying technique employed. In this context, Maa et al. (1999) established the competitive edge of spray-freeze-drying over spray drying for producing aerosols of different proteins such as recombinant human deoxyribonuclease (rhDNase), Immunoglobulin E (IgE) antibody and Anti-IgE antibody. Notably, the spray-freeze-dried powders showed larger median particle size, greater specific surface area and higher FPF than the spray-dried products. The relatively favorable aerosol behavior of SFD-protein powders was attributed to the absence of hot air drying as in spray drying (leads to shrinking of atomized droplets upon drying), due to which the atomized droplets retained their shape and size after the subsequent freezing and freeze-drying operations of SFD. The pores created during the sublimation process increased the SSA of SFD protein aerosols by 40-fold. The particle density of SFD product was $1/9^{th}$ of that of the spray-dried powder, based on which their aerodynamic diameters were found to be 2.7 μm and 3.5 μm, respectively (SFD < SD). Thus, SFD-protein powder was much inhalable than its spray-dried counterpart (Table 8.3).

The impactor analysis showed the increase in powder deposition in bottom stages (stages 3 and 4 and the filter) and reduced deposition in device and throat for the spray-freeze-dried powder. With respect to the powder deposition in each stage, the SFD powders consistently outweighed the spray-dried product (Fig. 8.8). Type of atomizer used had a significant influence on the aerodynamic properties of the SFD-protein powders. While ultrasonic atomization produced powders of larger particle size (32 μm) and smaller SSA (44.1 m^2/g), two-fluid atomization resulted in smaller particles with larger SSA: 19 μm/49.7 m^2/g and 5.9 μm/72.9 m^2/g, for atomizing flow rate of 600 g/L and 1050 g/L, respectively. The FPF decreased evidently with increase in particle size. Hence, to obtain inhalable protein powders with FPF > 30%, two-fluid atomization with atomization flow rate greater than 1000 L/h is appropriate (Maa et al., 1999).

The above study, being an initial investigation on the use of SFD for dry aerosol protein powders struck a comparison between the spray drying (SD) and SFD processes (Table 8.4). The authors suggested that the complexities involved in SFD such as longer processing time and higher operating cost compared to SD are counterbalanced by its higher production yield and excellent

Table 8.3 Physical and aerosol properties of protein powders: Spray-freeze-drying vis-à-vis spray drying (Modified from Maa et al., 1999)

Formulation	Method	Atomization	Particle size (μm)	Surface area (m^2/g)	FPF (%)
Excipient-free rhDNase	SD	Air at 1050 L/h	3.4	3.4	46
Excipient-free anti-Ig E antibody	SFDSD	Air at 1050 L/h	7.0 3.3	121.2 2.8	70 27
Anti-Ig E antibody:trehalose = 60:40	SFD	UltrasoundAir at 600 L/h Air at 1050 L/h	7.7 19 5.9	127.7 49.7 72.9	50 16 52

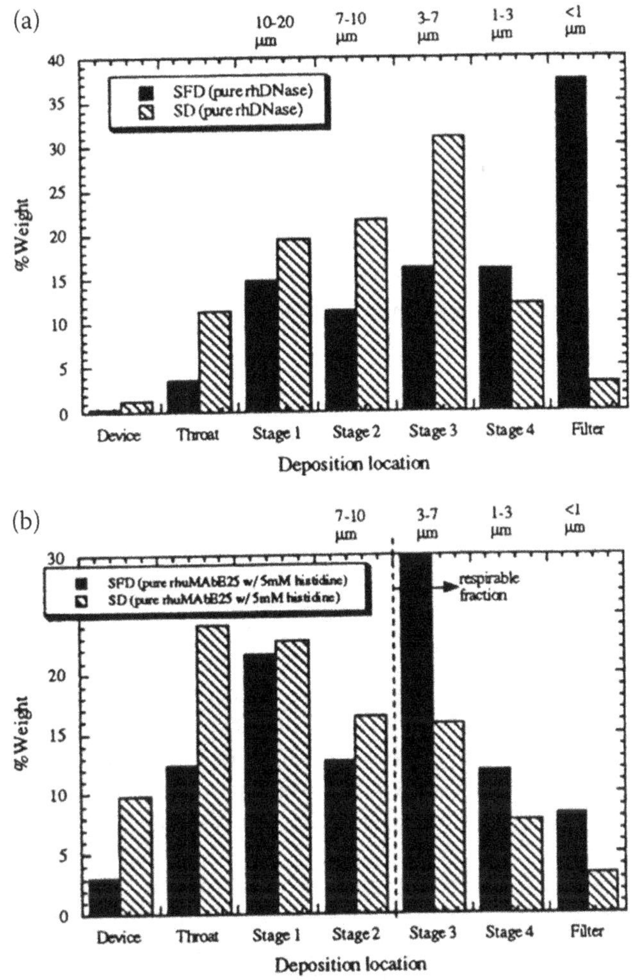

Figure 8.8 Comparison between the dispersibility of spray-freeze-dried and spray-dried protein aerosols (Maa et al., 1999).

Table 8.4 Process comparison between SFD and SD for production of dry aerosol powders of protein (Maa et al., 1999)

Factors	Spray drying	Spray-freeze-drying
Operation	Easy, fast, convenient	Time consuming, inconvenient to handle liquid N_2
Scalability	Straightforward	Comparable to SD, but more complicated due to freezing by liquid N_2
Operation cost	Low	High
Yield	50–70%	>95%
Aerosolization	Good	Excellent (at least 50% better)

powder aerosol performance (Table 8.4). Scalability of both the processes were found to be equivalent, due to the common rate/speed-limiting step, i.e., atomization. The conservation of protein stability against the stressful events of SFD was remarked as the major challenge (Maa et al., 1999), which could perhaps be addressed with the use of suitable excipients. The excipient

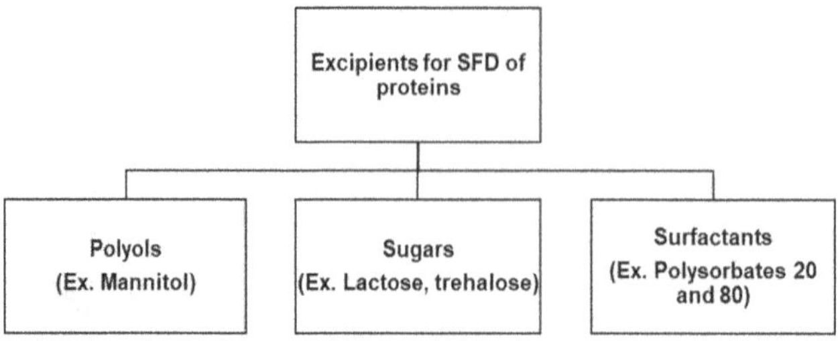

Figure 8.9 Classification of excipients used in the spray-freeze-drying of proteins for dry powder inhaler production.

usage approach was adopted in a recent study by Lo, Pan, & Lam (2021), the findings of which are presented in the forthcoming section.

Generally, one or a combination of stabilizing excipients are added to the feed formulation to address the shear and thermal stresses encountered by the protein during atomization and freeze-drying stages of SFD. Some of the established categories of excipients (Grasmeijer et al., 2013; Mensink et al., 2017; Depreter et al., 2013) are given in Fig. 8.9. Lo et al. (2021) used 2-hydroxypropyl-β-cyclodextrin (HPβCD), a hydroxyalkyl derivative of cyclodextrin as the excipient for preparing the dry powder aerosol of a model protein, i.e., bovine serum albumin (BSA). Cyclodextrin is an oligosaccharide that is known for its ability to protect proteins in dry form by different mechanisms such as water replacement, vitrification, amino acid complexation and surfactant-like effect (Milani et al., 2020; Pouya et al., 2018; Serno, Geidobler, & Winter, 2011).

In the study of Lo et al. (2021), the stock solutions of BSA and HPβCD were prepared at a concentration of 100 mg/mL, which were then mixed with ultrapure water to obtain a total solute mass of 120 mg before SFD. A two-fluid nozzle was used for atomization and the SFV/L mode of atomization was carried out, followed by vacuum freeze-drying for about 64 hours. The presence of HPβCD in feed formulation offered protection to the protein during SFD and improved its integrity (estimated by SDS-PAGE) and inhalation indices of the final product (TED: ~98.4%; FPF: 52.3–65.5%; MMAD: 1.4–2.5 μm), compared to the excipient-free formulations (TED: maximum of 96.3%; FPF: maximum of 79.4%; MMAD: maximum of 13.6 μm). However, addition of HPβCD did not lead to a complete eradication of protein aggregation. This implies that SFD is a robust drying method, wherein the aerosol properties are mainly governed by the processing conditions rather than the feed composition. However, there is a potential scope for future research to corroborate the above finding by conducting the study with a combination of excipients. Sugars such as trehalose or an oligosaccharide like cyclodextrin alone might not be fully proficient in tackling the shear and interfacial stresses posed by the SFD process. Hence, in combination with USFDA-approved surfactants like polysorbates (Serno et al., 2010) at relatively low concentrations, the improvement in protein stability caused by sugars may become more prominent (Mensink et al., 2017), especially when high atomization gas flow rate is required.

Contrarily, increasing the protein (BSA) concentration (2% to 10%) reduced the tendency to aggregate, due to the *'volume exclusion'* phenomenon of macromolecular crowding. At higher concentration of macromolecules, it is hypothesized that the protein unfolding is suppressed and its overall mobility is reduced due to thermodynamic stabilization (Minton 2005; Ohtake, Kita, & Arakawa, 2011). An auxiliary hypothesis for the same event is that, provided the interface is saturated with protein molecules, then, any further increase in the concentration of protein in bulk will reduce protein aggregation (Arsiccio & Pisano, 2020). This study demonstrated that HPβCD is a competent excipient in the preparation of spray-freeze-dried protein formulations with good aerosol properties (Lo et al., 2021). But, further studies must be extended in this line to come up with more stable and effective protein inhaler powders.

8.1.3.4 Drugs and APIs

Based on the investigations conducted thus far, it is evident that spray-freeze-drying is the method of choice for producing dry powder aerosols of drugs/APIs that are:

i. highly susceptible to oxidation upon contact with air;
ii. poorly soluble in aqueous solutions;
iii. prone to extensive first pass metabolism; and,
iv. characterized by low bioavailability limited by dissolution rate after oral administration.

Accordingly, Δ^9-tetrahydrocannabinol (THC) is a pharmacologically active constituent derived from *Cannabis sativa*, which exhibits broncho-dilative property. However, being a sticky resin with poor aqueous solubility, it is difficult to convert THC into a suitable dosage form. But, SFD has been successful in producing an inhalable solid dispersion powder of THC, using inulin as a stabilizing carrier. A 10 mg/ml THC in tertiary butyl alcohol (TBA) solution was mixed with aqueous inulin solution of various concentrations with a volume ratio for water/TBA at 6:4. The SFD process comprised of twin-fluid nozzle atomization followed by SFV/L into liquid nitrogen and vacuum freeze-drying for 48 h. For the above formulations, the drug load after spray-freeze-drying was found to be ranging between 4–30% (w/w).

The SFD-THC solid dispersions appeared as white powder, which did not change with time, thus indicating the effective stabilization of the sensitive THC within the inulin glass (Van Drooge et al., 2004). Depending on the total solid content of the feed solution, the final powder exhibited low bulk density (20 to 85 mg/cm^3), very high bulk porosity (94 to 99%) and high SSA (70 to 110 m^2/g). Notably, the powder whirled up easily, which is suggestive of its suitability for inhalation. Pure THC degraded completely within 50 days and 15 days of storage (exposed to air) at 20°C/45%RH and 60°C/8% RH. But, the SFD-THC remained stable with about 70% drug recovery after 390 days, which was attributed to the successful encapsulation of THC within the glassy inulin matrix. The aforesaid storage stability of SFD-THC was found to be true irrespective of the drug load.

Likewise, powders prepared from different feed formulations unanimously showed a high FPF (40–50%) (Fig. 8.10), which in this study has been defined as the sum of the third, fourth and the

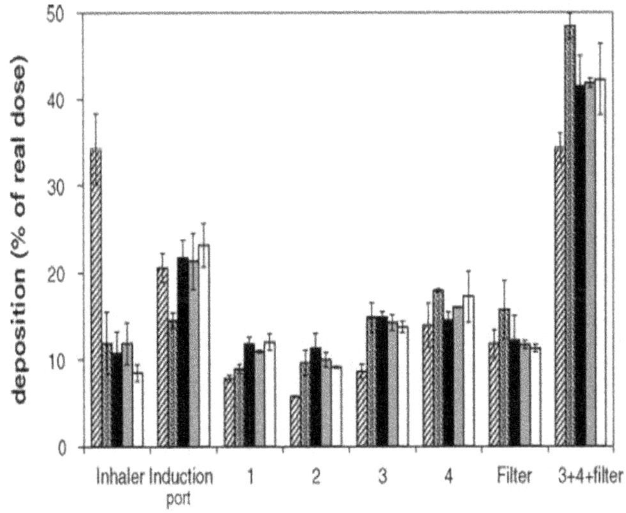

Figure 8.10 Cascade impactor analysis of SFD-THC powders with different drug loads, columns from right to left: 4%, 8%, 12%, 16% and the extreme right columns: 20% drug load in the solid dispersion (van Drooge et al., 2005).

filter stage relative to the total dose. This confirms the superior inhalation behavior of SFD-THC and its aptness for pulmonary delivery in combination with an air classifier based inhaler. The underlying reasons for the higher stability of spray-freeze-dried Δ^9-tetrahydrocannabinol were the higher cooling rate, greater surface area for heat transfer from the solution and the spontaneous freezing of droplets that shortened the available time for phase separation of either inulin or THC in the partially frozen solution. All the above enhanced the adequate incorporation of THC in the glassy inulin matrix. Thus, this study established that SFD is an optimum process to prepare stable solid dispersions of THC for a wide-range of drug load (van Drooge et al., 2005).

Spray-freeze-drying has also been successful in producing DPI aerosol of budesonide, a poorly water-soluble corticosteroid used in treating asthma and chronic obstructive pulmonary disease (Parsian et al., 2014). SFD-budesonide particles have been prepared both in the presence and absence of stabilizers. In the former case, micronized porous particles of budesonide in a mannitol-based dry powder formulation were produced using HPβCD and/or L-leucine as complementary excipients. While leucine had higher affinity towards particle surface (Chew et al., 2005) to form a protective shell and reduce the inter-particle cohesive forces and aggregation, HPβCD improved the sphericity of microparticles. A synergistic combination of leucine and HPβCD resulted in spherical porous particles of budesonide. Moreover, in the presence of leucine and HPβCD, the crystallinity of mannitol diminished and it was converted into a metastable δ polymorphic form during the freezing stage (Parsian et al., 2014), which is known to improve the inhalation properties of the final product (Niwa, Mizutani, & Danjo, 2012). The budesonide drug molecules at a low dose remained molecularly dispersed in the amorphous state within the homogeneous matrix.

Regarding the inhalation indices, the TED of different formulations containing different combinations of leucine, HPβCD and mannitol varied between 44.3% and 77.9%. Further, the FPF was at its maximum (~58%) in the powders containing both HPβCD and leucine. Higher amount of HPβCD in the feed induced favorable morphological changes in the microparticles that increased both the FPF as well as the dissolution rate in pure water (~94% in 30 min). The dissolution obtained in the presence of HPβCD was 61% higher than that of pure budesonide, in which merely ~30% of drug was released after 30 min of the test. On the other hand, presence of leucine was beneficial in terms of reducing the overall amount of drug deposited in the device and throat section of the apparatus, thus improving the TED (Parsian et al., 2014).

Contrastingly, a novel approach comprising a microfluidic reactor integrated with ultrasonic SFD eliminated the need for a homogenization step or stabilizer during the SFD of budesonide. Microfluidics is an effective alternative to the conventional bottom-up approaches for preparing fine particles. It uses small-sized reactors with an array of narrow channels that mix reagents to offer a stable reaction environment (Mansur et al., 2008; Thiele et al., 2011). Microfluidics has an edge over the other particle production techniques in terms of its small scale production, effective use of reagents and prospective scaling-up using an array of microfluidic reactors simultaneously and the use of less toxic base chemicals such as methanol, ethanol, acetone and water (Makgwane & Ray, 2014). Thus, a T-junction microfluidic reactor with narrow channels (100–110 μm) was used to produce budesonide particle suspension using an ethanol–water, methanol–water and an acetone–water system. The resultant suspension was atomized using an ultrasonic probe and spray-frozen in liquid nitrogen, followed by vacuum freeze-drying (Fig. 8.11). The resultant product was the fine crystalline budesonide powder with median diameter of 5.85 μm, 7.33 μm and 8.83 μm for the acetone, ethanol and methanol systems, respectively. After blending with lactose and dispersing using an Aerolizer at 100 L/min, the FPF of SFD-budesonide particles ranged between 48% and 55%. Thus, even in the absence of homogenization of feed solution and excipients, the inhalable drug particles produced by SFD exhibited good aerosol performance (Saboti et al., 2017).

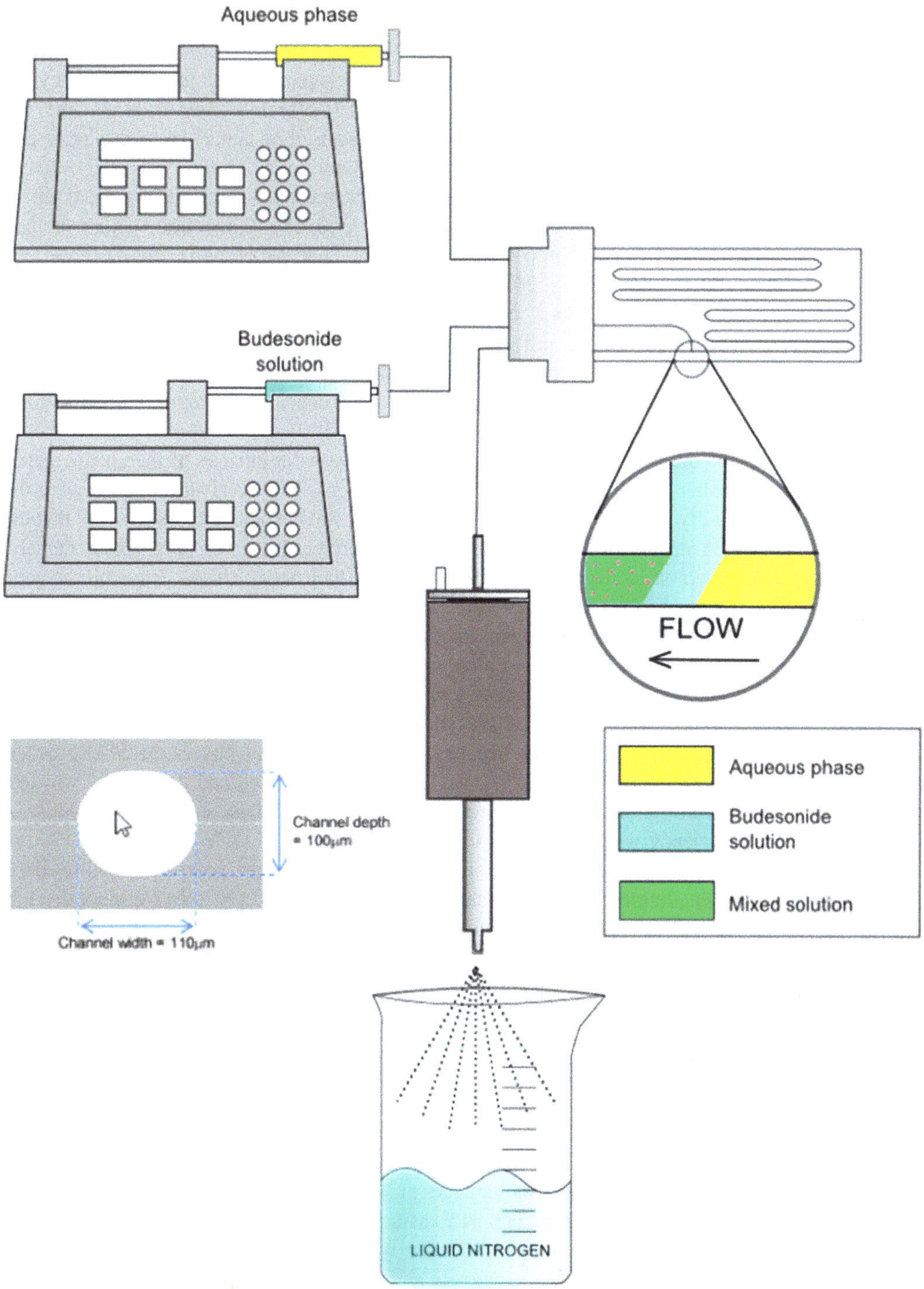

Figure 8.11 (A) Schematic diagram of the coupled microfluidic reactor with ultrasonic SFD system: 1 – syringe pump with a syringe, 2 – syringe filter, 3 – T-junction microfluidic chip (Source: Dolomite Centre ltd, UK), 4 – ultrasonic atomization probe, 5 – collecting beaker with liquid nitrogen; (B) Channel diameter in the microfluidic chip (Source: Dolomite Centre Ltd., UK) (Saboti et al., 2017).

8.1.3.5 Therapeutic genes and nucleic acids

Advent of pulmonary gene therapy is an alternative viable option in the treatment of several conditions, wherein the conventional methods are not completely effective (Jenkins et al., 2003). It is possible to treat several lung diseases such as cystic fibrosis and lung cancer by high-degree and prolonged expression of the related gene of interest (Bivas-Benita et al., 2004). Targeted delivery of genes to the lungs with the aid of nebulizers, pressurized metered-dose inhalers (pMDIs) or DPIs overcomes the major apprehensions associated with intravenous injection, which include non-targeted distribution and rapid degradation of genes by endonucleases in the systemic circulation (Mohri et al., 2010; Okuda et al., 2015). Thus, with pulmonary delivery, the gene stability is preserved until it reaches the lung. But, converting genes into powder form involves multiple complexities, as the chosen process should maintain their stability and biological activity, whilst producing free-flowing and dispersible powders (Lam, Liang, & Chan, 2012). Use of appropriate non-viral gene delivery vectors has been the commonly adopted approach to improve gene stability during SFD, as would be discussed subsequently.

The initial studies on the use of SFD for preparation of dry powder inhalers of therapeutic genes encountered various challenges and presented inherent limitations. Kuo & Hwang (2004) were the first to prepare inhalable dry plasmid DNA (pDNA) powders by SFD. These authors reported that the destabilization of pDNA during SFD was resolved by the protective effect of a vector namely, polyethyleneimine (PEI, a non-biodegradable polycation). But, their study did not focus on the gene transfection characteristics *in vivo* (Kuo & Hwang, 2004). As a development over the above said investigation, Mohri et al. (2010) were successful in establishing SFD as a process for the preparation of stable dry inhaler powders of pDNA. They employed chitosan (CS) to overcome the instability of pDNA and protect it against the physical stresses encountered during SFD. The SFD-pDNA-CS powders showed highly porous and spherical particles, with diameter in the range of 20–40 μm. Electrophoresis confirmed the intact ternary structure and integrity of pDNA in the SFD powder. However, in the absence of chitosan, the pDNA was completely degraded and its stability was lost during SFD. Although, chitosan showed good protective effect upon pDNA and high tolerability *in vivo,* its transfection efficiency was found to be lower than the PEI.

In contrast to the above studies, Okuda et al. (2015) synthesized the poly(aspartamide) derivatives with an ethylenediamine unit containing side chain (poly{N [N-(2-aminoethyl)-2-aminoethyl]aspartamide} (PAsp(DET)) and their block copolymers with poly(ethylene)glycol (PEG-PAsp(DET)) and used these as vectors to prepare inhalable dry gene powders with high gene transfection efficiency. Compared to PEI, the above vectors are biodegradable polycations with greater efficiency and least cytotoxicity. As expected, the vectors used by Okuda et al. preserved the integrity of pDNA during the SFD process for powder production. Further, L-leucine was used as a dispersibility enhancer in the feed formulation, to augment the inhalable property of spray-freeze-dried pDNA. The resultant product possessed spherical and porous particles with a diameter of 5–10 μm. PAsp(DET)- and PEG-PAsp(DET)-based dry gene powders showed 50-fold and 4.5-fold higher gene transfection efficiencies, respectively, in the lungs than the chitosan-based dry gene powders prepared previously. The advantage of adding L-leucine was assessed by determining two inhalation indices, i.e., TED and FPF, using an 8-stage Anderson cascade impactor. SFD dry inhaler powders produced with the addition of L-leucine showed exceptionally high values of TED and FPF at 98% and 62%, respectively and reduced MMAD at 1.2 μm (Okuda et al., 2015). Comparing the abovementioned inhalation indices obtained with L-leucine against other formulations used for the SFD of pDNA (Table 8.5) clearly confirmed its superiority in facilitating improved dispersibility and deep lung deposition.

Small nucleic acid molecules show great potential to treat respiratory diseases (asthma, influenza, respiratory syncytial virus (RSV) infection and tuberculosis) (DeVincenzo, 2012; Koli et al., 2014; Man et al., 2016; Qiu et al., 2016). For instance, the small interfering RNA (siRNA)

Table 8.5 Inhalation indices of SFD-pDNA: A comparative analysis (Okuda et al., 2015)

Formulation	TED (%)	FPF (%)	MMAD (μm)
Mannose + pDNA	74.3 ± 19.2	6.8 ± 1.3	15.6 ± 2.5
Mannose + L-leucine + pDNA	97.8 ± 0.9	62.3 ± 3.0	1.2 ± 0.2
PAsp(DET) + pDNA	95.5 ± 2.9	54.3 ± 6.2	2.4 ± 0.5
PEG-PAsp(DET) + pDNA	79.9 ± 13.2	11.2 ± 1.6	8.7 ± 1.5

functions by inhibiting specific gene expression through RNA interference (RNAi) (Fire et al., 1998). Liang et al. (2018) validated the potential of spray-freeze-drying as a feasible particle engineering technology to produce dry powder inhalers of small interfering RNA (siRNA) for pulmonary delivery. Herring sperm DNA was used as the model small nuclei acid therapeutic and mannitol was used as the bulking agent. SFD yielded 83–90% of powder. The concentration of solute (1% to 7.5%, w/v) in feed exerted major influence on the volumetric diameter and morphology of particles, at a constant DNA concentration of 1% (w/w). The median diameter of particles produced from different formulations ranged between 8.7 μm and 12.6 μm. But, the particle size increased and the size distribution turned narrower with lower span value as the solute concentration increased. The particles were more discrete and individual at high solute concentration, compared to the porous and fragile particles obtained from the feed with low solute concentration. Results showed that the solution concentration must be no less than 5% (w/v) to obtain robust particles suitable for inhalation. However, change in DNA concentration (0.25% to 2%, w/w) did not affect the particle characteristics, significantly and all the formulations prepared in this study showed a TED of >90%.

On the other hand, the aerodynamic diameter of particles increased and FPF reduced with increase in solute concentration, due to their low porosity and higher density. However, the above condition could be combated by increasing the DNA concentration. Accordingly, FPF of the 2% w/w DNA formulation was nearly 30%, which was the highest among the formulations with different DNA concentrations. Similarly, at high DNA concentration, the powder was predominantly deposited on Stage 1 of a 7-stage next generation cascade impactor (NGI), which had a cut-off diameter of 6.12 μm. This is still higher than the range of aerodynamic diameter suitable for deep-lung deposition (1–5 μm), which could be improved by increasing the DNA concentration further. Conversely, the particle diameter can be reduced by slowing the feed flow during atomization or by changing the frequency of ultrasonic nozzle (Costantino et al., 2000; Ishwarya et al., 2015). Nevertheless, using a higher concentration of the active ingredient is advantageous due to the reduced amount of excipient and the resultant reduction in the mass of powder to be delivered. This circumvents the need for frequent administration to obtain the same therapeutic dose. Further, in this study, the gel retardation and liquid chromatography assays showed that the siRNA remained intact after spray-freeze-drying even without using a delivery vector (Liang et al., 2018).

8.2 RATIONALE AND APPLICATIONS OF SPRAY-FREEZE-DRYING FOR THE DEVELOPMENT OF TRANSDERMAL DELIVERY SYSTEMS

8.2.1 Transdermal delivery systems – A prelude

Transdermal delivery system, also known as *'patches'* is a self-administrable dosage form that can be applied to intact skin to deliver an active ingredient to the systemic circulation at a controlled rate through the different epidermal layers (stratum corneum of thickness, 10–20 μm and viable epidermis of thickness, 50–100 μm) (Bhowmik et al., 2010; Divya et al., 2012; Kumar et al.,

2011). Protein-based drugs and therapeutics (antibodies, growth factors and interleukins) are mainly delivered through the transdermal route. Because, these proteins are highly susceptible to the negative action of several endogenous and exogenous factors, such as temperature, pH, enzymatic degradation and immune clearance. But, an AI delivered through skin bypasses the acidic pH and enzymatic activity in the gastrointestinal tract as well as the first-pass metabolism that occurs in the liver. Hence, transdermal route of administration maintains the active component at a uniform concentration in the plasma and enhances its bioavailability. Consequently, the duration of action is prolonged at a low dosage frequency of the AI (Peña-Juárez, Guadarrama-Escobar, & Escobar-Chávez, 2021). The added advantages of dry powder-mediated transdermal delivery are that it eliminates the need for trained persons for drug administration, needle/syringe and refrigerated storage. Also, transdermal delivery of vaccines requires much less antigen than the conventional route of administration, due to its targeted delivery to the epidermis (Dean & Chen 2004; Maa et al., 2004).

8.2.1.1 *Stages in transdermal delivery*

Transdermal delivery of any drug or therapeutic molecule from the skin to blood circulation occurs in a series of stages. Initially, the molecule penetrates the stratum corneum (SC; the outermost layer of skin) and then passes through the subsequent layers of epidermis and dermis, without accumulating in the dermal layer. Once the drug reaches the dermal layer, it becomes available for systemic absorption through the dermal microcirculation (Donnelly et al., 2012; Kretsos & Kasting, 2007). Dermal patches/films and hydrogels are the commonly employed dosage forms in transdermal delivery.

The efficacy of aforesaid methods/devices of transdermal delivery is limited by physical (hydrophobicity of stratum corneum) and chemical (presence of multiple enzyme types in the epidermal layer) barriers. Various physical modifications have been carried out to enhance the permeability of molecules through the skin layers, listed as follows: (1) thermal ablation; (2) iontophoresis (applying local electric current to deliver ionic therapeutic compounds into the systemic circulation); (3) sonophoresis (usage of ultrasound to enhance the absorption of topical compounds through the different layers of skin); (4) electroporation (generating small pores on the surface of stratum corneum using pulse voltage) and (5) microneedles (Fig. 8.12). Chemical penetration enhancers like dimethylsulfoxide (DMSO) or Azone, oleic acid, ethanol, propylene glycol, menthol and limonene are also used to enhance drug permeability. Nevertheless, both physical and chemical enhancers need to be used at high dose or high potency to achieve their intended purpose. But, this can lead to skin irritation and affect the skin barrier function.

The aforesaid challenges led to the advent of particle-mediated transdermal delivery of therapeutic agents without the aid of physical or chemical enhancers (Tomoda & Makino, 2014). Particles penetrate the skin layers through the trans-appendicular pathway comprising hair follicles and sweat glands (Prow et al., 2011). In this context, particles of sub-micron scale dimensions are more effective than their micron-sized counterparts, in penetrating through the follicles (Toll et al., 2004). The optimal ranges for particle density and injection velocity have been reported in the range of 800–1500 kg m^{-3} and 200–3000 m s^{-1}, respectively (Sarphie & Burkoth, 1999). The desirable particle diameter falls in the range of ~30–50 μm. As indicated by pre-clinical trials, the skin surface can reject or repel particles with an average MMAD of less than 20 μm. And, particles larger than 75 μm can cause tiny sub-dermal hemorrhages or petechiae, which are not associated with pain or bleeding, but are not cosmetically acceptable (Burkoth et al., 1999). When the particle diameter is about 40 μm, the injected powder can reach the target site, i.e., the epidermis/dermis (skin), without rupturing the 15–20 μm thick outer layer of stratum corneum that covers it (Lahm & Lee, 2006). Uniformly spherical or elliptical aerodynamic shape of the spray-freeze-dried particles is ideal for their application in needleless powder injection (Prestrelski, Burkoth & Maa, 2002; Ziegler, 2008).

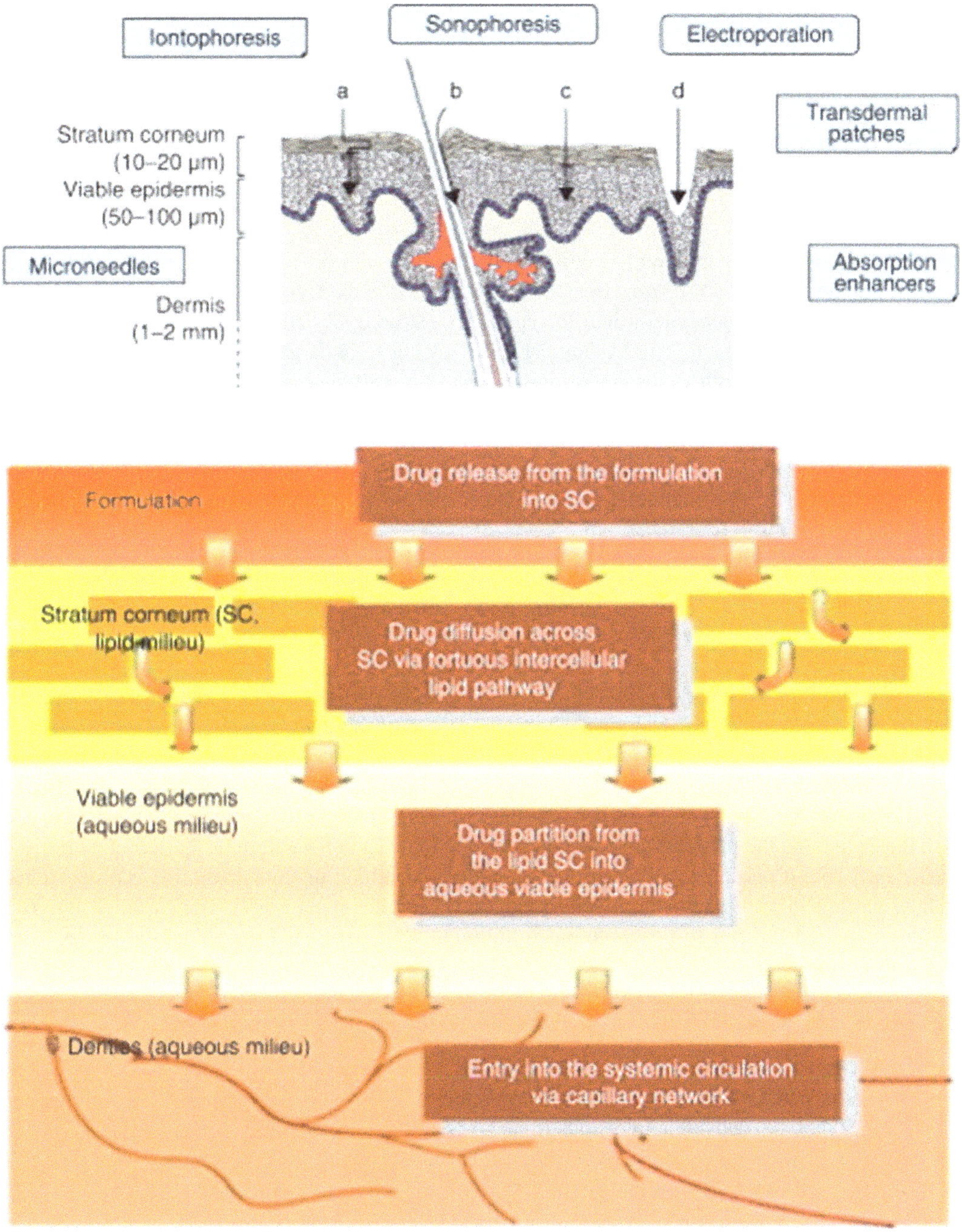

Figure 8.12 (a) Structure of human skin and methods of transdermal delivery approaches; (b) stages of drug absorption into the systemic circulation through transdermal delivery (Sonia & Sharma, 2014).

During transdermal delivery, the penetration of drug particles through the epidermal and sub-epidermal layers of skin is facilitated by accelerating the particles to an adequate momentum using high-velocity gas jets (Burkoth et al., 1999). The particles should attain a threshold momentum in the range of 7–12 kg m^{-1} s^{-1} before rupturing through the stratum corneum and delivered into the viable epidermis (Kendall et al., 2004). Generally, two types of particles are suitable for transdermal

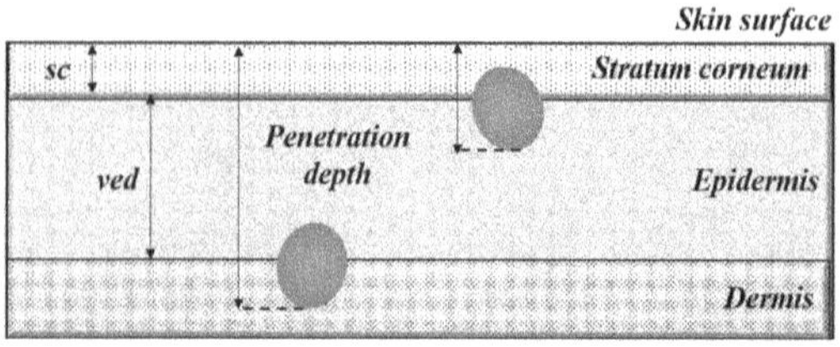

Figure 8.13 Measuring the skin penetration depth – A schematic representation (Adapted and redrawn from Weissmueller et al., 2015).

delivery: (i) small-sized particles (1–3 μm) with high density and (ii) low-density particles with large diameter in the range of 20–70 μm (Dean et al., 2003). While the former type provides the required momentum density for penetration into the epidermis (Dean et al., 2003; Kendall et al., 2004), the latter one demonstrates high drug-loading capability owing to its larger size.

Skin penetration depth is the important criteria of a particle-based formulation intended for needle-free ballistic injection (Kendall, Mitchell, & Wrighton-Smith, 2004). Penetration depth (D_p) of microparticles into the skin is a function of particle diameter (d_p; m), density (ρ_p; kg m^{-3}) and velocity at impact (v_p; m s^{-1}) (Mitchell, Kendall & Bellhouse, 2001). The concept of momentum density governs the aforesaid relationship, which can be expressed by the following equation (Burkoth et al., 1999).

$$D_p \propto M_D = \frac{v_p m_p}{A_p} \approx v_p d_p \rho_p \tag{8.2}$$

Where M_D is the momentum density (kg m^{-1} s^{-1}), m_p is the particle mass (kg) and A_p is the particle frontal area (m^2). It is quantified by measuring the length of the line connecting the deepest part of a particle in the tissue to the surface of stratum corneum and is perpendicular to the latter (Fig. 8.13) (Weissmueller et al., 2015). According to Eq. (8.2), particles with higher density and larger diameter would penetrate deeper into the skin at a given particle impact velocity. But, it is also important that the particles are adequately strong to resist deformation or shattering on impact. Sufficiently robust particles will penetrate the epidermis and dermis to a depth that is proportional to their particle impact parameter, given by Eq. (8.3) (Sarphie et al., 1997). The particles require a definite impact parameter to penetrate the stratum corneum and reach the epidermal tissue.

$$Partice\ impact\ parameter = v_p \frac{d_p}{2} \rho_p = v_p r_p \rho_p \tag{8.3}$$

For the reason stated above, microparticles are preferred over nanoparticles for ballistic delivery as the latter cannot be delivered into epidermal layer of human skin by direct powder injection. Nanoparticles require high impact velocity (Kendall et al., 2004), which is limited by the particle accelerating device. Generally, particles less than 100 nm in diameter penetrate the skin without causing pain and those larger than 200 nm do not penetrate into the epidermis (Prow et al., 2011; Tomoda et al., 2011). Therefore, microparticles with a penetration depth of about 20–100 μm are useful for ballistic drug delivery. In the Eq. (8.2) for particle penetration depth, the particle diameter and particle density are the key parameters as the particle velocity does not change significantly in a given injector assembly. When the particles are accelerated inside the injector device, their velocities are usually less than the gas velocity due to particle slippage. Thus, particle velocity is a function of the driver gas used, apart from particle size and density. While larger and heavier particles exert the optimal impact to breach the stratum corneum layer, smaller and lighter particles lead to higher fraction of the gas velocity and so an adequate acceleration in the injector device (Burkoth et al., 1999). Therefore, the ideal particle diameter of powders for ballistic delivery is a trade-off between obtaining higher acceleration in the injector device and greater penetration depth underneath the skin that is facilitated by a higher impact parameter (Ziegler, 2008).

Compared to conventional techniques such as spray drying and freeze-drying, spray-freeze-drying is more appropriate to produce particles for transdermal delivery due to the ability of its atomization step to control the particle size of final powder (Schiffter, 2007). Nevertheless, a major apprehension of SFD is that it often results in porous fragile particles with low density and poor mechanical robustness, which are not fit for needle-free ballistic delivery (Costantino et al., 2000 & 2002; Schiffter, 2007). But, the above problem can be overcome by adding high concentrations (~350 mg/mL) of sugars and polymers (ex. dextran or hydroxyethyl starch) in the feed formulation (Maa et al., 2004; Rochelle & Lee, 2007; Sonner et al., 2002).

8.2.2 Preparation of dry powders for transdermal delivery by spray-freeze-drying

The requisite characteristics of particles for transdermal delivery were already discussed in Section 8.2.1. Accordingly, a primary outcome that is expected of spray-freeze-drying when used as an approach to produce particles for transdermal delivery is the production of robust, physically and biologically stable particles with high loading of a poorly water-soluble drug. Density and mechanical robustness of the drug-loaded particles, in other words, the ***'ρvr'*** value for the manufactured powders are imperative. To accomplish the same, a rational choice of the following parameters are vital (Schiffter et al., 2010):

1. ***Atomizer type and atomization conditions:*** to produce large droplets and eventually large particles with diameter in the range of 50–100 μm;
2. ***Composition of feed formulation:*** use of excipients to enhance particle strength; and,
3. ***Conditions of freeze-drying:*** regulating the conditions of primary drying phase to increase the density of microparticles.
4. ***Maximum area loading factor:***The area covered by particles divided by the total skin area must be maximum; this parameter along with the target area decides the powder payload for a given particle size and density.

Schiffter et al. (2010) explored the potential of spray-freeze-drying to produce dense and mechanically robust insulin-loaded particles for needle-free ballistic delivery, having diameter of around 50 μm, monodispersed size distribution and high drug loading. Different excipients such as trehalose dehydrate (T), mannitol (M), dextran (D) 10 kDa, dextran 150 kDa were used in the SFD formulation. To obtain high density particles without complicating the process of atomization, which happens when the feed formulation turns viscous due to larger amounts of dense excipients, the pre-prepared insulin nanoparticles (obtained from SFD of 0.5% insulin solutions) were suspended at different concentrations (theoretical insulin content: 0–50%, w/w) in the 350 mg/g matrix solution of sugar polymers, i.e., TMDD (3:3:3:1). Ultrasonic atomization at an operational frequency of 25 kHz (1.5W) and 48 kHz (3.1W) were found advantageous over two-fluid nozzles with respect to obtaining larger droplet sizes (>40 μm) at low feed flow rates (0.25 and 0.50 ml min^{-1}). Also, ultrasonic nozzle was found to protect the stability of insulin during the atomization operation. The primary drying temperature during freeze-drying was −10°C.

The authors found that the insulin particles produced at the above temperature were irregularly shaped with contracted and wrinkled surface morphology, due to structural collapse. This occurred as the primary drying temperature was higher than the glass transition temperature (T_g) of the aforementioned TMDD formulation, which was found to be −29.4°C. A primary drying temperature of −30°C can prevent the collapse phenomenon. But, as the particles prepared were intended for needle-free ballistic powder injection, structural collapse appeared to be advantageous as it increased the mechanical robustness and particle density (662 kg m^{-3} at −30°C to 791 kg m^{-3}

at −10°C) due to the shrinkage of particles at higher drying temperatures. Hence, in the above study, −10°C was preferred as the primary drying temperature to prepare insulin microparticles for transdermal delivery. The insulin particles prepared from different formulations had median diameter in the range of 46.81 μm (0% insulin) and 61.67 μm (25% insulin). When the TMDD solid content was decreased to 250 mg/g, significant particle break-up occurred with reduction in mean volume diameter to 41 μm. Thus, it is evident that a high solid content of matrix solution before freeze-drying is necessary to provide sufficient mechanical stability and robustness of the particles.

The particle penetration property was tested by injecting the insulin-loaded drug particles into tissue mimicking 3% agar hydrogel phantoms using an 8 mm inline venturi device (Fig. 8.14) loaded with a 60 bar helium cylinder. Agar is a polysaccharide extracted from red algae, which contains about 70% agarose and 30% agaropectin. It mimics the mechanical material properties of human skin. Upon injecting the pure TMDD particles (0% theoretical insulin load), neither particles nor particle fragments were observed in the sliced agar targets, when visualized under the microscope. This is because the particles dissolved instantaneously due to the high aqueous content of agar hydrogel (97%). Fig. 8.15 shows the microscopic image of polydisperse insulin particles injected into agar hydrogel test beds. After injection, the particles depicted an elongated, disrupted and slightly porous appearance. Particulate clusters of very small particles were retained in the agar. The highly soluble TMDD matrix seemed to dissolve instantly, leaving behind the clusters of originally suspended small and poorly soluble insulin nanoparticles (Schiffter et al., 2010).

The mean penetration depth of insulin particles (25% insulin load) was 251.3 + 114.7 μm. Due to the polydispersity of SFD particles following a log-normal distribution, a large amount of small particles merely penetrated to depths less than 250 μm. But, still larger insulin-loaded particles were found to penetrate up to a maximum depth of 450 μm. This is relevant as, to be injected into the inside of forearm, the capillary region of dermis has been reported to commence more than 150–250 μm beneath the skin surface (Aulton, 2007). Thus, the mean penetration depth of polydisperse insulin-loaded SFD particles is well within the above range, which renders them suitable for ballistic injection. The authors ascribed the optimal penetration depth of insulin-loaded SFD

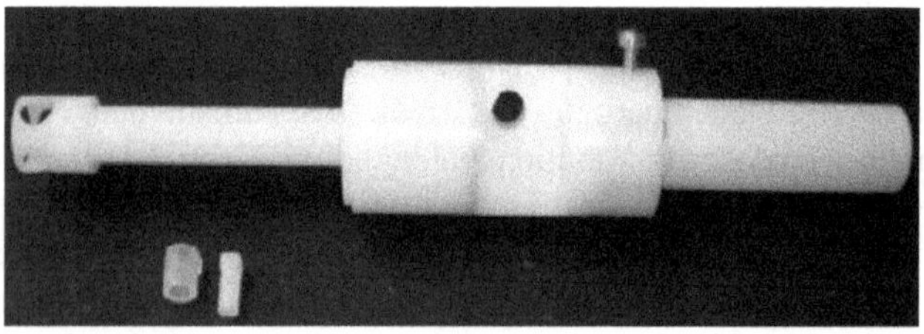

Figure 8.14 Photograph of the inline Venturi injection device with an 8 mm exit diameter used for the needle-free ballistic particle injection (Schiffter, Condliffe, & Vonhoff, 2010).

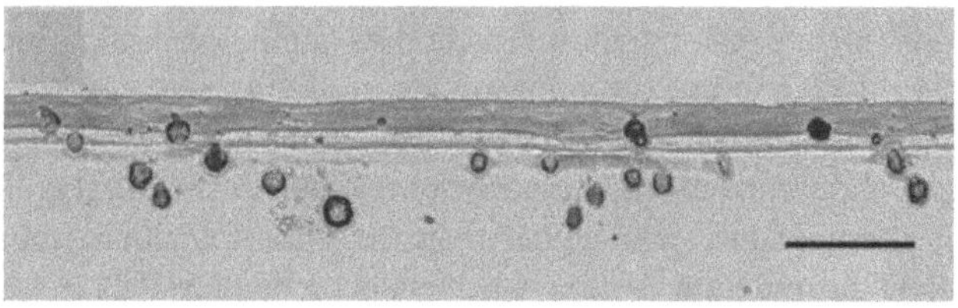

Figure 8.15 Light microscopy image of polydisperse particles of SFD-insulin prepared with TMDD (3:3:3:1) and 25% insulin (frequency of ultrasonic atomization: 48 kHz) injected into 3% agar hydrogel test beds; 10X magnification; scale bar: 300 μm (Schiffter, Condliffe, & Vonhoff, 2010).

particles to their superior physical stability ascertained by the size exclusion chromatography and Fourier transform infrared (FTIR) spectroscopy analyses. Results showed only an insignificant increase in the intermolecular β-sheet of 0.4% after SFD (Schiffter et al., 2010).

Like for the insulin, proteins such as bovine serum albumin ($BSA_{66.5\ kDa}$) and bovine carbonic anhydrase ($BCA_{30\ kDa}$ – an enzyme) were also converted to microparticles intended for ballistic injection by inducing structural collapse. Contrary to the previous study, besides adding sugar-based plasticizers and increasing the temperature of primary drying above the collapse temperature of the frozen feed matrix, an additional annealing step was performed. Because mere use of primary drying temperature at +10°C did not result in pronounced particle shrinkage, due to the high glass transition temperature of pure proteins (Ex. pure BSA: T_g at −11°C). But, the expected collapse phenomenon occurred upon adding either trehalose or sucrose to the protein feed and prolonging the duration of primary drying up to 2745 minutes at a shelf temperature of −12°C to −8°C. Result was the formation of shriveled and wrinkled particles of protein with reduced porosity and increased density that were appropriate for needle-free ballistic injection (Straller & Lee, 2017). The formulations that were spray-freeze-dried without plasticizers in the feed and subjected to shorter-duration primary drying (240 min) did not reveal shrinkage and showed lower density than the wrinkled particles. Nevertheless, when dried in combination with plasticizers, higher concentration of proteins prevented collapse even at a primary drying temperature close to or above their T_g (Johnson & Lewis, 2011). This is due to the reduced tendency of regularly freeze-dried cakes of amorphous proteins to collapse at higher concentrations. Accordingly, the wrinkled SFD particles as required were obtained only when the protein-to-plasticizer ratio (ex. BSA/sucrose) was in the order of 1:3, which exhibited T_g at −29.6 °C.

Interestingly, the above study by Straller & Lee (2017) showed that the degree of protein (BSA/BCA) aggregation during the 2745 min-long annealing step at −12°C/−8°C was lesser than that observed when the shelf temperature was increased from −12°C to −5°C, but conducting a short-duration primary drying for 240 min. An apparent reduction in monomers and increase in dimers were observed with the increase in shelf temperature, despite the fact that higher shelf temperature did not induce particle collapse. Thus, it is evident that the vulnerability of protein to aggregate is more sensitive to higher shelf temperature than prolonged duration of primary drying. Therefore, to prepare wrinkled, shriveled and dense SFD particles of pure protein (BSA or BCA) that are less damaged or aggregated, it is more preferable to manipulate the duration of primary drying step of freeze-drying rather than increasing the shelf temperature.

CONCLUSIONS

The information presented in this chapter established the edge of spray-freeze-drying as a technique to fabricate dry powder-based delivery vehicles for pulmonary and transdermal administration of active pharmaceutical ingredients. The high porosity, low density, smaller particle size with broader size distribution of spherical SFD particles were favorable towards obtaining optimal aerosolization behavior. On the contrary, the flexibility of SFD process to produce larger particles with wrinkled or spherical shape and higher density rendered it an appropriate technique to produce particles for transdermal delivery. For this particular application, the choice of atomizer and the conditions of atomization and freeze-drying steps were found to be the critical control points of SFD process. Use of ultrasonic atomizers and performing an annealing step before the primary drying stage transformed the physical characteristics of SFD dry powders to acquire the required aerosolization behavior and penetration depth into the skin. The concept of employing SFD powders as novel delivery systems is a truly interdisciplinary approach involving the expertise of process engineers and biologists. Further studies are required to establish precise and evidence-based guidelines on the choice of formulation and processing conditions for each category of active pharmaceutical ingredient discussed in this chapter. This will increase the future prospects of spray-freeze-drying being used as a commercial process for the production of aerosols and formulations for needle-free ballistic injection.

REFERENCES

Abu-Thabit, N. Y., & Makhlouf, A. S. (2018). Historical development of drug delivery systems: From conventional macroscale to controlled, targeted, and responsive nanoscale systems. In: A. S. Makhlouf & N. Y. Abu-Thabit (Eds.), *Stimuli Responsive Polymeric Nanocarriers for Drug Delivery Applications* (Volume 1), Woodhead Publishing, Duxford, United Kingdom, pp. 3–41.

Adhikari, B. R., Bērziņš, K., Fraser-Miller, S. J., Gordon, K. C., & Das, S. C. (2020). Co-amorphization of kanamycin with amino acids improves aerosolization. *Pharmaceutics*, 12(8), 715.

Ahmad, F. J. (2021). Novel drug delivery systems - Industrial advancements. Available from: https://www.pharmafocusasia.com/manufacturing/novel-drug-delivery-systems (Accessed 10 September 2021).

Ali, N., & Gary, P. M. (2015). *Pulmonary Drug Delivery: Advances and Challenges*, United Kingdom: John Wiley & Sons Inc.

Amorij, J.-P., Hinrichs, W. L. J., Frijlink, H. W., Wilschut, J. C., & Huckriede, A. (2010). Needle-free influenza vaccination. *The Lancet Infectious Diseases*, 10(10), 699–711.

Amorij, J.-P., Saluja, V., Petersen, A. H., Hinrichs, W. L. J., Huckriede, A., & Frijlink, H. W. (2007). Pulmonary delivery of an inulin-stabilized Influenza subunit vaccine prepared by spray-freeze drying induces systemic, mucosal humoral as well as cell-mediated immune responses in balb/c mice. *Vaccine*, 25(52), 8707–8717.

Arsiccio, A., & Pisano, R. (2020). The ice-water interface and protein stability: A review. *Journal of Pharmaceutical Sciences*, 109(7), 2116–2130.

Aulton, M. E. (2007). Pharmaceutics. In: M. E. Aulton & K. M. G. Taylor (Eds.), *The design and manufacture of medicines*, 3rd Edition, Churchill Livingston, London, UK, pp. 565–597.

Banga, A. K. (2015). *Therapeutic peptides and proteins: formulation, processing, and delivery systems* (2nd Edition). CRC Press, Boca Raton, Florida; London, England, New York.

Banker, G. S., & Rhodes, T. R. (2002). *Modern Pharmaceuticals*. Marcel Dekker, New York, NY, pp. 529–586.

Bhowmik, D., Chiranjib, Chandira, M., Jayakar, B., & Sampath, K. P. (2010). Recent advances in transdermal drug delivery system. *International Journal of PharmTech Research*, 2(1), 68–77.

Bi, R., Shao, W., Wang, Q., & Zhang, N. (2008). Spray-freeze-dried dry powder inhalation of insulin-loaded liposomes for enhanced pulmonary delivery. *Journal of Drug Targeting*, 16(9), 639–648.

Bivas-Benita, M., Romeijn, S., Junginger, H. E., & Borchard, G. (2004). PLGA-PEI nanoparticles for gene delivery to pulmonary epithelium. *European Journal of Pharmaceutics and Biopharmaceutics*, 58(1), 1–6.

Bosquillon, C., Lombry, C., Preat, V., & Vanbever, R. (2001). Comparison of particle sizing techniques in the case of inhalation dry powders. *Journal of Pharmaceutical Sciences*, 90(12), 2032–2041.

Burkoth, T. L., Bellhouse, B. J., Hewson, G., Longridge, D. J., Muddle, A. G., & Sarphie, D. F. (1999). *Transdermal and transmucosal powdered drug delivery. Critical Reviews™ in Therapeutic Drug Carrier Systems*, 16, 331–384.

Byron, P. R., Roberts, S. R. N., & Clark, A. R. (1986). An isolated perfused rat lung preparation for the study of aerosolized drug deposition and absorption. *Journal of Pharmaceutical Sciences*, 75(2), 168–171.

Byron, P. R. et al. (2010). In vivo-in vitro correlations: predicting pulmonary drug deposition from pharmaceutical aerosols. *Journal of Aerosol Medicine and Pulmonary Drug Delivery*, 23(Suppl. 2), S59–S69.

Cassetti, M. C., Katz, J. M., & Wood, J. (2006). Report of a consultation on role of immunological assays to evaluate efficacy of influenza vaccines. *Vaccine*, 24(5), 541–543.

Chakrabarty, R. K., Moosmüller, H., Garro, M. A., Arnott, W. P., Slowik, J. G., Cross, E. S., Han, J.-H., Davidovits, P., Onasch, T. B., & Worsnop, D. R., 2008. Morphology based particle segregation by electrostatic charge. *Journal of Aerosol Science*, 39, 785–792.

Chaurasiya, B., & Zhao, Y.-Y. (2021). Dry powder for pulmonary delivery: A comprehensive review. *Pharmaceutics*, 13(1), 31.

Chen, W., & Fryrear, D. W. (2001). Aerodynamic and GEOMETRIC diameters of airborne particles. *Journal of Sedimentary Research*, 71(3), 365–371.

Chew, N. Y. K., Shekunov, B. Y., Tong, H. H. Y., Chow, A. H. L., Savage, C., Wu, J., & Chan, H.-K. (2005). Effect of amino acids on the dispersion of disodium cromoglycate powders. *Journal of Pharmaceutical Sciences*, 94(10), 2289–2300.

Chew, N. Y., & Chan, H.-K. (2002). The Role of Particle Properties in Pharmaceutical Powder Inhalation Formulations. *Journal of Aerosol Medicine*, 15, 325–330.

Clements, M. L., Betts, R. F., Tierney, E. L., & Murphy, B. R. (1986). Serum and NASAL Wash ANTIBODIES associated with resistance to experimental challenge with influenza a wild-type virus. *Journal of Clinical Microbiology*, 24(1), 157–160.

Cole, R. B., & Mackay, A. D. (1990). Concepts of pulmonary physiology. In: *Essentials of Respiratory Disease*. Churchill Livingstone, New York, NY, pp. 49–60.

Cooper, C. D., & Alley, R. C. (1990). *Air Pollution Control: A Design Approach. Prospect Heights*, Waveland Press, IL.

Costantino, H. R., Firouzabadian, L., Hogeland, K., Wu, C. C., Beganski, C., Carrasquillo, K. G., Cordova, M., Griebenow, K., Zale, S. E., & Trace. (2000). Protein spray-freeze drying. Effect of atomisation conditions on particle size and stability. *Pharmaceutical Research*, 17(11), 1374–1383.

Costantino, H. R., Firouzabadian, L., Wu, C., Hogeland, K., C., Carrasquillo, K. G., Cordova, M., et al. (2002). Protein Spray freeze drying. 2. Effect of formulation variables on particle size and stability. *Journal of Pharmaceutical Sciences*, 91, 388–395.

Courrier, H. M., Butz, N., & Vandamme, T. F. (2002). Pulmonary drug delivery systems: Recent developments and prospects. *Critical Reviews in Therapeutic Drug Carrier Systems*, 19(4-5), 425–498.

de Boer, A. H., Hagedoorn, P., Westerman, E. M., Le Brun, P. P. H., Heijerman, H. G. M., & Frijlink, H. W. (2006). Design and in vitro performance testing of multiple air classifier technology in a new disposable inhaler concept (Twincer®) for high Powder doses. *European Journal of Pharmaceutical Sciences*, 28(3), 171–178.

de Boer, A. H., Gjaltema, D., Hagedoorn, P., & Frijlink, H. W. (2002). Characterization of inhalation aerosols: a critical evaluation of cascade impactor analysis and laser diffraction technique. *International Journal of Pharmaceutics*, 249, 219–231.

de Boer, A. H., Hagedoorn, P., Hoppentocht, M., Buttini, F., Grasmeijer, F., & Frijlink, H. W. (2017). Dry powder inhalation: past, present and future. *Expert Opinion on Drug Delivery*, 14, 499–512.

Dean, H. J., & Chen, D. (2004). Epidermal powder immunization against influenza. *Vaccine*, *23*(5), 681–686.

Dean, H. J., Fuller, D., & Osorio, J. E. (2003). Powder and particle-mediated approaches for delivery of DNA and protein vaccines into the epidermis. *Comparative Immunology, Microbiology and Infectious Diseases*, 26(5-6), 373–388.

Depreter, F., Pilcer, G., & Amighi, K. (2013). Inhaled proteins: Challenges and perspectives. *International Journal of Pharmaceutics*, 447(1-2), 251–280.

Desai, T. R., Hancock, R. E. W., & Finlay, W. H. (2002). A facile method of delivery of liposomes by nebulization. *Journal of Controlled Release*, 84(1-2), 69–78.

DeVincenzo, J. P. (2012). The promise, pitfalls and progress of RNA-interference-based antiviral therapy for respiratory viruses. *Antiviral Therapy*, 17, 213–225.

Dhand, C., Prabhakaran, M. P., Beuerman, R. W., Lakshminarayanan, R., Dwivedi, N., & Ramakrishna, S. (2014). Role of size of drug delivery carriers for pulmonary and intravenous administration with emphasis on cancer therapeutics and lung-targeted drug delivery. *RSC Advances*, 4(62), 32673–32689.

Divya, A., Rao, M. K., Gnanprakash, K., Sowjanya, A., Vidyasagar, N., & Gobinath, M. (2012). A review on current scenario of transdermal drug delivery system. *International Journal of Research in Pharmaceutical Sciences*, 3(4), 494–502.

Donnelly R. F., Singh T. R. R., Morrow D. I., & Woolfson A. D. (2012). *Microneedle-Mediated Transdermal and Intradermal Drug Delivery*. Wiley, Hoboken, NJ, USA.

Edwards, D. A., Ben-Jebria, A., & Langer, R. (1998). Recent advances in pulmonary drug delivery using large, porous inhaled particles. *Journal of Applied Physiology*, 85(2), 379–385.

Edwards, D. A., Hanes, J., Caponetti, G., Hrkach, J., Ben-Jebria, A., Eskew, M. L., Mintzes, J., Deaver, D., Lotan, N., & Langer, R. (1997). Large porous particles for pulmonary drug delivery. *Science*, 276(5320), 1868–1872.

El-Sherbiny I. M., Villanueva D. G., Herrera D., & Smyth H. D. C. (2011). Overcoming Lung Clearance Mechanisms for Controlled Release Drug Delivery. In: H. Smyth & A. Hickey (Eds.), *Controlled Pulmonary Drug Delivery. Advances in Delivery Science and Technology*, Springer, New York, NY, pp. 101–126.

Emami, F., Vatanara, A., Park, E., & Na, D. (2018). Drying technologies for the stability and bioavailability of biopharmaceuticals. *Pharmaceutics*, 10(3), 131.

ersjournals (*n.d.*). https://erj.ersjournals.com/highwire/markup/98668/expansion?width=1000&height=500&iframe=true&postprocessors=highwire_figures%2Chighwire_math%2Chighwire_inline_linked_media (Accessed 10 September 2021).

Fenton, C., Keating, G. M., & Plosker, G. L. (2003). Novolizer?? *Drugs*, 63(22), 2437–2445.

Ferro, T. J., Kern, J. A., Elias, J. A., Kamoun, M., Daniele, R. P., & Rossman, M. D. (1987). Alveolar macrophages, blood monocytes, and density-fractionated alveolar macrophages differ in their ability to promote lymphocyte proliferation to mitogen and antigen. *The American Review of Respiratory Disease*, 135(3), 682–687.

Fire, A., Xu, S., Montgomery, M., Kostas, S. A., Driver, S. E., & Mello, C. C. (1998). Potent and specific genetic interference by double-stranded RNA in *Caenorhabditis elegans*. *Nature*, 391, 806–811.

Friebel, C., & Steckel, H. (2010). Single-use disposable dry powder inhalers for pulmonary drug delivery. *Expert Opinion on Drug Delivery*, 7(12), 1359–1372.

Frijlink, H. W., & de Boer, A. H. (2005). Trends in the technology-driven development of new inhalation devices. *Drug Discovery Today: Technologies*, 2, 47–57.

Garmise, R. J., Staats, H. F., & Hickey, A. J. (2007). Novel dry powder preparations of whole inactivated influenza virus for nasal vaccination. *AAPS PharmSciTech*, 8(4), 2–10.

Geiser, M. (2002). Morphological aspects of particle uptake by lung phagocytes. *Microscopy Research and Technique*, 57(6), 512–522.

Gill, S., Löbenberg, R., Ku, T., Azarmi, S., Roa, W., & Prenner, E. J. (2007). Nanoparticles: Characteristics, mechanisms of action, and toxicity in pulmonary drug delivery—a review. *Journal of Biomedical Nanotechnology*, 3(2), 107–119.

Gong, J. L., McCarthy, K. M., Rogers, R. A., & Schneeberger, E. E. (1994). Interstitial lung macrophages interact with dendritic cells to present antigenic peptides derived from particulate antigens to T cells. *Immunology*, 81(3), 343–351.

Grasmeijer, N., Stankovic, M., de Waard, H., Frijlink, H. W., &; Hinrichs, W. L. J. (2013). Unraveling protein stabilization mechanisms: Vitrification and water replacement in a glass transition temperature controlled system. *Biochimica Et Biophysica Acta (BBA) - Proteins and Proteomics*, 1834(4), 763–769.

Guo, S., Jańczewski, D., Zhu, X., Quintana, R., He, T., & Neoh, K. G. (2015). Surface charge control for zwitterionic polymer brushes: Tailoring surface properties to antifouling applications. *Journal of Colloid and Interface Science*, 452, 43–53.

Hamedinasab, H., Rezayan, A. H., Mellat, M., Mashreghi, M., & Jaafari, M. R. (2020). Development of chitosan-coated liposome for pulmonary delivery of n-acetylcysteine. *International Journal of Biological Macromolecules*, 156, 1455–1463.

Her, J.-Y., Song, C.-S., Lee, S. J., & Lee, K.-G. (2010). Preparation of kanamycin powder by an optimized spray freeze-drying method. *Powder Technology*, 199, 159–164.

Heyder, J., & Rudolf, G. (1984). Mathematical models of particle deposition in the human respiratory tract. *Journal of Aerosol Science*, 15(6), 697–707.

Hickey, A. J. (2004). *Pharmaceutical Inhalation Aerosol Technology*. American Association of Pharmaceuticals Scientists, 2nd Edition, Marcel Dekker, Inc., New York, NY, USA.

Hillery, A. M., Lloyd, A. W., & Swarbrick, J. (2001). *Drug delivery and targeting for pharmacists and pharmaceutical scientists*. Taylor & Francis, Oxford.

Huang, J., Garmise, R. J., Crowder, T. M., Mar, K., Hwang, C. R., Hickey, A. J., Mikszta, J. A., & Sullivan, V. J. (2004). A novel dry powder influenza vaccine and intranasal delivery technology: Induction of systemic and mucosal immune responses in rats. *Vaccine*, 23(6), 794–801.

Ishwarya, S. P., & Anandharamakrishnan, C. (2015). Spray-Freeze-Drying approach for soluble coffee processing and its effect on quality characteristics. *Journal of Food Engineering*, 149, 171–180.

Ishwarya, S. P., Anandharamakrishnan, C., & Stapley, A. G. F. (2015). Spray-freeze-drying: A novel process for the drying of foods and bioproducts. *Trends in Food Science & Technology*, 41, 161–181.

Jaffari, S., Forbes, B., Collins, E., Barlow, D., Martin, G. P., & Murnane, D. (2013). Rapid characterisation of the inherent dispersibility of respirable powders using dry dispersion laser diffraction. *International Journal of Pharmaceutics,* 447, 124–131.

Jain, G. (1995). *Development of transdermal drug delivery system for vasoactive drugs.* Ph.D. Thesis, Submitted to the University of Delhi, India.

Jenkins, R. G., McAnulty, R. J., Hart, S. L., & Laurent, G. J. (2003). Pulmonary gene therapy. Realistic hope for the future, or false dawn in the promised land? *Monaldi Arch Chest Dis.*, 59(1), 17–24.

Johnson, R., & Lewis, L. (2011). Freeze-drying protein formulations above their collapse temperatures: Possible issues and concerns. *American Pharmaceutical Review*, 14(3), 50–54.

Kanig, J. L. (1963). Pharmaceutical aerosols. *Journal of Pharmaceutical Sciences*, 52(6), 513–535.

Kawabata, Y., Wada, K., Nakatani, M., Yamada, S., & Onoue, S. (2011). Formulation design for poorly water-soluble drugs based on biopharmaceutics classification system: basic approaches and practical applications. *International Journal of Pharmaceutics*, 420(1), 1–10.

Kendall, M., Mitchell, T., & Wrighton-Smith, P. (2004). Intradermal ballistic delivery of micro-particles into excised human skin for pharmaceutical applications. *Journal of Biomechanics*, 37(11), 1733–1741.

Kohler, D. (2003). Novolizer: the new technology for the management of asthma therapy. *Current Opinion in Pulmonary Medicine*, 9 (Suppl1), S11–S16.

Kohler, D. (2004). The Novolizer®: Overcoming inherent problems of dry powder inhalers. *Respiratory Medicine*, 98, S17–S21.

Koli, U., Krishnan, R. A., Pofali, P., Jain, R., & Dandekar, P. (2014). siRNA-Based therapies for pulmonary diseases. *Journal of Biomedical Nanotechnology*, 10(9), 1953–1997.

Kretsos, K., & Kasting, G. B. (2007). A geometrical model of dermal capillary clearance. *Mathematical Biosciences*, 208(2), 430–453.

Kumar, D., Sharma, N., Rana, A. C., Agarwal, G., & Bhat, Z. A. (2011). A review: transdermal drug delivery system: tools for novel drug delivery system. *International Journal of Drug Development & Research*, 3(3), 70–84.

Kuo, J.-H. S., & Hwang, R. (2004). Preparation of DNA dry powder for non-viral gene delivery by spray-freeze drying: Effect of protective agents (polyethyleneimine and sugars) on the stability of DNA. *Journal of Pharmacy and Pharmacology*, 56(1), 27–33.

Lahm, K., & Lee, G. (2006). Penetration of crystalline powder particles into excised human skin membranes and model gels from a supersonic powder injector. *Journal of Pharmaceutical Sciences*, 95(7), 1511–1526.

Lam, J. K.-W., Liang, W., & Chan, H.-K. (2012). Pulmonary delivery of therapeutic siRNA. *Advanced Drug Delivery Reviews*, 64(1), 1–15.

Lashkar, S. (2012). An overview on: pharmaceutical aerosols. *IRJP*, 3(9), 68–75.

Laube, B. L. (2005). The expanding role of aerosols in systemic drug delivery, gene therapy, and vaccination. *Respiratory Care*, 50(9), 1161–1176.

Liang, W., Chan, A. Y. L., Chow, M. Y. T., Lo, F. F. K., Qiu, Y., Kwok, P. C. L., & Lam, J. K. W. (2018). Spray freeze drying of small nucleic acids as inhaled powder for pulmonary delivery. *Asian Journal of Pharmaceutical Sciences*, 13(2), 163–172.

Lindsley, W. G., Green, B. J., Blachere, F. M., Martin, S. B., Law, B. F., & Jensen, P. A. (2017). Sampling and Characterization of Bioaerosols. NIOSH Manual of Analytical Methods (Fifth edition), National Institute for Occupational Safety and Health, Cincinnati.

Liu, D.-L. (2010). Particle deposition onto enclosure surfaces. In: Kohli, R., & Mittal, K. (Eds.), *Developments in Surface Contamination and Cleaning*. Elsevier/William Andrew, Oxford, UK, pp. 1–56.

Lo, J. C., Pan, H. W., & Lam, J. K. (2021). Inhalable protein powder prepared by spray-freeze-drying using hydroxypropyl-β-cyclodextrin as excipient. *Pharmaceutics*, 13(5), 615.

Maa, Y.-F., Nguyen, P.-A., Sweeney, T., Shire, S. J., & Hsu, C. C. (1999). Protein inhalation powders: spray drying vs spray freeze drying. *Pharmaceutical Research*, 16(2), 249–254.

Maa, Y. F., Ameri, M., Shu, C., Payne, L. G., & Chen, D. (2004). Influenza vaccine powder formulation development: Spray-freeze-drying and stability evaluation. *Journal of Pharmaceutical Sciences*, 93(7), 1912–1923.

Makgwane, P. R., & Ray, S. S. (2014). Synthesis of nanomaterials by continuous-flow microfluidics: A Review. *Journal of Nanoscience and Nanotechnology*, 14(2), 1338–1363.

Man, D. K. W., Chow, M. Y. T., Casettari, L., Gonzalez-Juarrero, M., & Lam, J. K. W. (2016). Potential and development of inhaled RNAi therapeutics for the treatment of pulmonary tuberculosis. *Advanced Drug Delivery Reviews*, 102, 21–32.

Mansur, E. A., Ye, M., Wang, Y., & Dai, Y. (2008). A state-of-the-art review of mixing in Microfluidic Mixers. *Chinese Journal of Chemical Engineering*, 16(4), 503–516.

Manufacturing Chemist (2019). *The advantages and challenges of spray drying*. Available from: https://manufacturingchemist.com/news/article_page/The_advantages_and_challenges_of_spray_drying/153451 (Accessed 12 September 2021).

McClements, D. J. (2012). Requirements for food ingredient and nutraceutical delivery systems. In N. Garti & D.J. McClements (Eds.), *Encapsulation Technologies and Delivery Systems for Food Ingredients and Nutraceuticals*. Elsevier Science, India, pp. 3–18.

Mehrani, P., Bi, H. T., & Grace, J. R. (2007). Electrostatic behavior of different fines added to a faraday cup fluidized bed. *Journal of Electrostatics*, 65(1), 1–10.

Mensink, M. A., Frijlink, H. W., van der Voort Maarschalk, K., & Hinrichs, W. L. J. (2017). How sugars protect proteins in the solid state and during drying (review): Mechanisms of stabilization in relation to stress conditions. *European Journal of Pharmaceutics and Biopharmaceutics*, 114, 288–295.

Milani, S., Faghihi, H., Najafabadi, R. A., Amini, M., Montazeri, H., & Vatanara, A. (2020). Hydroxypropyl beta cyclodextrin: A water-replacement agent or a surfactant upon spray freeze-drying of IgG with enhanced stability and aerosolization. *Drug Development and Industrial Pharmacy*, 46(3), 403–411.

Minton, A. P. (2005). Influence of macromolecular crowding upon the stability and state of association of proteins: Predictions and observations. *Journal of Pharmaceutical Sciences*, 94(8), 1668–1675.

Mitchell, T. J. , Kendall, M.A.F. , & Bellhouse, B. J. (2001). Microparticle penetration to the oral mucosa. *Proceedings of the 2001 Bioengineering Conference*, Snowbird, Utah, pp. 763–764.

Mohri, K., Okuda, T., Mori, A., Danjo, K., & Okamoto, H. (2010). Optimized pulmonary gene transfection in mice by spray-freeze dried powder inhalation. *Journal of Controlled Release*, 144(2), 221–226.

Momin, M. A. M., Sinha, S., Tucker, I. G., Doyle, C., & Das, S. C. (2017). Dry powder formulation of kanamycin with enhanced aerosolization efficiency for drug-resistant tuberculosis. *International Journal of Pharmaceutics*, 528(1-2), 107–117.

Murugappan, S., Patil, H. P., Kanojia, G., ter Veer, W., Meijerhof, T., Frijlink, H. W., Huckriede, A., & Hinrichs, W. L. J. (2013). Physical and immunogenic stability of spray freeze-dried influenza vaccine powder for pulmonary delivery: Comparison of inulin, dextran, or a mixture of dextran and trehalose as protectants. *European Journal of Pharmaceutics and Biopharmaceutics*, 85(3), 716–725.

Murugappan, S. (2014). *New strategies for simplifying influenza vaccination*. Ph.D. Thesis submitted to the University of Groningen, Netherlands.

National Center for Biotechnology Information (2021). PubChem Compound Summary for CID 2764, Ciprofloxacin. Retrieved September 12, 2021 from https://pubchem.ncbi.nlm.nih.gov/compound/Ciprofloxacin.

Niwa, T., Mizutani, D. and Danjo, K. 2012. Spray-freeze-dried porous microparticles of a poorly water soluble drug for respiratory delivery. *Chemical and Pharmaceutical Bulletin*, 60(7), 870–876.

Ohtake, S., Kita, Y., & Arakawa, T. (2011). Interactions of formulation excipients with proteins in solution and in the dried state. *Advanced Drug Delivery Reviews*, 63(13), 1053–1073.

Okamoto, H., & Danjo, K. (2008). Application of supercritical fluid to preparation of powders of high-molecular weight drugs for inhalation. *Advanced Drug Delivery Reviews*, 60(3), 433–446.

Okuda, T., Suzuki, Y., Kobayashi, Y., Ishii, T., Uchida, S., Itaka, K., Kataoka, K., & Okamoto, H. (2015). Development of biodegradable polycation-based inhalable dry gene powders by spray freeze drying. *Pharmaceutics*, 7(3), 233–254.

Paranjpe, M., & Müller-Goymann, C. C. (2014). Nanoparticle-mediated pulmonary drug delivery: a review. *International Journal of Molecular Sciences*, 15, 5852–5873.

Parsian, A. R., Vatanara, A., Rahmati, M. R., Gilani, K., Khosravi, K. M., & Najafabadi, A. R. (2014). Inhalable budesonide porous microparticles tailored by spray freeze drying technique. *Powder Technology*, 260, 36–41.

Patton, J. S., & Byron, P. R. (2007). Inhaling medicines: Delivering drugs to the body through the lungs. *Nature Reviews Drug Discovery*, 6(1), 67–74.

Peña-Juárez, M. C., Guadarrama-Escobar, O. R., & Escobar-Chávez, J. J. (2021). Transdermal delivery systems for Biomolecules. *Journal of Pharmaceutical Innovation*, 1–14.

Peng, T., Lin, S., Niu, B., Wang, X., Huang, Y., Zhang, X., Li, G., Pan, X., & Wu, C. (2016). Influence of physical properties of carrier on the performance of dry powder inhalers. *Acta Pharmaceutica Sinica B*, 6(4), 308–318.

Pepper, I. L. , Gerba, C. P., & Brendecke, J. W. (2011). *Environmental microbiology: A laboratory manual.* Elsevier Science.

Pouya, M. A., Daneshmand, B., Aghababaie, S., Faghihi, H., & Vatanara, A. (2018). Spray-Freeze Drying: A suitable method for aerosol delivery of antibodies in the presence of trehalose and cyclodextrins. *AAPS PharmSciTech*, 19(5), 2247–2254.

Prestrelski, S. J., Burkoth, T. L., & Maa, Y.-F. (2002). *Spray freeze-dried compositions.* Australian Patent, AU2002302814A1.

Prow, T. W., Grice, J. E., Lin, L. L., Faye, R., Butler, M., Becker, W., Wurm, E. M. T., Yoong, C., Robertson, T. A., Soyer, H. P., & Roberts, M. S. (2011). Nanoparticles and microparticles for skin drug delivery. *Advanced Drug Delivery Reviews*, 63(6), 470–491.

Puius, Y. A., Stievater, T. H., & Srikrishnan, T. (2006). Crystal structure, conformation, and absolute configuration of kanamycin A. *Carbohydrate Research*, 341(17), 2871–2875.

Qiu, Y., Lam, J., Leung, S., & Liang, W. (2016). Delivery of RNAi therapeutics to the airways – from bench to bedside. *Molecules*, 21(9), 1249.

Renegar, K. B., & Small, P. A. Jr (1991). Passive transfer of local immunity to influenza virus infection by IgA antibody. *Journal of Immunology*, 146(6), 1972–1978.

Rochelle, C., & Lee, G. (2007). Dextran or hydroxyethyl starch in spray freeze-dried trehalose/mannitol microparticles intended as ballistic particulate carriers for proteins. *Journal of Pharmaceutical Sciences*, 96(9), 2296–2309.

Rouwkema, J., Rivron, N. C., & van Blitterswijk, C. A. (2008). Vascularization in tissue engineering. *Trends in Biotechnology*, 26(8), 434–441.

Saboti, D., Maver, U., Chan, H.-K., & Planinšek, O. (2017). Novel budesonide particles for dry powder inhalation prepared using a microfluidic reactor coupled with ultrasonic spray freeze drying. *Journal of Pharmaceutical Sciences*, 106(7), 1881–1888.

Sahilhusen, I. J., & Mukesh, R. P. (2014). Pharmaceutical Controlled Release Drug Delivery Systems: A Patent Overview. *Aperito Journal of Drug Designing and Pharmacology*, 1(2), p. 22. 10.14437/AJDDP-1-107.

Saluja, V., Amorij, J.-P., Kapteyn, J. C., de Boer, A. H., Frijlink, H. W., & Hinrichs, W. L. J. (2010). A comparison between spray drying and spray freeze drying to produce an influenza subunit vaccine powder for inhalation. *Journal of Controlled Release*, 144(2), 127–133.

Sarphie, D. F., & Burkoth, T. L. (1999). *Method for providing dense particle compositions for use in transdermal particle delivery.* European Patent EP0912239B1.

Sarphie, D. F., Johnson, B., Cormier, M., Burkoth, T. L., & Bellhouse, B. J. (1997). Bioavailability following transdermal powdered delivery (TPD) of radiolabeled inulin to hairless Guinea Pigs. *Journal of Controlled Release*, 47(1), 61–69.

Schiffter, H. (2007). Spray-freeze-drying in the manufacture of pharmaceuticals. *European Pharmaceutical Review*, 12, 67–71.

Schiffter, H., Condliffe, J., & Vonhoff, S. (2010). Spray-freeze-drying of nanosuspensions: The manufacture of insulin particles for needle-free ballistic powder delivery. *Journal of the Royal Society Interface*, 7, S483–S500.

Sciarra, J. J., & Cutie, A. J. (1991). *Pharmaceutical aerosols* (Second edition), Varghese Publishing House, Bombay, India, pp. 589–618.

Serno, T., Carpenter, J. F., Randolph, T. W., & Winter, G. (2010). Inhibition of agitation induced aggregation of an IgG-antibody by hydroxypropyl-beta-cyclodextrin. *Journal of Pharmaceutical Sciences*, 99, 1193–1206.

Serno, T., Geidobler, R., & Winter, G. (2011). Protein stabilization by cyclodextrins in the liquid and dried state. *Advanced Drug Delivery Reviews*, 63(13), 1086–1106.

Shi, S., Dodds Ashley, E. S., Alexander, B. D., & Hickey, A. J. (2009). Initial characterization of micafungin pulmonary delivery via two Different Nebulizers and multivariate data analysis of Aerosol mass distribution profiles. *AAPS PharmSciTech*, 10(1), 129–137.

Silva, C. O., Rijo, P., Molpeceres, J., Figueiredo, I. V., Ascensão, L., Fernandes, A. S., Roberto, A., & Reis, C. P. (2015). Polymeric nanoparticles modified with fatty acids encapsulating betamethasone for anti-inflammatory treatment. *International Journal of Pharmaceutics*, 493(1-2), 271–284.

Sjoholm, P. E. T. R. I., Ingham, D. E. R. E. K. B., Lehtimaki, M. A. T. T. I., Perttu-Roiha, L. E. E. N. A., Goodfellow, H. O. W. A. R. D., & Torvela, H. E. I. K. K. I. (2001). Gas-cleaning technology. In H. Goodfellow & E. Tähti (Eds.), *Industrial Ventilation Design Guidebook*. Academic Press, San Deigo, USA, pp. 1197–1316.

Smith, D. J., Bot, S., Dellamary, L., & Bot, A. (2003). Evaluation of novel aerosol formulations designed for mucosal vaccination against influenza virus. *Vaccine*, 21(21-22), 2805–2812.

Sonia, T. A., & Sharma, C. P. (2014). *Oral delivery of insulin*. Woodhead Publishing, Cambridge, United Kingdom.

Sonner, C., Maa, Y. F., & Lee, G. (2002). Spray-freeze-drying for protein powder preparation: Particle characterization and a case study with trypsinogen stability. *Journal of Pharmaceutical Sciences*, 91(10), 2122–2139.

Stahlhofen, W., Koebrich, R., Rudolf, G., & Scheuch, G. (1990). Short-term and long-term clearance of particles from the upper human respiratory tract as function of particle size. *Journal of Aerosol Science*, 21, S407–S410.

Straller, G., & Lee, G. (2017). Shrinkage of spray-freeze-dried microparticles of pure protein for ballistic injection by manipulation of freeze-drying cycle. *International Journal of Pharmaceutics*, 532(1), 444–449.

Sweeney, L. G., Wang, Z., Loebenberg, R., Wong, J. P., Lange, C. F., & Finlay, W. H. (2005). Spray-freeze-dried liposomal ciprofloxacin powder for inhaled aerosol drug delivery. *International Journal of Pharmaceutics*, 305(1-2), 180–185.

Thepen, T., Hoeben, K., Brevé, J., & Kraal, G. (1992). Alveolar macrophages down-regulate local pulmonary immune responses against intratracheally administered T-cell-dependent, but not T-cell-independent antigens. *Immunology*, 76(1), 60–64.

Thiele, J., Windbergs, M., Abate, A. R., Trebbin, M., Shum, H. C., Förster, S., & Weitz, D. A. (2011). Early development drug formulation on a chip: Fabrication of nanoparticles using a microfluidic spray dryer. *Lab on a Chip*, 11(14), 2362–2368.

Toll, R., Jacobi, U., Richter, H., Lademann, J., Schaefer, H., & Blume-Peytavi, U. (2004). Penetration profile of microspheres in follicular targeting of terminal hair follicles. *Journal of Investigative Dermatology*, 123(1), 168–176.

Tomoda, K., & Makino, K. (2014). Nanoparticles for transdermal drug delivery system (TDDS). In: H. Ohshima & K. Makino (Eds.), *Colloid and Interface Science in Pharmaceutical Research and Development*, Elsevier, Amstersam, The Netherlands, pp. 131–147.

Tomoda, K., Terashima, H., Suzuki, K., Inagi, T., Terada, H., & Makino, K. (2011). Enhanced transdermal delivery of indomethacin-loaded PLGA nanoparticles by iontophoresis. *Colloids and Surfaces B: Biointerfaces*, 88(2), 706–710.

Van Drooge, D. J., Hinrichs, W. L. J., & Frijlink, H. W. (2004). Incorporation of lipophilic drugs in sugar glasses by lyophilization using a mixture of water and tertiary butyl alcohol as solvent. *Journal of Pharmaceutical Sciences*, 93(3), 713–725.

van Drooge, D.-J., Hinrichs, W. L. J., Dickhoff, B. H. J., Elli, M. N. A., Visser, M. R., Zijlstra, G. S., & Frijlink, H. W. (2005). Spray freeze drying to produce a stable δ9-tetrahydrocannabinol containing inulin-based solid dispersion powder suitable for inhalation. *European Journal of Pharmaceutical Sciences*, 26(2), 231–240.

Vanbever, R., Mintzes, J. D., Wang, J., Nice, J., Chen, D., Batycky, R., Langer, R., & Edwards, D. A. (1999). Formulation and physical characterization of large porous particles for inhalation. *Pharmaceutical Research*, 16(11), 1735–1742.

Vladykina, T., Deryagin, B., & Toporov, Y. P. (1985). The effect of surface roughness on triboelectrification of insulators. *Phys. Chem. Mech. Surf.*, 3, 2817–2821.

Wang, W. (2000). Lyophilization and development of solid Protein pharmaceuticals. *International Journal of Pharmaceutics*, 203(1-2), 1–60.

Weissmueller, N. T., Schiffter, H. A., Carlisle, R. C., Rollier, C. S., & Pollard, A. J. (2015). Needle-free dermal delivery of a diphtheria toxin crm197mutant on potassium-doped hydroxyapatite microparticles. *Clinical and Vaccine Immunology*, 22(5), 586–592.

Westerman, E. M., Heijerman, H. G., & Frijlink, H. W. (2007). Dry powder inhalation versus wet nebulisation delivery of antibiotics in cystic fibrosis patients. *Expert Opinion on Drug Delivery*, 4, 91–94.

Yang, W., Peters, J. I., & Williams-III, R. O. (2008). Inhaled nanoparticles - a current review. *International Journal of Pharmaceutics*, 356, 239–247.

Ye, T., Yu, J., Luo, Q., Wang, S., & Chan, H.-K. (2017). Inhalable clarithromycin liposomal dry powders using ultrasonic spray freeze drying. *Powder Technology*, 305, 63–70.

Zanen, P., Go, L. T., & Lammers, J. J. (1994). The optimal particle size for β-adrenergic aerosols in mild asthmatics. *International Journal of Pharmaceutics*, 107, 211–217.

Zeng, X. M., Martin, G. P., & Marriott, C. (2000). *Particulate interactions in dry powder formulation for inhalation* (1st Edition). CRC Press, Boca Rato, FL.

Zhanel, G. G., Ennis, K., Vercaigne, L., Walkty, A., Gin, A. S., Embil, J., Smith, H., & Hoban, D. J. (2002). A critical review of the fluoroquinolones. *Drugs*, 62(1), 13–59.

Zhu, K., Tan, R. B., Chen, F., Ong, K. H., & Heng, P. W. S. (2007). Influence of particle wall adhesion on particle electrification in mixers. *International Journal of Pharmaceutics*, 328, 22–34.

Ziegler, A. S. (2008). Needle-free delivery of powdered protein vaccines: A new and rapidly developing technique. *Journal of Pharmaceutical Innovation*, 3(3), 204–213.

CHAPTER 9

Spray-freeze-drying for the Production of Therapeutic Nanoparticles

In recent years, use of nanoparticles as vehicles for the targeted delivery of drugs and bioactives has gained increased attention. According to the International Organization for Standardization (ISO, 2017), nanoparticles are *materials with any external nanoscale dimension or having internal nanoscale surface structure*. Accordingly, their size falls within the range of 1–1000 nm. The transformed surface chemistry of nanoparticles is responsible for their unique properties (ex. different color, lower melting point, higher solubility), relative to the original bulk material (Surface energy, 2021). At decreasing dimensions of the nanoparticles, their surface-to-volume ratio (SA/V) increases enormously (Figure 9.1). The higher curvature of nanoparticles results in interesting biological, physicochemical and mechanical properties. This is because at sufficiently small size, properties of nanomaterials are governed by surface atoms, rather than bulk of the material.

In the above background, application of nanoparticles (NPs) as therapeutic carriers is an interesting area of investigation. The active drug is either physically distributed in the matrix of polymer-based or lipid-based nanoparticles or adsorbed onto the surface of NPs. The compounds loaded in nanoparticle-based carriers have shown a substantial improvement in their therapeutic efficiency. Besides, NP-based drug delivery systems overcome the drug-associated toxicity and facilitate targeted delivery (Hadavi & Poot, 2016) due to their ability to penetrate tissues. Nanostructures are easily taken up by cells (15–250 fold greater than that of microparticles in the 1–10 μm range), thereby permitting the effective delivery of drugs to the target sites of action. Also, the nanoparticles remain in the blood circulation for long duration and release the incorporated drug in a controlled manner. Further, nanostructures are capable of protecting the drugs against unfavorable conditions in the gastrointestinal tract. They facilitate the APIs to circumvent the first pass metabolism in the liver. In addition, nanoparticles enhance the oral bioavailability of poorly water-soluble drugs by specialized uptake mechanisms such as absorptive endocytosis. Thus, nanoparticles are promising delivery vehicles for challenging conventional drugs that are used for the management and treatment of lifestyle diseases (ex. diabetes) and chronic ailments (cancer, asthma, hypertension, HIV) (Ochubiojo et al., 2012).

From the above discussion, it is apparent that the use of NP-based systems is a one-stop solution to overcome the challenges posed by the larger size of active materials during drug delivery, which include poor dissolution, *in vivo* instability in the gastrointestinal tract, reduced bioavailability and impaired absorption in the body.

DOI: 10.1201/9781003019312-9

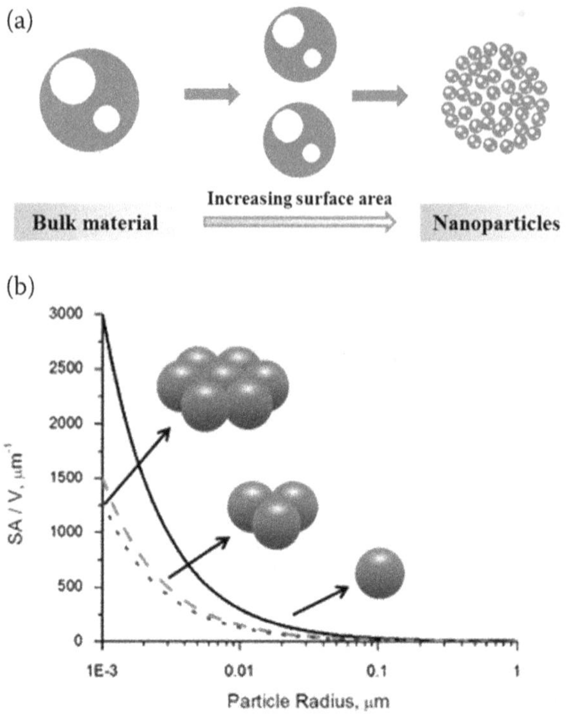

Figure 9.1 (a) Schematic illustration of the size reduction of bulk material to form NPs with high surface-to-volume ratio (Sun et al., 2017); (b) Progressive reduction in the surface area-to volume-ratio of a particle with decreasing particle size (Modified and redrawn from Akron Ascent Innovations, 2017).

9.1 SPRAY-FREEZE-DRYING AS THE TECHNIQUE FOR NANOPARTICLE PRODUCTION

The abovementioned advantages of nanoparticles are jeopardized by their chemical and/or physical instability in dispersion. This necessitates the removal of water or other solvent(s) from the dispersed system. Hence, drying of nanoparticle dispersions becomes an indispensable unit operation in the production of nanoparticles. Drying ensures long-lasting stability of nanoparticle dispersions and facilitates their conversion into solid dosage forms. A drying technique that is appropriate for nanoparticle production is the one which does not promote aggregation or instability of nanoparticles during the process. Spray-drying and lyophilization are conventionally employed for the production of nanoparticles in dry form. However, their major limitation is the impact on the biological activity of APIs, especially, proteins. Spray drying requires very high temperatures in the order of 150°C, to evaporate solvent from the atomized drug solution. The high temperature may accelerate degradation of the API. Moreover, drugs loaded in lipid-based carriers cannot be processed by spray drying due to the melting of lipids at its elevated operating temperature.

On the other hand, the freeze-dried products have a cake-like structure that pose limitations with respect to handling and further transformation into solid dosage forms. Minimal particle handling of APIs leads to simplified cleaning and sterilization procedures. The prolonged freezing times of bulk liquid nanodispersions may lead to the aggregation of nanoparticles before the drying is complete. Also, it is imperative that the particle size and biological activity of the dried nanoparticles are retained even after reconstitution into nanodispersion (Lek

Pharmaceuticals d.d., 2015). Therefore, the search began for a drying technique that can produce dry and free-flowing powder that can be converted into solid dosage forms such as tablets or capsules.

Based on the understanding obtained from previous chapters, it is evident that spray-freeze-drying is the apt candidate for the production of nanoparticles in dry form with good flowability. Indeed, SFD is a recent entrant among the drying techniques for nanoparticle production. Compared to the prolonged freezing time during conventional freeze-drying of bulk liquid nanodispersions, SFD's rapid freezing of atomized droplets may possibly prevent aggregation of nanoparticles dispersed in liquid. This may be attributed to events such as ice phase-separation and liquid freeze-concentration (Cui et al., 2003; Shaik et al., 2001). Spray-freeze-drying is a top-down approach for the production of nanoparticles. Physical mechanization is the basis of any top-down approach that achieves size reduction by applying one or a combination of the compression, impact and shear forces. Most top-down approaches involve either emulsification or emulsification followed by solvent evaporation, of which SFD belongs to the latter type. The atomization step of SFD subjects the feed emulsion to the necessary impact and shear required to form the nanoparticles eventually.

In general, top-down approaches are more commercially viable than the bottom-up methods that involve chemical-reaction-induced self-assembly and self-organization of molecules (Augustin & Sanguansri, 2009). The specific advantages of SFD process are its low-temperature operation and ability to produce nanoparticle aggregates with significantly high aqueous re-dispersibility. This testifies the appropriateness of SFD as the method to produce solid-dosage-forms of thermally sensitive nanoparticles (Cheow et al., 2011). This chapter would elaborate on the major applications of nanoparticles produced by spray-freeze-drying. Further, the principle of designing spray-freeze-drying trials for nanoparticle production will be explained with emphasis on the standardization of feed formulation and freezing conditions.

9.2 CUSTOMIZATION OF SFD PROCESS FOR NANOPARTICLE PRODUCTION

Tailoring the spray-freeze-drying process for fabricating nanoparticles can be done either in the stage of feed preparation (addition of adjuvants/excipients), atomization (choice of suitable atomizer) or freezing (selection of the appropriate freezing mode amongst SFL, SFV or SFV/L).

9.2.1 Use of freezing adjuvants in the feed formulation

Freezing adjuvants are agents that are added to improve the dissolution, flowability, aerosolization efficiency, bioavailability and controlled release characteristics of active pharmaceutical substances, especially, the inhalable drugs. The fundamental role of an adjuvant is to prevent the irreversible nanoparticle coalescences during freeze-drying, caused by mechanical stresses exerted by the ice crystal formation. An adjuvant prevails in its crystalline state during freeze-drying and it is capable of preventing ice nucleation and thereby the ice crystal formation that imposes mechanical stress on the nanoparticles (Abdelwahed et al., 2006). Thus, adjuvants aid in the instantaneous re-dispersibility of nanoparticle aggregates into primary nanoparticles in an aqueous continuous phase. Their ready disassociation into primary nanoparticles upon exposure to the lung interstitial fluid permits them to act against lung phagocytosis. Only those drug particles that are smaller than 1–2 μm are effective in accomplishing their intended therapeutic function (Chono et al., 2006).

Furthermore, inclusion of adjuvants in the formulation creates *'adjuvant bridges'* that connect the nanoparticles, hence preventing their inter-particle contact during freezing (Figure 9.2). Adjuvants with surfactant properties form a coating over the nanoparticle surface and stabilize it against the inter-nanoparticle contacts during freeze-drying. Thus, it is the dissolution rate of adjuvant bridges that exerts significant control over the aqueous re-dispersibility of SFD nanoparticles, rather than the

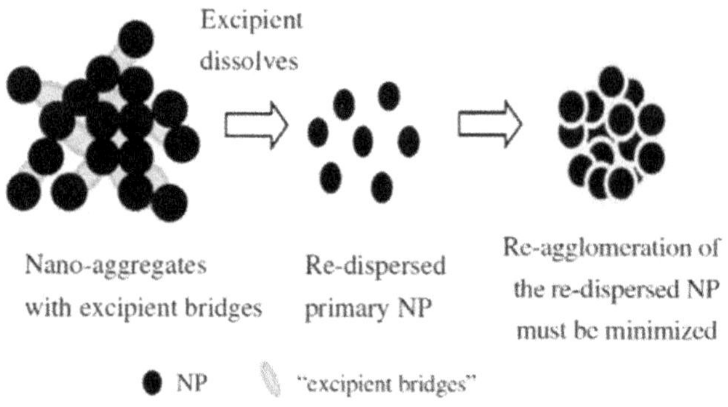

Figure 9.2 Role of *'adjuvant bridge'* in controlling the aqueous redispersibility of nano-aggregates (Kho & Hadinoto, 2010).

intrinsic attractive forces such as the van der Waals (Cheow et al., 2011). Besides, adjuvants act as cryoprotectants to protect the nanoparticles from the sub-zero temperature of spray-freeze-drying process. The commonly used adjuvants are high molecular weight polymers such as mannitol, trehalose, dextran and polyvinyl alcohol (PVA).

The suitability of a substance as an adjuvant is judged by spray-freeze-drying its aqueous solution in the absence of drug nanosuspension. To qualify as an adjuvant, a substance must not be hygroscopic, which would otherwise lead to highly cohesive particles with poor flowability after SFD. Further, the morphology of dried adjuvant particles must be intact and spherical with either smooth or porous surface. Cheow et al. (2011) assessed the suitability of four substances as adjuvants: leucine, lactose, mannitol and PVA. Among the four, mannitol and PVA were found appropriate as SFD of their aqueous solutions (7%, w/v) led to non-cohesive spherical particles with porous (Figure 9.3[b]) and smooth (Figure 9.3[c]) surface, respectively. On the other hand, the high hygroscopicity of lactose resulted in highly cohesive particles (Figure 9.3[a]) and leucine solution produced irregularly shaped and fragile fragments (Figure 9.3[d]). Thus, lactose and leucine were deemed inappropriate for use as adjuvants.

In the absence of adjuvants, the aqueous re-dispersibility of PCL nano-aggregates was poor, as only 20% of the nano-aggregates were recovered as primary nanoparticles. A majority of the nano-aggregates were recovered as aggregate fragments having size in the range of 1500–2500 nm. But, due to low particle density, the aerodynamic diameter of PCL nanoaggregates was in the range of ~ 4–5 μm, even without adjuvant inclusion, thus affirming their suitability for lung depositions. Further, mannitol and leucine were used as adjuvants during the SFD of drug-loaded poly(caprolactone) (PCL) nanoparticle suspension. The PCL nanoparticles were loaded with levofloxacin, an aerosol antibiotic formulation intended for lung biofilm infection therapy. Both the adjuvants led to nanoparticle aggregates that exhibited large, porous, and spherical morphologies (Figure 9.4) with good flowability and effective aerosolization behavior. While the mannitol-PCL nanoaggregates exhibited an aqueous re-dispersibility of ~85% (at 70% mannitol concentration), the PVA-PCL nanoaggregates showed a re-dispersibility of 90% (at 70% PVA concentration). The PVA-PCL nano-aggregate was superior to its mannitol-PCL counterpart in terms of its instantaneous and complete re-dispersion into primary nanoparticles at a lower adjuvant concentration. The variation in spontaneity of re-dispersion between mannitol and PVA was attributed to the difference in nanoparticle–adjuvant structures. Owing to its surfactant characteristics, PVA was observed to form a coating on the surface of PCL nanoparticles, thus preventing inter-particle contact during lyophilization. But, with mannitol as adjuvant, PCL nanoparticles were physically dispersed in the porous mannitol matrix (Cheow et al., 2011).

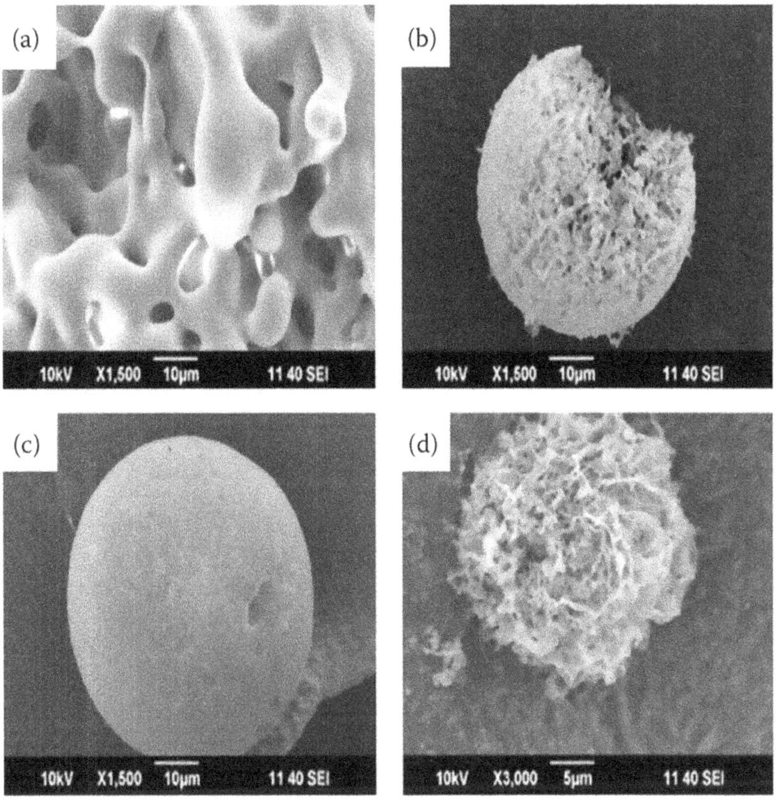

Figure 9.3 Particle morphologies of spray-freeze-dried (a) lactose; (b) mannitol; (c) PVA; and (d) leucine (Cheow et al., 2011).

9.2.2 Atomization: The use of ultrasonic nozzles and spray gun

In studies involving spray-freeze-drying for the production of nanoparticles, specialized atomizers are used rather than the conventional ones. In majority of the studies, ultrasonic nozzle was used. The working principle of ultrasonic atomizer was already discussed in Section 2.2.3 of Chapter 2. For instance, an ultrasonic nozzle powered by a 40 kHz ultrasonic generator was used for atomizing the aqueous solutions of mannitol, lysozyme, or bovine serum albumin (BSA). These were used in the production of inhalable porous nanoparticles that would function as carriers of cholesterol nanoparticles (D'Addio et al., 2013), intended for the treatment of tuberculosis. Ultrasonic atomization was also employed in other SFD trials involving the production of insulin nanoparticles for needle-free ballistic powder delivery (60 kHz; Schiffter, Condliffe & Vonhoff, 2010), aerosolizable mannitol carrier particles that encapsulate nanoparticles (40 kHz; D'Addio et al., 2013) and polyelectrolyte nanoparticles containing protein drug (25 kHz; Lek Pharmaceuticals d.d., 2015).

The vibration of ultrasonic nozzle is responsible for preventing agglomeration of nanoparticles that occurs during conventional air spray atomization (Figure 9.5). The electrical impulses generated from the control module produce continuous vibrations that maintain a uniform dispersion of nanoparticles in the suspension for several hours (Riemer, 2011). Frequency of nozzle vibration is the key control parameter of ultrasonic atomization that governs the droplet size. Smaller droplets can be obtained at higher ultrasonic frequency. The other atomization parameters include the surface tension and density of feed liquid and the rate at which it is fed to the ultrasonic nozzle.

Figure 9.4 (a) Large and spherical morphologies of PCL–PVA nano-aggregates and closely packed PCL nanoparticles exhibiting larger size due to PVA coatings; (b) Large, spherical and porous morphologies of PCL–mannitol nanoaggregates and physical dispersion of PCL nanoparticles in porous mannitol matrix (Cheow et al., 2011).

While the droplet size varies inversely with the density of feed liquid, it exhibits a direct relationship with the surface tension and flow rate of feed (Avvaru et al., 2006).

The main advantage of ultrasonic atomization is its discrete control over particle size and aerodynamic size, which are highly relevant parameters in the production of porous and inhalable drug nanoparticles. In a study performed by D'Addio et al. (2012), ultrasonic spray-freeze-drying resulted in large porous mannitol particles with a high nanoparticle loading (50% by mass) and fine particle fraction (63%) that exhibited excellent aerosol performance. Compared to conventional atomization, the other merits of ultrasonic atomization with respect to nanoparticle production include small droplet sizes with a homogeneous size distribution, high droplet sphericity, larger liquid flow passages and requirement for low liquid feed pressure (Rajan & Pandit, 2001).

Alternatively, a spray gun has been used as atomization device in a spray freeze-drying setup (Figure 9.6) for the production of nanocrystalline cellulose (NCC) (Kamal & Khoshkava, 2015). Spray gun is a pneumatic device that uses compressed air to atomize a liquid. The impingement between high-velocity turbulent air and the surface of liquid film causes their division of liquid

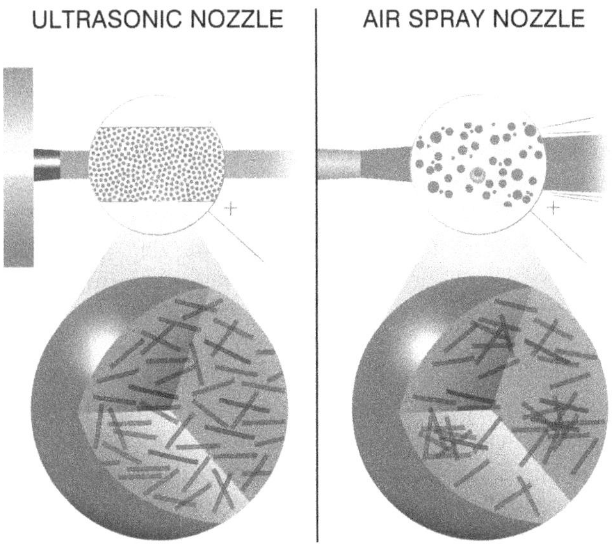

Figure 9.5 Comparison of droplet size and occurrence of droplet agglomeration phenomenon between ultrasonic atomization and conventional air spray atomization (Image courtesy: Sono-Tek CorporationSonoTek 2021).

feed into fine droplets of broad size range. In the abovementioned study, a dilute aqueous suspension of NCC was supplied to the spray gun using a peristaltic pump at a flow rate of 30 mL/min. The spray gun was in turn connected to an air pressure line, which provides the necessary energy for atomization. Initially, the spray gun was fixed at a distance of 30 cm above the surface of liquid nitrogen held in a container of 30 cm diameter and 50 cm height fitted with a magnetic stirring bar. The NCC suspension was sprayed into the liquid nitrogen. Then, the excess liquid nitrogen was boiled-off the frozen droplets, after which they were transferred to the lyophilizer. Freeze-drying was carried out at -52°C and 0.05 mbar for a period of 24–72 h, depending on the material quantity. The size of spray-freeze-dried NCC was same as the original size of particles (5–20 nm in diameter) used for the preparation of suspension. The ratio between the Brunauer-Emmett-Teller (BET) surface area of SFD-NCC and that of spray-dried NCC or freeze-dried NCC was more than or equal to about 10. This indicates the low density and high porosity of spray-freeze-dried nanoparticles that are advantageous in achieving the intended functionality of particles (Kamal & Khoshkava, 2015).

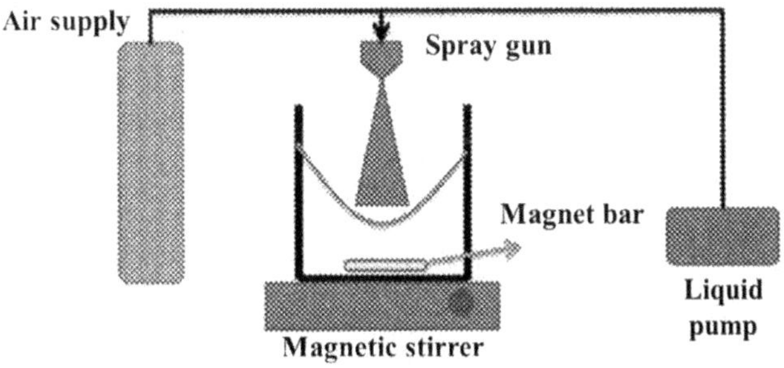

Figure 9.6 Use of spray gun in a spray-freeze-drying setup for the production of nanoparticles (Kamal & Khoshkava, 2015).

9.2.3 Freezing

Among the different spray-freezing modes of SFD process, the spray-freezing-into-liquid (SFL) is the most desirable arrangement for the production of nanoparticles. With other modes of freezing such as spray-freezing-into vapor (SFV) and spray-freezing-into vapor over liquid (SFV/L), substantial droplet aggregation occurs during the atomization step wherein the particles adsorb at the air–liquid interface as the droplets pass through the vapor. Due to their larger surface area, nanoparticles are prone to aggregation caused by the freezing and drying-induced stresses. Consequently, the stability of nanoparticles is lost, which may cause irreversible structural damage, denaturation (in the case of proteins), aggregation, and loss of activity (in the case of enzymes), upon rehydration (DePaz et al., 2002; Heller et al., 1999). However, SFL prevents the nanoparticle aggregation due to particle growth in the vapor gap, as the nanoparticle dispersion is directly atomized below the surface of cryogenic liquid. The emanation of feed liquid at a high velocity through a fine orifice nozzle and the friction created by directly passing the droplets through a cryogenic liquid leads to intense atomization of the droplets due to the liquid-liquid impingement between the feed solution and the cryogenic liquid (Yu et al., 2004).

Further, the subsequent ultra-rapid freezing shortens the time for which the nanoparticles (specifically of proteins) are exposed to the air–water interface during atomization. This limits the aggregation and/or diffusion of proteins to the water-air and water-ice interface, which are the typical sites of stability loss. The freezing is fast enough to trap the drug in an amorphous state without letting it to crystallize. The amorphous state of particles is relevant from the perspective of obtaining high dissolution rates (Hu et al., 2002; Rogers et al., 2002). Generally, amorphous powders exhibit higher solubility and dissolution rate than their crystalline counterparts, owing to the lesser energy barrier requirement to dissolve the molecules (Karagianni, Kachrimanis, & Nikolakakis, 2018). Collectively, SFL was found to result in highly porous and nanostructured microparticles with large surface areas, nanoscale dimensions (<500 nm) and a significant improvement in stability (Webb et al., 2002; Yu et al., 2004; Yu, Johnston & Williams, 2006). The edge of SFL over SFV and SFV/L is of high relevance in the production of low density microparticles for the pulmonary delivery of proteins and peptides to improve lung deposition (Dellamary et al., 2000; Edwards, 1997; Edwards, Ben-Jebria & Langer, 1998). Further, the uniform size distribution of SFL nanoparticles is expected to reduce the initial burst release (Webb et al., 2002).

9.3 NANOPARTICLES PRODUCED USING SPRAY-FREEZE-DRYING

A thorough analysis of the studies on nanoparticles produced using SFD reveals that their applications are either towards the targeted delivery and sustained release of drugs to the lungs or to improve the dissolution property of poorly water-soluble drugs for enhanced bioavailability. Nanoparticle-based carriers can successfully deliver the drug in dry powder inhalers (DPI) to the intercellular compartments of respiratory tract by avoiding the lung phagocytic and mucociliary clearance mechanisms. Consequently, therapeutic nanoparticles lead to prolonged drug residence time and hence greater bioavailability (Rogueda & Traini, 2007). Similarly, NP-based drugs exhibit improved dissolution rate than micron-scale drug particles, due to their higher SA/V. Accordingly, nanoparticles with size in the range of 5–50 nm show a significant increase in aqueous solubility and thereby lead to improved bioavailability (Leuenberger, 2002).

9.3.1 Spray-freeze-drying for the production of nanoparticle aggregates for inhaled drug delivery

As discussed in Chapter 8, NPs with size less than 50 nm are exhaled from the lungs due to low inertia (Rogueda & Traini, 2007). Moreover, their high surface energy lead to formulation instabilities (Hinds, 1998). To overcome the above challenges and attain optimal stability and aerosolization properties, the nanoparticles are formulated into micron-scale nanoparticle aggregates or nano-composite microparticles (NCMPs that include NPs inside a microcarrier). The nano-aggregates exhibit large geometric size and low density, thus leading to a reduced theoretical aerodynamic diameter between 1 and 5 μm, which is the desirable range for targeted delivery to the alveolar region (Figure 9.7[a]). Even in the absence of coarse carrier particles, nano-aggregates with geometric size greater than 5 μm are instantly aerosolized from the inhaler. The aggregates remain effectively deposited in the lung due to their smaller theoretical aerodynamic diameter (Sung et al., 2009). In addition, nano-aggregates are capable of reconstituting into individual nanoparticles after being deposited in the lung interstitial fluid. Reconstitution ability is important for the nanoparticles to bypass the lung clearance mechanisms and retain their intended therapeutic functions such as to improve the dissolution rate (Cheow & Hadinoto, 2010; Wang et al., 2012). On the other hand, the corrugated surface of

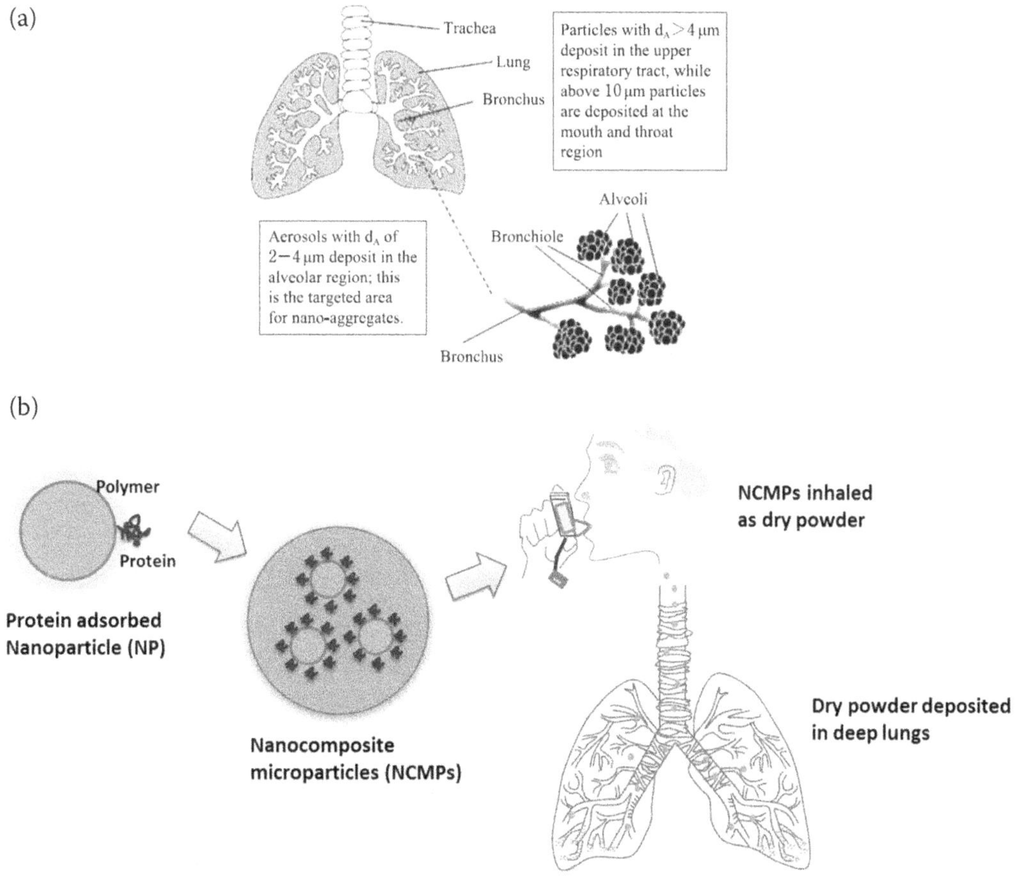

Figure 9.7 (a) Schematic of the target site of action of nano-aggregates in the respiratory tract (Cheow & Hadinoto, 2010); (b) Efficacy of nanoparticle-based carriers for pulmonary drug delivery: An example of protein nanocomposite microparticles (NCMPs) (Kunda et al., 2015).

NCMPs results in reduced contact points between particles that lead to enhanced aerosolization performance and deep deposition in the bronchial-alveolar region of the lungs ((Figure 9.7[b])) (Feng et al., 2011; Kunda et al., 2013; Sou et al., 2013).

Several techniques are available for the production of nanoparticle aggregates, including the flocculation of oppositely charged nanoparticles (Shi, Plumley & Berkland, 2007), dry powder coating (Yang, Cheow & Hadinoto, 2012) and spray drying (Jensen et al., 2012; Beck-Broichsitter et al., 2012). But, the therapeutic advantage of nano-aggregates or NCMPs produced by spray-freeze-drying confers an edge upon it and justifies its added cost and complexity of sterile process. The above reasoning can be explained as follows based on three different schools of thought: (1) mechanism of moisture removal/particle formation; (2) type of driving force for nanoparticle aggregation during drying, and (3) complexity of factors influencing product yield.

The efficacy of a DPI is a function of its aerodynamic performance, which has a substantial influence on its deposition in the lungs. In this regard, SFD-nanoaggregates exhibit low aerodynamic diameters despite their large geometric diameters, owing to their porous microstructure and the resulting low density (D'Addio et al., 2012; Maa et al., 1997; Wahjudi et al., 2013). This renders them suitable for effective lung depositions. But, spray dried particles have smaller size and non-porous morphology that lead to nanoparticle aggregation. Mechanism of moisture removal could influence the degree of nanoparticle aggregation during a drying process. Hence, variances in the mechanisms of moisture removal and particle formation between SFD and SD can be responsible for the difference in size, density, specific surface area and morphology of SFD- and SD-nanoaggregates/NCMPs.

The sequence of particle formation events during the SFD process to produce nanoaggregates is illustrated in Figure 9.8. The steps are essentially the same as that described in introductory chapter except that, here the feed is specifically a nanoparticle suspension rather than a simple solution or emulsion. Atomization involves channeling of the nanoparticle suspension through a suitable nozzle to form extremely fine droplets. The primary/sublimative drying phase of freeze-drying step is the control point in the SFD process for nanoaggregate production. Sublimation is

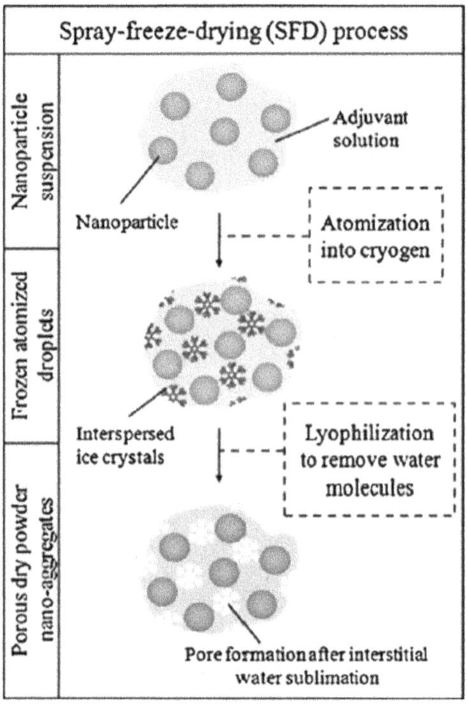

Figure 9.8 Schematic diagram of the formation of nano-aggregates during spray-freeze-drying process (Cheow et al., 2011).

carried out in a controlled manner below the triple point of water (6 mbar, 0.01°C) to prevent the melting of ice crystals into liquid water that destroys the solid structure of nano-aggregates. Nevertheless, the frozen droplets in which the ice crystals are sublimed interstitially to retain their size, without undergoing shrinkage (Cheow et al., 2011).

Consequently, SFD results in interconnected porous particles due to the sublimation-mediated removal of ice crystals that are distributed throughout the frozen droplet after spray-freezing (Maa et al., 1999; Amorij et al., 2007). This leads to effective disaggregation of nanoparticles in the SFD nanodispersions. Contrarily, during spray drying, evaporation of moisture from atomized feed droplets is extremely rapid upon contacting a stream of hot gas. This results in droplet shrinkage followed by solidification of the outer surface, leading to particles with smaller size and non-porous surface (Maa et al., 1997; Amorij et al., 2007).

The second school of thought is that the nanoparticles may aggregate during SFD due to the rapid formation of ice crystals, which cluster the nanoparticles to proximity as freeze-concentration shrinks the volume of amorphous matrix between the growing ice crystals (Zhang & Cooper, 2007). Contrarily, aggregation of nanoparticles during spray drying is caused by the increase in capillary pressure across the 'funicular' liquid bridges between nanoparticles with the increase in water loss (Urso et al., 2002), which pulls the nanoparticles together. However, the SFD-nanoaggregates disintegrate more readily after rehydration in aqueous medium than the SD-nanoaggregates, as the convective force originating from ice phase-separation and freeze concentration that causes NP aggregation in SFD is weaker than the capillary force of drying that draws the nanoparticles together during spray drying (Braig et al., 2019[a]). In the above context, the reconstitution ability of nanoparticles is judged by the change in size of nanoparticles after reconstitution. This is given by the ratio, S_i/S_f, where, S_i and S_f are the size of nanoparticles before and after redispersion, respectively. While $S_i/S_f = 1$ indicates complete reconstitution, $S_i/S_f > 1.5$ implies poor reconstitution (Kho & Hadinoto, 2010; Wang et al., 2012).

Moreover, the size and morphology of spray-dried nanoparticles are dependent on the relative rate of evaporation between the components (κ) and solute diffusion (D_i) (Broadhead, Edmond Rouan & Rhodes, 1992). Solid and spherical particles are produced only when diffusion phenomenon dominates. But, under rapid evaporation, the redistribution of solute or NP and their resultant enrichment at the droplet surface leads to particle buckling or crumpling after drying (Tsapis et al., 2002; Vehring, Foss, & Lechuga-Ballesteros, 2007). The significant difference in diffusion coefficients between the molecularly dissolved matrix constituents and the dispersed NPs results in separation of components in the final dried particle (Tsapis et al., 2002), which can promote irreversible aggregation of NPs during spray drying (Kho et al., 2010). Contrastingly, during spray-freeze-drying (Maa et al., 1999), freezing is much rapid due to which the solute is instantaneously trapped in the porous and homogenous structure of the frozen droplet. Therefore, before the freeze-drying step, the solute diffusion and its migration to the particle surface are arrested. Hence, the probability of NP aggregation is less in SFD. Also, the frozen droplet retains the initial volume of the aqueous droplet. Thus, unlike spray drying, the control of freezing and freeze-drying steps on the final particle size is decoupled, such that it is directly controlled by the size of the aqueous droplet (D'Addio et al., 2013).

The second aspect that demonstrates the superiority of SFD over SD for production of nanoaggregates is the yield of NPs. Compared to SD, SFD results in higher yield of nanoparticles with higher reconstitutability, flowability, emitted dose (ED), fine particle fraction (FPF) and lower mass median aerodynamic diameter (MMAD). Similar to particle formation, yield from SD is quite complicated as it is controlled by multiple factors (temperature, feed flow rate, particle collection) while that of SFD is governed by the spray-freezing step. In the latter, the productivity depends only on the number of sprayed droplets that are lost on the walls of liquid nitrogen (LN_2) vessel without reaching the cryogen. A simple solution of using a wider LN_2 container can improve the yield of NP from SFD (Wang et al., 2012). Thus, SFD is a competent technique for the preparation of stable nanoparticle-based carriers for pulmonary delivery. In fact, the major application of SFD in the pharmaceutical sector has been realized in the production of dry powder therapeutic nanoparticle aerosols for inhaled drug delivery. The forthcoming sections would highlight the major findings of pertinent investigations on the applications of SFD for the production of nanoaggregates/NCMPs for pulmonary delivery.

The superiority of SFD over spray drying (SD) for the production of DPI nanoparticle drugs was demonstrated in a study carried out by Ali & Lamprecht (2014). SFD and SD were used for the preparation of inhalable polymer- and lipid-based nanocarriers from the nanoparticle dispersions of various polymers (poly DL-lactide-coglycolide [PLGA], ethyl cellulose [EC] or Eudragit RL [EDRL]) and lipids (solid lipid nanoparticles [SLN] and lipid nanocapsules [LNC]), respectively. While maltodextrin was used as the cryoprotectant along with polymeric nanoparticles, trehalose was added to the lipid-nanocarrier formulations. NCMs were produced from SFD after twin-fluid nozzle atomization of the nanoparticle dispersions followed by SFV (-130°C) and vacuum freeze-drying (for 36 hours). SD was carried out by twin fluid nozzle atomization followed by drying at an inlet air temperature of 110°C. Production of NCMs from polymeric NPs was possible with both SFD and SD. But, NCM preparation from lipid nanocarriers was not feasible with SD due to its high temperature operation that led to melting of the lipid core and stickiness of the melt droplets to form a waxy mass. With respect to morphology, except for that prepared from lipid nanocapsules which showed an irregular shape, SFD-NCM prepared from polymeric and SLN-NPs exhibited spherical shape with larger particle size and a slightly rough surface (Figure 9.9[A-E]). But, SD-NCM were smaller with a slightly wrinkled smooth surface (Figure 9.10[A-C]). The other aspects demonstrating the superiority of SFD-NCM over SD-NCM in terms of producing particles suitable for pulmonary deposition are presented in Table 9.1.

In yet another study, D'Addio et al. (2013) demonstrated the potential of ultrasonic atomization for the successful production of nanoparticle aggregates (NA) of cholesterol by SFD and their comparative advantage over SD-NA. As discussed earlier, ultrasonic atomization generates uniform-sized droplets, wherein the droplet size is governed by the nozzle frequency rather than the properties of feed formulation (Sears et al., 1977). When the cholesterol NPs were loaded into mannitol carriers to form the nanoparticle aerosol carriers, the average size (21 ± 1.7 μm) and size distribution (Figure 9.11[a]) of the SFD particles were almost equal regardless of the nanoparticle loading (0%, 3% and 50%, w/w). The particles were porous (Figure 9.11[b]) with good aerosol properties (Figure 9.11[d]) and a FPF of 60%. On the contrary, the SD NP-loaded aerosol carriers (NPAC) were smaller in size (5–10 μm; but, smaller particles of size 1 μm were predominantly found) with denser particles (Figure 9.11[c]).

The release of NPs into aqueous medium after the dissolution of mannitol was determined by rehydrating the SFD and SD powder at a concentration of 10 mg/mL followed by shaking (30 s) and sonication (1 min). The size of native particles and that after shaking and sonication were measured by dynamic light scattering method and compared. SFD-NPAC with 50% NP loading, showed slight aggregation upon redispersion by hand agitation (Figure 9.11[e]) with an average size of 680 nm. But, with reduction in NP loading (3%), the size of redispersed NPs dropped to less than 400 nm with simple shaking. With further addition of energy by sonication, the size of redispersed particles was below 200 nm, irrespective of the NP loading (Figure 9.11[e]). Contrarily, the method of redispersion did not have any effect on the size of reconstituted spray dried particles. But, the reduction in NP loading reduced the aggregate size. After dissolution with shaking and sonication, SD-NPACs with 10% and 3% NP loading resulted in NP aggregates with sizes of 800 nm and 540 nm, respectively (Figure 9.11f]), which were significantly larger than those of their SFD counterparts. This proves the excellent redispersibility of SFD nanoparticle aggregates.

9.3.2 Spray-freeze-drying as an approach for the improved dissolution and oral bioavailability of poorly water-soluble drugs

A long-time apprehension in the product development cycle of a new drug is the poor dissolution property of potentially bioactive molecules. The dissolution rate of drugs depends on the

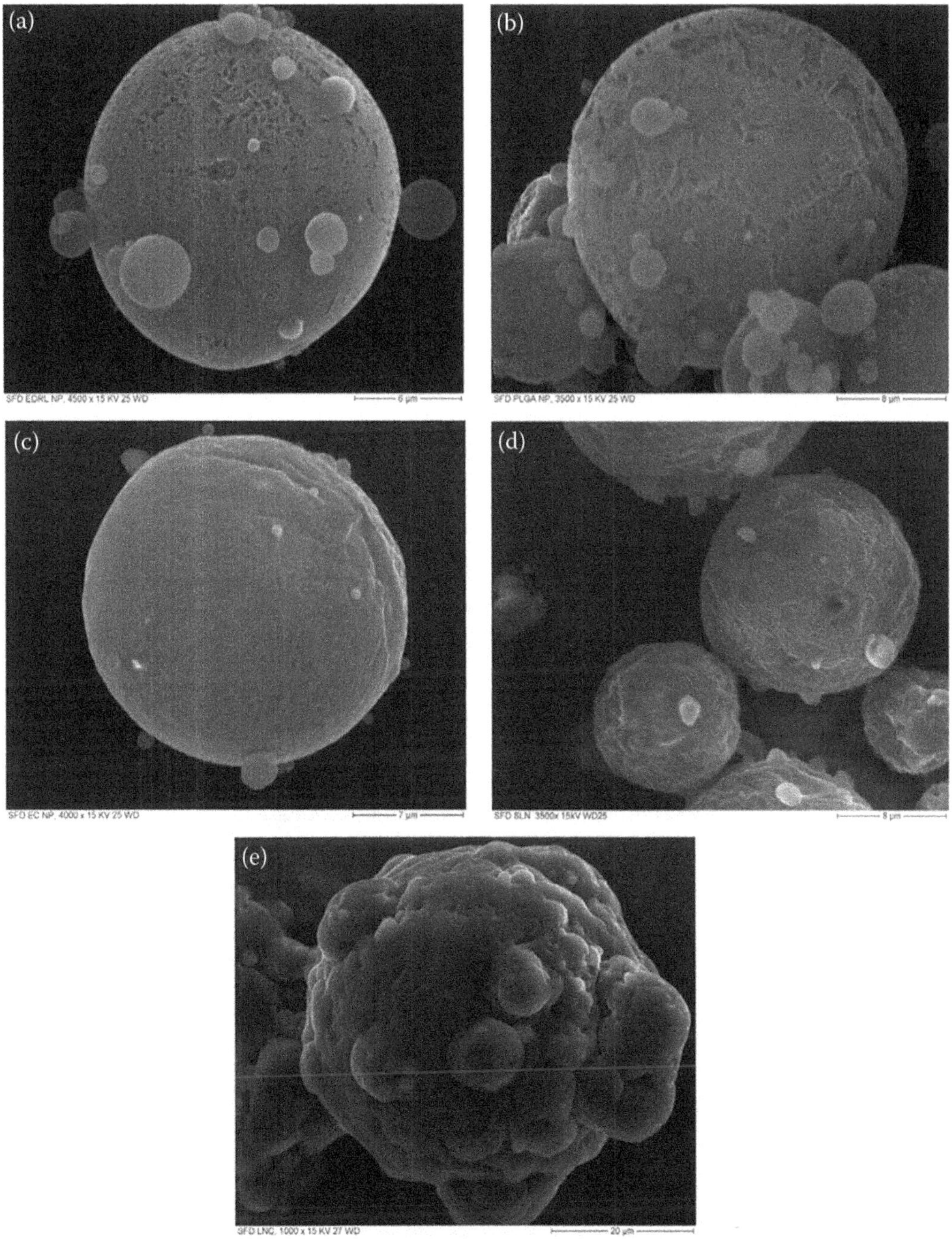

Figure 9.9 Morphology of SFD-NCM prepared from: (A) EDRL nanoparticles; (B) PLGA nanoparticles; (C) EC nanoparticles; (D) SLN; (E) LNC (Ali & Lamprecht, 2014).

concentration gradient of the solute, governed by the Fick's second law of diffusion. This led to the following Noyes-Whitney first-order equation (Eq. 9.1; Figure 9.12) (Smith, 2016), which demonstrates that the dissolution rate of an API is governed by its aqueous solubility.

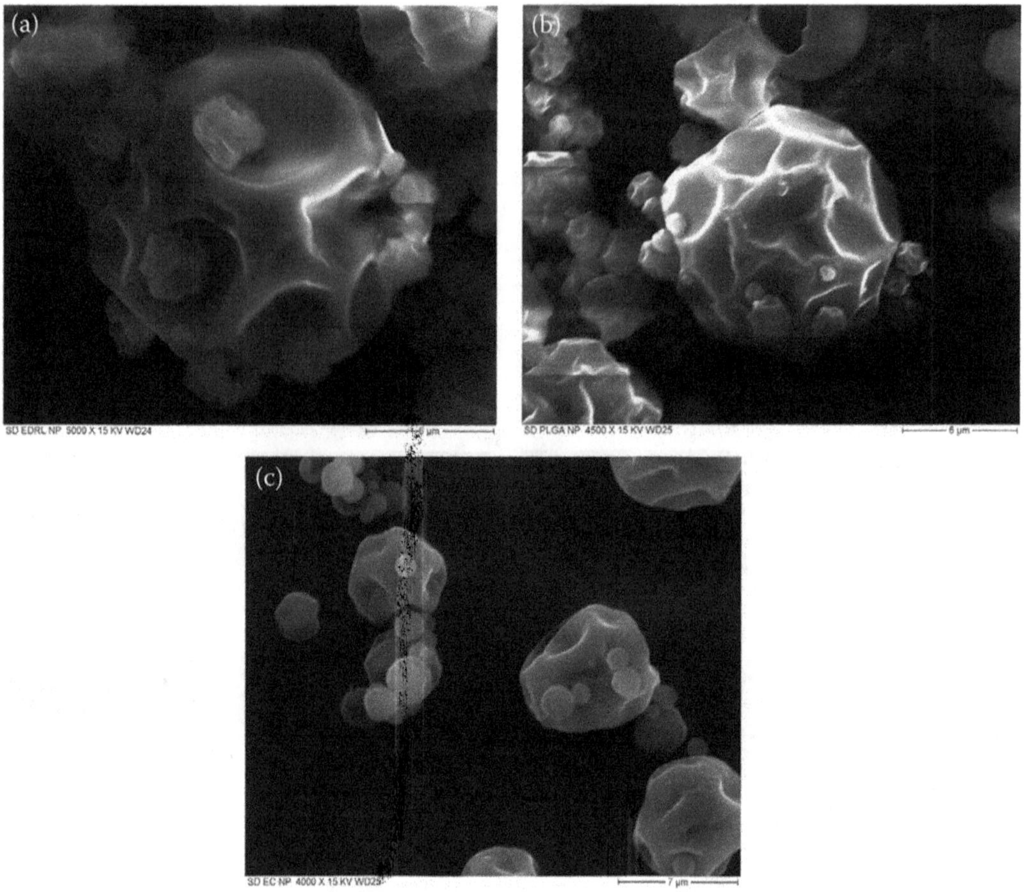

Figure 9.10 Morphology of SD-NCM produced from: (A) EDRL nanoparticles; (B) PLGA nanoparticles; (C) EC nanoparticles (Ali & Lamprecht, 2014).

$$\frac{dc}{dt} = k\,(C_s - C_b) \tag{9.1}$$

Where,

dc/dt = solute dissolution rate in terms of concentration (mol s^{-1})
k = dissolution constant
C_s = particle surface (saturation) concentration (mol L^{-1})
C_b = concentration in the bulk solvent or solution (mol L^{-1})

Due to their poor aqueous solubility, many of the prospective candidate molecules are rejected during the initial stages of drug development. Classified as BCS Class II drugs (Biopharmaceutics Classification System), these poorly water-soluble APIs exhibit low bioavailability and limited absorption in the gastrointestinal tract, when administered orally. The common examples of BCS Class II drugs include danazol and carbamazepine. The major factors that influence the dissolution property of APIs are the particle size, specific surface area and degree of crystallinity. Specific surface area of a drug is directly related to its dissolution rate and bioavailability (Lindenbaum et al., 1973). But, as mentioned earlier, submicron and nanoscale particles need to be stabilized both physically and chemically. Many techniques have been commonly used to improve the dissolution rate of poorly

Table 9.1 Spray-freeze-drying versus spray-drying in the production of DPI drug nanoparticles (data from Ali & Lamprecht, 2014)

Property	SFD-NCM	SD-NCM	Remarks
Particle yield (%)	91	69	• SFD resulted in 1.3-fold higher yield than SD.
Bulk density (g/cm^3)	0.02–0.07	0.24–0.27	• SFD-NCM exhibited 10-fold lower bulk density than the SD-NCM.
Carr index (CI) (flowability)	6–15	22–28	• Lower CI of SFD signifies excellent to good flowability, compared to the SD-NCM, which exhibited poor flowability. • Thus, more SFD powder can be expected to be dispensed from the capsules and inhaler than the SD powder.
Specific surface area (SSA)	60–77 m^2/g	1.8–2.4 m^2/g	• The higher SSA of SFD-NCM than the SD-NCM has been attributed to the difference in scheme of particle formation between the SFD and SD processes. • Spray-freezing of droplets leads to the formation of ice crystals throughout the frozen particle, which are then removed during freeze-drying step to form powder with high inter-particular porosity and a consequently larger SSA. • But, during SD, spraying of liquid feed into a stream of hot gas causes rapid evaporation of water, which results in shrinkage of the droplets. The outer surface solidifies to form smaller particles with non-porous surface and smaller SSA.
Mass median aerodynamic diameter (MMAD)	2.4–14.2	3.1–3.4	• The lower MMAD of SFD-NCM than SD-NCM was the result of its low density and porous morphology. • Compared to that of SD-NCM, the MMAD of SFD-NCM is within the range suitable for effective lung depositions.
Emitted fraction (EF) Fine particle fraction (FPF)	• Higher EF of SFD powder led to higher emptying of the SFD-NCM from the capsule and inhaler than the SD powder. • SFD-NCM and SD-NCM showed similar FPFs (≤5.2 μm), due to which some large particles were likely to exist in the emitted SFD powders which were deposited in the induction port and the pre-separator of the inhaler.		
S_f/S_i ratio; S_i: particle size before reconstitution; S_f: particle size after reconstitution	~ 1 for polymeric nanoparticles and <1.5 for SLNs	>1.5	• S_f/S_i ratio ~1 denotes complete reconstitution and S_f/S_i ratio >1.5 indicates poor reconstitution. • With SFD process, it was possible to maintain the particle size of the polymeric and lipid nanocarriers, even after aqueous reconstitution

water soluble APIs, including mechanical milling, spray drying, precipitation, and freeze-drying. But most of these techniques fail in preserving the stability of nanoparticles.

In this background, the advent of SFD technology came in as a boon to the pharmaceutical scientists to improve the dissolution of poorly water-soluble drugs, whilst retaining their stability. Niwa & Danjo (2013) reported that almost 100% of the SFD-azithromycin drug prepared from a nanodispersion of milled drug nanoparticles embedded in the polymeric network of polyvinylpyrrolidone (PVP) was dissolved in water within 5 min after immediate homogeneous dispersion. This onset speed was much greater than the release from spray-dried equivalents (Niwa, Miura & Danjo, 2011). Mechanism of the redispersion process of SFD particles is depicted in

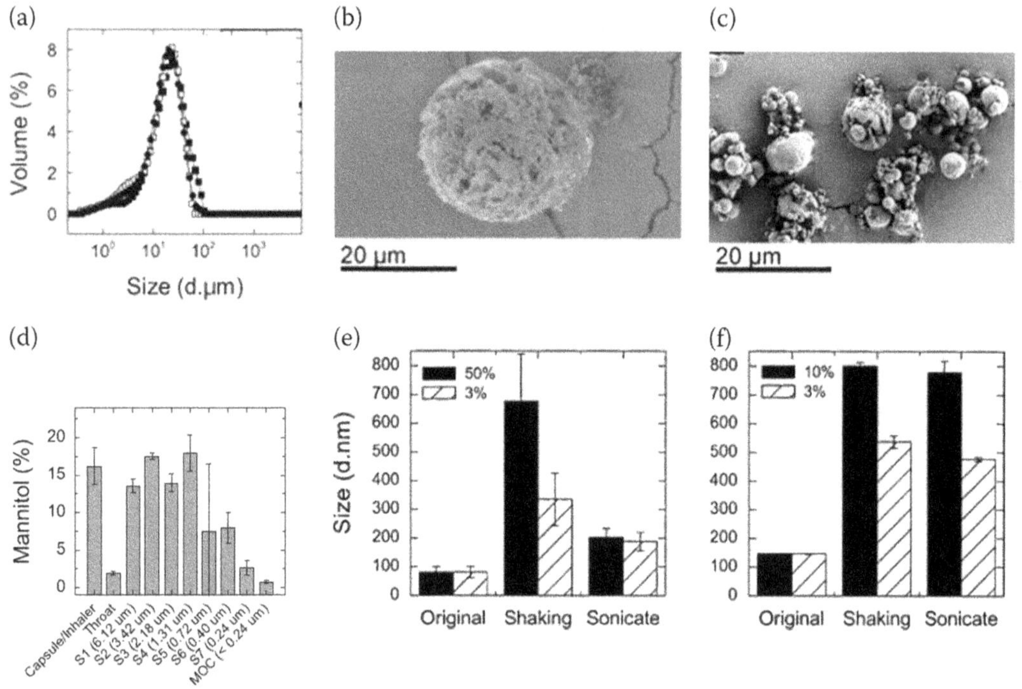

Figure 9.11 Characterization of nanoparticle-loaded aerosol carriers (NPAC): SFD vs. SD: (a) Volume weighted particle size distributions of NPAC prepared by SFD with different NP loadings (■) SFD01 (50%, w/w, cholesterol NPs), (□) SFD02 (3%, w/w, cholesterol NPs), and (●) SFD03 (0% cholesterol NPs); (b) SFD-NPAC with 50% NP loading; (c) SD-NPAC with 1:1 NP/mannitol; (d) Impaction data for SFD01 dispersed at 100 l min-1 in the Aerolizer®: plot showing the fraction of recovered mannitol dose in each stage of the impactor (S1 to S7 denote impactor stages 1 to 7 and MOC is the micro-orifice collector); (e) Z-average particle sizes of the resuspended NPs obtained from SFD with 50% NP loading (black bars) or 3% NP loading (hashed bars); (f) Z-average particle sizes of the resuspended NPs obtained from SD with 10% NP loading (black bars) or 3% NP loading (hashed bars) (D'Addio et al., 2013).

Figure 9.13. As soon as the SFD particle is placed in the aqueous medium, water instantaneously penetrates into the micropores within the particle. Subsequently, the hydrophilic polymeric network dissolves easily in a short duration to release the drug nanoparticles. Then, the NPs are dispersed spontaneously in the aqueous phase to restructure the nanosuspension. Thus, the porous structure of SFD particles plays a crucial role in their rapid dissolution.

Many studies have established the enhanced dissolution property of spray-freeze-dried drugs that have inherently poor water solubility. In this context, crystalline naproxen is a poorly water-soluble drug with a saturated solubility of 0.0159 mg/mL in pure water at 25°C (Yalkowsky & He, 2003). Braig et al. (2019[a]) prepared SFD-NPs of naproxen using lactose as bulking agent and compared it against the corresponding spray-dried product. Choice of lactose is justified by its ability to form amorphous state on freezing (Paterson Ripberger & Bridges, 2015) and reduce the coagulation of dried nanoparticles (Chaubal & Popescu, 2008). Initially, nanodispersions of naproxen at different pH (2.8, 6.2, and 7.0) were prepared by milling in the presence of an aqueous dispersion medium with hydroxypropyl cellulose. Results showed that the freezing-mediated aggregation of naproxen nanoparticles (at pH 2.8 of the nanodispersion) during SFD was largely reversible after rehydration. Redispersion via ultrasonication for 1 min and 5 min resulted in particles of size 300 nm and 160 nm, respectively, of which the latter was just 30 nm more than that of the original nanodispersion (after milling) containing non-aggregated nanoparticles (130 nm). But, redispersion of SFD-solids with stirring showed a large micron-sized population

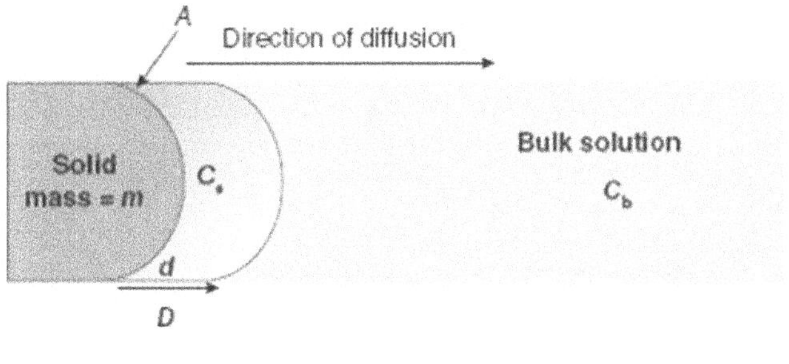

Figure 9.12 Schematic representation of the Noyes-Whitney parameters for drug dissolution rate (Pharmacy, 2016; https://basicmedicalkey.com/solubility-and-dissolution/).

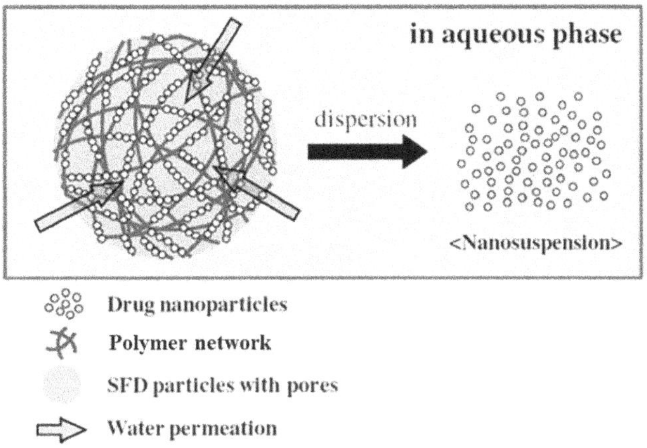

Figure 9.13 Schematic diagram of the re-dispersing process of SFD- nanosuspension in the dissolution medium (modified from Niwa & Danjo, 2013).

(up to 10 μm diameter) and a significant population of submicron aggregates (~300 nm). However, the redispersed spray-dried particles still exhibited a large population of micron-sized aggregates (≥1 μm) after sonication for 5 min. The underlying reasons for the superior disaggregation of SFD particles were already discussed in the earlier sections of this chapter.

Strikingly, the superior disaggregation of SFD-nanoparticles over the SD solids improved at the higher pH of the nanodispersions. At pH 6.2, redispersion with mere stirring resulted in a substantial population of sub-micron particles (~300 nm). Further, at pH 7.0, only stirring was adequate to yield a complete sub-micron population with a volumetric mean diameter of ~190 nm, which is just slightly greater than that of the original nanodispersion before SFD. It is likely that the improved disaggregation at higher pH initiates during the milling-based preparation of nanodispersions rather than in the SFD process. It can be attributed to the dependency of the drug's saturation solubility on the pH of medium utilized for the nanomilling process (Braig et al., 2019[b]). As naproxen is a weak acid (pKa = 4.18) (Avdeef et al., 2000), its saturation solubility in the milling medium will be more at higher pH. Thus, the original nanodispersion before SFD will contain a progressively greater proportion of drug that remains dissolved in the dispersion medium with increase in pH (Konnerth et al., 2017). At a fixed amount of naproxen, the portion of naproxen present as nanoparticles will be less at a higher pH (Braig et al., 2019[a]).

An alternative approach to overcome nanoparticle instability during SFD was proposed by Wei et al. (2017), which entailed the preparation of *'cage-like'* composite drug particles by combining the homogenization and SFD technologies. The authors used a diterpenoid lactone (code name: AGP) as

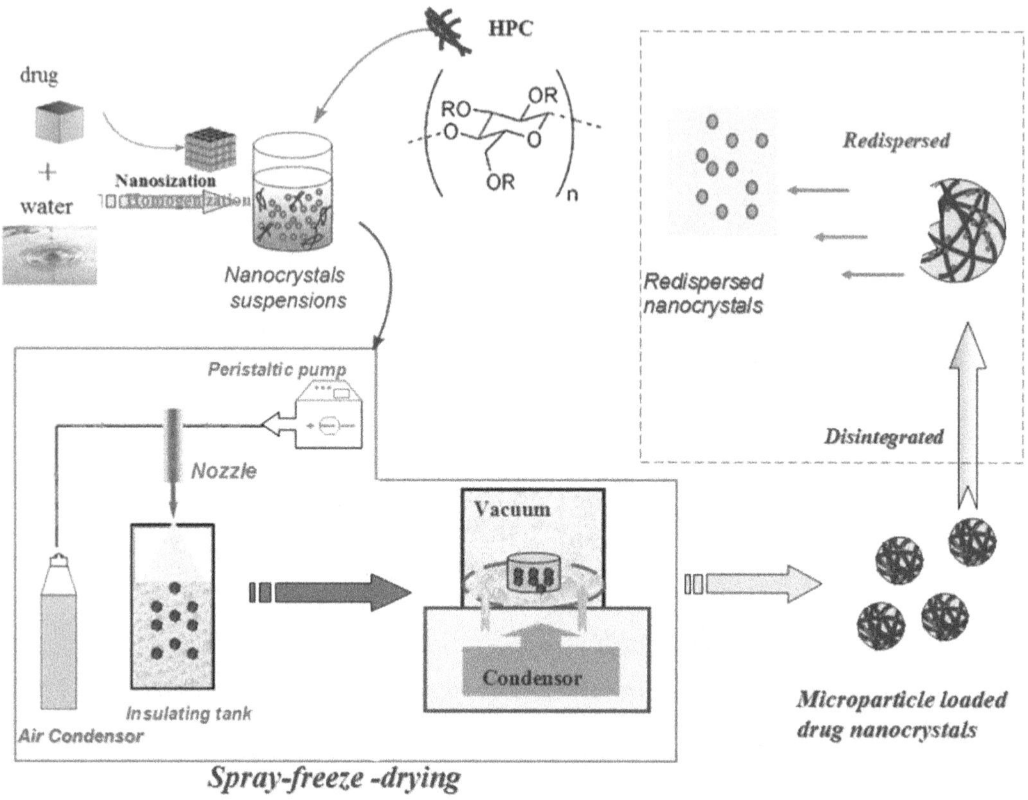

Figure 9.14 Schematic representation of the synergistic approach to prepare composite microparticles by homogenization-cum-SFD technology (Wei et al., 2017).

model drug. AGP is an anti-inflammatory and anti-viral drug, which is associated with low oral bioavailability (1.9%), high lipophilicity[#] (log P = 2.6) and low aqueous solubility (3.3μg/mL). Moreover, it is also a heat-labile compound. Thus, SFD turns out to be an ideal technology for the processing of AGP, due to its low-temperature operation and positive effect on aqueous solubility. Hydroxypropylcellulose (HPC) – a cellulosic polymer was used as the carrier. It immobilizes the drug nanocrystals in the cage-like structure and inhibits their aggregation during storage due to attractive inter-particle forces. The *'steric barrier effect'* of HPC is responsible for its stabilization effect on the drug nanocrystals suspension during homogenization. The molecules of HPC adsorb onto the surface of fragmented drug particles in the suspension and prevent the aggregation between primary particles by creating steric barriers (Pongpeerapat et al., 2008).

Initially, a suspension of HPC-stabilized AGP nanocrystals (AGP-NS) was prepared by homogenization technology, followed by its conversion into AGP microparticles with cage-like structure (AGP-CM) by spray-freeze-drying (Figure 9.14). Thus, the expectation of the above study was that, when the AGP nanoparticles are *'imprisoned'* within *'cage-like'* matrix structure of HPC, their free mobility would be impeded. Also, the *'cage-like'* structure of HPC could prevent the particle-particle

[#] Lipophilicity is a key parameter of the drug that influences its activity in the human body. The Log P value of a compound signifies the permeability of drugs to reach the target tissue in the body. Log P > 0 (or P > 1) indicates the lipophilic nature of drugs (Czyrski, 2019).

agglomeration of nanocrystals during storage and thereby lead to easy reconstitution of the AGP composite microparticles into nanosuspensions upon rehydration and improve their dissolution.

Initially, a coarse powder of AGP (1%, w/v) was dispersed in HPC solutions at different concentrations using a high-pressure homogenizer at a cooling temperature of 10°C. Then, the AGP-NS was converted into composite microparticles by SFD. SFV/L followed by vacuum freeze-drying was conducted using the following conditions and parameters: (1) atomizer: twin fluid nozzle; (2) cryogen: liquid nitrogen; (3) feed flow rate: 10 mL/min; (4) atomizing air flow: 30 mmHg; (5) feed solid content: 1% (w/v); (6) freezing at –40°C for 1 h followed by primary drying and secondary drying at –20°C and 10°C for 12 h, respectively. The results showed that the HPC-to-drug ratio is the key parameter that controlled the homogenization-mediated size reduction efficiency of the drug as well as the formation of a cage-like structure of the composite particles. Both the dissolution and redispersibility of the spray-freeze-dried composite particles were high, attributed to the particle size reduction of drug resultant from homogenization and the high porosity and surface area obtained after the SFD process.

The degree of size reduction shifted to submicron range (<1 μm) with narrow size distribution as the HPC-to-drug ratio increased from 1:10 to 2:1. However, at the ratio of 1:20, the size reduction was inadequate (D_{50} >1.5 μm). Further increase in the HPC-to-drug ratio (1:1 and 2:1) reduced the homogenization efficiency due to the increase in viscosity (from 2.1 mPa.s at 1:1 to 12.3 mPa.s at 2:1) that resulted from the excess HPC dispersed in water. High viscosity can weaken the collision power of cavitations during homogenization. Apart from HPC-to-drug ratio, homogenization pressure also exerted influence on the particle size of AGP nanocrystal suspension, which was completely of nanoscale (<1 μm) at a homogenization pressure of 1200 bar for 30 cycles, relative to other pressures (600, 800 and 1000 bar).

The redispersibility index of spray-freeze-dried AGP composite microparticles decreased with increase in the HPC-to-drug ratio, due to the corresponding increase in specific surface area (SSA) (Figure 9.15). At 1:2, 1:1 and 2:1 ratios, the particles showed superior redispersibility with their redispersibility index (RDI) close to 1, with larger SSA of 3000 m^2/g. But at low HPC-drug ratio (1:20, 1:10 and 1:5), the spray-freeze-dried drug particles exhibited poor redispersibility (RDI > 1.5). This was ascribed to the insufficient amount of HPC to prevent the aggregation of

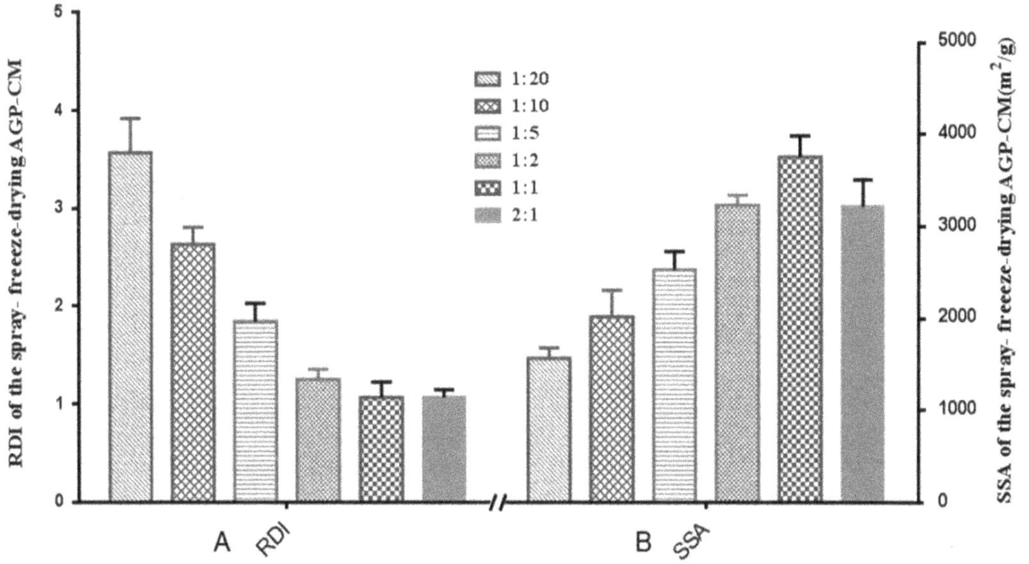

Figure 9.15 The redispersibility index (RDI) and specific surface area (SSA) of the spray-freeze-dried drug composite microparticles at different carrier-to-drug ratios (Wei et al., 2017).

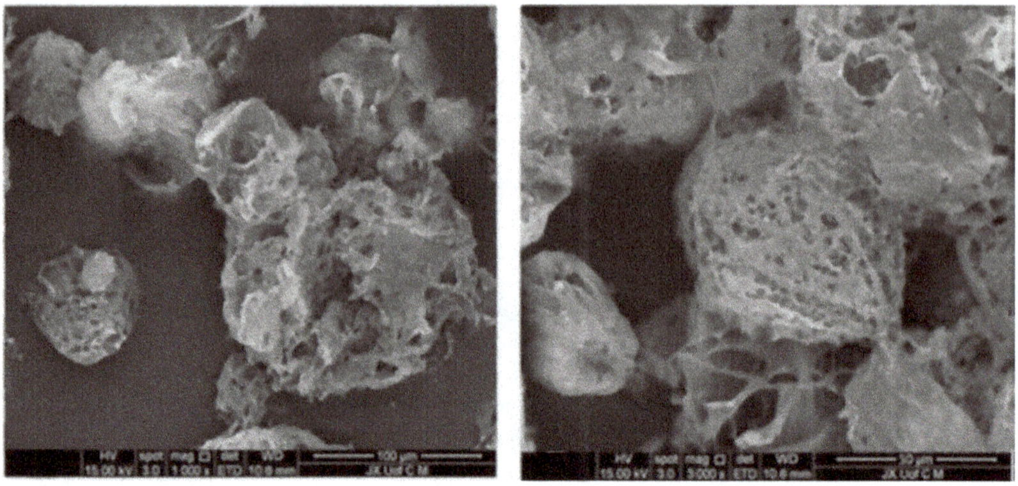

Figure 9.16 Morphology of SFD-AGP composite microparticles at HOC-to-drug ratio of: (a) 1:2; (b) 1:1, depicting the cage-like microstructure (Wei et al., 2017).

nanocrystals during the SFD process. All the aforesaid effects were attributed to the fact that the *'cage-like'* microstructure of SFD-AGP composite microparticles (Figure 9.16) were formed only at high HPC content in the ratio of 1:2 and 1:1, which was disintegrated into some hybrid structure containing agglomerates and large drug fragments at lower ratios (1:20, 1:10 and 1:5). The nanoscale AGP nanocrystals were embedded inside the cage-like structure of HCP (Wei et al., 2017).

Further, the SFD-AGP composite microparticles with higher HPC content (1:2 and 1:1) showed superior dissolution than the crude AGP and those with lower HPC content (1:10 and 1:5) (Figure 9.17). While only 33.35% of AGP was released from the coarse AGP within 30 min, almost 49.87% and 72.16% concentration of AGP was released from the SFD microparticles at lower ratios of HPC (1:10 and 1:5). Strikingly, the release attained 90% with higher ratios of HPC (1:2 and 1:1). Thus, it was evident that a high concentration of polymer (HPC) in the formulation can completely prevent the aggregation of AGP microparticles during SFD. The enhanced dissolution of SFD-AGP microparticles was attributed to their particle size reduction, larger SSA and porous microstructure (Wei et al., 2017).

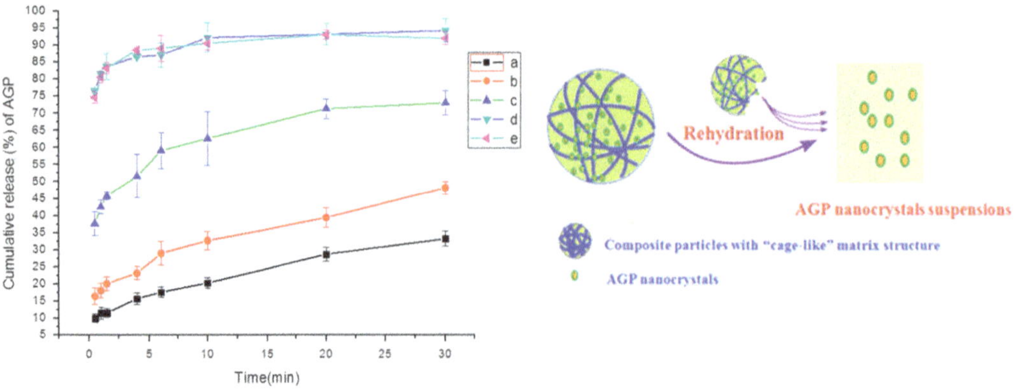

Figure 9.17 Dissolution kinetics of unprocessed drug and the drug-composite microparticles at different HPC-drug ratios (b 1:10, c 1:5, d 1:2, e 1:1) (modified from Wei et al., 2017).

Similar to the above study, Adeli (2017) demonstrated the enhanced dissolution of azithromycin, a poorly water-soluble drug with low bioavailability, by preparing its solid dispersions using SFD. The *in vitro* drug release profiles showed that the dissolution rate of the solid dispersion SFD sample was 8.9-fold greater than the unprocessed/pure drug.

CONCLUSIONS

The process capability of spray-freeze-drying for nanoparticle production was explained in this chapter. From the discussions, it is apparent that SFD has a clear competitive edge over the other top-down approaches for nanoparticle production as well as the bottom-up techniques. Nevertheless, the process is intricate with certain critical factors such as the use of adjuvants/stabilizers, choice of atomizer and freezing mode, and the freezing rate. Certainly, SFD is a promising approach to handle poorly water-soluble drugs and produce inhalable nanoaggregates and nanocomposite microparticles. The relatively high process cost of SFD for nanoparticle production is justified by the unique structure, high stability and aerosolization efficiency, enhanced dissolution and redispersibility characteristics of its final product. Yet, one caveat about SFD that has scope for improvement is the slight increase in the size of nanoparticles after redispersion. Since the role of atomization and freezing in the above lacuna is ruled out, further investigations can be conducted on the influence of freeze-drying step on the aggregation of SFD-nanoparticles.

REFERENCES

Abdelwahed, W., Degobert, G., & Fessi, H. (2006). Freeze-drying of nanocapsules: Impact of annealing on the drying process. *International Journal of Pharmaceutics*, 324(1), 74–82.

Adeli, E. (2017). The use of spray freeze drying for dissolution and oral bioavailability improvement of azithromycin. *Powder Technology*, 319, 323–331.

Akron Ascent Innovations (2017). Nanoscale phenomena: Importance of being small. Available from: http://www.akronascent.com/tech-blog/nanoscale2 (Accessed 16 September 2021).

Ali, M. E., & Lamprecht, A. (2014). Spray freeze drying for dry powder inhalation of nanoparticles. *European Journal of Pharmaceutics and Biopharmaceutics*, 87(3), 510–517.

Amorij, J.-P., Saluja, V., Petersen, A. H., Hinrichs, W. L. J., Huckriede, A., & Frijlink, H. W. (2007). Pulmonary delivery of an inulin-stabilized Influenza subunit vaccine prepared by spray-freeze drying induces systemic, mucosal humoral as well as cell-mediated immune responses in balb/c mice. *Vaccine*, 25(52), 8707–8717.

Augustin, M. A., & Sanguansri, P. (2009). Nanostructured materials in the food industry. *Advances in Food and Nutrition Research*, 58(4), 183–213.

Avdeef, A., Berger, C. M., & Brownell, C. (2000). pH-Metric Solubility 2: Correlation Between the Acid-Base Titration and the saturation shake-flask solubility pH methods. *Pharmaceutical Research*, 17(1), 85–89.

Avvaru, B., Patil, M. N., Gogate, P. R., & Pandit, A. B. (2006). Ultrasonic atomization: Effect of liquid phase properties. *Ultrasonics*, 44(2), 146–158.

Beck-Broichsitter, M., Schweiger, C., Schmehl, T., Gessler, T., Seeger, W., & Kissel, T. (2012) Characterization of novel spray-dried polymeric particles for controlled pulmonary drug delivery. *Journal of Controlled Release*, 158, 329–335.

Braig, V., Konnerth, C., Peukert, W., & Lee, G. (2019a). Can spray freeze-drying improve the re-dispersion of crystalline nanoparticles of pure naproxen? International *Journal of Pharmaceutics*, 564, 293–298.

Braig, V., Konnerth, C., Peukert, W., & Lee, G. (2019). Effects of pH of processing-medium on re-dispersion of spray dried, crystalline nanoparticles of pure naproxen. *International Journal of Pharmaceutics*, 558, 261–267.

Broadhead, J., Edmond Rouan, S. K., & Rhodes, C. T. (1992). The spray drying of pharmaceuticals. *Drug Development and Industrial Pharmacy*, 18(11-12), 1169–1206.

Chaubal, M. V., & Popescu, C. (2008). Conversion of nanosuspensions into dry powders by spray drying: A case study. *Pharmaceutical Research*, 25(10), 2302–2308.

Cheow, W.S., & Hadinoto, K. (2010). Enhancing encapsulation efficiency of highly watersoluble antibiotic in poly(lactic-co-glycolic acid) nanoparticles: modifications of standard nanoparticle preparation methods. *Colloids and Surfaces A: Physicochemical and Engineering Aspects*, 370, 79–86.

Cheow, W. S., Ng, M. L., Kho, K., & Hadinoto, K. (2011). Spray-freeze-drying production of thermally sensitive polymeric nanoparticle aggregates for inhaled drug delivery: Effect of freeze-drying adjuvants. *International Journal of Pharmaceutics*, 404(1-2), 289–300.

Chono, S., Tanino, T., Seki, T., & Morimoto, K. (2006). Influence of particle size on drug delivery to rat alveolar macrophages following pulmonary administration of ciprofloxacin incorporated into liposomes. *Journal of Drug Targeting*, 14(8), 557–566.

Cui, Z., Hsu, C.-H., & Mumper, R. J. (2003). Physical characterization and macrophage cell uptake of mannan-coated nanoparticles. *Drug Development and Industrial Pharmacy*, 29(6), 689–700.

Czyrski, A. (2019). Determination of the lipophilicity of ibuprofen, naproxen, ketoprofen, and flurbiprofen with thin-layer chromatography. *Journal of Chemistry*, 2019, 1–6.

D'Addio, S. M., Chan, J. G., Kwok, P. C., Prud'homme, R. K., & Chan, H.-K. (2012). Constant size, variable density aerosol particles by ultrasonic spray freeze drying. *International Journal of Pharmaceutics*, 427(2), 185–191.

D'Addio, S. M., Chan, J. G., Kwok, P. C., Benson, B. R., Prud'homme, R. K., & Chan, H.-K. (2013). Aerosol delivery of nanoparticles in Uniform MANNITOL carriers formulated by ultrasonic spray freeze drying. *Pharmaceutical Research*, 30(11), 2891–2901.

Dellamary, L. A., Tarara, T. E., Smith, D. J., Woelk, C. H., Adractas, A., Costello, M. L., Gill, H., & Weers, J. G. (2000). Hollow porous particles in metered dose inhalers. *Pharmaceutical Research*, 17(2), 168–174.

DePaz, R. A., Dale, D. A., Barnett, C. C., Carpenter, J. F., Gaertner, A. L., & Randolph, T. W. (2002). Effects of drying methods and additives on the structure, function, and storage stability of subtilisin: Role of protein conformation and molecular mobility. *Enzyme and Microbial Technology*, 31(6), 765–774.

Edwards, D. A. (1997). Large porous particles for pulmonary drug delivery. *Science*, 276(5320), 1868–1872.

Edwards, D. A., Ben-Jebria, A., & Langer, R. (1998). Recent advances in pulmonary drug delivery using large, porous inhaled particles. *Journal of Applied Physiology*, 85(2), 379–385.

Feng, A. L., Boraey, M. A., Gwin, M. A., Finlay, P. R., Kuehl, P. J., & Vehring, R. (2011). Mechanistic models facilitate efficient development of leucine containing microparticles for pulmonary drug delivery. *International Journal of Pharmaceutics*, 409(1-2), 156–163.

Hadavi, D., & Poot, A. A. (2016). Biomaterials for the treatment of Alzheimer's disease. *Frontiers in Bioengineering and Biotechnology*, 4, 49.

Heller, M. C., Carpenter, J. F., & Randolph, T. W. (1999). Conformational stability of lyophilized PEGylated proteins in a phase-separating system. *Journal of Pharmaceutical Sciences*, 88(1), 58–64.

Hinds, W.C. (1998). *Aerosol Technology: Properties, Behavior, and Measurement of Airborne Particles.* Wiley, New York

Hu, J., Rogers, T. L., Brown, J., Young, T., Johnston, K. P., & Williams III, R. O. (2002). Improvement of dissolution rates of poorly water soluble apis using novel spray freezing into liquid technology. *Pharmaceutical Research*, 19(9), 1278–1284.

ISO/TS 80004-1:2010, Nanotechnology – Vocabulary – Part 1: Core Terms. Geneva, Switzerland: International Organization for Standardization; 2010. [Jul 17; 2017]. Available from: https://www.iso.org/standard/51240.html (Accessed 22 September 2021).

ISO/TR 18401:2017(en)Nanotechnologies — Plain language explanation of selected terms from the ISO/IEC 80004 series. Available from: https://www.iso.org/obp/ui/#iso:std:62384:en (Accessed 14 December 2021).

Jensen, D. K., Jensen, L. B., Koocheki, S., Bengtson, L., Cun, D., Nielsen, H. M., & Foged, C. (2012). Design of an inhalable dry powder formulation of DOTAP-modified PLGA nanoparticles loaded with siRNA. *Journal of Controlled Release*, 157(1), 141–148.

Kamal, M., & Khoshkava, V. (2015). *Spray freeze-dried nanoparticles and method of use thereof.* United States Patent, US20150307692A1.

Karagianni, A., Kachrimanis, K., & Nikolakakis, I. (2018). Co-amorphous solid dispersions for solubility and absorption improvement of drugs: Composition, preparation, characterization and formulations for oral delivery. *Pharmaceutics*, 10(3), 98.

Kho, K., & Hadinoto, K. (2010). Aqueous re-dispersibility characterization of spray-dried hollow spherical silica nano-aggregates. *Powder Technology*, 198(3), 354–363.

Kho, K., Cheow, W. S., Lie, R. H., & Hadinoto, K. (2010). Aqueous re-dispersibility of spray-dried antibiotic-loaded polycaprolactone nanoparticle aggregates for inhaled anti-biofilm therapy. *Powder Technology*, 203(3), 432–439.

Konnerth, C., Braig, V., Ito, A., Schmidt, J., Lee, G., & Peukert, W. (2017). Formation of mefenamic acid nanocrystals with improved dissolution characteristics. *Chemie Ingenieur Technik*, 89(8), 1060–1071.

Kunda, N. K., Somavarapu, S., Gordon, S. B., Hutcheon, G. A., & Saleem, I. Y. (2013). Nanocarriers targeting dendritic cells for pulmonary vaccine delivery. *Pharmaceutical Research*, 30(2), 325–341.

Kunda, N. K., Alfagih, I. M., Dennison, S. R., Somavarapu, S., Merchant, Z., Hutcheon, G. A., & Saleem, I. Y. (2015). Dry powder pulmonary delivery of cationic pga-co-pdl nanoparticles with surface adsorbed model protein. *International Journal of Pharmaceutics*, 492(1-2), 213–222.

Lek Pharmaceuticals d.d. (2015). *Spray-Freeze Drying of Polyelectrolyte Nanoparticles Containing the Protein Drug*. Available from: https://patents.justia.com/patent/20170258726 (Accessed 16 September 2021).

Leuenberger, H. (2002). Spray Freeze drying e the process of choice for low water soluble drugs? *Journal of Nanoparticle Research*, 4, 111–119.

Lindenbaum, J., Butler Jr., V., Murphy, J., & Cresswell, R. (1973). Correlation of digoxintablet dissolution-rate with biological availability. *The Lancet*, 301(7814), 1215–1217.

Maa, Y.-F., Costantino, H. R., Nguyen, P.-A., & Hsu, C. C. (1997). The effect of operating and Formulation variables on the morphology Of Spray-Dried Protein Particles. *Pharmaceutical Development and Technology*, 2(3), 213–223.

Maa, Y.-F., Nguyen, P.-A., Sweeney, T., Shire, S.J., & Hsu, C.C. (1999). Protein inhalation powders: spray drying vs spray freeze drying. *Pharmaceutical Research*, 16, 249–254.

Niwa, T., & Danjo, K. (2013). Design of self-dispersible dry nanosuspension through wet milling and spray freeze-drying for poorly water-soluble drugs. *European Journal of Pharmaceutical Sciences*, 50(3-4), 272–281.

Niwa, T., Miura, S., & Danjo, K. (2011). Design of dry nanosuspension with highly spontaneous dispersible characteristics to develop solubilized formulation for poorly water-soluble drugs. *Pharmaceutical Research*, 28(9), 2339–2349.

Ochubiojo, M., Chinwude, I., Ibanga, E., & Ifianyi, S. (2012). Nanotechnology in Drug Delivery, Recent Advances. In: A.D. Sezer (Ed.), *Novel Drug Carrier Systems*, IntechOpen, DOI: 10.5772/51384. Available from: https://www.intechopen.com/chapters/40262 (Accessed 24 September 2021).

Paterson, A. H. J., Ripberger, G. D., & Bridges, R. P. (2015). Measurement of the viscosity of freeze dried amorphous lactose near the glass transition temperature. *International Dairy Journal*, 43, 27–32.

Pharmacy (2016). Solubility and Dissolution. Available from: https://basicmedicalkey.com/solubility-and-dissolution/ (Accessed 16 September 2021).

Pongpeerapat, A., Wanawongthai, C., Tozuka, Y., Moribe, K., & Yamamoto, K. (2008). Formation mechanism of colloidal nanoparticles obtained from probucol/PVP/SDS ternary ground mixture. *International Journal of Pharmaceutics*, 352(1-2), 309–316.

Rajan, R., & Pandit, A. B. (2001). Correlations to predict droplet size in ultrasonic atomisation. *Ultrasonics*, 39(4), 235–255.

Riemer, J. (2011). Ultrasonic spray coating of nanoparticles. *Global Solar Technology*. Available from: http://www.sono-tek.com/wp-content/uploads/2018/07/Ultrasonic-Spray-of-Nanoparticles.pdf (Accessed 16 September 2021).

Rogers, T. L., Hu, J., Yu, Z., Johnston, K. P., & Williams, R. O. (2002). A novel particle engineering technology: Spray-freezing into liquid. *International Journal of Pharmaceutics*, 242(1-2), 93–100.

Rogueda, P. G. A., & Traini, D. (2007). The nanoscale in pulmonary delivery. Part 1: Deposition, fate, toxicology and effects. *Expert Opinion on Drug Delivery*, 4(6), 595–606.

Schiffter, H., Condliffe, J., & Vonhoff, S. (2010). Spray-freeze-drying of nanosuspensions: The manufacture of insulin particles for needle-free ballistic powder delivery. *Journal of The Royal Society Interface*, 7, S483–S500.

Sears, J., Huang, K., Ray, S., & Fairbanks, H. (1977). Effect of liquid properties on the production of aerosols with ultrasound. *Ultrasonics Symposium*. 10.1109/ultsym.1977.196808.

Shaik, M. S., Ikediobi, O., Turnage, V. D., McSween, J., Kanikkannan, N., & Singh, M. (2001). Long-circulating monensin nanoparticles for the potentiation of immunotoxin and anticancer drugs. *Journal of Pharmacy and Pharmacology*, 53(5), 617–627.

Shi, L., Plumley, C. J., & Berkland, C. (2007). Biodegradable nanoparticle flocculates for dry powder aerosol formulation. *Langmuir*, 23(22), 10897–10901.

Smith, B.T. (2016). Solubility and dissolution (Chapter 3). In: Physical Pharmacy, Remington, Pharmaceutical Press, East Smithfield, London, UK, pp. 31–50.

SonoTek (2021). *Nanotechnology Coatings Deposited Using Ultrasonic Spray*. Available from: http://www.sonotek.com/nanotechnology-overview/ (Accessed 16 September 2021).

Sou, T., Kaminskas, L. M., Nguyen, T.-H., Carlberg, R., McIntosh, M. P., & Morton, D. A. V. (2013). The effect of amino acid excipients on morphology and solid-state properties of multi-component spray-dried formulations for pulmonary delivery of biomacromolecules. *European Journal of Pharmaceutics and Biopharmaceutics*, 83(2), 234–243.

Sun, X., Zhang, Y., Chen, G., & Gai, Z. (2017). Application of nanoparticles in enhanced oil recovery: A critical review of recent progress. *Energies*, 10(3), 345.

Sung, J. C., Padilla, D. J., Garcia-Contreras, L., VerBerkmoes, J. L., Durbin, D., Peloquin, C. A., Elbert, K. J., Hickey, A. J., & Edwards, D. A. (2009). Formulation and pharmacokinetics of self-assembled rifampicin nanoparticle systems for pulmonary delivery. *Pharmaceutical Research*, 26(8), 1847–1855.

Surface Energy (2021). Available from: https://chem.libretexts.org/@go/page/183363 (Accessed 22 September 2021).

Tsapis, N., Bennett, D., Jackson, B., Weitz, D. A., & Edwards, D. A. (2002). Trojan particles: Large porous carriers of nanoparticles for drug delivery. *Proceedings of the National Academy of Sciences of the United States of America*, 99, 12001–12005.

Urso, M. E., Lawrence, C. J., & Adams, M. J. (2002). A two-dimensional study of the rupture of funicular liquid bridges. *Chemical Engineering Science*, 57(4), 677–692.

Vehring, R., Foss, W. R., & Lechuga-Ballesteros, D. (2007). Particle formation in spray drying. *Journal of Aerosol Science*, 38(7), 728–746.

Wahjudi, M., Murugappan, S., van Merkerk, R., Eissens, A. C., Visser, M. R., Hinrichs, W. L. J., & Quax, W. J. (2013). Development of a dry, stable and inhalable acyl–homoserine–lactone–acylase powder formulation for the treatment of pulmonary pseudomonas aeruginosa infections. *European Journal of Pharmaceutical Sciences*, 48(4-5), 637–643.

Wang, Y., Kho, K., Cheow, W. S., & Hadinoto, K. (2012). A comparison between spray drying and spray freeze drying for dry powder inhaler formulation of drug-loaded lipid–polymer hybrid nanoparticles. *International Journal of Pharmaceutics*, 424(1-2), 98–106.

Webb, S. D., Golledge, S. L., Cleland, J. L., Carpenter, J. F., & Randolph, T.W. (2002). Surface adsorption of recombinant human interferon-c in lyophilized and spray-lyophilized formulations. *Journal of Pharmaceutical Sciences*, 91, 1474–1487.

Wei, S., Ma, Y., Luo, J., He, X., Yue, P., Guan, Z., & Yang, M. (2017). Hydroxypropylcellulose as matrix carrier for novel cage-like microparticles prepared by spray-freeze-drying technology. *Carbohydrate Polymers*, 157, 953–961.

Yalkowsky, S.H., & He, Y. (2003). *Handbook of aqueous solubility data: An extensive compilation of aqueous solubility data for organic compounds extracted from the aquasol database*. CRC Press LLC, Boca Raton, FL.

Yang, Y., Cheow, W. S., & Hadinoto, K. (2012). Dry powder inhaler formulation of lipid–polymer hybrid nanoparticles via electrostatically-driven nanoparticle assembly onto microscale carrier particles. *International Journal of Pharmaceutics*, 434(1-2), 49–58.

Yu, Z., Garcia, A. S., Johnston, K. P., & Williams, R. O. (2004). Spray freezing into liquid nitrogen for highly stable protein nanostructured microparticles. *European Journal of Pharmaceutics and Biopharmaceutics*, 58(3), 529–537.

Yu, Z., Johnston, K. P., & Williams, R. O. (2006). Spray freezing into liquid versus spray-freeze drying: Influence of atomisation on protein aggregation and biological activity. *European Journal of Pharmaceutical Sciences*, 27, 9–18.

Zhang, H., & Cooper, A. I. (2007). Aligned porous structures by directional freezing. *Advanced Materials*, 19(11), 1529–1533.

CHAPTER 10

Properties of Spray-freeze-dried Products and their Characterization

Spray-freeze-drying produces powders with unique physicochemical and structural characteristics that are not observed in conventionally dried products. It is important to quantify these characteristics to judge whether the products meet their intended specifications and functionality that are expected to result from the SFD process. The characteristics to be measured vary with the end-use of the spray-freeze-dried product (Figure 10.1). Research publications and reference books often focus on the results related to their study objectives. But, the relevance of quantifying a certain characteristic, instruments for measurement and the test specifications are seldom discussed. Hence, this chapter would explain the properties of different spray-freeze-dried products. For the above purpose, spray-freeze-dried products are classified into four major categories and the characteristics have been grouped under each product category (Figure 10.1). The subsequent sections of this chapter would discuss the objectives, principle, instrumentation and specifications of different test methods available for characterizing the spray-freeze-dried products. The concepts related to particle morphology and aerodynamic size are not discussed here as the same have already been discussed in Chapter 4 and Chapter 8, respectively.

10.1 CHARACTERIZATION OF SPRAY-FREEZE-DRIED PHARMACEUTICAL PROTEINS

10.1.1 Protein stability and integrity

Quantifying the stability of spray-freeze-dried proteins is imperative as proteins are susceptible to denaturation at the ice-water interface during the spray-freezing step, especially in the SFV/L mode. The rate of denaturation increases with freezing rate and interfacial surface area (Chang, Kendrick & Carpenter, 1996; Eckhardt, Oeswein & Bewley, 1991; Hsu et al., 1995; Strambini & Gabellieri, 1996). The impact of freezing on proteins is indicated by certain changes. For instance, rapid freezing leads to increased turbidity in the freeze-dried protein during storage. Also, freeze-dried proteins are vulnerable to structural changes (Griebenow, Santos, & Carrasquillo, 1999) and solid-state aggregation, which could affect its sustained release from the encapsulant matrices (Costantino et al., 1997). Degree of protein instability or denaturation is determined by quantifying the loss of protein solubility (Anandharamakrishnan, 2008).

10.1.1.1 Size exclusion chromatography

Size-exclusion chromatography is the commonly used technique to determine the degree of protein aggregation in the spray-freeze-dried powder. Here, the results are reported as reduction in

DOI: 10.1201/9781003019312-10

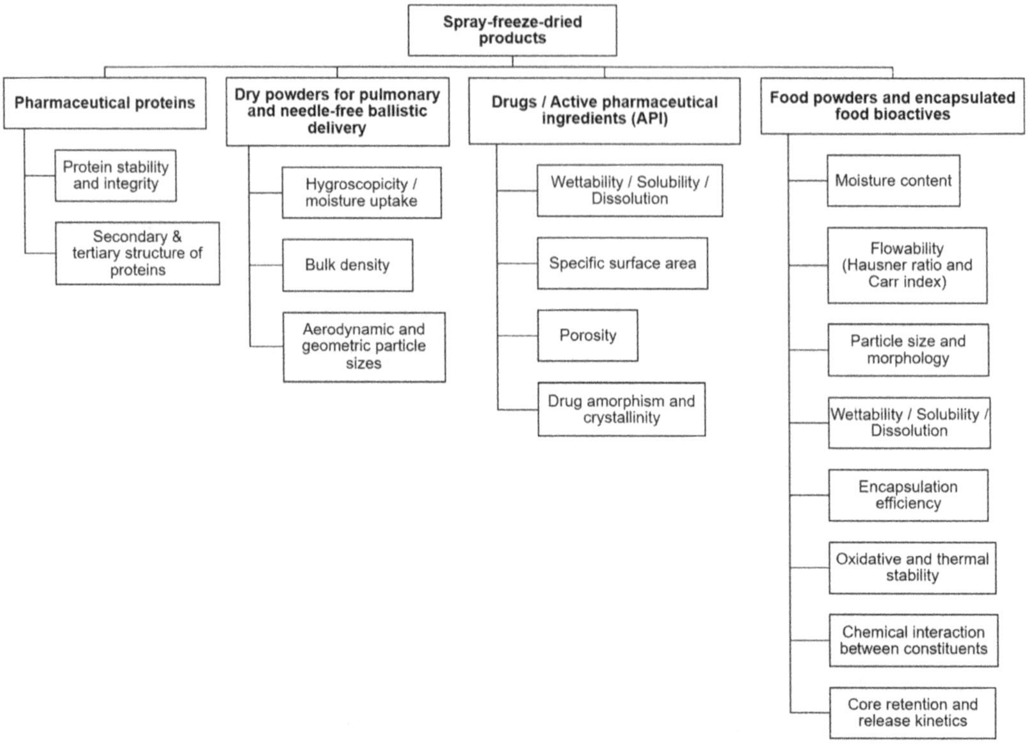

Figure 10.1 Classification of spray-freeze-dried products and their characteristics.

the percentage of monomers in the final product relative to that present in the protein before spray-freeze-drying. If the monomer content is found to be higher in the dried protein than the feed material, then the reduction in percentage monomers is reported as zero. In general, reduction in monomers occurs along with the increase in dimers and soluble aggregates (Figure 10.2). In other words, the number of dimers generated during SFD or rehydration of SFD protein powder is numerically equal to the monomer loss (Sonner, Maa & Lee, 2002). Further, monomer loss is inversely related to the particle size and directly related to the specific surface area of SFD protein particles (Costantino et al., 2000).

10.1.1.2 Reverse phase high performance liquid chromatography (RP-HPLC)

Alternatively, the effect of spray-freeze-drying on protein denaturation has been investigated using the RP-HPLC approach (Anandharamakrishnan, 2008), in which the mobile phase is more polar than the stationary phase (Mant & Hodges, 1991; Meyer, 1999). In this method, protein instability is reported as percentage solubility loss of individual (constituent) proteins. This method is relatively simple with the ability to resolve individual proteins within a short analytical time. Kjeldahl method is commonly employed to determine protein solubility based on the calculation of total nitrogen content. But, with this method, there are chances for further denaturation of the sample during the acid digestion phase of the analysis. But, the above apprehension is avoided in the RP-HPLC based approach, as the estimation is based on hydrophobic interactions that constitute the major stabilization forces of the three-dimensional structure of proteins (Ferreira et al., 2001). The proportions of native α-lactalbumin and β-lactoglobulin in the soluble fraction of spray-freeze-dried whey protein were estimated using the RP-HPLC method at pH 4.6 (the isoelectric

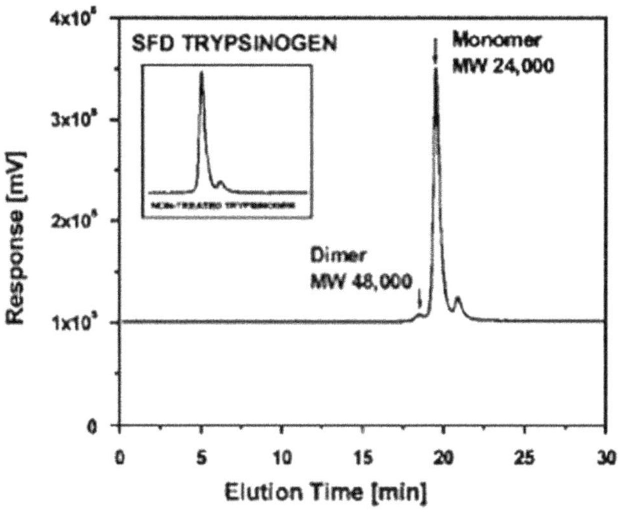

Figure 10.2 Size exclusion chromatogram of SFD-trypsinogen (Sonner, Maa & Lee, 2002).

point of whey protein) (Anandharamakrishnan, 2008). A dispersion of the spray-freeze-dried whey protein was subjected to its isoelectric point (pH 4.6) for a defined period (say 30 min), after which it was diluted appropriately and centrifuged. The resultant supernatant was then filtered, diluted and centrifuged again in a microfuge to remove any insoluble material before injecting it to the RP-HPLC system (Figure 10.3) (Parris & Baginski, 1991; Ferreira et al., 2001; Ferreira & Caçote, 2003).

The chromatograms of α-lactalbumin, β-lactoglobulin, mixture of these standards and the spray-freeze-dried whey protein (Figure 10.4[a-d]) showed that these native proteins were evidently resolved with distinct retention times at 20 min, 24 min and 25 min for α-lactalbumin, β-lactoglobulin B (P-IgB) and β-lactoglobulin A (0-IgA), respectively. The mean values of protein content determined from the HPLC analyses were used in the calculation of its degree of denaturation (Eq. 10.1):

$$\%\ Loss\ of\ solubility = \left(1 - \frac{SP_e}{SP_u}\right) \times 100 \tag{10.1}$$

Where, SP_e and SP_u are the amounts of soluble protein in the spray-freeze-dried whey protein and untreated whey protein, respectively. These values are obtained after normalizing the protein mass by accounting for the moisture content of dried product.

10.1.1.3 Gel electrophoresis

Apart from the chromatography techniques, the aggregation behavior of spray-freeze-dried proteins has also been ascertained by subjecting the rehydrated SFD powder to gel electrophoresis (Sodium Dodecyl Sulfate – Polyacrylamide Gel Electrophoresis, also known as SDS-PAGE) and ultraviolet (UV) spectrophotometry. With gel electrophoresis, the presence of an intense monomer band, dimer band and some aggregates in the SDS gel lanes signifies the occurrence of protein aggregation during freezing (Figure 10.5). The dimers and minimal aggregates in rehydrated SFD are the results of protein inactivation caused by its partial and irreversible unfolding to an inactive

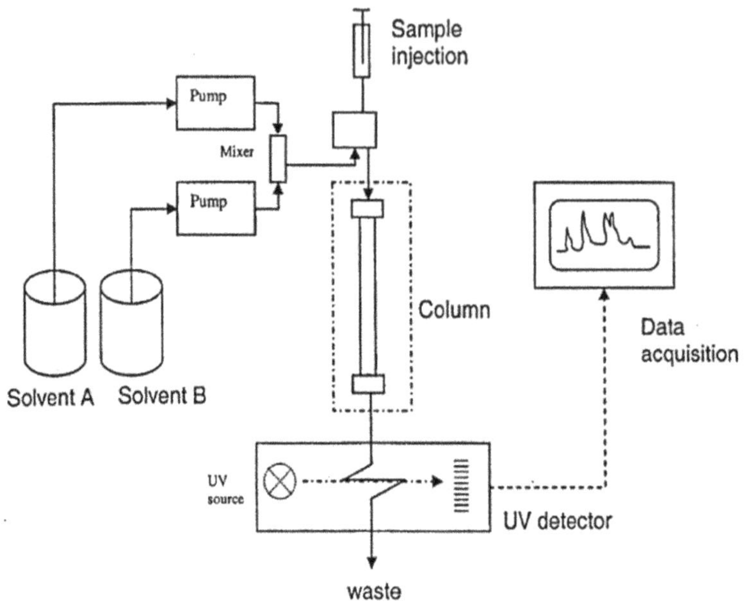

Figure 10.3 Schematic diagram of an RP-HPLC unit (Anandharamakrishnan, 2008).

molten globule state. The above inactivation events occur during the atomization stage, wherein the protein adsorbs and unfolds at the large liquid/air interface of the spray droplets. These unfolded proteins do not refold on rehydration and therefore continues to be inactive, but are not aggregated. Thus, studies have ascertained that protein aggregation is minimal during SFD/rehydration (Sonner, Maa & Lee, 2002).

10.1.1.4 Ultraviolet spectroscopy

Ultraviolet spectroscopy of spray-freeze-dried proteins provides insight to the degree of drying-induced protein aggregation. The ultraviolet spectrum of SFD protein is acquired by taking its rehydrated solution (at a concentration of, say, 1 mg/mL) in a quartz cuvette of path length 1 cm and measuring its absorbance in the region between 240 and 350 nm, using a UV spectrophotometer. Then, the index of protein aggregation (*AI*) is calculated from the absorbance values (A_n; where n denotes the wavelength at which the absorbance is measured) using the following equation:

$$AI = \frac{A_{340} \times 100}{(A_{280} - A_{340})} \tag{10.2}$$

Higher *AI* values correspond to the presence of large aggregates in the processed protein powder. With spray-freeze-dried immunoglobulin G (IgG), the *AI* values of protein powders produced from feed containing only IgG and a combination of IgG with trehalose and arginine as excipients were very high (Table 10.1) compared to the native IgG (0.6). Arginine with a positively charged guanidium group induced the aggregation of IgG and destabilized the protein's tertiary structure (Faghihi et al., 2014; Thakkar et al., 2012), during the SFD process. On the other hand, the *AI* was extremely low in the range of 0.1–1.7 for the SFD-IgG formulations prepared by adding a combination of trehalose and amino acids such as leucine (Leu), phenylalanine (Phe), cysteine (Cys)

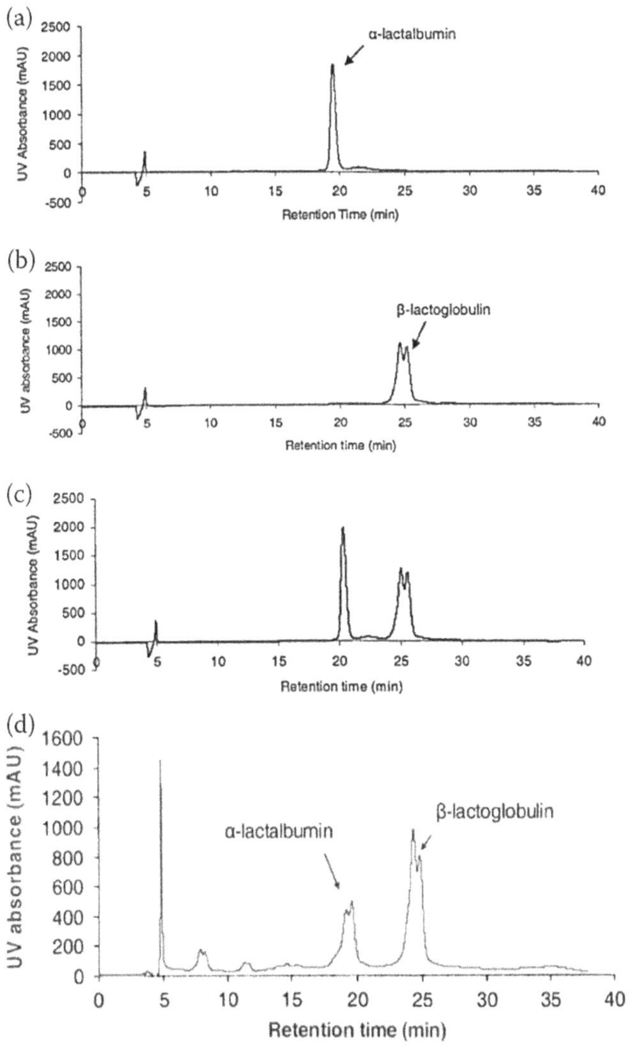

Figure 10.4 Chromatogram pattern of protein standards (a) α-lactalbumin; (b) β-lactoglobulin; (c) mixture of α-lactalbumin and β-lactoglobulin (1:2); (d) spray-freeze-dried whey protein (Anandharamakrishnan, 2008).

and glycine (Gly) as excipients. Specifically, Phe with a hydrophobic side chain resulted in the maximum stabilization of IgG during SFD and even after storage at 40°C and 75% RH for 2 months (Emami et al., 2018). Thus, the influence of adding excipients in feed formulation on the stability of spray-freeze-dried proteins was ascertained from their UV absorption spectra in the amide region.

10.1.2 Secondary and tertiary structure of proteins

10.1.2.1 Fourier-transform infrared (FTIR) spectroscopy

Fourier-transform infrared (FTIR) spectroscopy is an ideal technique for the non-invasive determination of secondary structure of proteins in solution as well as in the amorphous dry form

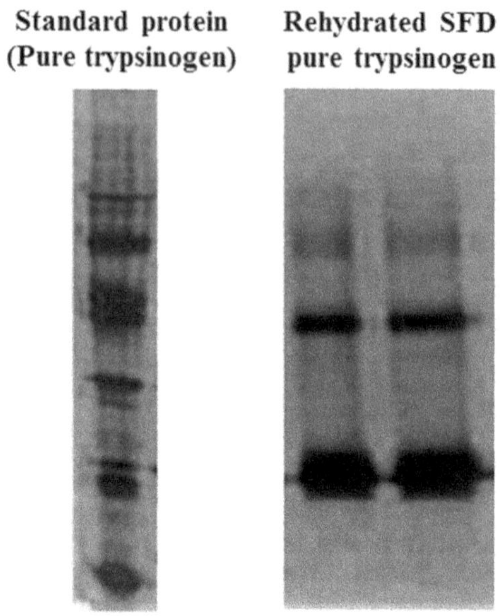

Figure 10.5 SDS-PAGE gel of rehydrated SFD pure trypsinogen solution (50 mg/mL) (Sonner, Maa & Lee, 2002).

(Costantino et al., 1998; Carrasquillo, Cordero et al., 1998; Carrasquillo, Costantino et al., 1999; Costantino, Nguyen & Hsu, 1996). From the IR spectra, it is possible to detect drying-induced conformational changes in the secondary structure of spray-freeze-dried proteins. Initially, the conditions of FTIR spectroscopy are standardized such that artifacts are not induced in the protein secondary structure during sample preparation or investigation. Costantino et al. (2000) proposed the following conditions for the spectroscopic analysis of spray-freeze-dried bovine serum albumin (BSA): (i) number of scans: 256; (ii) scan resolution: 2 cm^{-1}; (iii) pelletization of protein powder with IR-grade potassium bromide (KBr): by pressing ~1 mg of protein with 200 mg of KBr at a pressure of 5 kpsi.

After obtaining the IR spectrum by applying the above conditions, the difference in α-helix and β-sheet contents between the native protein and spray-freeze-dried powder is estimated by Gaussian curve-fitting the Fourier self-deconvoluted spectra in the amide-I region (Herbert et al., 1998). The freeze-drying stage of SFD process causes a reduction in the α-helix content and increase in the β-sheet structure of proteins relative to their native form. Notably, the α-helix content of BSA was reduced from 54% (in native form) to 29% after spray-freeze-drying along with an increase in β-sheet (Costantino et al., 2000). However, the FTIR spectroscopy technique is not sensitive to detect the transitional structural changes in the protein that lead to its instability. The molten globule state of proteins discussed in the previous section is an example for the transitional structural change that retains the native secondary structure, but has a non-native tertiary structure (Bychkova & Ptitsyn, 1993). Also, FTIR spectroscopy is not useful when the mechanism of protein aggregation (ex. thiol-disulfide interchange observed in the case of recombinant human albumin) is independent of its overall secondary structure (Costantino et al., 1995).

10.1.2.2 Circular dichroism spectroscopy

Conformational changes in spray-freeze-dried proteins have also been determined by performing circular dichroism (CD) spectroscopy in the far ultraviolet (UV) region of the electromagnetic spectrum (190–250 nm). A spectropolarimeter with quartz cell having a path length of 0.1 cm at 20°C is employed for this purpose. Dilute solutions of the native protein and SFD protein

Table 10.1 Aggregation index and content (%) of secondary structure elements in different IgG formulations (Modified from Emami et al., 2018)

Samples	IgG formulations for SFD	AI	α-Helix	β-Sheet	Turn	Random
Native IgG	–	0.6	6.0	72.4	5.4	16.2
F_1	• IgG: 200 mg • Trehalose: - • Amino acids: -	25.9	2.4	84.0	0.9	12.6
F_2	• IgG: 200 mg • Trehalose: 200 mg • Amino acids: -	0.1	6.9	70.0	8.3	14.8
F_3	• IgG: 200 mg • Trehalose: 200 mg • Amino acids: Leu 25 mg	0.2	5.6	73.2	6.4	14.7
F_4	• IgG: 200 mg • Trehalose: 200 mg • Amino acids: Leu 50 mg	1.0	4.6	76.8	4.6	14.1
F_5	• IgG: 200 mg • Trehalose: 200 mg • Amino acids: Phe 25 mg	0.6	7.2	73.7	3.0	16.0
F_6	• IgG: 200 mg • Trehalose: 200 mg • Amino acids: Phe 50 mg	0.7	7.6	70.5	5.0	16.9
F_7	• IgG: 200 mg • Trehalose: 200 mg • Amino acids: Arg 25 mg	58.9	0.3	83.2	0	16.5
F_8	• IgG: 200 mg • Trehalose: 200 mg • Amino acids: Arg 50 mg	61.3	0	83.2	0	16.8
F_9	• IgG: 200 mg • Trehalose: 200 mg • Amino acids: Cys 25 mg	0.3	5.7	68.1	7.3	18.8
F_{10}	• IgG: 200 mg • Trehalose: 200 mg • Amino acids: Cys 50 mg	1.7	5.3	67.8	6.5	20.3
F_{11}	• IgG: 200 mg • Trehalose: 200 mg • Amino acids: Gly 25 mg	0.6	7.5	71.4	6.8	14.3
F_{12}	• IgG: 200 mg • Trehalose: 200 mg • Amino acids: Gly 50 mg	0.3	7.4	71.4	7.2	14.0

samples (1 mg/mL) in phosphate buffer (pH 7.4) are filtered using a syringe filter (0.45 μm pore size). The spectra of protein solutions (Figure 10.6) are then obtained at a scanning rate of 200 nm/min and wavelength interval of 0.5 nm, with five scans accumulated. In the CD spectra, zero intensity observed at 210 nm and the extreme positive and negative peaks at 202 and 218 nm are typical of the high β-sheet content of native protein.

As an example, the secondary structure elements of native IgG and different formulations of spray-freeze-dried IgG are presented in Table 10.1. The proportion of secondary structure elements (%) in the protein samples was calculated after subtracting the spectra of buffer/additives from the sample spectrum. The varying SFD formulations (F_1-F_{12}) contained IgG (F_1) and its combination with trehalose and amino acids (F_2-F_{12}) as excipients to preserve the physicochemical stability of the antibody during drying (Emami et al., 2018). Sugars and amino acids are known to have lyoprotective effects (stability against the lyophilization process) upon the proteins via mechanisms such as water replacement, immobilization of proteins in a solid matrix and preservation of

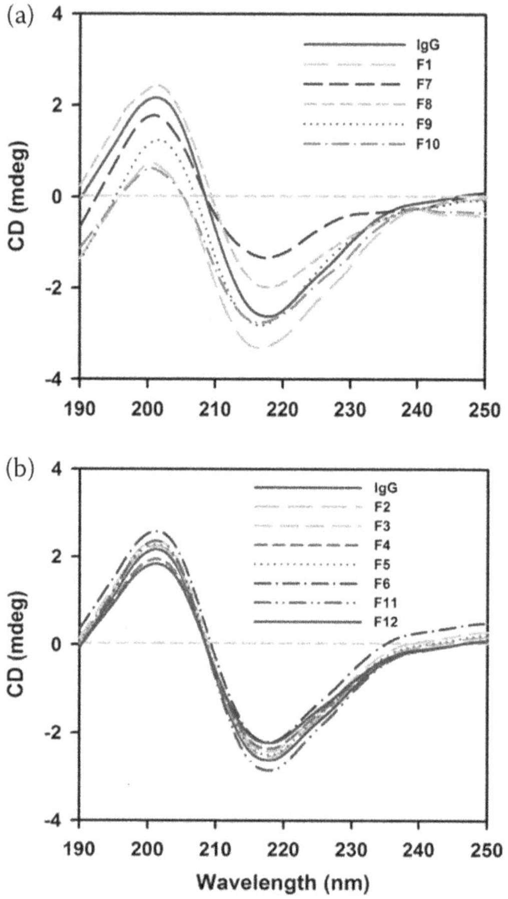

Figure 10.6 CD spectra of spray-freeze-dried IgG from feed formulations containing different combinations of sugar and amino acid excipients (F1-F12), compared to that of unprocessed IgG (Emami et al., 2018).

proteins in an amorphous state, with high glass transition temperature (T_g) (Depreter et al., 2013; Tonnis et al., 2014; Kanojia et al., 2016). Accordingly, five different amino acids under the following four categories were chosen as excipients to be used in combination with trehalose: (i) hydrophobic (leucine, phenylalanine); (ii) hydrophilic (arginine); (iii) polar sulfur-containing (cysteine) or (iv) nonpolar side chains (glycine) (Faghihi et al., 2014).

From the table, it can be noted that most of the spray-freeze-dried IgG formulations (F_2, F_3, F_4, F_5, F_6, F_{11} and F_{12}) followed similar positive and negative peak patterns as that in the secondary structure of native IgG (Figure 10.6[b]). But, the formulations that contained only IgG (F_1) and those prepared using arginine (F_7 and F_8) or cysteine (F_9 and F_{10}) as excipients depicted significant differences compared to the CD spectra of native IgG and other formulations. A substantial shift in the maximum (202 nm) and minimum peaks (218 nm) of abovementioned formulations was indicative of major alterations in the secondary structure of IgG upon spray-freeze-drying (Figure 10.6[a]).

10.1.2.3 Fluorescence spectroscopy

Fluorescence spectroscopy is a sensitive technique to determine the hydrophobicity and alterations in the secondary structure of proteins. It provides the aforesaid information based on the change in fluorescence intensity and shift in emission maxima (λ_{max}) of an extrinsic fluorescent dye (ex. 8-Anilino-1-naphthalene sulfonic acid ammonium salt or ANS) bound to the protein. In the case of SFD-IgG formulations given in Table 10.1, the λ_{max} was detected at about 472 nm and there was no measurable variation in fluorescence intensity between the native and stable compositions (F_2, F_3, F_4, F_5, F_6, F_{11} and F_{12}). But, the extrinsic fluorescence spectra of SFD-IgG formulations containing arginine and cysteine as excipients (F_7, F_8, F_9 and F_{10}) showed significant differences in the fluorescence intensity (Figure 10.6). Those formulations that contained cysteine (F_9 and F_{10}) had an apparently stronger ANS fluorescence intensity with a positive shift of wavelength from 472 to 480 nm. Whereas, the intensity of SFD feed with arginine (F_7 and F_8) reduced significantly with a marked blue shift from 472 to 464 nm (Figure 10.7). The above observations confirmed the conformational changes caused by spray-freeze-drying process in the secondary and tertiary structures of IgG, relative to its native form (Emami et al., 2018).

10.2 CHARACTERIZATION OF DRY POWDERS FOR PULMONARY AND NEEDLE-FREE BALLISTIC DELIVERY

10.2.1 Hygroscopicity

Hygroscopicity is the tendency of dry powders to absorb moisture from a high relative humidity environment (Bhusari et al., 2014). It leads to poor physical stability and reduced flowability of the solid product (Sonner, Maa & Lee, 2002). Hygroscopicity of powders is controlled by their physical and structural properties such as surface area and particle morphology (Bhandari, 2013). SFD process results in porous particles with large specific surface area that are hygroscopic.

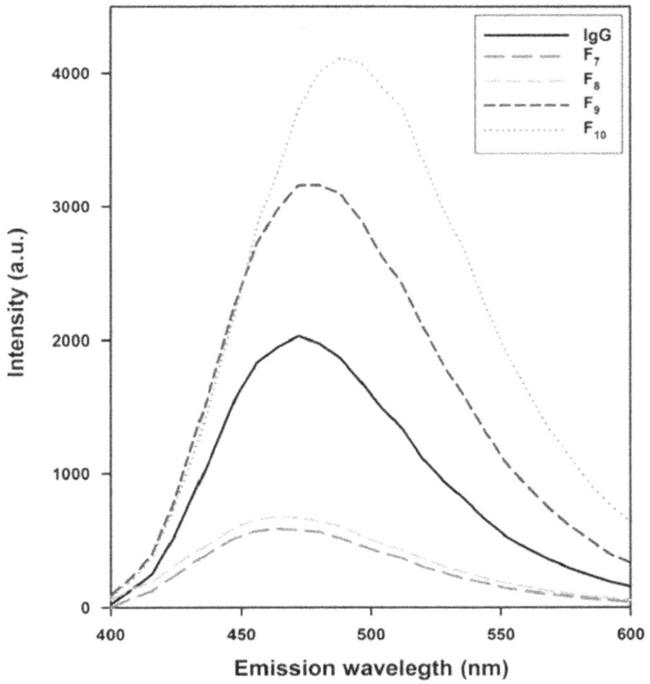

Figure 10.7 Fluorescence spectra of native IgG and spray-freeze-dried IgG obtained from formulations containing arginine (F_7 and F_8) and cysteine (F_9 and F_{10}) as lyoprotective excipients (Emami et al., 2018).

The hygroscopicity of spray-freeze-dried powders is determined by assessing their moisture uptake kinetics. A small sample of the powdered product is transferred to a highly-sensitive analytical balance maintained under humidified conditions in an air-tight container using saturated sodium chloride solution. The spray-freeze-dried powder takes up moisture under the above conditions. The resultant increase in powder weight is measured over a defined time period (in the order of 1 to 2 hours to a maximum of 4 weeks).

For instance, spray-freeze-dried trehalose powder absorbed about 4.5% moisture after 1 h, when stored at 33% relative humidity (RH) and 25°C. On the contrary, the spray-dried trehalose powder took up just 1.5% moisture after 2 hours, under the same storage conditions. Despite their ten-fold higher particle size (20–90 µm) than the spray-dried powder (<5 µm), the SFD-trehalose showed more hygroscopicity due to its higher internal porosity and the resultant larger specific surface area for moisture uptake (Sonner, Maa & Lee, 2002). Nevertheless, Rajam & Anandharamakrishnan (2015) reported on the lower hygroscopicity of spray-freeze-dried probiotic (*Lactobacillus plantarum)* microcapsules (6.50–7.33 g/100 g) than their freeze-dried counterparts (11.19–13.58 g/100g), when about 2 g of powder was exposed to 75% relative humidity at room temperature for seven days (Fritzen-Freire et al., 2012). The authors attributed the lower hygroscopicity of SFD probiotic powder to atomization and the use of film-forming wall materials such as WPI, denatured WPI and sodium alginate. During atomization, the size reduction of feed solution might create hydrophobic surfaces as the amorphous materials rupture interfaces with the weaker hydrophobic bond that is extremely susceptible to water sorption (Leuenberger, Plitzko, & Puchkov, 2006). On the other hand, the skin formed by the film-forming wall materials partially covered the pores on their outer surface. Consequently, the exposed area was reduced, which suppressed the water adsorption at high relative humidity (Tonon et al., 2011). But, the freeze dried microencapsulates had relatively larger pores that were left behind after the sublimation of bigger ice crystals formed after slow freezing of the process. The larger pores increased the moisture uptake. This study also established the influence of moisture content and wall material type on the hygroscopicity of SFD product. Higher residual moisture content and inherent hygroscopic nature of wall materials (ex. fructooligosaccharide) increase the hygroscopicity of final encapsulate (Rajam & Anandharamakrishnan, 2015).

Reducing the hygroscopicity of spray-freeze-dried powders is imperative, mainly to permit its ready conversion into solid pharmaceutical dosages for ballistic drug delivery (Lek Pharmaceuticals d.d., 2015). Certain approaches have been found useful in reducing the hygroscopicity of spray-freeze-dried powders. Applying an annealing step at the start of the primary drying stage by initiating the reduction of chamber pressure was found to reduce the total moisture uptake of spray-freeze-dried trehalose from about 4.5 to 3.5% after 1 h exposure to 33% RH. In the above study, annealing was carried out for 2 hours at a shelf temperature of 0°C. However, extending the duration of annealing beyond 2 hours to about 6 hours did show any further reduction in the moisture uptake (Figure 10.8) (Sonner, Maa & Lee, 2002). Annealing prevents hygroscopicity by causing a partial structural collapse and shrinkage of particles to reduce their specific surface area and thereby, moisture uptake (Nguyen, Herberger, & Burke, 2004; Sonner, Maa & Lee, 2002). Use of hydrophobic amino acids like L-leucine, L-isoleucine or L-phenylalanine as excipients at a high concentration is effective in alleviating the hygroscopicity of SFD powders (Otake, Okuda, & Okamoto, 2016). This approach is known to promote the applications of spray-freeze-dried therapeutics for pulmonary and transdermal delivery applications.

Alternatively, surfactants are added as excipients in the feed formulations to prevent hygroscopicity in spray-freeze-dried protein powders. The use of surfactants is based on the postulated tendency of surfactant molecules to adsorb at the air–liquid interface of droplets in an atomized mist. The adsorbed surfactants replace the protein molecules at the air–liquid interface and thus prevent exposure of the protein to surface unfolding, denaturation and the resultant hygroscopicity. Among the different categories of surfactants, the non-ionic candidates are more appropriate for

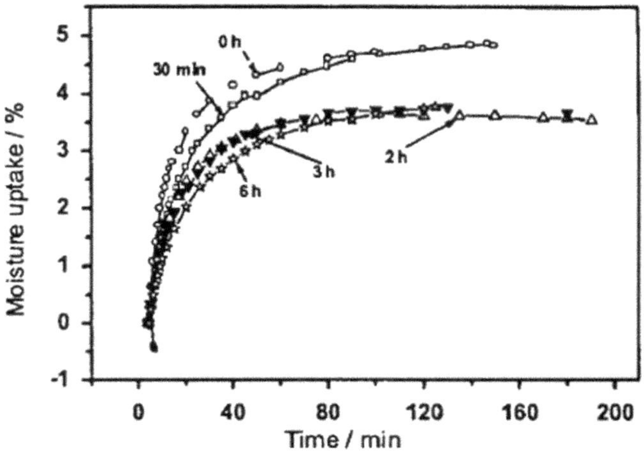

Figure 10.8 Moisture uptake kinetics of spray-freeze-dried trehalose particles (stored at 33% RH and 25°C) annealed for various times at a shelf temperature of 0°C during primary drying (Modified from Sonner, Maa & Lee, 2002).

the avoidance of drug–surfactant interaction within the formulation (Pikal, 2002 & 2003). Further, it was hypothesized that hydrophilic surfactants such as Pluronic F68, which exist as solid at room temperature, may form a less hygroscopic layer on the surface of the SFD particle. This could potentially improve the flowability and physical stability of particles, based on the abovementioned principle of the preferential adsorption of surfactant at the air-liquid interface (Maa et al., 2004).

10.2.2 Bulk density

Bulk density can be defined as the ratio between the mass of solid particles including the moisture content and the volume occupied by them. It is an important physical property that influences the transport, handling and storage of food powders (Caparino et al., 2012; Quispe-Condori et al., 2011), especially, instant products such as soluble coffee. Any non-compliance in the prescribed value of bulk density will lead to either an under net weight pack or a pack that seems to lack the product, despite of having the correct net weight (Barbosa-Canovas et al., 2005). The two types of bulk density are the *as-poured* bulk density and tapped density. *As-poured* bulk density is the density of particles, when heaped in a volumetric container without any compacting forces. This depends on the size, shape and surface characteristics of microcapsules (Rajam & Anandharamakrishnan, 2015). Tapped density is nothing but the bulk density of particles added to a volumetric vessel with the aid of compaction forces (Al-Hakim, 2004).

The bulk (ρ_B) and tapped densities (ρ_T) are usually determined with a graduated glass measuring cylinder of known volume (10 mL, 25 mL, 50 mL, 100 mL or 250 mL) and inner diameter. Initially, the measuring cylinder is weighed to the nearest 0.1 g (m_1) and then the cylinder is filled with the powder to a defined volume (V_B). Then, the cylinder and its contents are weighed to the nearest 0.1 g (m_2). ρ_B is calculated using the Eq. 10.3. Subsequently, the cylinder is either tapped manually (hand tapping) or using a tap density tester for a defined number of times as prescribed in the national or international standards and then the compacted volume (V_T) is noted. For instance, soluble coffee particles are hand-tapped 300 times before noting the compacted volume (IS 16033, 2012). Then, ρ_T is calculated using Eq. 10.4. The tapped bulk density of powders is governed by the structural strength and packed characteristics of the loosely packed sample (Mohammadi & Harnby, 1997).

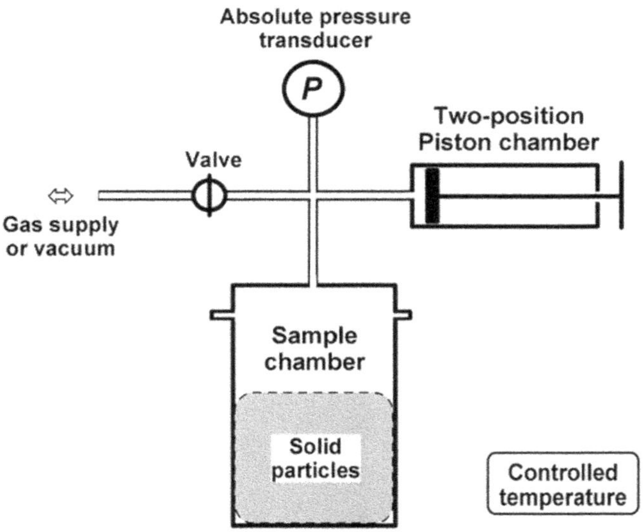

Figure 10.9 Diagram of a gas pycnometer (Tamari & Aguilar-Chávez, 2005).

$$\rho_B = \frac{m_2 - m_1}{V_B} \tag{10.3}$$

$$\rho_T = \frac{m_2 - m_1}{V_T} \tag{10.4}$$

Conversely, some studies on spray-freeze-dried particles have reported the true density of particles, wherein the open pores are omitted in the calculation. This is usually determined by a gas pycnometer (Figure 10.9), using a sample cup of suitable size (say 1 mL or 0.1 mL). It works on the principle of detecting the change in pressure resultant from the displacement of gas by a solid substance. Before placing the sample, a specific quantity of gas at known pressure is allowed to expand inside an empty chamber. Then, the gas pressure is measured to set the baseline. Afterwards, the sample is kept in the chamber, which is then resealed. Then, for the second time, the same quantity of gas at same pressure is expanded inside the sample chamber and the pressure is recorded. The difference between the two pressure values along with the known volume of the empty sample chamber is used to calculate the volume of the sample, based on the gas law (Bio-Equip, *n.d.*). Helium is the most commonly used gas for pycnometry-based density measurements. Alternatively, liquid pycnometry can also be used to determine the true density of particles, in which case, toluene replaces helium.

In many investigations, spray-freeze-drying has been observed to result in denser particles than freeze-drying (Ishwarya & Anandharamakrishnan, 2015; Parthasarathi & Anandharamakrishnan, 2016; Rajam & Anandharamakrishnan, 2015), due to the following reasons: (1) smaller particle size, spherical shape and the larger contact surface areas per unit volume of spray-freeze-dried powders; (2) uniform size and shape of SFD particles compared to the freeze-dried product, that reduces the external voids and lead to lower bulk volume and higher bulk density (Caparino et al., 2012). The typical bulk, true and tapped densities of spray-freeze-dried products are compiled in Table 10.2.

Table 10.2 Density of spray-freeze-dried products

Spray-freeze-dried product	Bulk density (BD) (or) True density (TD)	Tapped density	Reference
Soluble coffee	BD: 0.612 ± 0.007 g/mL	0.679 ± 0.008 g/mL	Ishwarya & Anandharamakrishnan (2015)
Probiotic (*Lactobacillus plantarum*) microcapsules • WPI + SA • WPI + FOS • DWPI + SA • DWPI + FOS	BD: • 169.37 ± 3.56 kg/m^3 • 321.43 ± 12.86 kg/m^3 • 193.49 ± 7.82 kg/m^3 • 367.81 ± 7.98 kg/m^3	• 176.76 ± 4.35 kg/m^3 • 368.30 ± 2.66 kg/m^3 • 216.89 ± 5.90 kg/m^3 • 401.22 ± 7.89 kg/m^3	Rajam & Anandharamakrishnan (2015)
Vitamin E microcapsules	BD: 266 ± 2.40 kg/m^3	321.34 ± 2.57 kg/m^3	Parthasarathi & Anandharamakrishnan (2016)
Baccharis dracunculifolia leaf power (BdDE) with D-mannitol (DMan) and modified corn starch (MCS) as drying adjuvants	BD: • 0.25 g/mL (BdDE + DMan) • 0.45 ± 0.03 g/mL (BdDE + MCS) • 0.40 ± 0.02 g/mL (BdDE + DMan + MCS)	• 0.34 ± 0.01 g/mL (BdDE + DMan) • 0.49 ± 0.03 g/mL (BdDE + MCS) • 0.44 ± 0.03 g/mL (BdDE + DMan + MCS)	Teixeira et al. (2017)
Trypsin/trehalose composite microparticles • Trypsin • Try + Tre (3:1) • Try + Tre (2:1) • Try + Tre (1:1) • Try + Tre (1:5) • Try + Tre (1:9) • Trehalose	BD: • 18 ± 1 kg/m^3 • 15 ± 3 kg/m^3 • 17 ± 1 kg/m^3 • 16 ± 1 kg/m^3 • 14 ± 1 kg/m^3 • 16 ± 1 kg/m^3 • 24 ± 1 kg/m^3	• 35 ± 1 kg/m^3 • 36 ± 2 kg/m^3 • 30 ± 2 kg/m^3 • 31 ± 2 kg/m^3 • 29 ± 1 kg/m^3 • 33 ± 1 kg/m^3 • 34 ± 1 kg/m^3	Zhang et al. (2018)
Polyethylene glycol polymeric particles (By Helium pycnometry: Helium gas at 19.5 psig)	TD: 0.5 – 1.5 g/mL	-	Barron et al. (2003)
Lysozyme (protein) spheres (TD: By Helium pycnometry)	**With BSA as stabilizer:** 0.1% (w/v) 0.0874 g/cc 1% (w/v) 0.0613 g/cc 10% (w/v) 0.2559 g/cc **With dextran as stabilizer:** 0.1% (w/v) 0.106 g/cc 1% (w/v) 0.1076 g/cc 10% (w/v) 0.2437 g/cc **With Polyvinylpyrollidone (PVP) as stabilizer:** 0.1% (w/v) 0.1253 g/cc 1% (w/v) 0.1298 g/cc 10% (w/v) 0.358 g/cc	-	Eggerstedt et al. (2012)

(Continued)

Table 10.2 *(Continued)* Density of spray-freeze-dried products

Spray-freeze-dried product	Bulk density (BD) (or) True density (TD)	Tapped density	Reference
Microbial Transglutaminase	• TD (by liquid pycnometry, using toluene): 1351.43 ± 8.80 kg/m^3 • BD: 152.3 ± 0.08 kg/m^3	244.32 ± 1.50 kg/m^3	Isleroglu et al. (2018)
• SFD-insulin (5 mg/mL) at pH 2.0 • SFD-insulin, 5 mg/mL + trehalose 5 mg/ml atpH 2.0	• TD (from helium pycnometry): 1.322 g/cc; BD: 3.23 x 10^{-3} g/cc • TD: 1.429 g/cc; BD: 3.45 x 10^{-3} g/cc	• 4.36 ± 10^{-3} g/cc • 4.63 x 10^{-3} g/cc	Schiffter et al. (2010)

10.3 CHARACTERIZATION OF SPRAY-FREEZE-DRIED DRUGS/API

10.3.1 Dissolution property

Dissolution has been defined in two ways: (i) process by which a solid substance enters into a solvent to form a solution (Agilent Technologies, 2015); (ii) the rate at which the active ingredient (s) in a solid or semi-solid pharmaceutical or nutraceutical formulation is released into the liquid medium under standard conditions of temperature, liquid/solid interface and media composition. According to the United States Pharmacopeia/National Formulary (USP/NF), dissolution testing is an important quality indicator and a definitive tool to characterize the performance characteristics of pharmaceutical products in their solid dosage form (Rosanske & Brown, 1996). Dissolution testing is not just limited to oral dosage forms such as tablets and capsules, but also to evaluate products such as suspensions and powders, transdermal patches and dry powders for inhalation (DPI). The methodology and apparatus for dissolution testing has continuously evolved and improved across the last seven decades to provide suitable conditions for the performance testing of wide-ranging products (Agilent Technologies, 2015). This section is intended to present the different methods of dissolution tests and the kind of apparatuses that have been employed for the spray-freeze-dried products.

In the USP/NF general chapter, seven apparatuses have been defined for dissolution testing, which mainly vary with their mixing accessory type. Of the above, Apparatus I and Apparatus II are commonly used for solid oral dosage forms. Both these apparatuses comprise a single vessel with hemispherical bottom, which holds the dissolution medium and a spindle positioned in its center (Figure 10.10). The vessel is immersed inside a constant temperature water bath for the precise maintenance of temperature. In Apparatus I, the dosage form is placed inside a basket that is attached to the spindle's end and the contents are mixed at a rotational speed of 100 rpm (Long & Chen, 2009). The medium of dissolution or solvent enters the basket from bottom in the axial direction, flows upward and exits via the sides (D'Arcy et al., 2006). Thus, Apparatus I is more suitable for dosage forms such as capsules that might float. The Apparatus II (Figure 10.10[b]) employs a paddle to agitate the contents at a rotational speed of 50–75 rpm. In this case, the drug to be tested is dropped directly into the vessel containing dissolution medium. Choice of dissolution medium is a critical factor for dissolution testing, which is governed by the objective of the study and the API under investigation. For research and development purpose, it is preferable to use media similar to biological fluids. Phosphate buffers comprising a group of proteins, lipids and

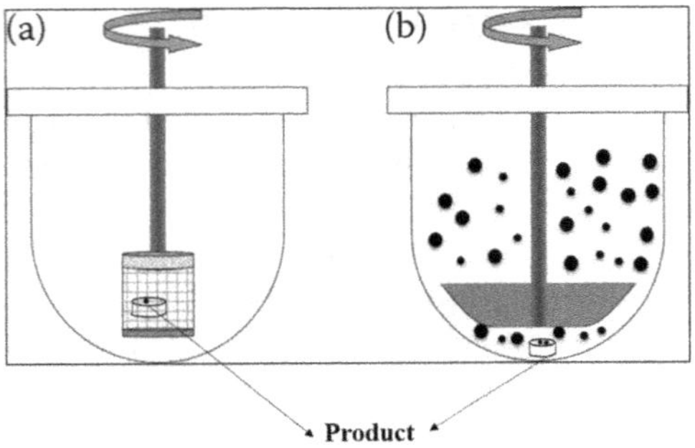

Figure 10.10 Schematic representations of (a) USP Apparatus I (Basket) and (b) USP Apparatus II (Paddle) (Slightly modified from Long & Chen, 2009).

surfactants (ex. Dipalmitoylphosphatidylcholine [DPPC], a pulmonary surfactant) are normally used as simulated lung fluids (Arora et al., 2010; Griese & Stiftung, 1999; May et al., 2012; Marques, Loebenberg, & Almukainzi, 2011; Sakagami & Lakhani, 2012; Salama et al., 2008). However, for routine quality control analysis, simple dissolution media such as buffers in the pH range of 6.8–7.4 are used for the reasons of higher reproducibility, lower costs and ease of preparation (Marques et al., 2011).

In an earlier study by Rogers et al. (2003), the authors used a USP 24 Type II Apparatus for the dissolution testing of spray-freeze-dried danazol. Danazol is an orally administered synthetic androgen used in the treatment of endometriosis, fibrocystic breast disease and hereditary angioedema (Brayfield, 2013; Morton & Hall, 2012). About 10 mg of SFD powder containing 4 mg of the active ingredient, Danazol was placed into 900 mL of dissolution media that was prepared by dissolving 150 g of sodium lauryl sulfate (SLS) and 242 g of tris(hydroxymethyl) aminomethane (Tris) in approximately 18 L of purified water. The pH of dissolution media was adjusted to 9.0 with 1 N HCl. Under constant stirring, its volume was made up to 20 L with purified water. The dissolution test was conducted at a constant paddle speed and bath temperature of 50 rpm and 37°C, respectively. Sink condition of the drug was maintained throughout the duration of dissolution testing. Intermittently, 5 mL of samples were collected at the 2nd, 5th, 10th, 20th, 30th and 60th min, using an autosampler and the percentage of dissolved API (danazol) was estimated using HPLC. Thus, this study determined dissolution according to the first definition mentioned in the beginning of Section 10.3.1.

Figure 10.11 shows the typical dissolution profile of spray-freeze-dried danazol powders obtained from a feed solution containing danazol and excipients such as polyvinyl alcohol (PVA), polaxamer and polyvinylpyrrolidone. The feed solution was spray-frozen in the SFL mode

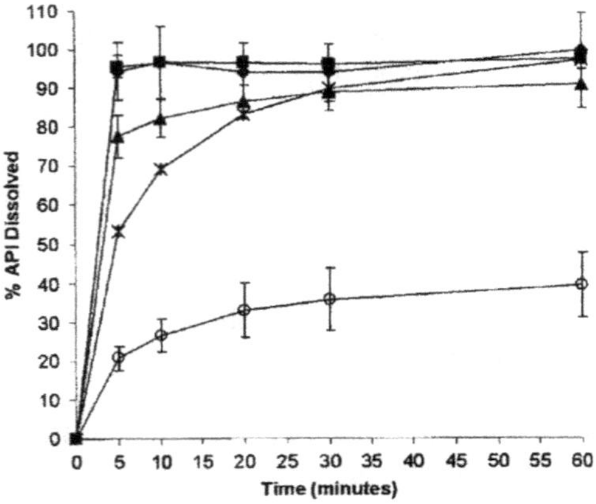

Figure 10.11 Aqueous dissolution profiles of the co-ground physical mixture (o), bulk danazol (*), slowly-frozen control (▲), atmospheric freeze-dried and micronized spray-freezing-into liquid (SFL) powder (♦) and vacuum freeze-dried micronized SFL powder (■) (Rogers et al., 2003).

followed by freeze-drying under atmospheric conditions (ATMFD) and vacuum freeze-drying (VFD). Three control samples were considered for comparison. The first was the bulk or un-processed danazol (BD) without excipients. The second control sample was a co-ground physical mixture (CGPM) containing the above constituents of SFD feed in their dry form. Third one, referred to as the 'slowly frozen control' (SFC) was the sample obtained after conventional freeze-drying of the spray-frozen feed solution same as that used for ATMFD. From Figure 10.11, the dissolution rate in media was found to increase in the following order: CGPM < BD < SFC < ATMFD < VFD. Notably, the spray-freeze-dried danazol powders dissolved completely (100%) in the dissolution medium within 5 min, against 78% of SFC within the same duration (Rogers et al., 2003).

The faster dissolution of VFD and ATMFD danazol powders was attributed to their amorphous nature and larger specific surface area. Moreover, the amorphous danazol was uniformly distributed within the hydrophilic excipient matrices of VFD and ATMFD powders, which facilitated their rapid and complete aqueous dissolution. The excipient, polaxamer is amphiphilic because of which it could adsorb hydrophobic danazol onto its surface, whilst interacting with the aqueous dissolution medium. The outcome of this simultaneous interaction was the enhanced dissolution of danazol. The crystalline nature of danazol in the SFC powder did not permit it to wet and dissolve as instantaneously as the amorphous VFD and ATMFD powders. The unexpected dissolution behavior was that of the co-ground physical mixture that dissolved more slowly and to a lesser extent than the other control formulations, even lower than the bulk danazol. When placed in the aqueous dissolution medium, the excipients present in the co-ground physical mixture hydrated and swelled to trap the danazol crystals within the gel layer. This hampered the complete dissolution within the stipulated dissolution testing period (Rogers et al., 2003).

Another study by Adeli (2017) quantified dissolution of azithromycin from spray-freeze-dried powder as the rate of its release from the solid formulation into the liquid medium. Azithromycin, belonging to the category of macrolide antibiotics, is an azalide derived from erythromycin. It exhibits bacteriocidal and bacteriostatic activities (National Center for Biotechnology Information, 2021). The same USP Apparatus II as that used in the previous study was employed in this study as well. The temperature and agitation conditions were 37°C and 100 RPM, respectively. 900 mL of phosphate buffer at pH 6.0 was used as the dissolution medium. As mentioned in the previous study, the physical mixture of azithromycin was considered as the control, for which, defined amounts of pure Azithromycin and PVA were mixed using a mortar and pestle for 5 min. To analyze the dissolution behavior, about 100 mg of samples (pure drug, solid dispersions and physical mixture) were used. The samples were withdrawn in intervals of 5 min, whilst making up for the liquid volume in the apparatus with fresh medium. The samples were filtered and analyzed by a UV spectrophotometer at 215 nm for the dissolved API. Dissolution efficiency (DE) was determined from the area under the dissolution curve at time '*t*' and expressed as a percentage of the area of the rectangle, given by 100% dissolution at the same time (Eq. 10.5).

$$\%DE = \left(\frac{\int_0^t y.\, dt}{y_t^{100}} \right) \times 100 \tag{10.5}$$

Where y is the percentage drug dissolved at any time t (Adeli, 2017). Results showed that the *in vitro* dissolution rate and cumulative drug release were maximum for the spray-freeze-dried azithromycin and minimum for its pure form. For a test period of 60 min, the maximum drug release was 87.7% from the SFD powder and 22% for the physical mixture (Figure 10.12). Similar trend was observed for the %DE (Table 10.3). The amorphous form of azithromycin and the local solubilizing action of carrier material, i.e., PVA (Chiou & Riegelman, 1971; Sekiguchi, Obi &

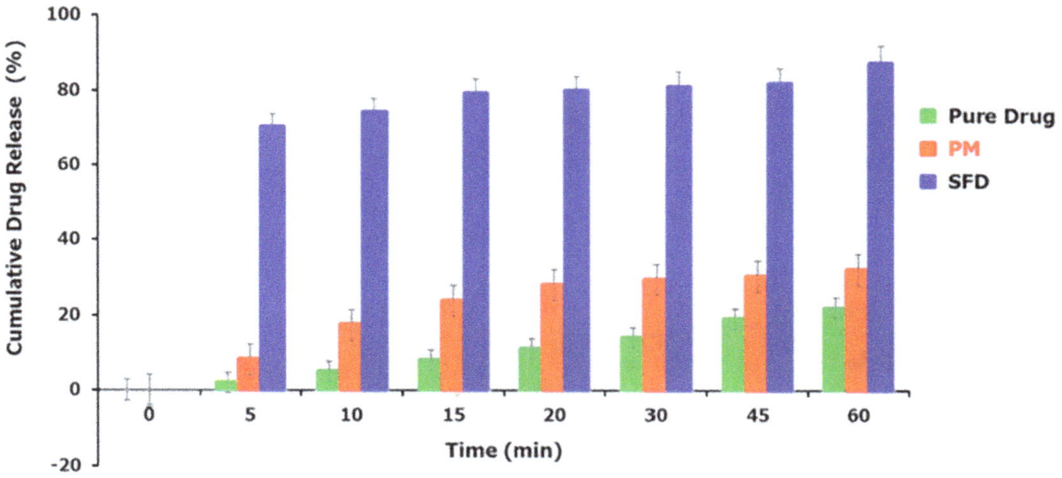

Figure 10.12 *In vitro* dissolution drug release of pure Azithromycin, SFD-drug and the physical mixture (Adeli, 2017).

Table 10.3 Dissolution parameters of pure Azithromycin, SFD-drug and the physical mixture (Adeli, 2017)

Sample	Drug: carrier (w/w)	DE_{10} (%)	DE_{30} (%)
Pure drug	1:0	4.50 ± 1.12	14.32 ± 0.99
SFD-drug	1:1	46.11 ± 0.66	80.40 ± 1.83
Physical mixture	1:1	7.19 ± 1.18	17.88 ± 1.39

Ueda, 1964; Adeli, 2017) were stated as reasons for the enhanced dissolution rate of spray-freeze-dried powders.

In yet another investigation by Niwa, Mizutani, & Danjo (2012), Ciclosporin – a poorly water-soluble model drug for respiratory delivery was loaded in mannitol (a carrier to improve dissolution) and then spray-freeze-dried. Different from the above studies, these researchers adopted two different methodologies to quantify the release rate of ciclosporin from the SFD composite particles: (1) soaking method in a large volume of medium and (2) filter permeation method in a small volume of medium. Probably, the above variations were studied since the systemic availability of dry powder inhalers (DPIs) is usually limited by the poor dissolution of their constituent API in limited physiological volumes of aqueous media (Labouta & Schneider, 2010). Also, the release pattern of API from the SFD particles were examined in a simulated respiratory fluid (phosphate-buffer solution containing 0.1% of Tween 80 at pH 6.6). The first method of dissolution testing involved a simple procedure in which a defined weight (in the order of milligrams) of the drug was placed into 100 mL of the dissolution medium at 37°C and then agitated at 300 rpm using magnetic stirrer. The latter was carried out in a specially designed system (Figure 10.13), which mimics the release process from particles that cling to the surface of bronchus and pulmonary mucosa (Niwa et al., 2012).

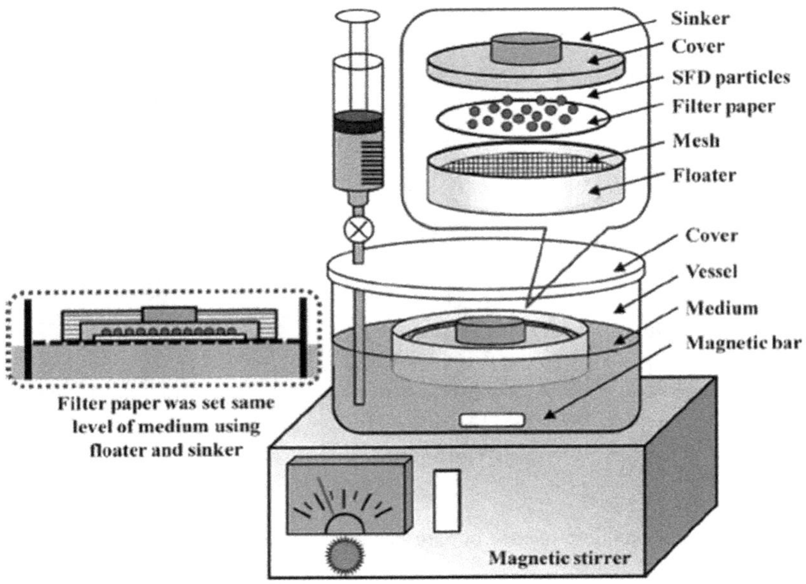

Figure 10.13 Schematic diagram of the permeation-type release test apparatus through a wet fiber filter (Niwa, Mizutani, & Danjo, 2012).

In the second method of dissolution test, a fiber filter covering the surface of medium (100 mL) was allowed to be completely wetted by the latter. A definite weight of SFD-ciclosporin particles were uniformly spread above the filter to simulate deposition of inhaled particles on the respiratory mucosa. The filter paper was attached to a mesh and floated on the dissolution medium using a floater. A sinker was employed to maintain the filter at the same level of the dissolution medium during the test. For both the methods, aliquots were withdrawn at defined time intervals, diluted appropriately in methanol and analyzed for the released API using HPLC. Figure 10.14 shows the faster release rate of SFD-ciclosporin particles than the particles produced without mannitol (Cic100%) and the bulk particles that exhibited poor solubility. Specifically, with the fiber filter method, the release curve of Cic100% particles flattened at ~15%. This implied that even the effectively aerosolized drug particles after reaching the respiratory region may not be necessarily available at levels required for the treatment efficacy. On the other hand, the dissolution rate of SFD-ciclosporin particles was high, which increased with the increase in the mannitol content of particles, even with a low volume of dissolution medium in the filter permeation method. The hypothesis was that the mannitol enhanced the entry of medium into the particles and promoted a substantial increase in the effective surface area for dissolution (Niwa, Mizutani, & Danjo, 2012).

Different from the above approaches, an imaging based method was adopted to study the dissolution characteristics of spray-freeze-dried pullulan particles in water (Xu et al., 2018). The SFD trials were carried out at different pullulan concentrations (5%, 10%, 15% and 20%, w/w), in two different cryogenic milieu: liquid nitrogen (−196°C) and cold air (−15°C). The prepared particles were placed in a container filled with water, without stirring. A high-speed camera was used for dissolution testing, by maintaining a constant distance between the camera lens and the liquid surface. Figs. 10.15[a & b] show the images of single pullulan particles at different stages of dissolution for a period of 40 min. With feed solutions having low concentration of pullulan and spray-frozen in liquid nitrogen, the resultant particles showed rapid swelling upon dissolution in water, while their surface layer dissolved gradually. Subsequently, the entire particle collapsed into

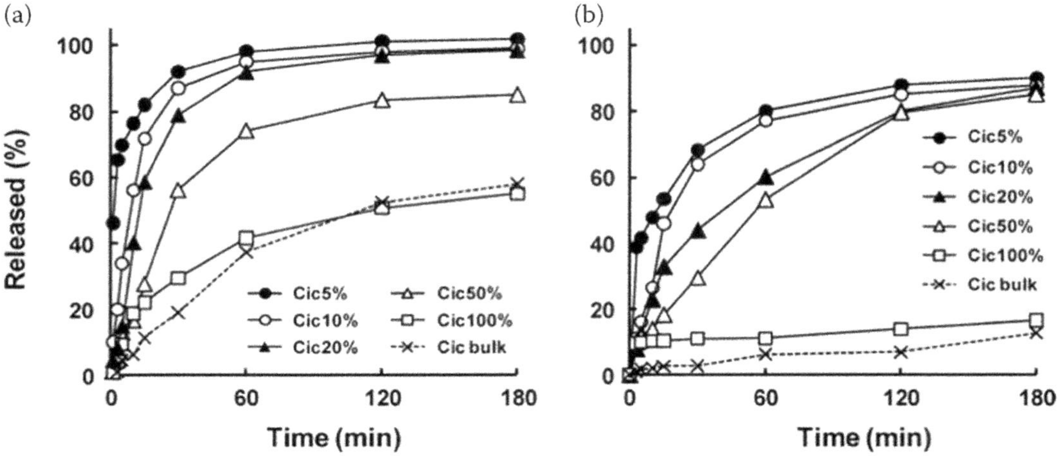

Figure 10.14 Release Profiles of Ciclosporin from the SFD Composite Particles with Varying Ciclosporin Content in Simulated Bronchial Medium (pH 6.6): (A) Dissolution test with a large volume of medium: dissolution from the immersed samples in the medium. (B) Dissolution test with a small volume of medium: permeation of drug from wet samples through a wet filter. Key: (●) Cic5%, (○) Cic10%, (▲) Cic20%, (Δ) Cic50%, (□) Cic100%, (x) ciclosporin bulk (Niwa, Mizutani, & Danjo, 2012).

multiple small particles and continued to dissolve rapidly till the solid particles disappeared. But, the pullulan particles prepared by freezing in cold air did not undergo swelling in water. Instead, they turned smaller as their surface layer dissolved slowly. Despite having a smaller size than those prepared in liquid nitrogen, the particles prepared in cold air showed a lower dissolution rate than the latter.

Further, initial concentration of pullulan in the feed solution had a major influence on the dissolution of SFD particles. At higher concentration (20%), irrespective of the cryogenic environment, the particle swelling rate in water was slower and the surface layer dissolution was not as apparent as in particles obtained after spray-freeze-drying of feed solutions with lower pullulan concentration (5%, 10% and 15% w/w) (Figure 10.15[c]). The above results were attributed to the highly dense particle surface of SFD particles prepared by freezing in cold air, which hindered the entry of water into the particles. Contrarily, the porous surface morphology of particles prepared after rapid freezing of pullulan solutions at low concentration in liquid nitrogen followed by freeze-drying was ascertained to be responsible for their good dissolution behavior. Further, the larger specific surface area of particles prepared in the LN_2 environment (5%: 7.45 m^2/g; 10%: 11.78 m^2/g; 15%: 5.10 m^2/g and 20%: 4.54 m^2/g) was also responsible for its shorter dissolution time than their cold air counterpart (5%: 5.26 m^2/g; 10%: 3.29 m^2/g; 15%: 3.44 m^2/g and 20%: 3.20 m^2/g) (Xu et al., 2018).

10.3.2 Specific surface area and porosity

Specific surface area (SSA) of solids is defined as the total surface area of a material per unit mass, measured in units of $m^2\ kg^{-1}$ or mg g^{-1} (IUPAC, 1997). Particularly, it is that portion of the total surface area, which is available for adsorption (Kuila & Prasad, 2013). The SSA of a particle is a function of its porosity, pore size distribution, shape, size and roughness (Amador & Martin de Juan, 2016), wherein porosity is the ratio between total pore volume and apparent volume after excluding the inter-particle voids (Salamon, 2014). SSA is an important parameter in material

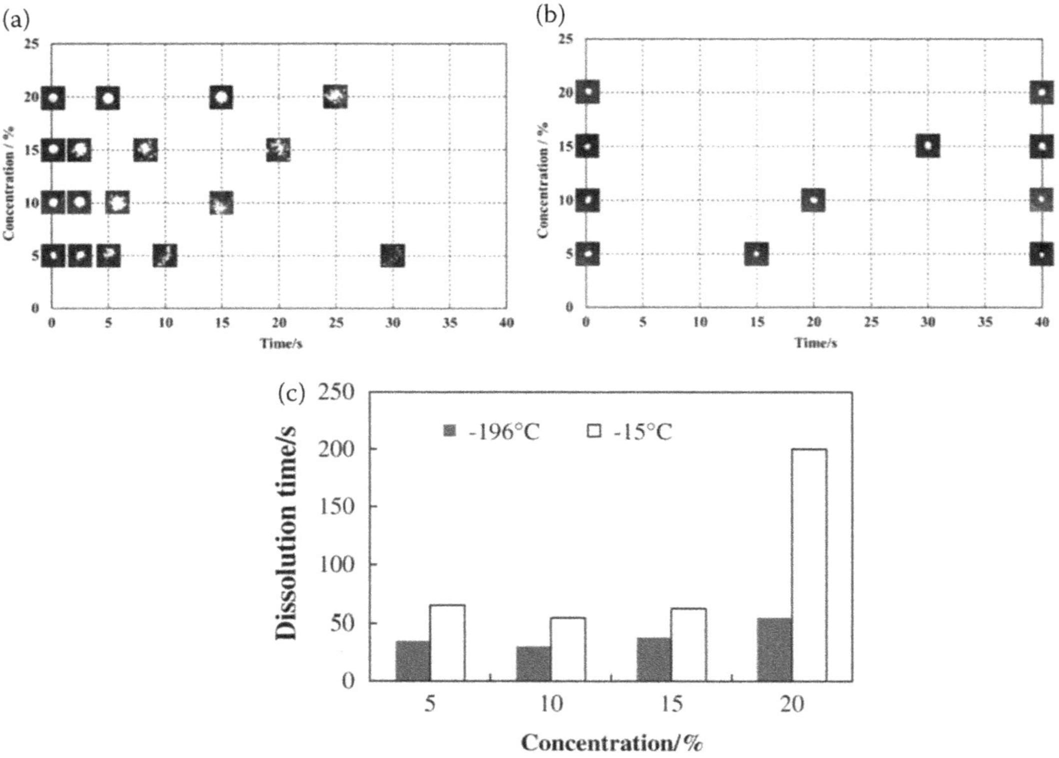

Figure 10.15 Dissolution steps of pullulan particles from different concentration solutions at freezing temperatures of (a) −196 °C; (b) −15 °C; (c) Influence of feed concentration on the dissolution time of particles prepared after spray-freezing in two different cryogenic environments (Xu et al., 2018).

characterization as it is related to several properties such as dissolution rates, bioavailability, catalytic activity, moisture retention and shelf life of dry powders (Anton Paar, *n.d.*). The uniqueness of spray-freeze-drying is that it results in particles of high surface area and porosity (Nguyen, Herberger, & Burke, 2004), due to its atomization step. This is because, additional surfaces are created when the feed solution is divided into smaller droplets during atomization, thereby increasing their surface area and eventually of the particles. Likewise, the surface area increases when pores are created after the sublimation phase of freeze-drying. The gas adsorption method is an established technique to measure the specific surface area and pore size distribution of materials. This method is based on the typical absorption properties of gas on solid surfaces (Fu, Lin, & Xu, 2017). It uses the well-known BET (Brunauer, Emmett and Teller) equation to calculate the specific surface area, the theory of which is explained subsequently.

10.3.2.1 Surface area

10.3.2.1.1 Theory and instrumentation of BET surface area determination

The BET concept of SSA calculation is based on the adsorption, i.e., adhesion of the atoms or molecules of an unreactive gas onto the solid particles under study. It is based on the assumptions that the gas molecules would physically adsorb on a solid in layers, infinitely and that the layers are non-interactive such that the BET theory is applicable to each layer (Raja & Barron, 2021).

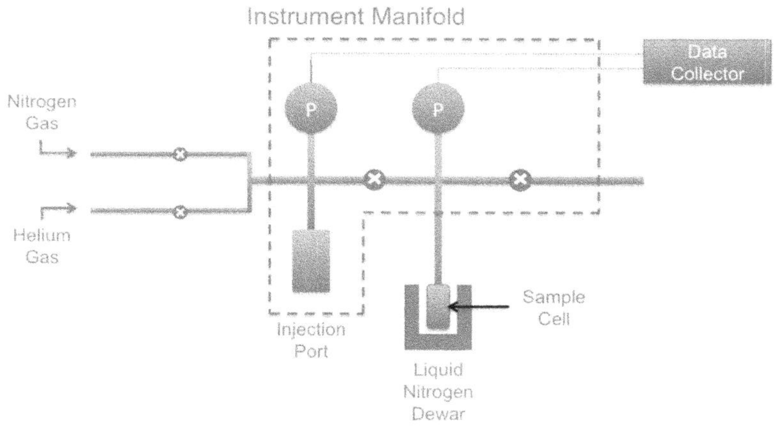

Figure 10.16 Schematic representation of the BET instrument (Raja & Barron, 2021).

A typical set-up of the BET instrument is shown in Figure 10.16. Before the gas adsorption analysis of spray-freeze-dried powders, a known quantity of the sample solid material (≤ 0.5 g) is placed in a glass cell and degassed for 24 hours or a minimum of 16 hours under helium at a sufficiently high temperature to remove unwanted vapors and gases from its surface. This step is critical as incomplete degassing of the sample can lead to underestimation of the reported SSA value. The selected degassing temperature must be the highest possible one that promotes water vapor removal within a shorter time, but without damaging the sample's structure. Nevertheless, for SFD powders, it has been observed that the degassing step is not adequate for the complete removal of adsorbed moisture. This is because of the low glass transition temperature of SFD powders, due to which higher degassing temperatures cannot be applied without damaging the solid sample. Hence, an additional preliminary step of drying the powders for 24 h under vacuum is practiced (Brunaugh et al., 2019). Then, the sample is shifted to the analysis port and cooled by a cryogenic liquid to enhance its interaction with the gas. Subsequently, a known amount of gas (adsorbate) is released into the sample cell with the aid of a calibrated piston. Nitrogen is the commonly used adsorbate owing to its high purity and strong interaction with most solids. Prior to the injection of nitrogen gas and also after each measurement, a gas such as helium that does not adsorb onto the sample is injected for a blank run and to calibrate the dead volume in the sample cell (Raja & Barron, 2021).

Under isothermal conditions of the solid sample, the pressure or concentration of the adsorbing gas is gradually increased. Concurrently, a partial vacuum is applied to attain relative pressures lower than the atmospheric pressure. With intensification of relative pressure, increasing number of molecules adsorb onto the surface to form a thin single or mono-layer that covers the entire surface (represented by dark red circles in Figure 10.17). Once the saturation pressure is reached, the gas adsorption ceases irrespective of further increase in pressure. Sophisticated transducers with high precision and accuracy are used to monitor the change in pressure during the gas adsorption process. Post the formation of adsorption layers, the solid sample is removed from the nitrogen atmosphere and then heated to release the adsorbed nitrogen to be quantified (desorption of gas). The adsorption–desorption cycles are repeated until the difference between consecutive surface area measurements is less than 5% (Brunaugh et al., 2019). Then, the surface area isotherm is plotted between the volume adsorbed/desorbed onto/from the sample (y-axis) and the relative pressure of gas (x-axis).

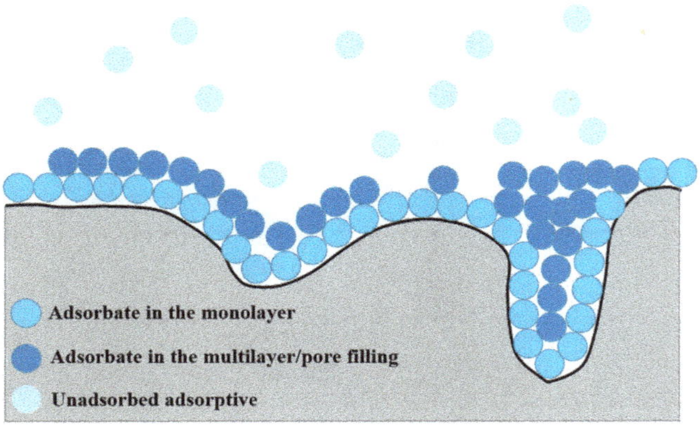

Figure 10.17 Schematic representation of monolayer formation (Adapted and redrawn from Anton Paar, *n.d.*).

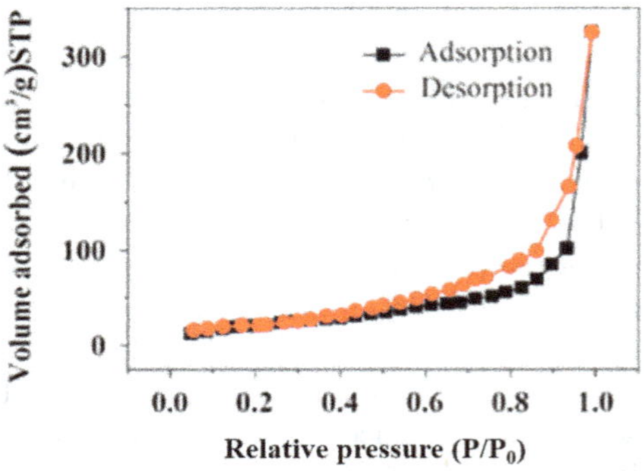

Figure 10.18 A typical surface area isotherm (Li et al., 2015).

A typical surface isotherm is depicted in Figure 10.18. Subsequently, the area of accessible surface is calculated from the isotherm as the cross-sectional area of the adsorbate is known. Nevertheless, the relationship between gas adsorption and relative pressure of gas is not linear (Figure 10.18), which necessitates the use of an appropriate mathematical model to calculate the surface area. The BET equation was developed for the above purpose. Since its advent in 1938, the BET equation (Brunauer et al., 1938) (Eq. 10.6) is widely used to calculate the number of molecules or atoms of a gas required to form a monolayer (X_m) of adsorbed gas onto a solid surface (Thommes et al., 2015). It provides an estimate of the number of gas molecules adsorbed (X) at a given relative pressure (P/P_0).

$$\frac{1}{X\left[(P_0/P)-1\right]}=\frac{1}{X_m C}+\frac{C-1}{X_m C}\left(\frac{P}{P_0}\right) \tag{10.6}$$

In the above equation, the parameter C is a constant related to the heat of adsorption and X_m is the monolayer capacity, which is the volume of gas adsorbed at standard temperature and pressure (STP) conditions (273 K and 1 atm), given by Eq. 10.7.

$$X_m = \frac{1}{slope + intercept} \tag{10.7}$$

The values of slope and y-intercept can be calculated using least squares regression. In most cases wherein nitrogen gas is used, the linear relationship is confined only to a limited region of the adsorption isotherm, usually in the P/P_0 range of 0.05 to 0.35 (Figure 10.18). With this understanding of the BET theory, the total surface area (S) is calculated using the below equation.

$$S = \frac{X_m\, L_{av}\, A_m}{M_v} \tag{10.8}$$

Where A_m is the cross-sectional area of the adsorbate (Lowell et al., 2004), which is a known quantity that equals 0.162 nm^2 for an absorbed nitrogen molecule, L_{av} is the Avogadro's number and M_v is the molar volume that is equal to 22414 mL. The value of total surface area thus obtained is then divided by the sample weight to obtain the specific surface area (Brunaugh et al., 2019).

Ideally, for an accurate determination of surface area using the BET equation, data points in the P/P_0 range of 0.025 to 0.30 must be considered. Otherwise, a minimum of three data points is mandatory. This is referred to as the multi-point BET method. The rationale of choosing the above range of relative pressure is that the onset of capillary condensation occurs at $P/P_0 > 0.5$ and at very low relative pressures, the gas adsorption process stops with monolayer formation. Also, the BET plot must be linear with a positive slope to deem that the BET method is appropriate to obtain the surface area. Conversely, a single-point BET calculation can be adopted, wherein the intercept is set to zero and the value of C is neglected. In general, the data point at $P/P_0 = 0.3$ is the best match to a multipoint BET (Raja & Barron, 2021). The test conditions for BET analysis applied for SFD products and the resultant SSA are compiled in Table 10.4.

10.3.2.2 Porosity

Porosity (ε) is an important characteristic of spray-freeze-dried powders, which is controlled by various processing parameters such as spray velocity, diameter of nozzle orifice and the temperature and flow rate of cryogenic liquid (Barron et al., 2003). It is the ratio of total pore volume to the volume of particle (Thommes et al., 2015). The porous characteristics of particles, i.e., porosity, pore volume and pore size distribution can also be determined by the gas or liquid sorption method. The total pore volume (Eq. 10.9) is obtained by quantifying the vapor adsorbed at a relative temperature close to unity. The above calculation is based on the assumption that the pores are filled with the adsorbate. Once the pore volume is determined, the mean pore radius (r_p) is calculated from it by assuming that the pores possess a cylindrical geometry (Eq. 10.10) (BET-TPX-Chemi-reading-material, *n.d.*).

$$V_{liq} = \frac{P_a\, V_{ads}\, V_m}{RT} \tag{10.9}$$

(Where, V_{ads} = volume of gas adsorbed; V_{liq} = volume of liquid Nitrogen in the pores; V_m = molar vol. of liquid adsorbate; P_a = ambient pressure; T = ambient temperature; R = universal gas constant)

$$r_p = \frac{2V_{liq}}{S} \tag{10.10}$$

(Where, S is the surface area)

Table 10.4 Conditions for the BET analysis of spray-freeze-dried powders and the resultant values of specific surface area – An overview

S.No.	Conditions employed	Type of SFD product	SSA (m^2/g)	Reference
1.	• Adsorbate: Nitrogen • Temperature: 77.4 K	• Danazol/danazol/ polyvinylpyrrolidone (PVP) K-15 powder • Danazol/PVP K-15/Sodium Lauryl Sulfate powder • Danazol/PVP K-5/ polaxamer powder	• 79.9 • 48.0 • 30.0	Hu, Johnston, & Williams (2004)
2.	• Degassed under helium at 40°C for 24 h prior to gas adsorption analysis • Drying of powders for 24 h under vacuum • Single-point method	Respirable protein powders containing lysozyme with different combinations of excipients such as sucrose, D-mannitol, histidine and polysorbate 80	4.9 - 13.0	Brunaugh et al. (2019)
3.	• Adsorbate: Nitrogen • Temperature: 77 K • Multi-point method	Polymeric nanoparticles of: • Poly DL-lactide-co-glycolide (PLGA) • Ethyl cellulose (EC) • Eudragit RL (EDRL)	• 77.1 ± 1.9 • 77.6 ± 0.9 • 67.8 ± 1.9	Ali & Lamprecht (2014)
4.	• Adsorbate: Argon	• Phenytoin (Phe), a poorly water-soluble drug • Phe + Eudracit-L (1:5) • Phe + hydroxypropyl methyl = 1:5	• 14.5 • 38.4 • 50.9	Niwa et al. (2009)
5.	Adsorbate: Nitrogen	Recombinant human epithelial growth factor liposomal dry powders (ultrasonic spray-freeze-drying) with Sucrose to phospholipid ratio: • 1:1 • 2:1 • 3:1 • 4:1	• 72.5 ± 4.87 • 108.7 ± 7.34 • 54.2 ± 4.26 • 43.5 ± 3.22	Yin et al. (2014)
6.	• Adsorbate: Nitrogen • Temperature: • 77 K • Multi-point (5-point) method	Δ^9-tetrahydrocannabinol containing inulin-based solid dispersion powder suitable for inhalation	70–110	van Drooge et al. (2005)
7.	• Adsorbate: Nitrogen • Temperature: 77 K • Multi-point method	Influenza vaccine powder for pulmonary delivery with following protectants • Inulin • Dextran • Mixture of dextran and trehalose	• 100 • 120 • 75	Murugappan et al. (2013)
8.	• Adsorbate: Nitrogen • Temperature: 77 K • Multi-point method	Whole inactivated virus influenza vaccine with inulin as stabilizer	> 60	Patil et al. (2015)
9.	• Purged with nitrogen overnight at 25 °C.	Influenza subunit vaccine powder for inhalation with inulin as stabilizer	76.33	Saluja et al. (2010)

Table 10.4 ***(Continued)*** **Conditions for the BET analysis of spray-freeze-dried powders and the resultant values of specific surface area – An overview**

S.No.	Conditions employed	Type of SFD product	SSA (m^2/g)	Reference
	• Adsorbate: Nitrogen • Temperature: 77 K			
10.	• Sample size: • 300 - 800 mg • Degassing under vacuum for 12 h at 30 °C • Adsorbate: Nitrogen • Cryogen: Liquid nitrogen	Pure bovine erythrocyte serum albumin and bovine erythrocyte carbonic anhydrase for needle-free ballistic injection	• 26.2 ± 0.3 for product obtained after primary drying at −15°C/ 240 min • 9.46 ± 0.2 for product obtained after primary drying at −10°C/ 2715 min • 8.30 ± 0.2 for product obtained after primary drying at −8°C/ 2715 min	Straller & Lee (2017)
11.	• Adsorbate: Argon • Temperature: −195.6°C	Aqueous feed solution containing tolbutamide bulk powder and hydroxypropylmethylcellulose (HPMC) dissolved in 1% aqueous ammonia, to form a spray solution containing TBM:HPMC at the ratio of 1: 5 an then converted to following concentrations: • 0.9% • 1.8% • 9.0% • 13.5%	• 18.07 • 22.13 • 28.32 • 27.47	Kondo et al. (2009)
12.	• Sample size: ~200 mg • Degassing for no less than 3 h before analysis	• Bulk danazol (Dan) • Dan/PVA/Polaxamer (Pol)/ PVP physical mixture • Slowly frozen Dan/PVA/ Pol/PVP • Lyophilized SFL Dan/PVA/ Pol/PVP • Atmospheric freeze-dried SFL Dan/PVA/Pol/PVP	• 0.52 • 1.92 • 3.14 • 8.90 • 5.72	Rogers et al. (2003)
13.	• Adsorbate: Nitrogen	BSA (w/v): • 0.1% • 1% • 10% Dextran (w/v): • 0.1% • 1% • 10% PVP (w/v): • 0.1% • 1% • 10%	• 21.97 • 19.41 • 9.52 • 20.4 • 19.78 • 9.34 • 20.87 • 20.64 • 5.69	Eggerstedt et al. (2012)

(Continued)

Table 10.4 *(Continued)* Conditions for the BET analysis of spray-freeze-dried powders and the resultant values of specific surface area – An overview

S.No.	Conditions employed	Type of SFD product	SSA (m^2/g)	Reference
14.	• Adsorbate: Nitrogen • Temperature: 77 K • Relative pressure: 0.05 to 0.2	Darbepoetin Alfa powders: • Obtained after ultrasonic atomization at 120 kHz • Obtained after ultrasonic atomization at 120 kHz	• 29 • 24	Nguyen et al. (2004)
15.	• Degassing at 40°C for 16 h • Single-point method	Microbial transglutaminase (ultrasonic spray-freeze-drying)	0.65	Isleroglu et al. (2018)

Alternatively, porosity (ε) of spray-freeze-dried powders has also been calculated using several other approaches. For instance, ε has been estimated based on its relationship with the different types of density. The relationship between ε, particle density (ρ_p) (calculated from Helium pycnometry) and tap density (ρ_t) (Jinapong et al., 2008) is given in Eq. (10.11).

$$\varepsilon = \frac{(\rho_p - \rho_t)}{\rho_t} \times 100 \qquad (10.11)$$

Accordingly, Schiffter et al. (2010) computed the porosity of SFD-insulin particles for needle-free ballistic delivery to be 99.7%. Using the same method, Isleroglu et al. (2018) found the porosity of USFD-microbial transglutaminase to be 88.7%. Barron et al. (2003) calculated the intragranule porosity (IP) of polyethylene glycol (PEG) polymeric particles of albuterol sulfate drug from the true density (ρ_t) and granule density (ρ_g) using the equation: IP = 1 - ρ_g/ρ_t. Eggerstedt et al. (2012) calculated the porosity of SFD-protein spheres (lysozyme) from the ratio of their bulk density to particle density, which ranged between 73.2% and 95.8%, depending on the type and concentration of stabilizers (BSA, dextran and PVP) used in the feed formulation.

Besides, Scanning Electron Microscopy (SEM) is a qualitative method adopted to visualize particle porosity, which is the most commonly adopted approach in the case of spray-freeze-dried powders. Yet another method is the mercury intrusion porosimetry (Liang et al., 2018), which utilizes the non-wetting properties of mercury to gather information on the porous characteristics of particles. Prior to the mercury porosimetry analysis, the powder sample is degassed or evacuated to remove moisture from its porous structure. Then, the sample is completely immersed in mercury under a pressure cycle. The rationale of applying pressure is to force the intrusion of mercury through the smaller pores in the sample, which happens readily through the larger pores at low pressure. Thus, the information obtained from mercury porosimetry is a measure of the intraparticulate pores that open to the outside of the particle, such that they can be penetrated by the mercury (Sonner, Maa, & Lee, 2002).

By using a broad range of pressure from vacuum up to a maximum of 400 MPa, it is possible to measure a wide range of pore size from 4 nm to 800 μm using mercury porosimetry (solids-solutions, n.d.). For instance, Straller & Lee (2017) employed mercury porosimetry to determine the porosity of spray-freeze-fried microparticles of pure protein (BSA and BCA). 50 mg of sample was placed in the sample holder of the apparatus and subjected to a maximum intrusion pressure of mercury at 2000 bar (200 MPa). Under different temperature conditions of the primary drying

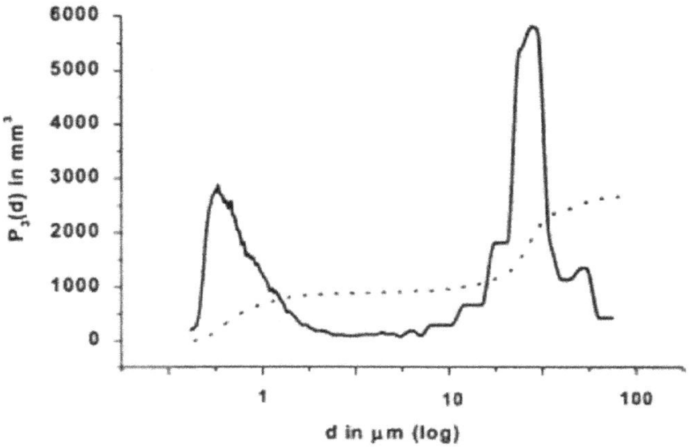

Figure 10.19 A model plot of mercury porosimetry (Sonner, Maa, & Lee, 2002).

stage of freeze-drying, they could detect pore diameter in the range of 840 µm to 41300 µm (41.3 mm) and corresponding total pore volume between 6.4 and 1.2 cm^3/g.

In the above manner, mercury porosimetry provides information on both intra-particle and inter-particle porosity, which in turn can be used to calculate different types of density such as bulk density, apparent density and true density and specific surface area. The result obtained from mercury porosimetry is a graphical representation comprising the mercury intrusion curve and extrusion curve and the corresponding data on pore size distribution, pore volume, pore size, density and porosity (solids-solutions, *n.d.*). A sample plot obtained after the mercury porosimetry analysis of SFD-trehalose (Sonner et al., 2002) is shown in Figure 10.19.

10.3.3 Drug amorphism and crystallinity

Depending on the production method, the molecular arrangement of food and pharmaceutical powders is either amorphous, crystalline or a combination of both (Figure 10.20). While a random alignment of molecules lead to amorphous structure, molecules packed in a specific order result in crystalline structure (Einfalt, Planinšek & Hrovat, 2013). Amorphous nature of dry powders is advantageous in terms of obtaining enhanced dissolution. But, crystalline forms have better chemical stability (Smyth & Hickey, 2005). Also, owing to their thermodynamically non-equilibrium state, amorphous powders are susceptible to physically undesirable changes during handling, storage and processing. Often, acceptable properties of food powders are obtained by structural transition from crystalline to amorphous and vice versa (Bhandari, 2013). Therefore, amorphism and crystallinity of dry powders is always a trade-off and it is important to quantify the amorphous/

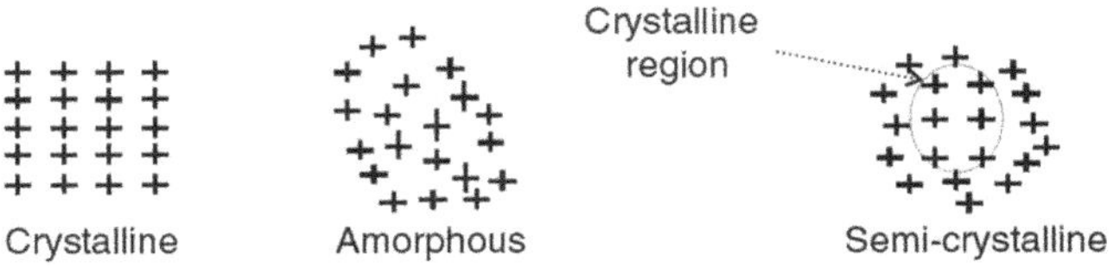

Figure 10.20 Schematic representation of the molecular arrangement in crystalline, amorphous and mixed structural powders (Bhandari & Roos, 2012).

crystalline fractions in food powders. This is relevant to control the powder quality during production and storage (Ho, Truong, & Bhandari, 2017). Several methods are available to quantitatively and qualitatively determine the molecular arrangement of powders, listed as follows (Einfalt et al., 2013).

- X-ray diffraction (XRD);
- Microscopy: SEM and polarized light microscopy;
- Thermal analysis: Differential scanning calorimetry (DSC) and Thermogravimetry (TGA);
- Spectroscopy: Fourier transform infrared (FTIR) spectroscopy, Nuclear magnetic resonance (NMR) spectroscopy and Raman spectroscopy;
- Dynamic mechanical analysis; and,
- Dissolution analysis.

Amongst the above, XRD and DSC are the commonly employed techniques to determine the degree of crystallinity or amorphism of spray-freeze-dried powders and hence would be substantiated in this section with pertinent case-studies.

10.3.3.1 XRD as a tool to determine the crystallinity of SFD powders

XRD is a non-invasive method to determine the relative proportions of amorphous and crystalline forms in food/drug powders. The construction of an X-ray diffractometer includes X-ray source, a monochromator, a detector and a sample cell. The working principle of XRD is that when X-rays strikes a powdered sample, they are diffracted from the series of repeating atomic planes present in the crystal lattice, which are separated from each other by a d distance (d-spacing) (Figure 10.21). The X-ray diffraction from powdered materials is governed by the Bragg's law, given by the following equation.

$$n\lambda = 2d \sin \theta \tag{10.12}$$

Where n is an integer number that gives the order of reflection; λ is the incident wavelength of X-ray beam, d is the distance between atomic planes and θ is the reflection angle of the incident X-ray. A photographic film or an electronic counter is used to detect the constructive interference of the diffracted X-ray beam (Pecharsky & Zavalij, 2009).

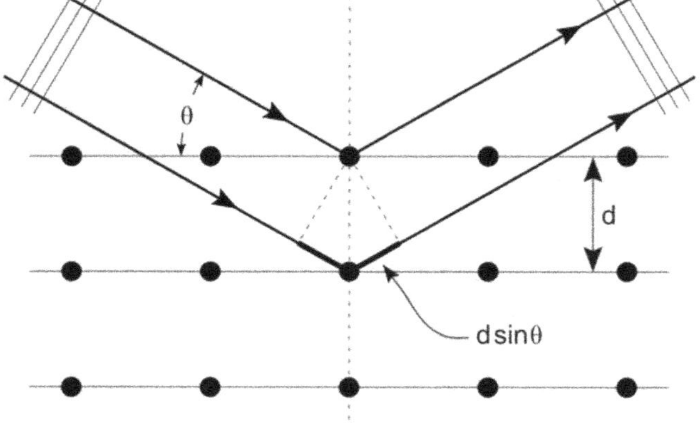

Figure 10.21 Schematic representation of Bragg's law (https://en.wikipedia.org/wiki/X-ray_crystallography#/media/File:Bragg_diffraction_2.svg; Accessed 2 July 2021).

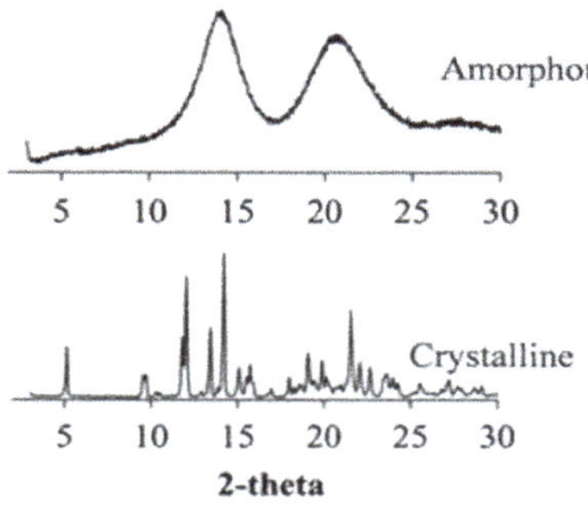

Figure 10.22 Model X-ray patterns of amorphous and crystalline powders (Ho et al., 2015).

In an X-ray diffractometer, the powder samples are scanned over a specific 2θ (°) range at a programmed rate (°/min). The difference in molecular arrangement between amorphous and crystalline powders is detected based on the variations in X-ray diffraction patterns. X-ray light that is incident on the lattice planes of crystalline materials scatters only in specific directions and results in narrow peaks of high intensity. Whereas, in the case of amorphous powders, the random orientation of atoms causes the incident X-ray light to be scattered in random directions to result in broad peaks (Figure 10.22). Quantification of powder crystallinity is possible by determining the height or intensity of the characteristic peak, provided, the curves are symmetrical (Saffari & Langrish, 2014).

Lavanya et al. (2020) scanned the SFD-bromelain aerosol powders prepared with different core-to-wall ratios and its constituent core (bromelain), wall (maltodextrin) and absorption enhancer (chitosan) in the 2θ range of 10°-80° at a scanning rate of 2°/min. XRD patterns (Figure 10.23[a]) of all the above formulations showed intense and sharp crystalline peaks. Besides

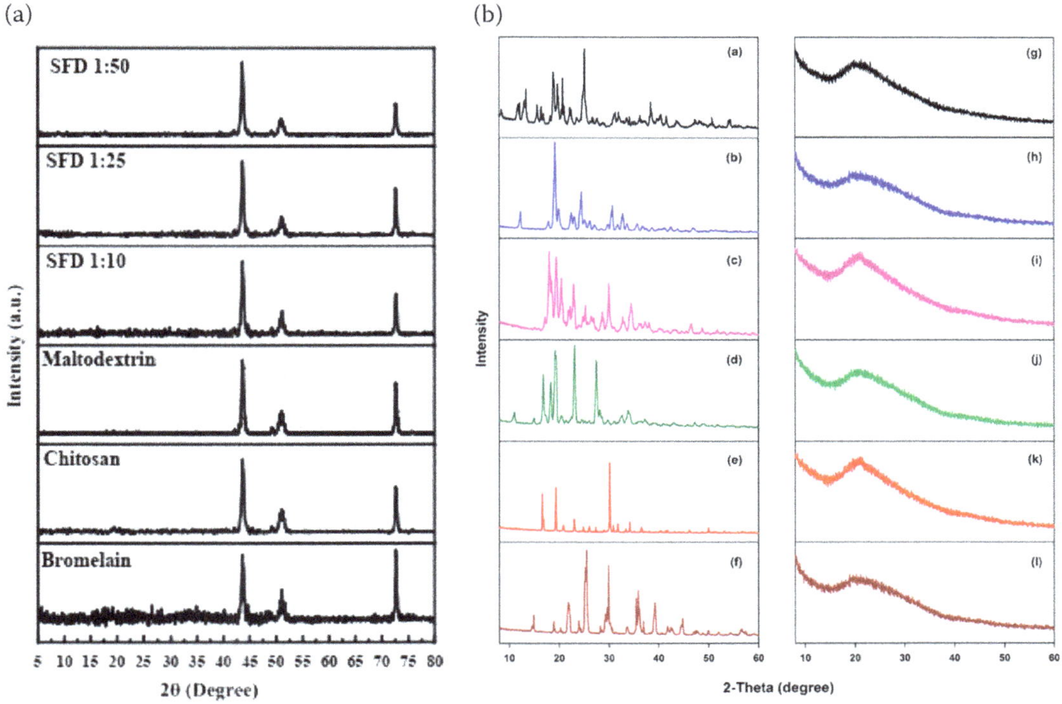

Figure 10.23 X-ray diffractograms of (a) SFD-bromelain aerosol formulations and its constituent components (Lavanya et al., 2020); (b) pure excipients (a: trehalose, b: Leu, c: Phe, d: Arg, e: Cys, f: Gly) and different spray freeze-dried powder formulations of IgG (g–l) (Emami et al., 2018).

several minor peaks, bromelain exhibited sharp peaks at [43.94°, 51.32° and 72.51°]. Maltodextrin and chitosan showed sharp peaks at [42.04°, 43.65°, 49.15°, 50.82°, 72.60°] and [43.87°, 50.73° and 72.66°], respectively. The SFD-bromelain aerosol powders retained the sharp, characteristic crystalline peaks of bromelain and maltodextrin in the 2θ range of 43°-72°, which signified retention of the native chemical structure of bromelain. Thus, SFD process did not alter the crystallinity of bromelain significantly, which was useful in retaining the chemical stability of the dry powder inhaler (DPI). However, the intensity of crystalline peaks was reduced in the SFD powders (Lavanya et al., 2020).

On the contrary, Emami et al. (2018) showed the SFD process-induced crystalline rearrangement of amino acids and sugar-based stabilizers. Here, the SFD-IgG powder and the excipients used to enhance the protein stability (trehalose, leucine, phenylalanine, arginine, cysteine or glycine) were scanned from 10° to 60° at a scanning rate of 5°/min. Diffractograms of pure excipients revealed sharp peaks, indicating the crystalline nature of those amino acids and sugar. But, the SFD-processed formulations exhibited broad peaks, thus confirming the amorphous nature of the microparticles (Figure 10.23[b]). Collectively, the X-ray diffractograms of SFD-IgG formulations were different from those of the pure excipients, implying potential interactions between IgG and the sugar/amino-acids. Excipients such as sugars, amino acids and biopolymers aided in retaining the amorphous state and thereby the stability of proteins such as IgG (Depreter et al., 2013; Kanojia et al., 2016; Tonnis et al., 2014). This is relevant as protein instability during the atomization and spray-freezing steps of SFD process is a major apprehension.

10.3.3.2 DSC as a tool to obtain structural information of SFD powders

DSC is an established thermal analysis technique to examine the structural and phase transitions, based on the differences in evaporation behavior of water/volatile components between the amorphous and crystalline fractions of powders. Further, it can also be used to quantify the amorphous and crystalline contents of powders. The characteristic DSC thermograms for amorphous and crystalline powders scanned with open and closed pans (i.e., with and without moisture loss, respectively) are shown in Figure 10.24. While the open-pan analysis is limited to qualitative analysis of the powder structure, the closed-pan analysis is capable of quantifying the percentage crystallinity in powders based on the enthalpies related to phase transitions. In the open pan analysis, amorphous powders with porous and open structure allows easy diffusion of water/volatile molecules. Consequently, their DSC thermograms exhibit merely a big endothermic hump over a wide temperature range (Figure 10.24). Contrarily, in crystalline powders, the DSC scan shows one or several sharp endothermic peaks, attributed to their highly packed structure that restrains the water/volatile molecules from escaping. In the closed pan condition, amorphous powders display an endothermic peak corresponding to the phase transition from glassy to rubbery state, followed by an exothermic peak that signifies crystallization (Figure 10.24) (Gill, Moghadam, & Ranjbar, 2010). Several investigations on spray-freeze-dried drugs/APIs have employed DSC as a tool to detect and quantify the amorphous or crystalline state of the final powders. The test conditions used in different studies are tabulated in Table 10.5. In a DSC thermogram, while an endothermic peak corresponds to melting and indicates crystallinity of powders, an exothermic peak represents amorphism.

Ye at al. (2017) studied the DSC profiles of ultrasonic SFD-clarithromycin (CLA) liposomal dry powders prepared with one or a combination of lyoprotectants (sucrose, mannose) in comparison with raw-CLA, blank liposome, raw-sucrose, USFD-sucrose, raw-mannose and USFD-mannose (Figure 10.25[a]). Notably, the thermogram of USFD-blank liposome did not show any thermal peak. But, raw CLA showed an endothermic peak at 228.12°C, which was absent in the DSC thermograms of USFD-clarithromycin (CLA) liposomal dry powder formulations. Both the unprocessed and USFD-mannose showed an endothermic peak at 164–165 °C related to melting,

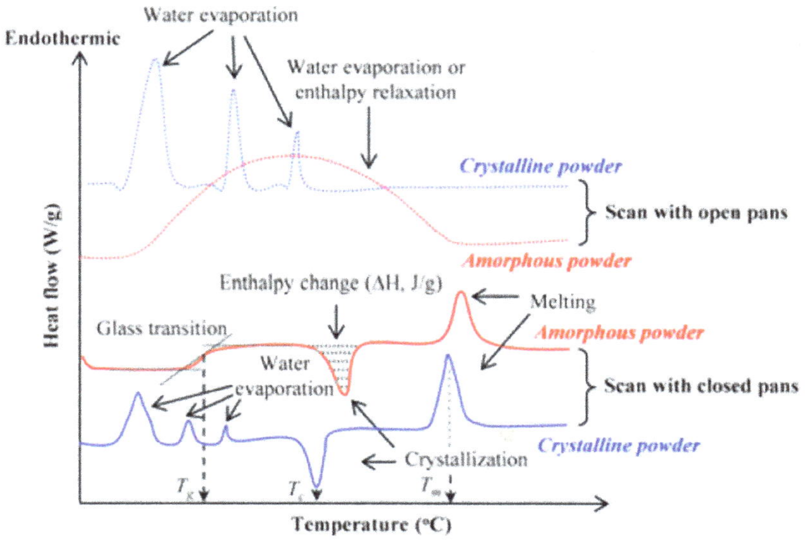

Figure 10.24 Characteristic DSC thermograms for amorphous and crystalline powders scanned with open and closed pans (Ho et al., 2017).

thus indicating the retention of its crystalline structure during USFD. Differently, USFD-sucrose had an additional exothermic peak at ~90 °C, followed by an endothermic melting peak, which hinted that the sucrose was obtained in the amorphous form after the USFD process. Thus, based on the DSC results, the authors ascertained mannose as the suitable lyoprotectant due to its ability to retain its crystalline structure during the SFD process and remain non-hygroscopic with the required moisture protection ability. The DSC thermograms of USFD-CLA liposomes were significantly distinct from those of its constituents (Figure 10.25[b]), which was indicative of possible interactions between the lipids and lyoprotectant. In contrast to the above observations, the DSC thermograms of SFD-IgG prepared with different amino acid and sugar-based excipients did not show any prominent endotherm or exothermic peak. This confirmed the amorphous nature of the microparticles. However, the pure excipients showed sharp endothermic peaks indicative of crystallinity (Figure 10.25[b]). Thus, in this case, SFD process caused the crystalline rearrangement of trehalose and amino acids (Emami et al., 2018).

10.4 CHARACTERIZATION OF FOOD POWDERS AND ENCAPSULATED FOOD BIOACTIVES

10.4.1 Moisture content

Moisture content plays an important role in the handling of any powdered product. Specifically, in the case of instant dehydrated products such as coffee, it is a critical aspect that is associated with the cohesiveness and formation of inter-particle liquid bridges. In such products, excessive moisture content can lead to severe caking and reduced flowability (Barbosa-Canovas et al., 2005). The final moisture content of spray-freeze-dried powder is mainly governed by its freeze-drying step. Lower moisture content of spray-freeze-dried products can be achieved by warming the drying medium or shelf during the secondary drying phase of conventional (vacuum) freeze-drying (Khwanpruk et al., 2008). Reasonable prolonging of the drying time can also lead to reduced moisture content of SFD powders (Ly et al., 2019). In the case of fluidized-bed freeze-drying, the

Table 10.5 Conditions of DSC analysis to determine the amorphous and crystalline contents of spray-freeze-dried powders

Product	DSC test conditions	Reference
Theophylline oxalic acid cocrystal (for inhalation)	• Weight of sample: 5 mg • Heating range: 25°C to 300°C • Heating rate: 5°C/min • Gas: Nitrogen • Gas purge rate: 30 mL/min	Tanaka et al. (2020)
Respirable protein (lysozyme) powders	• Weight of sample: 3 mg • Cooling range: Room temperature to −40°C • Cooling rate: 10°C/min • Heating range: −40 to 200°C • Heating rate: 2°C/min	Brunaugh et al. (2019)
Immunoglobulin G (IgG)	• Weight of sample: 10 mg • Sample holder: Aluminum pan • Heating range: −50°C to 300°C • Heating rate: 10°C/min	Daneshmand et al. (2018)
IgG in sugar-based matrices	• Weight of sample: 5–10 mg • Sample holder: Aluminum pan • Heating range: −20°C to 300°C • Heating rate: 10°C/min	Emami et al. (2018)
Inhalable clarithromycin liposomal dry powders	• Sample holder: Aluminum pan (hermetically sealed) • Heating range: 30°C to 200°C • Heating rate: 10°C/min • Gas: Nitrogen • Gas purge rate: 250 cm^3/min	Ye et al. (2017)
Inhalation phage powders	• Weight of sample: 5 ± 1 mg • Sample holder: Aluminium crucible crimped to a perforated lid • Heating range: 30°C to 300°C • Heating rate: 10°C/min • Gas: Nitrogen • Gas purge rate: 250 cm^3/min	Leung et al. (2016)
Salmeterol xinafoate microparticles	• Weight of sample: 5–10 mg • Sample holder: Aluminum pan Heating range: 25°C to 300°C • Heating rate: 10°C/min	Rahmati et al. (2013)
Salbutamol sulphate (for inhalation)	• Weight of sample: 4–6 mg • Sample holder: Aluminium pan (fitted with non-hermetic aluminium lid) • Heating range: Room temperature to 200°C • Heating rate: 2°C/min • Gas: Nitrogen • Gas purge rate: 50 mL/min	Mueannoom et al. (2012)

inlet gas temperature is known to influence the residual moisture content of SFD products. Faster drying rates were observed at higher gas temperatures to produce a powdered product of relatively low moisture content (Table 10.6). This is justified by the higher transport coefficients and pure ice vapor pressure at higher temperatures (Anandharamakrishnan, Rielly & Stapley, 2010). Studies pertaining to spray-freeze-drying have mainly employed the gravimetric method to determine the moisture content of final powder, based on the mass loss on drying. Hot air oven, vacuum oven, halogen moisture analyzer and thermogravimetric analyzer have also been used for estimating the moisture content of SFD powders. While the moisture content of SFD-food powders range between ~2% and 15%, that of pharmaceutical formulations fall in the range of 1.1–7.5% (Table 10.6).

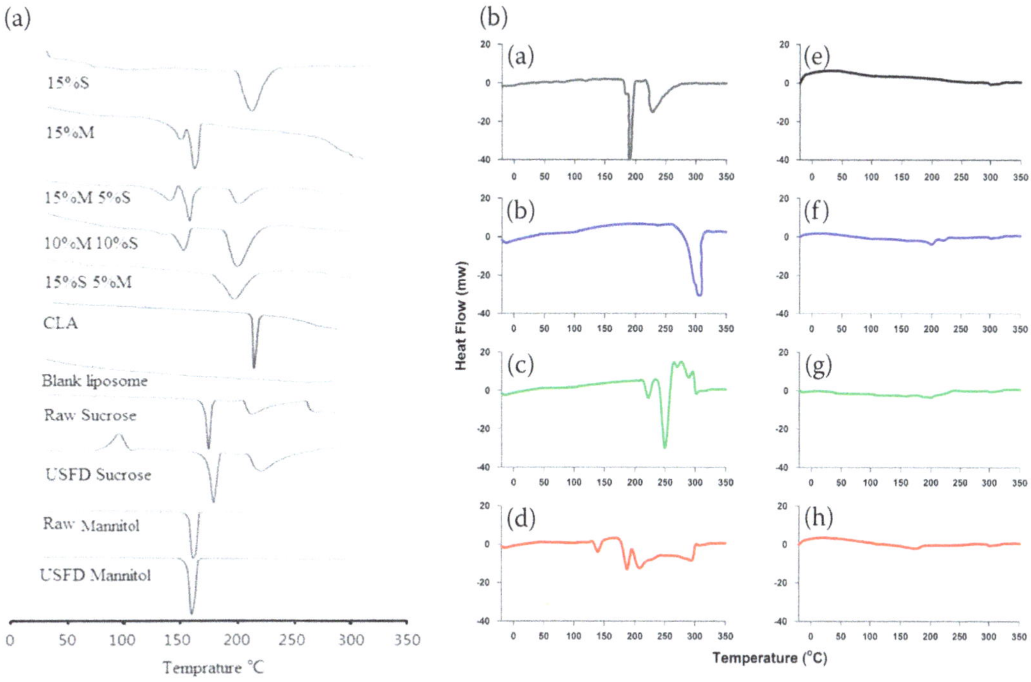

Figure 10.25 DSC thermograms of (a) clarithromycin (CLA) liposomal dry powders obtained by ultrasonic SFD with single and combination lyoprotectants, in comparison with CLA blank liposome, raw-sucrose, USFD-sucrose, raw-mannose and USFD-mannose (Ye et al., 2017); (b) pure excipients (a: trehalose, b: Leucine, c: Phenylalanine, d: Arginine e: Cysteine, f: Glycine) and different spray freeze-dried powder formulations of IgG (g–l) (Emami et al., 2018).

10.4.2 Flowability

Flowability or powder flow is defined as the relative movement of a bulk of particles amongst the adjacent particles or along the wall surface of a container (Peleg, 1977). It refers to the ability of a powder to flow in a preferred mode within the specific part of an equipment (Prescott & Barnum, 2000). Flowability is a vital parameter of product quality that promotes better manufacturing efficiency in industries that handle powders. An optimal powder flow circumvents capacity shortfall, stoppage of production line and thereby the production-related downtime. For instance, food industry may incur loss in production capacity and product damage due to irregular powder flow out of vertical bins or silos used for storage (Barbosa-Cánovas et al., 2005). Likewise, in pharmaceutical industry, inappropriate powder flow behavior can affect manufacturing efficiency and product quality with respect to its weight and content homogeneity (Barbosa-Canovas et al., 2005; Lumay et al., 2019). The above factors necessitate the determination of powder flowability, which is essentially a multidimensional property. Hence, it cannot be expressed as a single value or index estimated from a single and simple test method. Accordingly, in most of the studies wherein spray-freeze-drying has been employed for the production of food or pharmaceutical powders, flowability has been characterized as a function of bulk and tapped densities by two indices namely, Carr's index and Hausner ratio.

Carr compressibility index (CI) (Eq. 10.13) and Hausner ratio (HR) (Eq. 10.14) signify the tendency of powder to be compressed and describe interparticle interactions and settling behavior of powders. The interparticle interactions are usually insignificant for free-flowing powders, due to their relatively close values of bulk and tapped densities. On the other hand, poor-flowing powders

Table 10.6 Moisture content of spray-freeze-dried powders – An overview

Spray-freeze-dried product	Method/conditions of moisture content determination	Moisture content (%, wet basis)	Reference
Soluble coffee (Vacuum freeze-dried)	Hot air oven method • Weight of sample: 1 g • Temperature: 95° ± 2°C • Time: 2 hours	8.665 ± 0.001	IS 2791 (1992) Ishwarya & Anandharamakrishnan (2015)
Whey protein (Fluidized bed freeze-drying)	Vacuum oven method • Weight of sample: 0.5 g • Temperature: 105°C • Time: 12 hours	• 8.1 ± 2.27 (at −10°C) • 9.5 ± 0.43 (at −15°C) • 14.0 ± 0.61 (at −30°C)	Anandharamakrishnan, Rielly & Stapley (2010)
Vitamin E microcapsules with whey protein isolate as wall material	Halogen moisture analyzer • Weight of sample: 0.5 g • Sample holder: Aluminum pan • Temperature: 105°C	5.41 ± 0.24	Parthasarathi & Anandharamakrishnan (2016)
• Coffee • Maltodextrin	Hot air oven method • Weight of sample: 0.5 g • Sample holder: Aluminum pan • Temperature: 105°C • Time: 2 hours	• 15% (drying time: 1.75 hours) • 11.5% (drying time: 2 hours)	Khwanpruk et al. (2008)
Baccharis dracunculifolia leaf power (BdDE) with D-mannitol (DMan) and modified corn starch (MCS) as drying adjuvants	Halogen moisture analyzer • Weight of sample: 0.5 g	• 2.14 ± 0.01 (BdDE + DMan) • 2.41± 0.07 (BdDE + MCS) • 2.18 ± 0.02 (BdDE + DMan + MCS)	Teixeira et al. (2017)
Fish oil microcapsules	Hot air oven method • Temperature: 105°C • Time: 7–8 hours	3.38 ± 0.09	Pang et al. (2017)
Docosahexaenoic acid (DHA+WPI) microcapsules	Hot air oven method • Weight of sample: 0.5 g • Temperature: 105°C • Time: 12 hours	3.667 ± 0.014	Karthik & Anandharamakrishnan (2013)
Vanillin microcapsules with WPI and β-cyclodextrin (β-cyd) as wall materials	Hot air oven method • Weight of sample: 0.5 g • Sample holder: Aluminum pan • Temperature: 105°C • Time: 12 hours	• 6.63 ± 0.02 (Vanillin + WPI) • 4.15 ± 0.13 (Vanillin + β-cyd)	Hundre, Karthik & Anandharamakrishnan (2015)
Probiotic microcapsules (*Lactobacillus plantarum*)	Hot air oven method • Weight of sample: 0.5 g • Temperature: 105°C • Time: 12 hours	• 5.57 ± 0.11 (with whey protein isolate [WPI] and sodium alginate [SA] as wall materials) • 5.72 ± 0.12 (WPI and fructooligosaccharide [FOS] as wall materials) • 5.95 ± 0.07 (denatured WPI [DWPI] + SA as wall materials)	Rajam & Anandharamakrishnan (2015)

Table 10.6 *(Continued)* Moisture content of spray-freeze-dried powders – An overview

Spray-freeze-dried product	Method/conditions of moisture content determination	Moisture content (%, wet basis)	Reference
		• 6.32 ± 0.16 (DWPI + FOS as wall materials)	
Dry inhalable power of salbutamol sulphate (SS)	Thermogravimetric method: • Weight of sample: 4 – 6 mg • Sample holder: Ceramic crucible • Heating range: From 40°C to 200°C • Heating rate: 10°C/min • Gas: Nitrogen • Gas purge rate: 20 mL/min	• 1.1026 ± 0.15 (5% SS) • 1.8743 ± 0.12 (10% SS) • 1.1046 ± 0.26 (15% SS)	Mueannoom et al. (2012)
Mannitol and sucrose dry powders for pulmonary delivery	Thermogravimetric method: • Weight of sample: 4.5 – 12 mg • Heating range: From 40°C to 400°C • Gas: Nitrogen	• Mannitol: 5.5% • Sucrose: 7.5%	Babenko et al. (2019)
Porous powderformulation of naked smallinterfering RNA (siRNA) for inhalation	Thermogravimetric method: • Weight of sample: 1 – 2 mg • Heating range: From 25°C to 160°C • Heating rate: 10°C/min • Gas: Nitrogen • Gas purge rate: 25 mL/min • Loss in mass at 105 °C was calculated, which corresponds to the residual moisture evaporated from the sample during heating	• 0.88 – 1.32%	Liang et al. (2018)
Phage D29 powder (with trehalose and mannitol as excipients) [Atmospheric SFD]	Vacuum oven method • Weight of sample: 100 mg • Temperature: 100°C • Pressure: 3 kPa • Time: 8 hours	• 4.9 ± 0.1% (drying time: 6 hours) • 4.6 ± 0.1% (drying time: 7 hours)	Ly et al. (2019)

show typically high degree of interparticle interactions and thereby a larger difference between bulk and tapped. Carr index is given by the following equation:

$$CI = \left(\frac{\rho_T - \rho_B}{\rho_T} \right) \times 100 \tag{10.13}$$

$$HR = \frac{\rho_T}{\rho_B} \tag{10.14}$$

Powder flowability can be ascertained from the values of Carr index and Hausner ratio based on the guidelines that were already discussed in Chapter 6 (see Table 6.1).

Spray-freeze-dried food and pharmaceutical powders have been found to exhibit good to excellent flow properties. For example, the values of CI and HR for spray-freeze-dried soluble coffee powder prepared from a 40% (w/w) feed solution were 10 and 1.11, respectively. A study by Wang et al. (2012) categorized the particles as free-flowing and having poor-flowability, with Carr index value of $\leq$ 25 and $\geq$ 40, respectively. Accordingly, spray-freeze-dried dry powder inhaler containing drug-loaded (levofloxacin) polymer-lipid (poly(lactic-co-glycolic acid) + lecithin) hybrid nanoparticles (60%, w/w) with leucine and mannitol (1:1.7) as excipients exhibited free-flowing behavior with CI < 25. The above range was found to be suitable for pulmonary delivery. The flow behavior indices of other SFD products are tabulated in Table 10.7.

10.4.3 Encapsulation efficiency

Encapsulation efficiency (EE) is the fraction of active ingredient that is successfully encapsulated within the encapsulant, out of its initial concentration in the feed solution (Eq. 10.15) (Anandharamakrishnan & Ishwarya, 2015).

$$\textit{Encapsulation efficiency} = \frac{\textit{Weight of encapsulated bioactive}}{\textit{Initial weight of bioactive in the feed}} \times 100 \qquad (10.15)$$

While the parameter in denominator is known, the weight of encapsulated core is determined by extracting the same from the microcapsules using a suitable solvent with affinity for the core, followed by its quantification using HPLC or UV-visible spectrophotometry. The above formula varies slightly in the case of lipophilic core substances such as specialty oils or fat-soluble vitamins, as given in Eq. (10.16). Both the total oil and surface or superficial oil contents are considered for the calculation of %EE.

$$\%EE = \frac{(\textit{Total oil content} - \textit{Surface oil content})}{\textit{Total oil content}} \times 100 \qquad (10.16)$$

The surface oil content is determined by briefly (in the order of seconds to minutes) washing/shaking the microcapsules with a lipophilic solvent such as hexane. Then, the solvent mixture is passed through filter paper to collect the unencapsulated core after removing the hexane by vacuum evaporation. On the other hand, the total oil content is determined by Soxhlet extraction of the same powders for a prolonged period in the order of hours (Hundre, Karthik, Anandharamakrishnan, 2015; Tan, Chan, & Heng, 2005). Conversely, the %EE can also be calculated as the ratio between the weights of encapsulated oil and oil added to the feed. The weight of encapsulated oil is calculated by obtaining the oil from the microcapsules after a series of steps involving extraction (with a combination of polar (ethanol, water) and non-polar (hexane) solvents), centrifugation (at high speed to separate the organic solvent layer) and solvent evaporation (using nitrogen gas) (Beaulieu et al., 2002; Karthik & Anandharamakrishnan, 2013). The typical encapsulation efficiencies of SFD-microcapsules are charted in Table 10.8.

In the case of spray-freeze-dried live cells such as probiotics, the encapsulation efficiency is calculated by determining the cell viability after completely releasing the cells from the microcapsules by dispersing in 0.85% saline (NaCl, pH 7.0). Hydrocolloid-based wall materials such as WPI are entirely solubilized (98%) at neutral pH. The rehydrated SFD powders are then kept in a shaker (100 rpm) at 37°C for 30 min to release the cells from the microcapsules (Sinha, Shukla, Lal, & Ranganathan, 1982). 1 mL of rehydrated samples are then pour plated on MRS agar plates

Table 10.7 Hausner ratio and Carr index of spray-freeze-dried powders

Product	Hausner ratio	Carr index (%)	Reference
β-carotene aerosols	• 1.21 ± 0.085 (for wall-to-core ratio of 1:10) • 1.17 ± 0.33 (for wall-to-core ratio of 1:25) • 1.19 ± 0.27 (for wall-to-core ratio of 1:50)	• 17.5 ± 2.84 • 15.1 ± 0.57 • 16.6 ± 2.8	Lavanya et al. (2020)
Probiotic (*Lactobacillus plantarum)* microcapsules	• 1.04 ± 0.01 (with whey protein isolate [WPI] and sodium alginate [SA] as wall materials) • 1.15 ± 0.05 (WPI and fructooligosaccharide [FOS] as wall materials) • 1.12 ± 0.02 (denatured WPI [DWPI] + SA as wall materials) • 1.09 ± 0.02 (DWPI + FOS as wall materials)	• 4.17 ± 1.15 • 12.72 ± 3.63 • 10.80 ± 1.73 • 8.32 ± 1.46	Rajam & Anandharamakrishnan (2015)
Vitamin E microcapsules (with WPI as wall material)	1.19	16	Parthasarathi & Anandharamakrishnan (2016)
Polymeric (PLGA, ethyl cellulose and Eudragit®) nanoparticles and liposomes (soybean lecithin)	< 1.11	≤ 10	Ali & Lamprecht (2014)
Transglutaminase (Ultrasonic spray-freeze-drying)	• 48 kHz: 1.22 – 1.58 • 120 kHz: 1.36 – 1.51	• 17.95 – 32.4 • 26.32 – 33.9	Isleroglu et al. (2018)
Baccharis dracunculifolia leaf power (BdDE) with D-mannitol (DMan) and modified corn starch (MCS) as drying adjuvants	• 1.36 ± 0.03 (BdDE + DMan) • 1.09 ± 0.01 (BdDE + MCS) • 1.09 ± 0.01 (BdDE + DMan + MCS)	• 26.3 ± 1.77 (BdDE + DMan) • 8.71 ± 0.54 (BdDE + DMan) • 8.01 ± 0.45 (BdDE + DMan)	Teixeira et al. (2017)

after 10-fold serial dilutions in 0.85% saline solutions. Subsequently, the colony forming units (CFU) are enumerated after incubation at 37°C for 24 h and the %EE is calculated using the below formula:

$$\%EE = \frac{N}{N_0} \times 100 \tag{10.17}$$

Where N_0 and N are the log cell number (CFU) of viable cells in the microcapsules before and after drying, respectively.

10.4.4 Chemical interaction between constituents

Any encapsulation process is deemed appropriate provided there is no chemical interaction between the core and wall. Fourier Transform Infrared Spectroscopy (FTIR) is the commonly

Table 10.8 Encapsulation efficiency of spray-freeze-dried microcapsules

Core	Wall	Encapsulation efficiency (%)	Reference
Docosahexaenoic acid (DHA)	Tween-40	71	Karthik & Anandharamakrishnan (2013)
Vanillin	Whey protein isolate (WPI)/ β-cyclodextrin	70	Hundre, Karthik & Anandharamakrishnan (2015)
Vitamin E	WPI	89.3	Parthasarathi & Anandharamakrishnan (2016)
Fish oil	Acacia gum and sodium alginate	90.8	Pang et al. (2017)
Bromelain	Hydroxypropyl β-cyclodextrin (HPβCD)	72.3 to 85.4 (at different core-to-wall ratios: 1:10, 1:25 and 1:50)	Lavanya et al. (2020)
Probiotics (*Lactobacillus plantarum*)	• Whey protein isolate (WPI) with sodium alginate (SA) • WPI with fructooligosaccharide (FOS) • Denatured-WPI (DWPI) with SA • DWPI with FOS	87.92–94.86	Rajam & Anandharamakrishnan (2015)

employed technique to confirm the above. The FTIR spectra of the core compound, wall material and encapsulate are obtained in the spectral range of 4000–400 cm^{-1} and compared. The absence of core-wall interaction is confirmed when the core compound's characteristic functional groups are present intact in the encapsulate. For instance, in the case of spray-freeze-dried vanillin microcapsules, spectral peaks (absorbance) were detected at 1756.6 cm^{-1} and 1100.8 cm^{-1}, which correspond to the aldehyde and ether groups of pure vanillin (Figure 10.26). Thus, it was confirmed that there was no chemical interaction between vanillin and WPI/β-cyclodextrin, which were used as wall materials for its encapsulation (Hundre, Karthik, Anandharamakrishnan, 2015). Similarly, strong absorption at 3013.4 cm^{-1} confirmed the presence of DHA in all the microencapsulates and pure DHA oil.

10.4.5 Oxidative and thermal stability

Similar to the chemical stability, the oxidative and thermal stability of encapsulates are also imperative, especially when the core is lipophilic or heat-sensitive, respectively. The oxidative stability of microcapsules over a defined storage period can be determined by quantifying the products of primary (peroxides) and secondary oxidation (aldehydes/ketones). The standard procedures for the determination of peroxide value (PV), p-anisidine value and thiobarbituric acid reactive substances (TBARS) are available in many reference materials and hence not discussed here. Strikingly, Pang et al. (2017) determined the propanal content of SFD-fish oil microcapsules, using a gas chromatograph integrated with a static headspace sampler. Propanal is one among the major volatile compounds formed during the oxidative breakdown of omega-3 fatty acids and

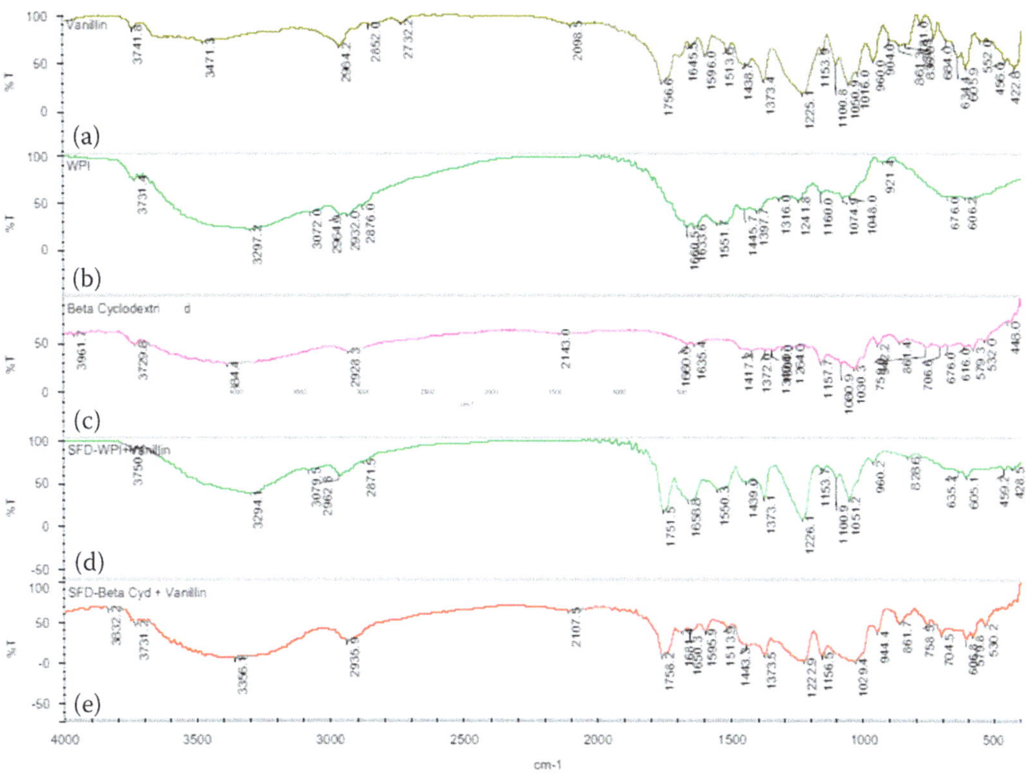

Figure 10.26 FTIR spectra: (a) vanillin; (b) WPI; (c) b-cyclodextrin; (d) SFD-whey protein isolate + vanillin; (e) SFD-b-cyclodextrin + vanillin (Hundre et al., 2015).

hence considered a good indicator of oxidative stability of fatty foods (Ramakrishnan et al., 2014). Owing to their lower surface oil content, SFD-fish oil microcapsules were found to have lower propanal content than the corresponding spray-dried and freeze-dried products, during a storage period of 60 days (Figure 10.27).

Thermal stability of encapsulates is analyzed using either a thermogravimetric analyzer (TGA). TGA is defined as a technique, wherein the sample is exposed to a controlled temperature program under a controlled atmosphere and its mass is monitored as a function of temperature or time (PerkinElmer, Inc., 2015). With respect to microcapsules, the weight loss signifies the release of core from the encapsulant matrix or the decomposition of constituents in the formulation. One has to be familiar with the rationale of defining values for the below parameters concerning TG analysis: (1) weight of sample (mg); (2) temperature range (°C) for the heating program; (3) heating rate (°C/min); (4) type of atmosphere (air/nitrogen); (5) gas purging rate (mL/min). Generally, for TG analysis, the sample size is in the range of 2 – 50 mg. In case of low sample availability, a minimum of 1 mg can be used. To ensure reproducibility with one sample and comparability between different samples, same sample weight must be loaded in the sample pan during each experiment. It is important to ensure that a large surface area of the sample is exposed to the purge. Choice of heating range requires prior knowledge about the thermal transition events

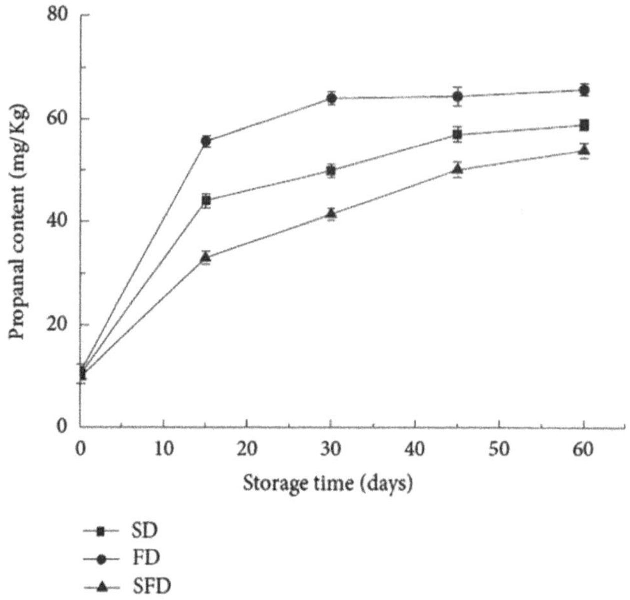

Figure 10.27 Oxidative stability of spray-freeze-dried fish oil microcapsules vis-à-vis spray-dried and freeze-dried counterparts (Pang et al., 2017).

of the material of interest. Then, a trial run can be carried out that starts and ends at 100°C below and above the transition of interest, respectively, at a heating rate of 20°C per minute. The actual heating program for the samples can be finalized based on the observations obtained from the trial run. The selection of heating rate depends on the expected resolution of thermal transition. If a higher resolution is required, then the scanning should be done at a slower rate, typically in the order of 10°C to 20°C per minute. A larger scanning rate (50°C/min) can be adopted when the temperature transition is not of interest (PerkinElmer, Inc., 2015). The latter is usually not practiced with the SFD samples. The typical conditions employed for the TG analysis of spray-freeze-dried products are compiled in Table 10.9.

The curve resultant from TGA is termed as *'thermogram'*, which provides information on the temperature at which the powder formulation begins to decompose. A model thermogram of SFD-inhalation phage powder prepared with trehalose as excipient is shown in Figure 10.28. According to the thermogram, the constituents of powder formulation begin to decompose at around 250°C, marked by the commencement of a steep decline in weight percentage that was constant at T<250°C. The above was attributed to the onset of decomposition or oxidation of trehalose in the formulation (Leung et al., 2016), as degradation events are always associated with a weight loss.

10.4.6 Release kinetics of encapsulated bioactives

Determining the release kinetics of core from the encapsulant matrix is relevant as it governs the bioaccessibility and bioavailability of the bioactive component. Bioaccessibility is that fraction of the bioactive compound, which is released from the food matrix into the gastrointestinal tract for intestinal absorption. Bioavailability is the portion of ingested bioactive, which reaches the circulatory system and subsequently becomes available for storage and biological processes (Parada

Table 10.9 Conditions employed for thermogravimetric analysis of SFD powders

SFD product	TGA conditions	Reference
Inhalation phage powders	• Weight of sample: 5 ± 1 mg • Heating range: 30°C to 400°C • Heating rate: 10°C/min • Gas: Nitrogen	Leung et al. (2016)
Vanillin	• Heating range: 30°C to 500°C • Heating rate: 20°C/min • Gas: Nitrogen	Hundre et al. (2015)
Salbutamol sulphate (for inhalation)	• Weight of sample: 4 – 6 mg • Heating range: Room temperature to 200°C • Heating rate: 10°C/min • Gas: Nitrogen • Gas purge rate: 20 mL/min	Mueannoom et al. (2012)

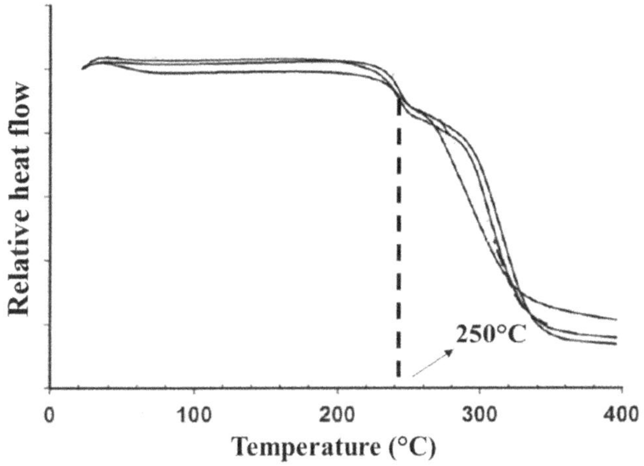

Figure 10.28 Thermogram of SFD-phage powder (modified from Leung et al., 2016).

& Aguilera, 2007; Tenore et al., 2013). The release of core from the encapsulant matrix can occur by different mechanisms and depends on various factors such as wall material type, pH and presence of enzymes. Microcapsules with hydrocolloid-based wall materials such as whey proteins and alginate gel, imbibe water and swell. This promotes diffusion and glassy-to-rubbery state transition in the microcapsules. Consequently, the polymer chains relax and facilitates the diffusion of core (Argin, Kofinas, & Lo, 2014; Mesquita et al., 2015). In spray-freeze-dried powders, core has been observed to diffuse from the polymeric matrix and released through the microporous structure (Mahalakshmi et al., 2020).

In addition to the moisture-controlled glass-to-rubbery transition as discussed above, core release can also be pH-dependent (Argin, Kofinas, & Lo, 2014; Siepmann & Siepmann, 2008). Nevertheless, the degree of polymer relaxation and dissolution at low pH is inadequate to promote release in the case of core with poor aqueous solubility (Prajapati et al., 2014). Presence of ionic species in the medium influences the pH value to maximize swelling. For instance, in the presence of sodium chloride, microcapsules swell at a lower pH value due to increase in osmotic pressure between the aqueous environment and core compound (Argin et al., 2014). The third major type of release mechanism is the *'burst effect'*, which is defined as the uncontrolled leakage of encapsulated core during the initial stage of release. Burst release often occurs in conjunction with the swelling of wall material (Bae et al., 2015; Chessa et al., 2014). Here, the release percentage is as high as 30% to 60% (Mesquita et al., 2015; Ye et al., 2010). But, after the rapid release, the cumulative release can either flatten into a plateau (Park et al., 2013) or shift to a gradual, steady-state release (Ye et al., 2010).

The burst release is instigated by the surface accumulation of non-encapsulated bioactives (Robert et al., 2012) or due to the saturation of interface with the entrapped hydrophilic peptides during the encapsulate production (Rawat & Burgess 2011; Siepmann & Siepmann, 2012). With lipophilic wall materials, the burst effect occurs after the encapsulate is acted upon by lipase during simulated intestinal digestion (Park et al., 2013; Sari et al., 2015). While burst release is desirable in certain food applications, it is not favored in pharmacological applications as the sudden release of core within 24 h might lead to overdosing, side-effects and eventually reduce the treatment efficiency (Bae et al., 2015; Park et al., 2013). With SFD-API microcapsules, wherein burst release is not desired, the same has been avoided by using high molecular weight wall materials in combination with polymers. This prolongs the release duration due to the increase in microsphere diameter (Carrasquillo et al., 2001).

In most cases, spray-freeze-dried encapsulates have been found to exhibit the burst release effect or an initial burst release followed by a slow-release towards the end. Encapsulates with larger size have been found to exhibit a higher initial burst due to their higher surface and internal porosities. The porous microstructure of SFD particles plays a major role in the rapid release of active ingredient trapped in the layers near the surface (Wong, Wang, & Wang, 2001). Also, porous particles can dissolve more easily and lead to rapid drug release than other particles (Rahmati et al., 2013). An investigation by Pang et al. (2017) demonstrated the higher release rate of fish oil from SFD products (Figure 10.29) compared to spray-dried and freeze-dried powders and found it to be advantageous for intestinal absorption. Generally, the initial burst is higher in

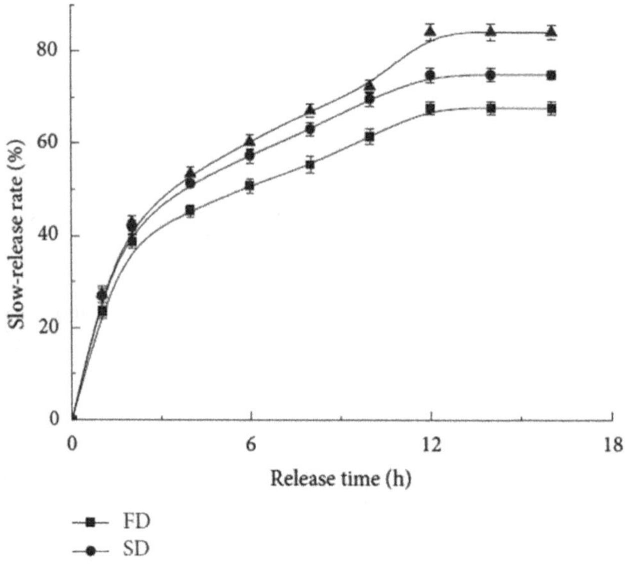

Figure 10.29 Typical bust release profiles of SFD-fish oil microcapsules (Pang et al., 2017).

Table 10.10 Studies on the core release kinetics of spray-freeze-dried microcapsules – An overview

Core	Release medium	Conditions to stimulate core release	Method to extract the released core	Method of quantifying the released core	Release rate/pattern	Reference
Fish oil	• Simulated gastric fluid (SGF) (pH 1) • Simulated intestinal fluid (SIF) (pH 7.1) containing 0.025 mol/L of Na_2HPO_4 and 0.025 mol/L of KH_2PO_4	• Fish oil microcapsules were placed in 100 ml of SGF and incubated for 2 h in a constant temperature oscillator at 90 RPM and 37°C • Followed by incubation in SIF	-	-	84.2% within 12 h	Pang et al. (2017)
IgG (40 mg)	Phosphate buffer saline (PBS) at pH 7.2: composed of 0.1 M sodium phosphate, 0.15 M sodium chloride and 4 mg/L of gentamicin sulfate (an antibacterial agent)	IgG spheres suspended in 1 mL of PBS and maintained at 37°C in a circulating water bath (100 rpm)	Centrifugation for 10 min at 4000 rpm	HPLC system	• Low initial burst of ~35% for PLA microspheres and 12% for PLGA microspheres in 1 day. • Subsequent period of decreasing release rate for almost 3 months.	Wong, Wang & Wang (2001)
Bovine serum albumin (10 mg)	10 mM phosphate buffer (PB) (pH 7.3)	Microspheres were placed in 2 ml of PB and incubated at 37°C	Centrifugation for 1 min at 500 rpm	UV-visible spectroscopy at 280 nm	• BSA in PLG microspheres: >80% of the protein was released rapidly within only 2 days • BSA + PLG + Polaxamer: Burst was reduced from 45% to a lowest of 15%	Bittner et al. (1998); Carrasquillo et al. (2001)
Salmeterol Xinafoate (5 mg)	Deionized water	Microcapsules were placed in 50 mL of freshly prepared deionized water at 37°C in a horizontal shaker at 100 rpm	Filtered using 200 nm inline syringe filter to remove suspended particles	HPLC analysis at 252 nm	~90% drug release in 30 min	Rahmati et al. (2013)

(*Continued*)

Table 10.10 *(Continued)* Studies on the core release kinetics of spray-freeze-dried microcapsules – An overview

Core	Release medium	Conditions to stimulate core release	Method to extract the released core	Method of quantifying the released core	Release rate/pattern	Reference
Bromelain (100 mg)	• Membrane dialysis in simulated lung fluid (SLF) at pH 7.4 • Egg shell membrane used as a membrane bag for the *in vitro* release study	Incubation in 37 mL of SLF at 37°C and 100 rpm	Centrifugation at 5000 RPM for 5 min	UV-visible spectroscopy at 280 nm	• 55.8, 50.18 and 46.91% release at core-to-wall (maltodextrin) ratio of 1:10, 1:25 and 1:50, respectively, after 12 hours. • Around, 24 h was required for 100% release of bromelain from microparticles.	Lavanya et al. (2020)
Bromelain (10 mg)	• Membrane dialysis in SLF at pH 7.4.	Incubation in 37 mL of SLF at 37°C and 100 rpm	Centrifugation at 5000 RPM for 5 min	UV-visible spectroscopy at 280 nm	• Approximately 74%, 49% and 44% release after 12 hours at core-to-wall (hyroxypropyl β-cyclodextrin) ratio of 1:10, 1:25 and 1:50, respectively.	Lavanya et al. (2020)
IgG (30–40 mg)	Phosphate buffer solution	Incubated in PBS at 37°C in a horizontally shaking water bath at 110 rpm for 2 hours	Centrifugation	Size exclusion chromatography - HPLC	~ 18%, 35% and 40% release at drug loading of 5%, 15% and 26% respectively, at the end of 28 days	Wang et al. (2004)
Procaine hydrochloride	Phosphate buffer solution (pH 7.2)	Drug was dispersed at a concentration of about 0.005 g/mL in PBS and maintained at room temperature	Passed through a 0.2 μm filter to remove particles	UV spectrometry	Controlled release from the PLA microcapsules was observed over 9 days, after an initial burst release for a short time	Yin & Yates (2009)
Human epithelial growth factor (rhEGF)	Phosphate-buffered saline (pH 7.4)	• Dialysis method • 5 mL of the reconstituted solution of SFD-rhEGF liposomal colloid solution placed into a dialysis bag with a cutoff of 12,000–14,000 Da • Bag was immersed in 250 mL of PBS at 37°C and 50 rpm	-	Human EGF ELISA kit.	Burst release of ~50% rhEGF within 30 min followed by sustained-release profile	Yin et al. (2014)

microcapsules with higher drug loading. Higher loading increases the diffusion gradient to enhance the diffusion rate of drug from the microsphere (Wang et al., 2004).

Release kinetics is usually determined by *in vitro* methods by exposing a defined quantity (in the order of milligrams) of microcapsules to a simple release matrix such as phosphate buffer or to simulated fluids selected based on the intended delivery site of the active ingredient (ex. gastric fluid (SGF), intestinal fluid (SIF), lung fluid (SLF)). The *in vitro* experiment is carried out under controlled conditions of mixing, temperature and pH. The aliquots are collected at regular intervals. The released core is then separated from the medium, usually by centrifugation and quantified using analytical techniques such as high performance liquid chromatography or UV-visible spectroscopy. The quantity of supernatant removed for analysis is replenished by adding a fresh aliquot of release medium to the microcapsules. The details of studies on the core release kinetics of selected spray-freeze-dried products and their outcomes are presented in Table 10.10.

CONCLUSIONS

Thus, this chapter provided a comprehensive understanding of the characteristics of spray-freeze-dried powders, with emphasis on the relevance to their functionality and methods of characterization. The insights provided in this chapter are expected to be a ready-reckoner for the readers, who intend to study the characteristics of their spray-freeze-dried products. Not limited to that, the pros and cons of methods discussed here would enable arriving at a logical design of experiments and reliable results. With the continuous technical advancements in this field, advent of new and sophisticated characterization techniques are expected in the near future.

REFERENCES

Adeli, E. (2017). The use of spray freeze drying for dissolution and oral bioavailability improvement of azithromycin. *Powder Technology*, 319, 323–331.

Agilent Technologies (2015). *Agilent Dissolution Seminar Series: The Fundamentals of Dissolution*. Available from: https://www.agilent.com/cs/library/flyers/Public/Dissolution_Seminar_Series.pdf (Accessed 17 September 2021).

Al-Hakim, K. (2004). *An Investigation of Spray-Freezing and Spray-Freeze-Dryings*. PhD Thesis, Loughborough University, United Kingdom.

Ali, M. E., & Lamprecht, A. (2014). Spray freeze drying for dry powder inhalation of nanoparticles. *European Journal of Pharmaceutics and Biopharmaceutics*, 87(3), 510–517.

Amador, C., & Martin de Juan, L. (2016). Strategies for Structured particulate systems design. *Computer Aided Chemical Engineering*, 39, 509–579.

Anandharamakrishnan, C. & Ishwarya, S. P. (2015). *Spray Drying Techniques for Food Ingredient Encapsulation*, John Wiley & Sons, Ltd., Chichester, UK, and the Institute of Food Technologists, Chicago, IL.

Anandharamakrishnan, C. (2008). *Experimental and computational fluid dynamics studies on spray-freeze-drying and spray-drying of proteins*. Ph.D. thesis. UK: Loughborough University.

Anandharamakrishnan, C., Rielly, C. D., & Stapley, A. G. F. (2010). Spray-freeze-drying of whey proteins at sub-atmospheric pressures. *Dairy Science and Technology*, 90, 321–334.

Anton Paar, (*n.d.*). *Solid Density of Pharmaceuticals: Assessing Flow Properties of Powders and Open Porosity of Tablets*. Available from: https://www.anton-paar.com/corp-en/services-support/document-finder/application-reports/solid-density-of-pharmaceuticals-assessing-flow-properties-of-powders-and-open-porosity-of-tablets/ (Accessed 17 September 2021).

Argin, S., Kofinas, P., & Lo, Y. M. (2014). The cell release kinetics and the swelling behavior of physically crosslinked xanthan–chitosan hydrogels in simulated gastrointestinal conditions. *Food Hydrocolloids*, 40, 138–144.

Arora, D., Shah, K. A., Halquist, M. S., & Sakagami, M. (2010). In vitro aqueous fluid-capacity-limited dissolution testing of respirable aerosol drug particles generated from inhaler products. *Pharmaceutical Research*, 27(5), 786–795.

Babenko, M., Peron, J.-M. R., Kaialy, W., Calabrese, G., Alany, R. G., & ElShaer, A. (2019). 1H NMR quantification of spray dried and spray freeze-dried saccharide carriers in dry powder inhaler formulations. *International Journal of Pharmaceutics*, 564, 318–328.

Bae, K. H., Lee, F., Xu, K., Keng, C. T., Tan, S. Y., Tan, Y. J., Chen, Q., & Kurisawa, M. (2015). Microstructured dextran hydrogels for burst-free sustained release of PEGylated protein drugs. *Biomaterials*, 63, 146–157.

Barbosa-Cánovas, G. V., Ortega-Rivas, E., Juliano, P., & Yan, H. (2005). Bulk properties. In: *Food Powders: Physical Properties, Processing, and Functionality*. Kluwer Academic/Plenum Publishers, New York, pp. 55–88.

Barron, M. K., Young, T. J., Johnston, K. P., & Williams, R. O. (2003). Investigation of processing parameters of spray freezing into liquid to prepare polyethylene glycol polymeric particles for drug delivery. *AAPS Pharm Sci Tech*, 4(2), 1–13, Article 12.

Beaulieu, L., Savoie, L., Paquin, P., & Subirade, M. (2002). Elaboration and characterization of whey protein beads by an emulsification/cold gelation process: Application for the protection of retinol. *Biomacromolecules*, 3(2), 239–248.

BET-TPX-Chemi-reading-material (*n.d.*). Available from: https://www.iitk.ac.in/che/pdf/resources/BET-TPX-Chemi-reading-material.pdf (Accessed 17 September 2021).

Bhandari, B. (2013). Introduction to food powders. In: B. Bhandari, N. Bansal, M. Zhang, & P. Schuck, *Handbook of Food Powders*, Woodhead Publishing, Cambridge, UK, pp. 1–25.

Bhandari, B., & Roos, Y. H. (2012). Food materials science and engineering: an overview. In: B. Bhandari & Y.H. Roos (Eds.), *Food Materials Science and Engineering*, Wiley-Blackwell, Oxford, pp. 1–23.

Bhusari, S. N., Muzaffar, K., & Kumar, P. (2014). Effect of carrier agents on physical and microstructural properties of spray dried tamarind pulp powder. *Powder Technology*, 266, 354–364.

Bio-Equip (*n.d.*). Available from: https://bio-equip.cn/enshow1equip.asp?equipid=35995 (Accessed 17 September 2021).

Bittner, B., Morlock, M., Koll, H., Winter, G., & Kissel, T. (1998). Recombinant human erythropoietin (rhepo) loaded poly(lactide-co-glycolide) microspheres: Influence of the encapsulation technique and polymer purity on microsphere characteristics. *European Journal of Pharmaceutics and Biopharmaceutics*, 45(3), 295–305.

Brayfield, A., (2013). *"Danazol". Martindale: The Complete Drug Reference*. Pharmaceutical Press, London, p. 2050.

Brunauer, S., Emmett, P. H., & Teller, E. (1938). Adsorption of gases in multimolecular layers. *Journal of the American Chemical Society*, 60(2), 309–319.

Brunaugh, A. D., Wu, T., Kanapuram, S. R., & Smyth, H. D. (2019). Effect of particle formation process on characteristics and aerosol performance of respirable protein powders. *Molecular Pharmaceutics*, 16(10), 4165–4180.

Bychkova, V. E., & Ptitsyn, O. B. (1993). Stostoianie rasplavlennoĭ globuly belkovykh molekul stanovitsia skoreepravilom, chem iskliucheniem [The state of unfolded globules of protein molecules is more quickly becoming a rule, rather than an exception]. *Biofizika*, 38(1), 58–66.

Caparino, A., Tang, J., Nindo, C. I., Sablani, S. S., Powers, J. R., & Fellman, J. K. (2012). Effect of drying methods on the physical properties and microstructures of mango (Philippine 'Carbao' var.) powder. *Journal of Food Engineering*, 111, 135–148.

Carrasquillo, K. G., Costantino, H. R., Cordero, R. A., Hsu, C. C., & Griebenow, K. (1999). On the structural preservation of recombinant human growth hormone in a dried film of a synthetic biodegradable polymer. *Journal of Pharmaceutical Sciences*, 88(2), 166–173.

Carrasquillo, K. G., Stanley, A. M., Aponte-Carro, J. C., De Jésus, P., Costantino, H. R., Bosques, C. J., & Griebenow, K. (2001). Non-aqueous encapsulation of excipient-stabilized spray-freeze dried BSA into poly(lactide-co-glycolide) microspheres results in release of native protein. *Journal of Controlled Release*, 76(3), 199–208.

Carrasquillo, K. G., Cordero, R. A., Ho, S., Franquiz, J. M., & Griebenow, K. (1998) Structure-guided encapsulation of bovine serum albumin in poly(DL-lactic-co-glycolic)acid. *Pharmacy and Pharmacology Communications*, 4, 563–571.

Chang, B. S., Kendrick, B. S., & Carpenter, J. F. (1996). Surface-induced denaturation of proteins during freezing and its inhibition by surfactants. *Journal of Pharmaceutical Sciences*, 85(12), 1325–1330.

Chessa, S., Huatan, H., Levina, M., Mehta, R. Y., Ferrizzi, D., & Rajabi-Siahboomi, A. R. (2014). Application of the dynamic gastric model to evaluate the effect of food on the drug release characteristics of a hydrophilic matrix formulation. *International Journal of Pharmaceutics*, 466(1-2), 359–367.

Chiou, W. L., & Riegelman, S. (1971). Pharmaceutical applications of Solid Dispersion Systems. *Journal of Pharmaceutical Sciences*, 60(9), 1281–1302.

Costantino, H. R., Chen, B., Griebenow, K., Hsu, C. C., & Shire, S. J. (1998). Fourier transform infrared spectroscopic investigation of the secondary structure of aqueous and dried recombinant human deoxyribonuclease I. *Pharmacy and Pharmacology Communications*, 4, 391–395.

Costantino, H. R., Firouzabadian, L., Hogeland, K., Wu, C. C., Beganski, C., Carrasquillo, K. G., Cordova, M., Griebenow, K., Zale, S. E., & Trace, (2000). Protein spray-freeze drying. Effect of atomisation conditions on particle size and stability. *Pharmaceutical Research*, 17(11), 1374–1383.

Costantino, H. R., Griebenow, K., Mishra, P., Langer, R., & Klibanov, A. M. (1995). Fourier-transform infrared spectroscopic investigation of protein stability in the lyophilized form. *Biochimica Et Biophysica Acta (BBA) - Protein Structure and Molecular Enzymology*, 1253(1), 69–74.

Costantino, H. R., Liauw, S., Mitragotri, S., Langer, R., Klibanov, A. M., & Sluzky, V. (1997). The pharmaceutical development of insulin. Historical perspectives and future directions. In: Z. Shahrokh, V. Sluzky, J.L. Cleland, S.J. Shire, & T.W. Randolph (Eds.), *Therapeutic Protein and Peptide Formulation and Delivery*, American Chemical Society, Washington, DC, pp. 29–66.

Costantino, H. R., Nguyen, T. H. & Hsu, C. C. (1996). Fourier transform infrared spectroscopy demonstrates that lyophilization alters the secondary structure of recombinant human growth hormone. *Pharm. Sci.*, 2, 229–232.

Daneshmand, B., Faghihi, H., Amini Pouya, M., Aghababaie, S., Darabi, M., & Vatanara, A. (2018). Application of disaccharides alone and in combination, for the improvement of stability and particle properties of spray-freeze dried IgG. *Pharmaceutical Development and Technology*, 24(4), 439–447.

D'Arcy, D. M., Corrigan, O. I., & Healy, A. M. (2006). Evaluation of hydrodynamics in the basket dissolution apparatus using computational fluid dynamics - dissolution rate implications. *European Journal of Pharmaceutical Sciences*, 27(2-3), 259–267.

Depreter, F., Pilcer, G., & Amighi, K. (2013). Inhaled proteins: Challenges and perspectives. *International Journal of Pharmaceutics*, 447(1-2), 251–280.

Eckhardt, B. M., Oeswein, J. Q., & Bewley, T. A. (1991). Effect of freezing on aggregation of human growth hormone. *Pharmaceutical Research*, 08(11), 1360–1364.

Eggerstedt, S. N., Dietzel, M., Sommerfeld, M., Süverkrüp, R., & Lamprecht, A. (2012). Protein spheres prepared by drop jet freeze drying. *International Journal of Pharmaceutics*, 438, 160–166.

Einfalt, T., Planinšek, O., & Hrovat, K. (2013). Methods of amorphization and investigation of the amorphous state. *Acta Pharmaceutica*, 63(3), 305–334.

Emami, F., Vatanara, A., Park, E., & Na, D. (2018). Drying technologies for the stability and bioavailability of biopharmaceuticals. *Pharmaceutics*, 10(3), 131.

Faghihi, H., Vatanara, A., Najafabadi, A. R., Ramezani, V., & Gilani, K. (2014). The use of amino acids to prepare physically and conformationally stable spray-dried igg with enhanced aerosol performance. *International Journal of Pharmaceutics*, 466(1-2), 163–171.

Ferreira, I. M. P. L. V. O., & Caçote, H. (2003). Detection and quantification of bovine, ovine and caprine milk percentages in protected denomination of origin cheeses by reversed-phase high-performance liquid chromatography of beta-lactoglobulins. *Journal of Chromatography A*, 1015(1-2), 111–118.

Ferreira, I. M., Mendes, E., & Ferreira, M. A. (2001). HPLC/UV analysis of proteins in dairy products using a Hydrophobic Interaction Chromatographic Column. *Analytical Sciences*, 17(4), 499–501.

Fritzen-Freire, C. B., Prudêncio, E. S., Amboni, R. D. M. C., Pinto, S. S., Negrão-Murakami, A. N., & Murakami, F. S. (2012). Microencapsulation of bifidobacteria by spray drying in the presence of prebiotics. *Food Research International*, 45(1), 306–312.

Fu, F., Lin, L., & Xu, E. (2017). Functional pretreatments of natural raw materials. In: M. Fan & F. Fu (Eds.), *Advanced High Strength Natural Fibre Composites in Construction*, Woodhead Publishing, Duxford, United Kingdom, pp. 87–114.

Gill, P., Moghadam, T. T., & Ranjbar, B. (2010). Differential scanning calorimetry techniques: applications in biology and nanoscience. *Journal of Biomolecular Techniques*, 21(4), 167–193.

Griebenow, K., Santos, A. M., & Carrasquillo, K. G. (1999). Secondary structure of proteins in the amorphous dehydrated state probed by FTIR spectroscopy. Dehydration induced structural changes and their prevention. *The Internet Journal of Vibrational Spectroscopy (IJVS)*, www.ijvs.com, 3, 1, 3.

Griese, M., & Stiftung, S. W. (1999). Pulmonary surfactant in health and human lung diseases: State of the art. *European Respiratory Journal*, 13(6), 1455–1476.

Herbert, P., Murphy, K., Johnson, O. F., Dong, N., Jaworowicz, W., Tracy, M. A., Cleland, J. L., & Putney, S. D. (1998). A large-scale process to produce microencapsulated proteins. *Pharmaceutical Research*, 15(2), 357–361.

Ho, T. M., Howes, T., & Bhandari, B. R. (2015). Characterization of crystalline and spray-dried amorphous α-cyclodextrin powders. *Powder Technology*, 284, 585–594.

Ho, T. M., Truong, T., & Bhandari, B. R. (2017). Methods to characterize the structure of food powders – a review. *Bioscience, Biotechnology, and Biochemistry*, 81(4), 651–671.

Hsu, C. C., Nguyen, H. M., Yeung, D. A., Brooks, D. A., Koe, G. S., Bewley, T. A., & Pearlman, R. (1995). Surface denaturation at solid-void interface--a possible pathway by which opalescent particulates form during the storage of lyophilized tissue-type plasminogen activator at high temperatures. *Pharmaceutical Research*, 12(1), 69–77.

https://en.wikipedia.org/wiki/X-ray_crystallography#/media/File:Bragg_diffraction_2.svg (Accessed 2 July 2021).

Hu, J., Johnston, K. P., & Williams, R. O. (2004). Rapid dissolving high potency danazol powders produced by spray freezing into liquid process. *International Journal of Pharmaceutics*, 271, 145–154.

Hundre, S. Y., Karthik, P., & Anandharamakrishnan, C. (2015). Effect of whey protein isolate and β-cyclodextrin wall systems on stability of microencapsulated vanillin by spray–freeze drying method. *Food Chemistry*, 174, 16–24.

IS 2791 (1992) (Reaffirmed 2009). *Soluble coffee powder [FAD 6: Stimulant Foods] (Third revision)*. Bureau of Indian Standards, New Delhi.

IS 16033 (2012). *Instant Coffee – Determination of Free-Flow and Compacted Bulk Densities [FAD 6: Stimulant Foods]*. Bureau of Indian Standards, New Delhi.

Ishwarya, S. P., & Anandharamakrishnan, C. (2015). Spray-freeze-drying approach for soluble coffee processing and its effect on quality characteristics. *Journal of Food Engineering*, 149, 171–180.

Isleroglu, H., Turker, I., Tokatli, M., & Koc, B. (2018). Ultrasonic spray-freeze drying of partially purified microbial transglutaminase. *Food and Bioproducts Processing*, 111, 153–164.

IUPAC (1997). *Compendium of Chemical Terminology*, 2nd Edition (the "Gold Book").

Jinapong, N., Suphantharika, M., & Jamnong, P. (2008). Production of instant soymilk powders by ultrafiltration, spray drying and fluidized bed agglomeration. *Journal of Food Engineering*, 84(2), 194–205.

Kanojia, G., Willems, G., Frijlink, H. W., Kesrten, G., Soema, P., & Amorij, J.-P. (2016). A Design of Experiment approach to predict product and process parameters for a spray dried influenza vaccine. *International Journal of Pharmaceutics*, 511(2), 1098–1111.

Karthik, P., & Anandharamakrishnan, C. (2013). Microencapsulation of docosahexaenoic acid by spray-freeze-drying method and comparison of its stability with spray-drying and freeze-drying methods. *Food and Bioprocess Technology*, 6, 2780–2790.

Khwanpruk, K., Anandharamakrishnan, C., Rielly, C. D., & Stapley, A. G. F., (2008). Volatiles retention during the sub-atmospheric spray freeze drying of coffee and maltodextrin. In: *Proceedings of the 16th International Drying Symposium* (IDS2008), 9–12 November 2008, Hyderabad, India, pp. 1066–1072.

Kondo, M., Niwa, T., Okamoto, H., & Danjo, K. (2009). Particle characterization of poorly water-soluble drugs using a spray freeze drying technique. *Chemical and Pharmaceutical Bulletin*, 57(7), 657–662.

Kuila, U., & Prasad, M. (2013). Specific surface area and pore-size distribution in clays and shales. *Geophysical Prospecting*, 61(2), 341–362.

Labouta, H. I., & Schneider, M. (2010). Tailor-made biofunctionalized nanoparticles using layer-by-layer technology. *International Journal of Pharmaceutics*, 395(1-2), 236–242.

Lavanya, M. N., Preethi, R., Moses, J. A., & Anandharamakrishnan, C. (2020). Production of bromelain aerosols using spray-freeze-drying technique for pulmonary supplementation. *Drying Technology*, 39(3), 358–370.

Lek Pharmaceuticals d.d. (2015). *Spray-Freeze Drying of Polyelectrolyte Nanoparticles Containing the Protein Drug*. Available from: https://patents.justia.com/patent/20170258726 (Accessed 17 September 2021).

Leuenberger, H., Plitzko, M., & Puchkov, M. (2006). Spray freeze drying in a fluidized bed at normal and low pressure. *Drying Technology*, 24, 711–719.

Leung, S. S., Parumasivam, T., Gao, F. G., Carrigy, N. B., Vehring, R., Finlay, W. H., Morales, S., Britton, W. J., Kutter, E., & Chan, H.-K. (2016). Production of inhalation phage powders using spray freeze drying and spray drying techniques for treatment of respiratory infections. *Pharmaceutical Research*, 33(6), 1486–1496.

Li, X., Jiang, L., Zhou, C., Liu, J., & Zeng, H. (2015). Integrating large specific surface area and high conductivity in hydrogenated $NiCo_2O_4$ double-shell hollow spheres to improve supercapacitors. *NPG Asia Materials*, 7(3), e165.

Liang, W., Chow, M. Y. T., Chow, S. F., Chan, H.-K., Kwok, P. C. L., & Lam, J. K. W. (2018). Using two-fluid nozzle for spray freeze drying to produce porous powder formulation of Naked Sirna for inhalation. *International Journal of Pharmaceutics*, 552, 67–75.

Long, M., & Chen, Y. (2009). Dissolution testing of solid products. In: Y. Qiu, Y. Chen, G. G. Z. Wang (Eds.), *Developing Solid Oral Dosage Forms*, Academic Press, London, UK, pp. 319–340.

Lowell, S. , Shields, J. E. , Thomas, M. A., &Thommes, M. (2004). *Characterization of Porous Solids and Powders: Surface Area, Pore Size and Density (First Edition)*. Springer, Dordrecht, The Netherlands.

Lumay, G., Tripathi, N. M., & Francqui, F. (2019). How to gain a full understanding of powder flow properties, and the benefits of doing so. *ONdrugDelivery Magazine*, 102, 42–46.

Ly, A., Carrigy, N. B., Wang, H., Harrison, M., Sauvageau, D., Martin, A. R., Vehring, R., & Finlay, W. H. (2019). Atmospheric spray freeze drying of sugar solution with Phage D29. *Frontiers in Microbiology*, 10.

Maa, Y. F., Ameri, M., Shu, C., Payne, L. G., & Chen, D. (2004). Influenza vaccine powder formulation development: Spray-freeze-drying and stability evaluation. *Journal of Pharmaceutical Sciences*, 93(7), 1912–1923.

Mahalakshmi, L., Leena, M. M., Moses, J. A., & Anandharamakrishnan, C. (2020). Micro- and nano-encapsulation of β-carotene in zein protein: Size-dependent release and absorption behavior. *Food & Function*, 11(2), 1647–1660.

Mant, C. T., & Hodges, R. S. (1991). *High-performance liquid chromatographý of peptides and proteins*, CRC Press, London.

Marques, M. R., Loebenberg, R., & Almukainzi, M. (2011). Simulated biological fluids with possible application in dissolution testing. *Dissolution Technologies*, 18(3), 15–28.

May, S., Jensen, B., Wolkenhauer, M., Schneider, M., & Lehr, C. M. (2012). Dissolution techniques for in vitro testing of dry powders for inhalation. *Pharmaceutical Research*, 29(8), 2157–2166.

Mesquita, P. C., Oliveira, A. R., Pedrosa, M. F., de Oliveira, A. G., & da Silva-Júnior, A. A. (2015). Physicochemical aspects involved in methotrexate release kinetics from biodegradable spray-dried chitosan microparticles. *Journal of Physics and Chemistry of Solids*, 81, 27–33.

Meyer, V. R. (1999). *Practical high-performance liquid chromatography*, Third Edition, John Wiley & Sons, England.

Mohammadi, M. S., & Harnby, N. (1997). Bulk density modelling as a means of typifying the microstructure and flow characteristics of cohesive powders. *Powder Technology*, 92(1), 1–8.

Morton, I. K., & Hall, J. M. (2012). *Concise Dictionary of Pharmacological Agents: Properties and Synonyms*. Springer Science & Business Media.

Mueannoom, W., Srisongphan, A., Taylor, K. M. G., Hauschild, S., & Gaisford, S. (2012). Thermal ink-jet spray freeze-drying for preparation of excipient-free salbutamol sulphate for inhalation. *European Journal of Pharmaceutics and Biopharmaceutics*, 80(1), 149–155.

Murugappan, S., Patil, H. P., Kanojia, G., ter Veer, W., Meijerhof, T., Frijlink, H. W., Huckriede, A., & Hinrichs, W. L. J. (2013). Physical and immunogenic stability of spray freeze-dried influenza vaccine powder for pulmonary delivery: Comparison of inulin, dextran, or a mixture of dextran and trehalose as protectants. *European Journal of Pharmaceutics and Biopharmaceutics*, 85(3), 716–725.

National Center for Biotechnology Information (2021). *PubChem Compound Summary for CID 447043, Azithromycin*. Available from: https://pubchem.ncbi.nlm.nih.gov/compound/Azithromycin (Accessed 17 September 2021).

Nguyen, X. C., Herberger, J. D., & Burke, P. A. (2004). Protein powders for encapsulation: a comparison of spray-freeze drying and spray drying of Darbepoetin Alfa. *Pharmaceutical Research*, 21(3), 507–514.

Niwa, T., Shimabara, H., Kondo, M., & Danjo, K. (2009). Design of porous microparticles with single-micron size by novel spray freeze-drying technique Using four-fluid nozzle. *International Journal of Pharmaceutics*, 382(1-2), 88–97.

Niwa, T., Mizutani, D., & Danjo, K. (2012). Spray-freeze-dried porous microparticles of a poorly water soluble drug for respiratory delivery. *Chemical and Pharmaceutical Bulletin*, 60(7), 870–876.

Otake, H., Okuda, T., & Okamoto, H. (2016). Development of spray-freeze-dried powders for inhalation with high inhalation performance and antihygroscopic property. *Chemical and Pharmaceutical Bulletin*, 64(3), 239–245.

Pang, Y., Duan, X., Ren, G., & Liu, W. (2017). Comparative study on different drying methods of fish oil microcapsules. *Journal of Food Quality*, 2017, 1–7.

Parada, J., & Aguilera, J. M. (2007). Food microstructure affects the bioavailability of several nutrients. *Journal of Food Science*, 72(2), R21–R32.

Park, S. H., Kwag, D. S., Lee, U. Y., Lee, D. J., Oh, K. T., Youn, Y. S., & Lee, E. S. (2013). Highly porous poly(lactide-co-glycolide) microparticles for sustained tiotropium release. *Polymers for Advanced Technologies*, 25(1), 16–20.

Parthasarathi, S., & Anandharamakrishnan, C. (2016). Enhancement of oral bioavailability of vitamin E by spray-freeze drying of whey protein microcapsules. *Food and Bioproducts Processing*, 100, 469–476.

Parris, N., & Baginski, M. A. (1991). A rapid method for the determination of whey protein denaturation. *Journal of Dairy Science*, 74(1), 58–64.

Patil, H. P., Murugappan, S., de Vries-Idema, J., Meijerhof, T., de Haan, A., Frijlink, H. W., Wilschut, J., Hinrichs, W. L. J., & Huckriede, A. (2015). Comparison of adjuvants for a spray freeze-dried whole inactivated virus influenza vaccine for pulmonary administration. *European Journal of Pharmaceutics and Biopharmaceutics*, 93, 231–241.

Pecharsky, V. K., & Zavalij, P. Y. (2009). *Fundamentals of Powder Diffraction and Structural Characterization of Materials*, 2nd edition. Springer, New York.

Peleg, M. (1977). Flowability of food powders and methods for its evaluation – A review. *Journal of Food Process Engineering*, 1, 303–328.

PerkinElmer, Inc. (2015). Thermogravimetric Analysis (TGA). Available from: https://resources.perkinelmer.com/lab-solutions/resources/docs/FAQ_Beginners-Guide-to-Thermogravimetric-Analysis_009380C_01.pdf (Accessed 17 September 2021).

Pikal, M.J. (2002). Freeze drying. In: *Encyclopedia of Pharmaceutical Technology*, Marcel Dekker, New York, pp. 1299–1326.

Pikal, M.J. (2003). Heat and mass transfer in low pressure gases: applications to freeze drying. In: Pikal, M.J. (Ed.), *Freeze Drying of Pharmaceuticals*. Lecture script, University of Connecticut, Storrs.

Prajapati, V. D., Jani, G. K., Moradiya, N. G., Randeria, N. P., Maheriya, P. M., & Nagar, B. J. (2014). Locust bean gum in the development of sustained release mucoadhesive macromolecules of aceclofenac. *Carbohydrate Polymers*, 113, 138–148.

Prescott, J. K., & Barnum, R. A. (2000). *On powder flowability. Pharmaceutical Technology*, 60–84.

Quispe-Condori, S., Saldaña, M. D. A., & Temelli, F. (2011). Microencapsulation of flax oil with zein using spray and freeze drying. *LWT - Food Science and Technology*, 44(9), 1880–1887.

Rahmati, M. R., Vatanara, A., Parsian, A. R., Gilani, K., Khosravi, K. M., Darabi, M., & Najafabadi, A. R. (2013). Effect of formulation ingredients on the physical characteristics of salmeterol xinafoate microparticles tailored by spray freeze drying. *Advanced Powder Technology*, 24(1), 36–42.

Raja, P. M. V., & Barron, A. R. (2021). *Principles of Gas Chromatography*. Available from: https://chem.libretexts.org/@go/page/55860 (Accessed 17 September 2021).

Rajam, R., & Anandharamakrishnan, C. (2015). Spray freeze drying method for microencapsulation of lactobacillus plantarum. *Journal of Food Engineering*, 166, 95–103.

Ramakrishnan, S., Ferrando, M., Aceña-Muñoz, L., Mestres, M., & De Lamo-Castellví, S., & Güell, C. (2014). Influence of emulsification technique and wall composition on physicochemical properties and oxidative stability of fish oil microcapsules produced by spray drying. *Food and Bioprocess Technology*, 7(7), 1959–1972.

Rawat, A., & Burgess, D. J. (2011). USP apparatus 4 method for in vitro release testing of protein loaded microspheres. *International Journal of Pharmaceutics*, 409(1-2), 178–184.

Robert, P., García, P., Reyes, N., Chávez, J., & Santos, J. (2012). Acetylated starch and inulin as Encapsulating agents of gallic acid and their release behaviour in a hydrophilic system. *Food Chemistry*, 134(1), 1–8.

Rogers, T.L., Nelsen, A.C., Sarkari, M., Young, T.J., Johnston, K.P., & Williams, R.O. (2003). Enhanced aqueous dissolution of a poorly water soluble drug by novel particle engineering technology: spray-freezing into liquid with atmospheric freeze-drying. *Pharmaceutical Research*, 20, 485–493.

Rosanske, T. W., Brown, C. K. (1996). Dissolution studies. In: C.M. Riley, & T.W. Rosanske (Eds.), *Progress in Pharmaceutical and Biomedical Analysis*, Elsevier, Volume 3, pp. 169–184.

Saffari, M., & Langrish, T. (2014). Effect of lactic acid in-process crystallization of lactose/protein powders during spray drying. *Journal of Food Engineering*, 137, 88–94.

Sakagami, M., & Lakhani, A. D. (2012) Understanding dissolution in the presence of competing cellular uptake and absorption in the airways. *Respiratory Drug Delivery*, 1, 185–192.

Salama, R., Traini, D., Chan, H., & Young, P. (2008). Preparation and characterisation of controlled Release CO-SPRAY DRIED drug–polymer Microparticles for inhalation 2: Evaluation of in vitro release profiling methodologies for controlled release respiratory aerosols. *European Journal of Pharmaceutics and Biopharmaceutics*, 70(1), 145–152.

Salamon, D. (2014). Advanced Ceramics. In: J. Z. Shen & T. Kosmač (Eds.), *Advanced Ceramics for Dentistry*, Butterworth-Heinemann, Oxford, United Kingdom, pp. 103–122.

Saluja, V., Amorij, J.-P., Kapteyn, J. C., de Boer, A. H., Frijlink, H. W., & Hinrichs, W. L. J. (2010). A comparison between spray drying and spray freeze drying to produce an influenza subunit vaccine powder for inhalation. *Journal of Controlled Release*, 144(2), 127–133.

Sari, T. P., Mann, B., Kumar, R., Singh, R. R. B., Sharma, R., Bhardwaj, M., & Athira, S. (2015). Preparation and characterization of nanoemulsion encapsulating curcumin. *Food Hydrocolloids*, 43, 540–546.

Schiffter, H., Condliffe, J., & Vonhoff, S. (2010). Spray-freeze-drying of nanosuspensions: The manufacture of insulin particles for needle-free ballistic powder delivery. *Journal of the Royal Society Interface*, 7, S483–S500.

Sekiguchi, K., Obi, N., & Ueda, Y. (1964). Studies on absorption of eutectic mixture. II. Absorption of fused conglomerates of chloramphenicol and urea in rabbits. *Chemical and Pharmaceutical Bulletin*, 12(2), 134–144.

Siepmann, J., & Siepmann, F. (2008). Mathematical modeling of drug delivery. *International Journal of Pharmaceutics*, 364(2), 328–343.

Siepmann, J., & Siepmann, F. (2012). Modeling of diffusion controlled drug delivery. *Journal of Controlled Release*, 161(2), 351–362.

Sinha, R. N., Shukla, A. K., Lal, M. A. D. A. N., & Ranganathan, B. (1982). Rehydration of freeze-dried cultures of Lactic Streptococci. *Journal of Food Science*, 47(2), 668–669.

Smyth, H. D., & Hickey, A. J. (2005). Carriers in drug powder delivery. *American Journal of Drug Delivery*, 3(2), 117–132.

solids-solutions (n.d.). Available from: https://www.solids-solutions.com/rd/porosity-and-surface-area-analysis/porosity-analysis-by-mercury-porosimetry/ (Accessed 14 December 2021).

Sonner, C., Maa, Y.-F., & Lee, G. (2002). Spray freeze drying for protein powder preparation. *Journal of Pharmaceutical Sciences*, 91(10), 2122–2139.

Straller, G., & Lee, G. (2017). Shrinkage of spray-freeze-dried microparticles of pure protein for ballistic injection by manipulation of freeze-drying cycle. *International Journal of Pharmaceutics*, 532(1), 444–449.

Strambini, G. B., & Gabellieri, E. (1996). Proteins in frozen solutions: Evidence of ice-induced partial unfolding. *Biophysical Journal*, 70(2), 971–976.

Tamari, S., & Aguilar-Chávez A. (2005). Optimum design of gas pycnometers for determining the volume of solid particles. *Journal of Testing and Evaluation*, 33(2), 12674.

Tan, L. H., Chan, L. W., & Heng, P. W. (2005). Effect of oil loading on microspheres produced by spray-drying. *Journal of Microencapsulation*, 22, 253–259.

Tanaka, R., Hattori, Y., Otsuka, M., & Ashizawa, K. (2020). Application of spray freeze drying to theophylline-oxalic acid cocrystal engineering for inhaled dry powder technology. *Drug Development and Industrial Pharmacy*, 46(2), 179–187.

Tenore, G. C., Campiglia, P., Ritieni, A., & Novellino, E. (2013). In vitro bioaccessibility, bioavailability and plasma protein interaction of polyphenols from Annurca apple (*M. pumila Miller* cv *Annurca*). *Food Chemistry*, 141(4), 3519–3524.

Teixeira, C. C., Cabral, T. P., Tacon, L. A., Villardi, I. L., Lanchote, A. D., & Freitas, L. A. (2017). Solid state stability of polyphenols from a plant extract after fluid bed atmospheric spray-freeze-drying. *Powder Technology*, 319, 494–504.

Thakkar, S. V., Joshi, S. B., Jones, M. E., Sathish, H. A., Bishop, S. M., Volkin, D. B., & Middaugh, R. C. (2012). Excipients differentially influence the conformational stability and pretransition dynamics of two IgG1 monoclonal antibodies. *Journal of Pharmaceutical Sciences*, 101(9), 3062–3077.

Thommes, M., Kaneko, K., Neimark, A. V., Olivier, J. P., Rodriguez-Reinoso, F., Rouquerol, J., & Sing, K. S. W. (2015). Physisorption of gases, with special reference to the evaluation of surface area and pore size distribution (IUPAC TECHNICAL REPORT). *Pure and Applied Chemistry*, 87(9-10), 1051–1069.

Tonnis, W. F., Amorij, J.-P., Vreeman, M. A., Frijlink, H. W., Kersten, G. F., & Hinrichs, W. L. J. (2014). Improved storage stability and immunogenicity of hepatitis b vaccine after spray-freeze drying in presence of sugars. *European Journal of Pharmaceutical Sciences*, 55, 36–45.

Tonon, R. V., Grosso, C. R., & Hubinger, M. D. (2011). Influence of emulsion composition and inlet air temperature on the microencapsulation of flaxseed oil by spray drying. *Food Research International*, 44(1), 282–289.

Turchiuli, C., Fuchs, M., Bohin, M., Cuvelier, M.E., Ordonnaud, C., & Peyrat-Maillard, M.N. (2005). Oil encapsulation by spray drying and fluidized bed agglomeration. *Innovative Food Science and Emerging Technologies*, 6, 29–35.

van Drooge, D.-J., Hinrichs, W. L. J., Dickhoff, B. H. J., Elli, M. N. A., Visser, M. R., Zijlstra, G. S., & Frijlink, H. W. (2005). Spray freeze drying to produce a stable δ9-tetrahydrocannabinol containing inulin-based solid dispersion powder suitable for inhalation. *European Journal of Pharmaceutical Sciences*, 26(2), 231–240.

Wang, G., Liu, S. J., Ueng, S. W. N., & Chan, E. C. (2004). The release of cefazolin and gentamicin from biodegradable PLA/PGA beads. *International Journal of Pharmaceutics*, 273, 203–212.

Wang, Y., Kho, K., Cheow, W. S., & Hadinoto, K. (2012). A comparison between spray drying and spray freeze drying for dry powder inhaler formulation of drug-loaded lipid–polymer hybrid nanoparticles. *International Journal of Pharmaceutics*, 424(1-2), 98–106.

Wong, H. M., Wang, J. J., & Wang, C.-H. (2001). In vitro sustained release of human Immunoglobulin G from biodegradable microspheres. *Industrial & Engineering Chemistry Research*, 40(3), 933–948.

Xu, Q., Yao, Y., Zhao, T., Shi, Q., Li, Z., & Tian, W. (2018). Dissolution characteristics of freeze-dried Pullulan Particles affected by solution concentration and freezing medium. *International Journal of Food Engineering*, 14(7-8), pp. 20180073.

Ye, M., Kim, S., & Park, K. (2010). Issues in long-term protein delivery using biodegradable microparticles. *Journal of Controlled Release*, 146(2), 241–260.

Ye, T., Yu, J., Luo, Q., Wang, S., & Chan, H.-K. (2017). Inhalable clarithromycin liposomal dry powders using ultrasonic spray freeze drying. *Powder Technology*, 305, 63–70.

Yin, F., Guo, S., Gan, Y., & Zhang, X. (2014). Preparation of redispersible liposomal dry powder using an ultrasonic spray freeze-drying technique for transdermal delivery of human epithelial growth factor. *International Journal of Nanomedicine*, 9, 1665–1676.

Yin, W., & Yates, M. Z. (2009). Encapsulation and sustained release from biodegradable microcapsules made by emulsification/freeze drying and spray/freeze drying. *Journal of Colloid and Interface Science*, 336(1), 155–161.

Zhang, S., Lei, H., Gao, X., Xiong, X., Wu, W. D., Wu, Z., & Chen, X. D. (2018). Fabrication of uniform enzyme-immobilized carbohydrate microparticles with high enzymatic activity and stability via spray drying and spray freeze drying. *Powder Technology*, 330, 40–49.

CHAPTER 11

Computational Fluid Dynamics Modeling of Spray-freeze-drying Process

Spray-freeze-drying (SFD) is an intricate process that is governed by a number of factors (Table 11.1) (Al-Hakim, Wigley, & Stapley, 2006). Design and operation of a SFD process needs information on the particle behavior in the chamber, which include, temperature, velocity, and residence time (Anandharamakrishnan et al., 2010). Further, these details are essential to improve the process in terms of throughput and product quality. As discussed in the previous chapters, various experimental approaches have been used in combination with instruments such as thermocouples, Phase Doppler Anemometry (PDA) system and velocimeter, to measure the above particle parameters. However, several limitations are imposed by the SFD process on the real-time measurements. For instance, measuring the cold gas temperature inside the spray-freezing chamber is hindered by the deposition of frozen particles on the thermocouples. The limitations encountered during PDA measurements were already explained in Chapter 2.

In this background, Computational Fluid Dynamics (CFD) modeling serves as an effective tool in predicting the particle behavior during spray-freezing process. Studies have shown that it is possible to computationally model the physical fluid phenomena, which are expensive and time consuming to be measured experimentally. CFD modeling outweighs the experimental approach as it considers the actual flow domain under realistic operating conditions while the latter confines itself to a lab-scale prototype under a limited range of operational conditions (Anderson, 2013). Nevertheless, experimental validation of the CFD model must always be done before drawing conclusions (Anandharamakrishnan & Ishwarya, 2015). Thus, CFD is perceived as a third dimension of fluid dynamics, in addition to the first two dimensions of pure experiment and pure theory. Food processing applications of computational fluid dynamics (CFD) are well-established with respect to unit operations such as drying, pasteurization, and baking. With respect to SFD, influence of different chamber designs or equipment configurations and operating conditions on the particle characteristics have been studied *in silico* using CFD modeling.

This chapter would deal with the CFD modeling of spray-freeze-drying process, particularly, the spray-freezing operation. The initial sections of this chapter would explain the theory of CFD modeling. The latter part would present a case-study on its application to predict the gas flow pattern and particle histories (temperature, velocity profile, particle trajectory, and particle impact on the wall) during the SFD process.

11.1 THEORY OF CFD MODELING

Computational Fluid Dynamics (CFD) is a simulation tool that uses a combination of powerful computers and applied mathematics to model the fluid flow and aid in the optimization of industrial processes (Anandharamakrishnan, 2003 & 2013). CFD modeling is governed by the equations of

DOI: 10.1201/9781003019312-11

Table 11.1 Factors governing the spray-freezing process

Factor	Relevance
Fluid mechanics of the spray	Controls the formation and motion of individual droplets as regards each other and the gas
Local conditions such as gas temperature, droplet temperature and droplet-gas slip velocity	Regulates the heat transfer between the gas and the droplets
Freezing mechanism and crystallization of ice inside the droplets	Determines the drying rate and the microstructure and porosity of particles after freeze-drying

fluid dynamics, derived from the laws of conservation of continuity, momentum, and energy (Table 11.2), collectively known as the *'Navier-Stokes transport equations'* (Anderson, Jr., 2009). These equations are completed by adding other algebraic equations from thermodynamics such as the equation of state for density and a constitutive equation to describe rheology (Anandharam-akrishnan, 2013; Fletcher, 2000). CFD modeling solves these equations inside a defined geometry using numerical methods to predict the velocity, temperature and pressure profiles. Finally, the graphics interface is used to generate the three-dimensional (3D) images of the results on fluid flow behavior (Scott & Richardson, 1997).

11.2 STEPS IN THE CFD MODELING

CFD analysis comprises the following three steps (Xia & Sun, 2002) (Figure 11.1):

I. *Pre- processing:* This step involves defining the material to be studied by specifying the element nature and the real constants of the problem, creation of geometry, meshing to form a computational domain eventually. Geometry creation and meshing are usually done using computer aided design (CAD) software platforms. Meshing refers to dividing a computational domain into discrete volumes. It can be 2-dimensional (2D; that lie in single plane, generally in XY plane) or 3-dimensional (3D; that are not limited to lie in a single plane). The commonly used 2D (triangles or rectangles) and 3D (tetrahedral, hexahedral, pyramid or prism) mesh elements are shown in Figure 11.2. Apart from shape, fineness of mesh or its refinement is also a key aspect that needs to be defined according to the regions of the geometry. Mesh refinement is generally done on a trial-and-error basis in the following sequence: initial meshing → reanalysis with different possible element types → analysis of variance in the solutions for the geometry under consideration. Significant variations in the solutions between trials indicate that the mesh needs more refinement. Then, the above sequence is repeated until convergence is attained (Anandharamakrishnan & Ishwarya, 2015).

II. *Processing or solving:* During this stage, the type of analysis (steady state or unsteady state (transient)) is defined and the boundary conditions are defined. Then, the solution is obtained using commercially available CFD software packages, which solve the mathematical equations of fluid flow.

III. *Post-Processing:* The last step involves analysis of simulation results that are time-specific in the case of unsteady state problems. The results are compiled in the form of contour plots, vector displays or tables (Madenci & Guven, 2006).

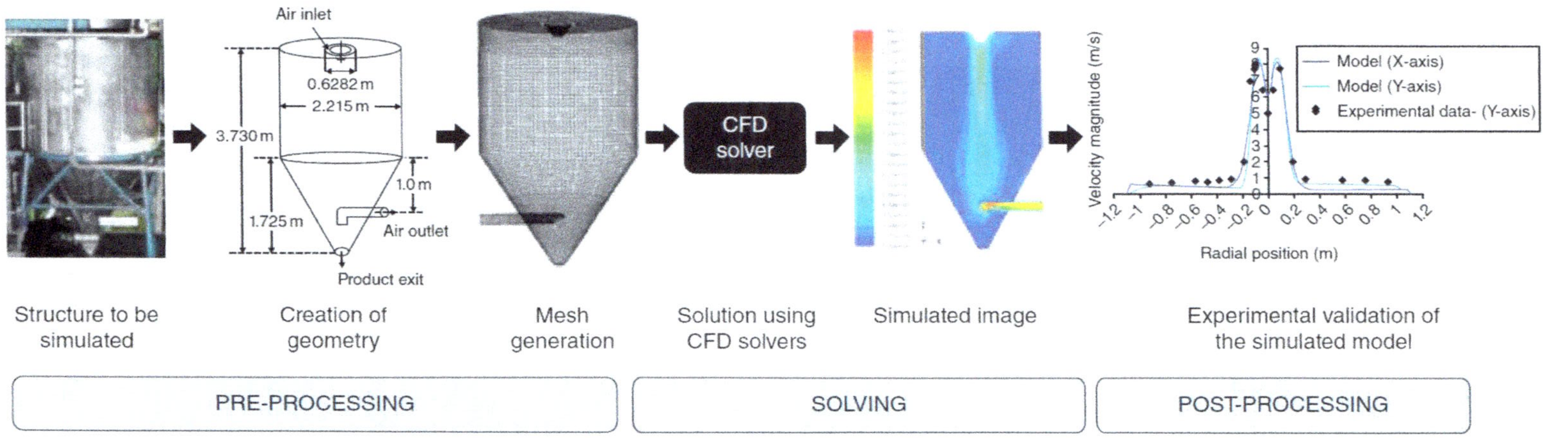

Figure 11.1 Steps involved in CFD modeling (Anandharamakrishnan & Ishwarya, 2015).

Table 11.2 Equations of the laws of conservation

Equation	Significance
Conservation of mass (continuity equation) $\nabla \cdot \underline{\mathbf{v}} = 0$ where, ∇ has the dimensions of reciprocal length:$\nabla = \frac{\partial}{\partial x} i + \frac{\partial}{\partial y} j + \frac{\partial}{\partial z} k$	Describes the rate of change of density at a fixed point resulting from the divergence in the mass velocity vector ρv.
Conservation of momentum (Newton's second law of motion) $\rho_g \frac{D\underline{\mathbf{v}}}{Dt} = -\nabla p + \nabla \cdot \underline{\underline{\tau}} + \rho_g \underline{g}$ The terms on the left hand side are the convection terms and those on the right are the pressure gradient (p), source terms of gravitational force (g) and stress tensor ($\underline{\underline{\tau}}$), which is responsible for diffusion of momentum.	This law is an application of the Newton's Second Law of Motion to an element of fluid. It signifies that a small volume of element moving with the fluid is accelerated due to the force acting upon it.
Conservation of energy $\frac{\partial}{\partial t}(\rho E) + \nabla . [\underline{\mathbf{v}} (\rho E + p)] = \nabla . [k_{eff} \nabla T - \sum_j h_j \underline{J}_j + (\underline{\underline{\tau}} . \underline{\mathbf{v}})]$	This law is an application of the first law of thermodynamics, which states that the rate of the sum of change in internal energy and kinetic energy is equal to the difference between the rates of heat transfer and work done by the system.

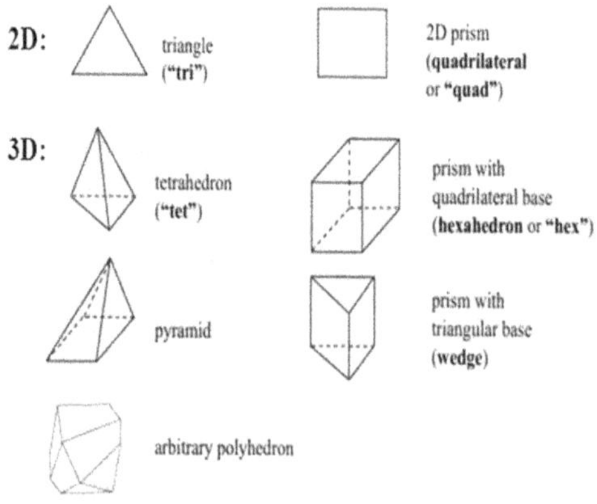

Figure 11.2 Classification of mesh elements (Anandharamakrishnan & Ishwarya, 2019).

11.3 NUMERICAL SIMULATIONS OF GAS-PARTICLE INTERACTIONS

The main objective of CFD modeling with respect to SFD is the simulation of gas-particle flows and interactions. Hence, the primary focus in this case is the coupling phenomena of mass, momentum, and energy equations between gas and droplets (Figure 11.3), which include:

- mass transfer from droplet to gas via evaporation;
- momentum exchange through drag; and,
- energy coupling by heat transfer.

The convective heat transfer between gas phase and droplets reduces the gas temperature, which in turn influences its viscosity and density and thereby effects the gas flow field. Consequently, the droplet trajectories and heat transfer rate between droplets and the gas are also

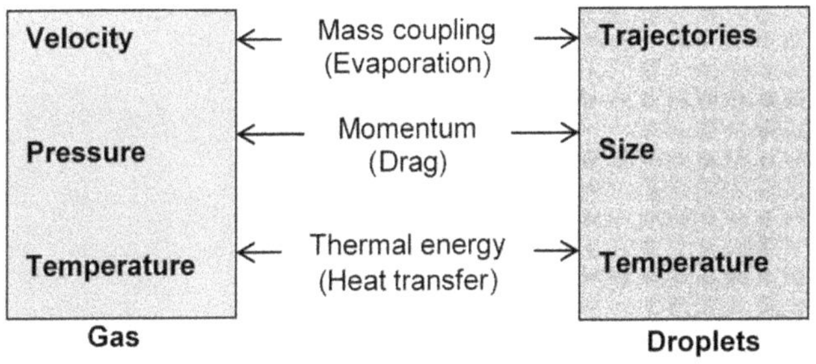

Figure 11.3 Gas-droplet coupling phenomena (Modified from Crowe et al., 1977; Anandharamakrishnan, 2008).

affected (Crowe et al., 1977). Thus, it is evident that all the three equations of conservation are mutually dependent and must be considered while simulating the gas-droplet interactions.

11.4 DISCRETIZATION

Discretization can be defined as the transformation of Navier-Stokes equations into distinct elements, such that they are suitable for numerical evaluation and easy to be solved digitally by the CFD software. The well-known computational approaches for the discretization of Navier-Stokes equations are:

- Finite Element Method (FEM)
- Finite Volume Method (FVM)
- Finite Difference Method (FDM)

The above approaches are collectively known as the *'numerical methods in CFD modeling'*, which is an alternative to finding an analytical solution to the partial differential equations (PDEs). In this approach, the PDEs are approximated to numerical model equations, after which they are solved using numerical methods. Indeed, the resultant solution is an approximation of the real solution to the PDEs.

11.4.1 Finite element method (FEM)

The term *'finite element'* refers to dividing a computational domain (geometry) into minute but finite-sized elements of geometrically simple shapes that constitute the *'finite element mesh'*. The response for each element is expressed in terms of a defined number of degrees of freedom (DOF), which is characterized as the value of an unknown function(s), at a set of nodal points. In finite element analysis, DOF refers to the minimum number of independent coordinates required to define the position of every mass in the system distinctively.

11.4.2 Finite volume method (FVM)

Like FEM, in the Finite volume method (FVM), the flow geometry is initially divided into infinitesimal but finite-sized and non-overlapping *'control volumes'* or *'cells'*. Here, the variable of interest is positioned at the centroid of the control volume. The governing equations are approximated over the control volumes as a function of the nodal values. Finally, numerical solution of the discretized equations are sought.

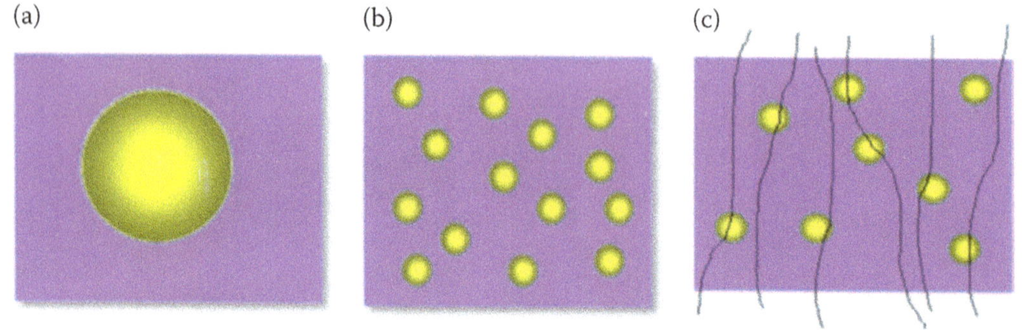

Figure 11.4 Reference frames: (a) Volume of fluid; (b) Eulerian-Eulerian; (c) Eulerian-Lagrangian (Anandharamakrishnan, 2013).

11.4.3 Finite difference method (FDM)

Unlike FEM and FVM, the finite difference method is a direct approach of approximating the PDEs. Here, a point in space is defined, where the continuum representation of equations is considered and substituted with a set of discrete equations, called *'finite-difference equations'*. The requirement for FDM method is a structured mesh defined on a regular grid. Consequently, FDM is preferred for rectangular or block-shaped models, but not useful for irregular geometries (Anandharamakrishnan & Ishwarya, 2019).

11.5 REFERENCE FRAMES IN CFD MODELING

'Reference frames' determine the pattern of interaction between the dispersed (droplets) and continuous (surrounding gas medium) phases in the computational domain of spray-freezing operation. The phases are identified based on their volume. Accordingly, continuous phase is the one in larger volume and the dispersed phase is that present in smaller volume. Depending on the interaction between continuous and dispersed phases, reference frames can be classified into four types, described as follows (Anandharamakrishnan & Ishwarya, 2015).

11.5.1 Volume of fluid (VOF)

The volume of fluid (VOF) reference frame (Figure 11.4[a]) is applicable for two or more immiscible fluids. VOF solves one set of momentum equations and monitors the volume fraction of each fluid throughout the domain. Since the fluids are immiscible, each computational cell contains only one of the pure fluids or the interface between two (or more) fluids. VOF modeling is useful in predicting the disintegration of a liquid jet (as in atomization of the spray-freezing operation), movement of large bubbles in a liquid, and the steady or unsteady state tracing of any liquid-gas interface (ANSYS Fluent user guide, 2006).

11.5.2 Eulerian–Eulerian (E–E)

According to the Eulerian–Eulerian (E–E) model (Figure 11.4[b]), the dispersed and continuous phases totally interpenetrate each other, such that each of the two phases is treated as a continuum (Huang, Kumar & Mujumdar, 2006). This approach is inexpensive with respect to its computational requirements as it arrives at the solution with one set of transport equations. But, this holds true only when the dispersed phase can be defined by a single particle diameter. On the contrary, in the case of multimodal particle size distribution, distinct sets of transport equations are needed for different particle sizes to obtain the solution. Therefore, the E–E approach is appropriate under the following conditions:

1. For flows with a monomodal particle size distribution;
2. When the droplet and gas phases are present in similar proportions;
3. When the emphasis is not on the particle properties (Mostafa & Mongia, 1987; Jakobsen et al., 1997).

Under the E–E approach, the following two cases are considered:

1. Each computational cell contains definite proportions of the dispersed and continuous phases. The means of averaged transport equations of motion is evaluated in the computational domain. The Navier–Stokes equation is solved individually for each phase, and then coupled by applied pressure and inter-phase exchange systems. The transport equations are expressed such that the volume fractions sum up to one (unity).
2. Each computational cell consists only one of the phases. Hence, the transport equations for the two phases are simplified to a single-phase system.

11.5.3 Eulerian–Lagrangian (E–L)

Different from the E–E approach, the Eulerian–Lagrangian approach (Figure 11.4[c]) is categorized under the discrete phase modeling (DPM), which considers droplets and surrounding gas medium as two separate entities. With this reference frame, it is possible to track individual particles through the geometry between their injection point and site of escape from the domain (Nijdam et al., 2006) (Figure 11.5). The E–L model overcomes the limitation associated with the E–E approach with respect to handling droplets with polydisperse size distribution. It is also computationally cheaper than the latter in dealing with the broad size distribution of droplets. But, E–L approach can be expensive when a large number of particles have to be tracked. Thus, it is

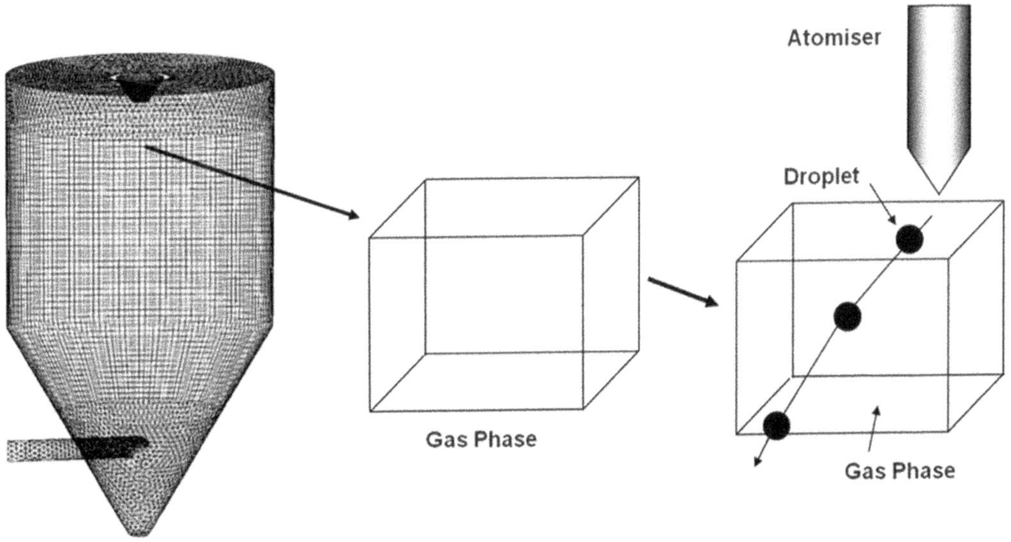

Figure 11.5 Schematic representation of the Eulerian-Lagrangian approach (Anandharamakrishnan & Ishwarya, 2015).

appropriate to adopt the E–L approach when the volume of dispersed phase is less than 10% by volume of the mixture in any region (Marshall & Bakker, 2002). Also, this model can provide additional details on the behavior and residence times of individual particles and result in an accurate prediction of heat and mass transfer.

11.5.4 Turbulence models

Turbulence models in CFD aim at including the effect of disordered motion of fluid flows in the simulation. Turbulent flows are commonly encountered in industrial and engineering applications. The following turbulence models are commonly used in the simulation of systems involving sprays (Anandharamakrishnan, 2008):

a. Standard k-ε (k: turbulence kinetic energy and ε: turbulence dissipation rate): This model focusses on the mechanisms that influence the turbulent kinetic energy. Owing to its robustness and considerable accuracy, this model is used over a widespread range of turbulent flows.
b. RNG k-ε: This model is derived from the Navier-Stokes equations based on a mathematical method called *'renormalization group'* (RNG). The analytical derivation leads to a model having constants that are distinct from those in the standard k-ε model with additional terms and functions in the transport equations for k and ε. The uniqueness of this model is that it considers the effect of swirl on turbulence, which improves the accuracy for swirling flows.
c. Realizable k-ε: This model fulfills some of the mathematical constraints on normal stresses, consistent with the physics of turbulent flows. It facilitates better prediction for flows comprising rotation, boundary layers under strong adverse pressure gradients, separation, and recirculation (Anderson, 1984; Fletcher, 2000).
d. Reynolds Stress Model (RSM): This model exhibits the same form as the instantaneous Navier-Stokes equations, with respect to the velocities and other time-averaged solution variables. It is appropriate for problems involving highly swirling flows, wherein anisotropy of turbulence exerts a significant influence on the mean flow (Fluent, 2005).

11.6 CASE-STUDY: CFD MODELING TO PREDICT THE INFLUENCE OF SPRAY-FREEZING CHAMBER DESIGN AND SPRAY PATTERNS ON PARTICLE CHARACTERISTICS

11.6.1 Step-1: Pre-processing

11.6.1.1 Definition of the problem

In this case-study (Anandharamakrishnan, 2008), the objective of CFD modeling was to simulate the spray-freezing-into-vapor (SFV) operation and evaluate the design improvement as a function of spray pattern and chamber design. The design improvement was judged in terms of the enhanced product collection efficiency. The study considered the influence of two different spray patterns produced by (1) solid cone spray pressure nozzle atomizer (spraying angle: 30°) and (2) hollow cone spray nozzle (Figure 11.6) and modified chamber design obtained by increasing the diameter of the chamber outlet. A solid cone spray is composed of drops that are homogeneously distributed throughout its volume. On the other hand, in a hollow cone spray, majority of the drops are concentrated at the periphery of a conical spray configuration (Lefebvre, 1989). The solid cone spray is characterized by coarse atomization, wherein the droplets at the spray center are larger than those at the edge (Tate, 1969). It distributes the droplets over a circular area. However, hollow cone nozzles offer a comparatively better atomization as they generate a circular liquid screen that disintegrates into fine droplets

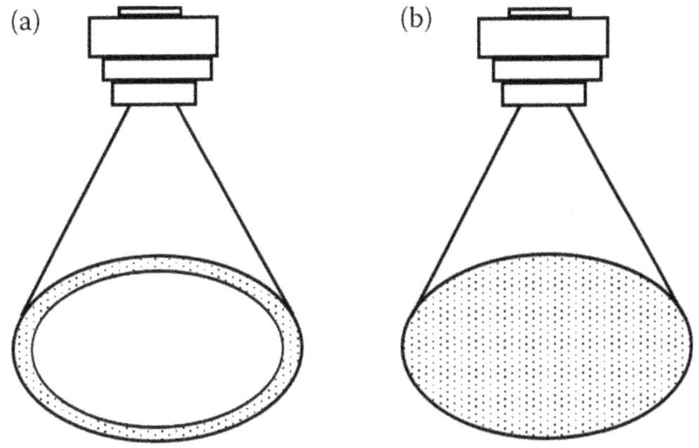

Figure 11.6 Schematic representations of (a) hollow cone and (b) solid cone sprays.

immediately after leaving the nozzle orifice (Anandharamakrishnan, 2008; Lefebvre, 1989). Accordingly, the following three cases were considered (Anandharamakrishnan et al., 2010):

- ***Case A***: Solid cone spray with unmodified design of spray-freezing chamber
- ***Case B***: Hollow cone spray with unmodified design of spray-freezing chamber
- ***Case C***: Solid cone spray with modified spray-freezing chamber design

The output parameters considered were the gas temperature profiles and the axial velocities, temperatures, residence time distribution, trajectories, impact positions and collection efficiency of particles.

11.6.1.2 Geometry creation

The experimental system and geometry used for the CFD simulation pertaining to Case A and Case B are shown in Figs. 11.7[a & b]. The model considered a cylindrical chamber of height: 1.5 m and diameter: 0.8 m, with a solid cone spray pressure nozzle atomizer housed at the roof of the chamber and the freezing gas (liquid nitrogen) entering through the annular region. The geometry of modified spray-freezing chamber design (Case C) obtained after increasing the chamber outlet diameter from 0.15 m (Figure 11.7[b]) to 0.45 m, is shown in Figure 11.8.

11.6.1.3 Meshing

A hexahedral mesh, typically of size 0.001 m was used for the simulation. The preliminary investigations showed that 180000 cells were adequate to obtain grid-independent solutions for the mean velocity field. Thus, the same number of grid cells were retained in the final meshed geometry (Figure 11.9[a]). To maintain the accuracy of solution, a fine mesh was used near the nozzle (displayed with dark color in Figure 11.9[a]), relative to a coarse grid in the other positions of chamber design. Figure 11.9[b] shows the measurement and predicted levels in axial positions under the nozzle.

11.6.2 Solving

An *inert particle* model was chosen for the simulation of spray-freezing operation. Inert particle is a discrete phase element (particle, droplet, or bubble), governed by the inert heating or

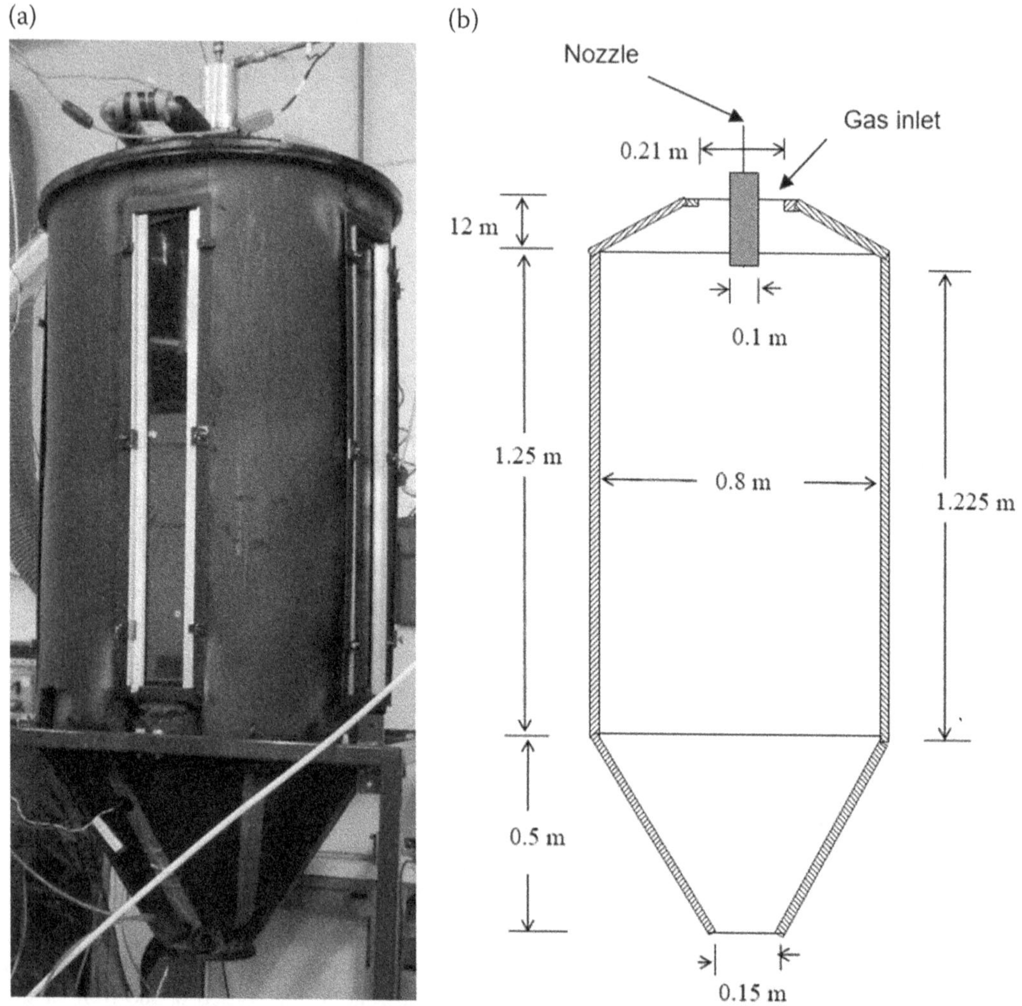

Figure 11.7 (a) Photograph of the experimental SFV apparatus; (b) Geometry of spray-freezing chamber (Case-A and Case-B): cross-section showing the dimensions (Anandharamakrishnan, 2008).

cooling law, which is applied when the particle temperature (T_p) is less than the vaporization temperature (T_{vap}) (Release 12.0 © ANSYS, Inc., 2009). Hence, in the case of spray-freezing, the Newton's law of cooling (Eq. 11.1) was applied to determine the heat transfer between droplet and cold gas.

$$m_p c_p \frac{dT_p}{dt} = hA_p(T_g - T_p) \tag{11.1}$$

Where, m_p is the mass of the droplet, c_p is the specific heat of droplet, T_p is the droplet temperature, T_g is the gas temperature, and A_p is the surface area of the droplet. The heat transfer coefficient (h) was calculated from the Ranz-Marshall (1952) equation, given by Eq. 11.2.

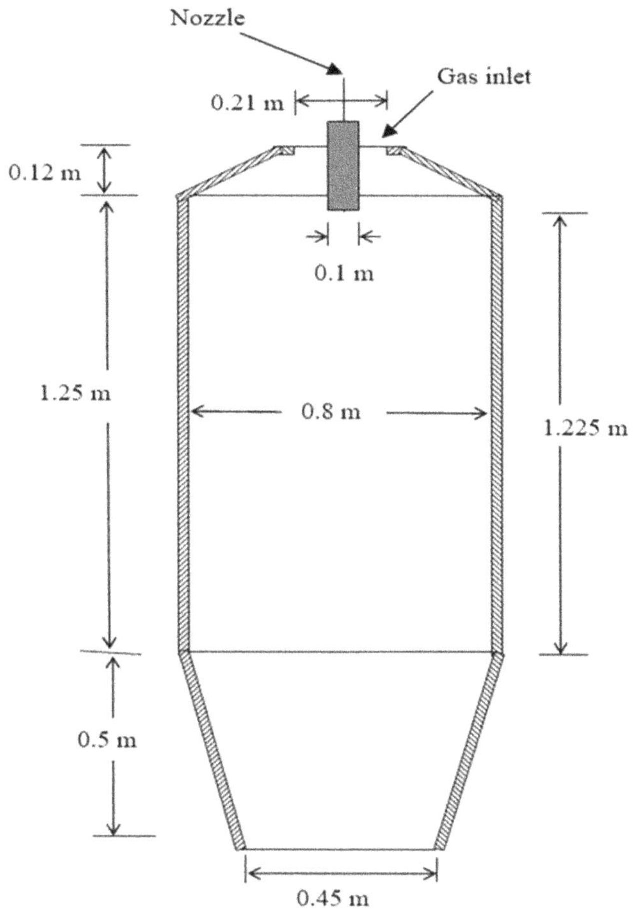

Figure 11.8 Geometry of the modified spray-freezing chamber (Case-C) (Anandharamakrishnan, 2008).

$$Nu = \frac{hd_p}{k_{ta}} = 2 + 0.6(Re_d)^{1/2}(Pr)^{1/3} \tag{11.2}$$

Where Nu is the dimensionless Nusselt number, d_p is the droplet diameter, k_{ta} is the thermal conductivity, Re_d and Pr are the dimensionless Reynolds and Prandtl number, respectively. The influence of latent heat of fusion during the recalescence and crystal growth phases of freezing was included in the discrete phase model by assuming that solidification took place in the temperature range between 0 and –10°C. During the aforesaid phases of freezing, values for the pseudo specific heat (c_p) of particles were defined as a piece-wise linear function of temperature that accounts for the additional latent heat (value of latent heat of ice divided by 10 K) (Table 11.3).

11.6.2.1 Boundary conditions

Details on the boundary conditions applied in the CFD model of spray-freezing are given in Table 11.4. The PDEs were solved by the finite volume method after applying the semi-implicit pressure-linked equations (SIMPLE) method for pressure-velocity coupling and a second-order upwind scheme to interpolate the variables on the surface of the control volume. The initial droplet

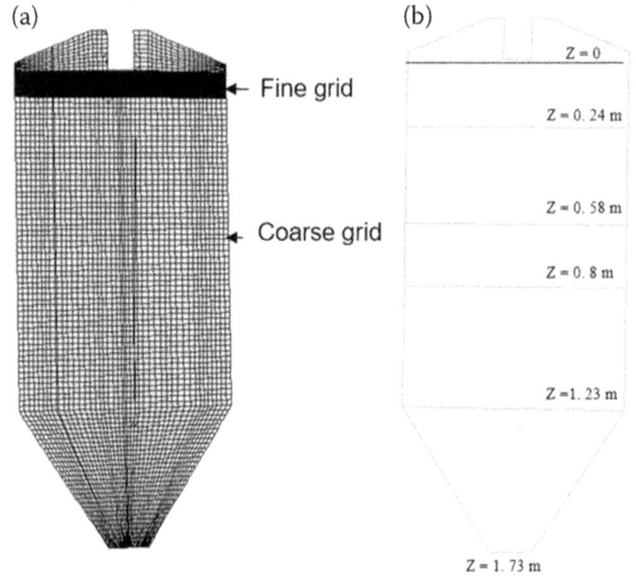

Figure 11.9 (a) Mesh used in the spray-freezing simulation; (b) Measurement and prediction of axial positions (Anandharamakrishnan, 2008).

Table 11.3 Particle specific heat values (Anandharamakrishnan, 2008)

Point	Temperature (K)	Specific heat (c_p) (J/kg K)
1	0	2093 (ice)
2	262.9	2093 (ice)
3	263	35343 (phase change)
4	273	35343 (phase change)
5	273.1	4185 (water)
6	373	4185 (water)

size distribution was derived from the fit of Rosin-Rammler (RR) distribution to PDA measurements of the solid cone spray droplet size distribution at z = 38 mm below the nozzle. 1600 particles were tracked and then the Rosin-Rammler droplet distribution was divided into 16 bins of particle size ranging between 0.5 and 250 μm. Two-way coupling between the cold gas and droplets was achieved using the discrete phase model (DPM). The stochastic effects of turbulence on the particle trajectories were incorporated via an eddy interaction model. The standard *k-ε* turbulence model was used with the *escape* wall boundary condition, which considered that the particles are lost from calculation at the point of impact with the wall (Anandharamakrishnan et al., 2010; Anandharamakrishnan, 2008).

11.6.3 Post-processing: Simulation results and experimental validation

11.6.3.1 Simulation without spray injection

In the initial part of the simulation, the gas flow pattern was obtained in the absence of spray injection (Figure 11.10[a]). From the flow pattern, it was evident that the axial velocity of gas was

Table 11.4 Boundary conditions used for spray-freezing simulation (Anandharamakrishnan, 2008)

Inlet Air	
Air inlet temperature (K)	203
Air mass flow rate (kg/s)	0.336
Air velocity magnitude (m/s)	0.9897
Outlet Condition	
Outlet	Outflow
Turbulence model	
Turbulence *k*-value (m^2/s^2)	3.588×10^{-3}
Turbulence -value (m^2/s^3)	3.21×10^{-4}
Liquid spray from nozzle	
Liquid feed rate (spray rate) (kg/s)	0.0125
Feed Temperature (°C)	20
Spray angle (full angle)	30°
Minimum droplet diameter (μm)	0.5
Maximum droplet diameter (μm)	250
Average droplet diameter $\bar{d}$ (μm)	141
Droplet velocity at nozzle exit (m/s)	28
Rosin-Rammler spread parameter	3.21
Chamber wall conditions	
Chamber wall thickness (m)	0.001
Wall material	Steel
Wall-heat transfer co-efficient (W/m^2K)	0.001
Interaction between wall and droplet	Escape

nearly uniform in the chamber. However, higher velocity was detected at the bottom of the chamber owing to the reduced area of chamber cone outlet. In addition, recirculation zone was not noticed in the spray-freezing chamber due to the low inlet gas velocity (0.09 m/s). When the gas axial velocity profiles were plotted as a function of distance from the nozzle (z = 0.24 m and z = 1.73 m), it was found that at the lower axial depth, the air flow patterns were almost symmetric in both the X and Y planes throughout the spray-freezing chamber (Figure 11.11). But, at a greater depth from the nozzle, the axial velocity increased to 1.5 m/s as the air stream moved downwards in the chamber. The abovementioned increase in axial velocity was attributed to the chamber's bottom cone (Figure 11.10[b]). Also, in the absence of spray, the simulated chamber temperatures were similar to the inlet gas temperature (203 K), due to the use of extremely low value for wall heat transfer coefficient, which indicated the effective insulation of the system (Anandharamakrishnan, 2008).

11.6.3.2 Simulation with spray injection

11.6.3.2.1 Case A: Solid (or) full cone spray

11.6.3.2.1.1 Axial velocity and temperature of gas Spray injection from a solid cone spray nozzle caused an upsurge in the axial velocities of gas (Figure 11.12) compared to those in its absence (Figure 11.10). Further, gas recirculation was observed at the conical bottom of the chamber in a region of relatively stagnant gas that encloses the rapidly moving core formed by the spray. This is in contrary to the simulation results obtained in the absence of spray, wherein, there

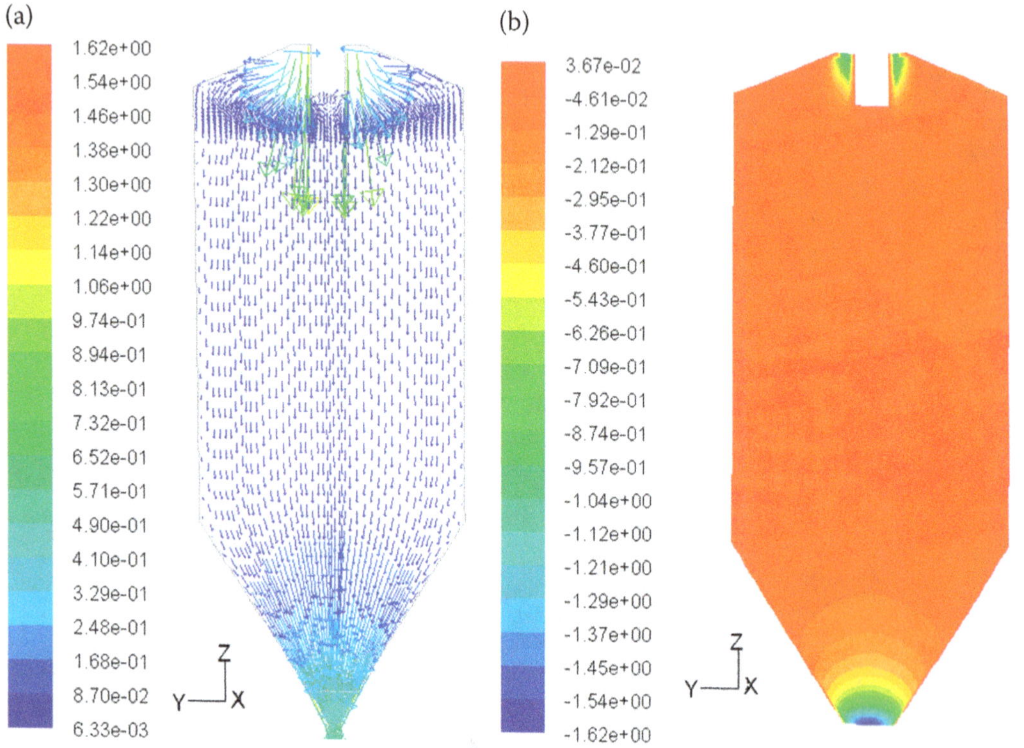

Figure 11.10 Gas velocity without spray injection: (a) velocity magnitude vector (m/s) and (b) contours of axial velocity (m/s) (Anandharamakrishnan, 2008).

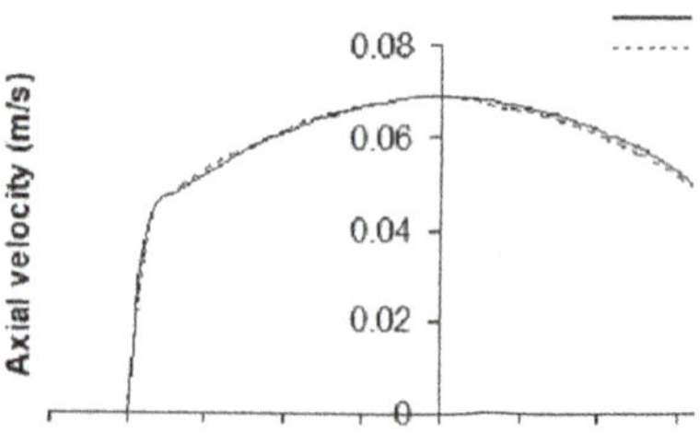

Figure 11.11 Gas axial velocity profiles without spray condition at z = 0.24 m axial distance (Anandharamakrishnan, 2008).

was no recirculation at the bottom. Gas recirculation reduces the particle temperature by increasing the residence time of some particles. Besides, it instigates the cold particles to be transported towards the cone wall or bottom of the cylindrical wall region. This leads to increased deposition of particles on the conical wall. Figs. 11.13[a-c] depict the gas axial velocity profiles at different axial distances below the nozzle (z = 0.24 m, 0.8 m and 1.23 m). The simulation results revealed a higher axial velocity (8 m/s) in the core region at z = 0.24 m (Figure 11.13[a]), relative to the outer regions of the chamber. This behavior is typical of the solid cone spray. As the spray passes down the chamber, the centerline axial velocity falls and exhibits an upward motion (negative velocity) near the wall region due to gas recirculation.

The simulation results on gas temperature profiles at different axial positions (z = 0.24 m, 0.58 m, 0.8 m, and 1.23 m) showed a good correlation with the experimental observations except in the core spray region (Figs. 11.14[a-d]). The deviations may arise due to the drops of liquid nitrogen flowing into the core of the chamber from the inlet (Figure 11.14[a]). However, at a farther distance from the

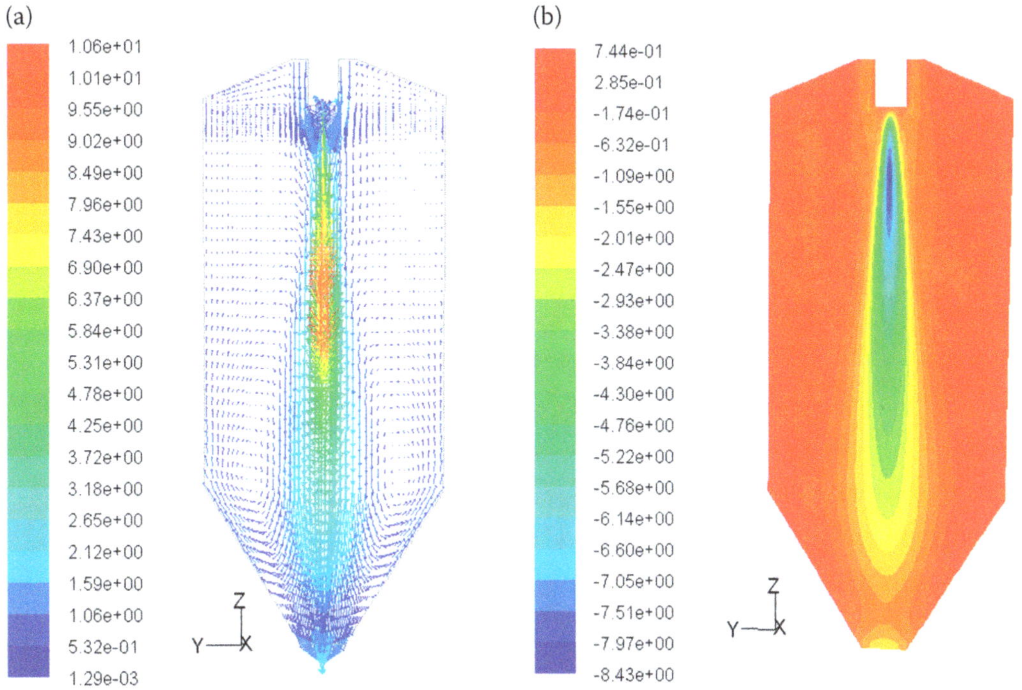

Figure 11.12 Gas velocity with spray injection: (a) velocity magnitude vector (m/s) (b) contours of axial velocity (m/s) (Anandharamakrishnan, 2008).

nozzle (z = 0.8 m and 1.23 m), due to the evaporation of liquid nitrogen, the spray fanned out to flatten the temperature profiles and widen the core region of higher temperature. Besides, the temperature contours demonstrated that the solid cone spray pattern caused most of the freezing to occur in the core region. Furthermore, the temperature of gas outside the core was found to be almost uniform and it seemed that most of the droplets did not penetrate this zone (Figure 11.15) (Anandharamakrishnan, 2008).

11.6.3.2.1.2 Particle histories (Velocity, temperature, and residence time distributions) To simulate the particle histories as a function of particle size, four size ranges of 17 µm, 50 µm, 100 µm and 150 µm particles were chosen. On the other hand, the experimental measurement of droplet axial velocities was done using Phase Doppler Anemometry (PDA) along the center-line of the spray at various distances vertically below the nozzle orifice: 0.038 m, 0.068 m and 0.108 m, at a chamber temperature of –42°C. Then, the data was extracted for droplets having size of 20 µm, 50 µm, 100 µm and 150 µm, so as to be correlated with the simulation data. The simulation results were plotted along with the experimental results and presented in Figure 11.16 (Anandharamakrishnan, 2008).

i. **Axial velocity of droplets**

Experimental validation of simulation results established that the larger particles travel faster than smaller size particles, as the latter slow down more rapidly when they approach their terminal velocities. This is because the terminal velocity varies directly with the square of droplet diameter (Papadakis & King, 1988). While the simulation results for larger diameter particles (100 and 150 µm) matched the experimental results obtained from PDA, the smaller diameter particles (17 and 50 µm)

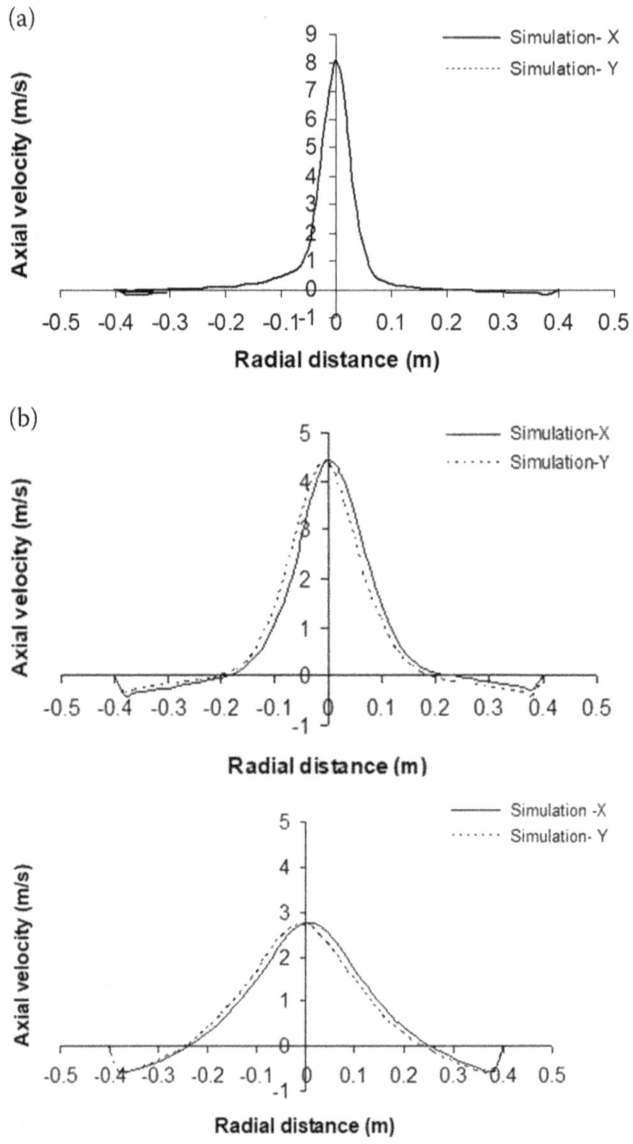

Figure 11.13 Gas axial velocity profiles at axial distances of (a) z = 0.24 m; (b) z = 0.8 m and (c) z = 1.23 m below the nozzle (Anandharamakrishnan, 2008).

did not. Minor errors in the experimental measurement of gas and liquid flow rates were stated as possible reason for the above deviation (Anandharamakrishnan, 2008).

ii. **Temperature profiles**

The temperature data was extracted from the vertical axis centre-line. Smaller sized particles were colder than their larger counterparts that exhibited higher thermal inertia and smaller specific surface area. The gas temperature raised until an axial distance of z = 0.4 m, yet close to the temperature of smallest particles (Figure 11.17). The simulated radial profiles of particle temperature at z = 0.8 m and 1.23 m below the nozzle are shown in Figs. 11.18[a] and [b]. Outside the

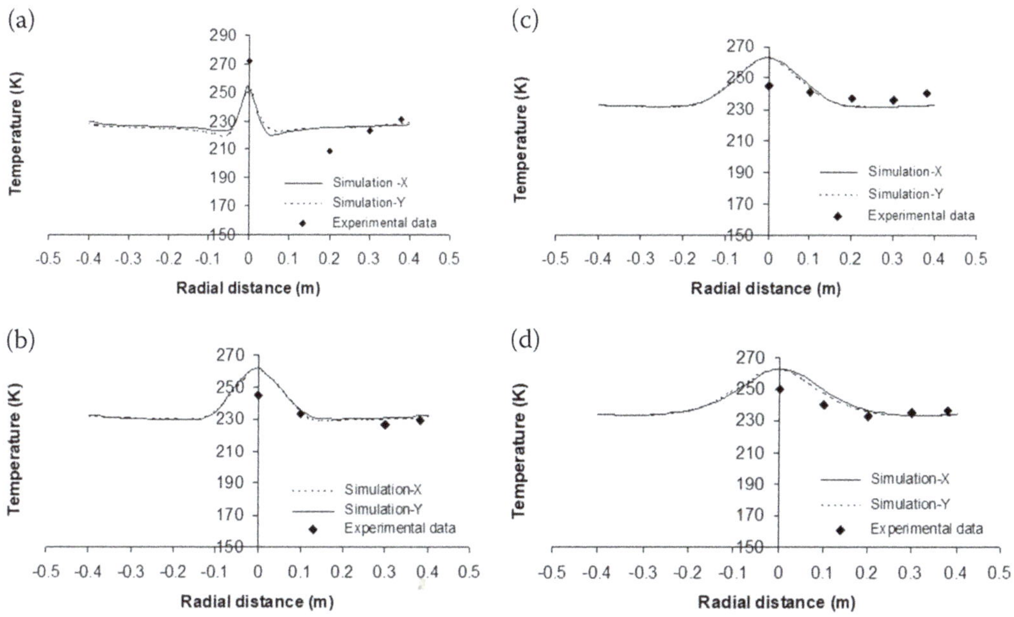

Figure 11.14 Comparison of gas temperature profiles between measurements and simulation at (a) z = 0.24 m (b) z = 0.58 m (c) z = 0.8 m and (d) z = 1.23 m from the nozzle spray point (Anandharamakrishnan, 2008).

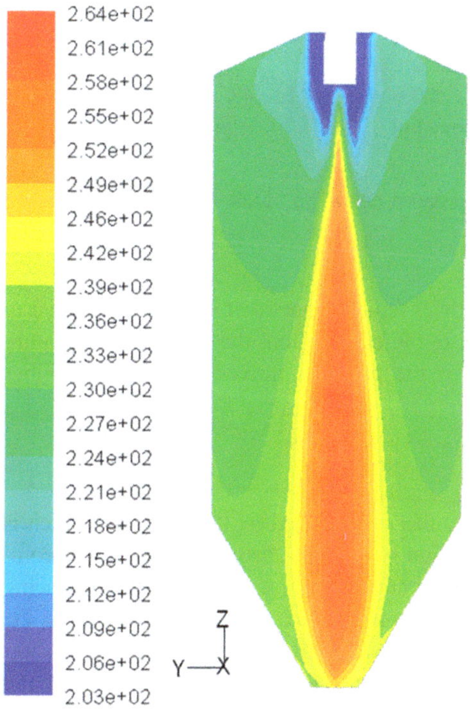

Figure 11.15 Gas temperature (K) contours of solid cone spray condition (Anandharamakrishnan, 2008).

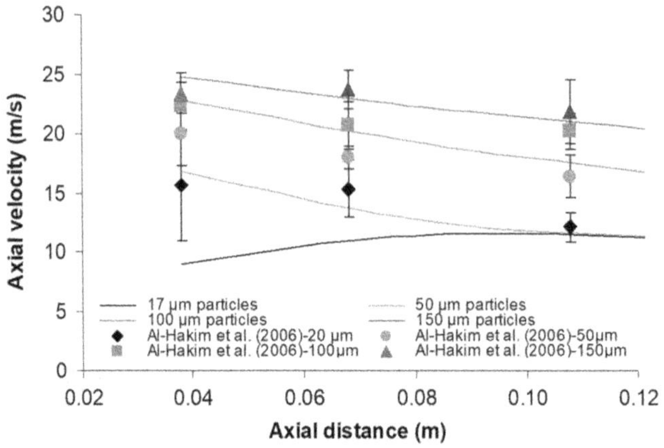

Figure 11.16 Comparison of axial velocity between measurements and simulation at an axial distances of z = 0.038 m, 0.068 m and 0.108 m from the nozzle for various particle sizes (Anandharamakrishnan, 2008).

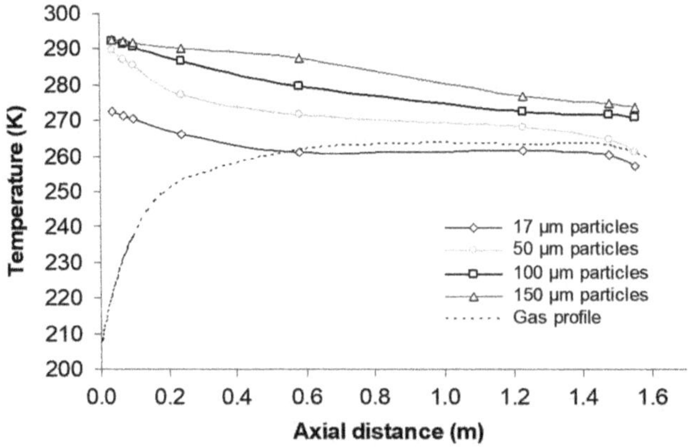

Figure 11.17 CFD simulated particle temperatures at axial distance below the nozzle (Anandharamakrishnan, 2008).

core region (0.2 m < r < 0.4 m), the droplets showed lower temperatures that were almost equal to that of gas, implying complete freezing. Nevertheless, in the core region, the particle temperature was higher than the gas temperature and decreased as the particles moved down the chamber (Figure 11.18[a & b]) (Anandharamakrishnan, 2008).

iii. **Particle residence time distributions**

The residence time of a particle inside the spray chamber is governed by its trajectory, which is nothing but the path followed by particles under the action of air flow pattern. Thus, in the computational model, the primary particle residence time distribution (RTD) was determined by tracing multiple particles through the flow domain, after which the time of each particle to travel from the atomizer to a wall or to a product outlet was recorded. The simulation results revealed the short residence time (RT: 1.5 to 5.4 s) of most of the particles in the trajectories (Figure 11.19[a]) and indicated that the smaller particles were recirculated by the gas phase (Figure 11.19[b]). Further,

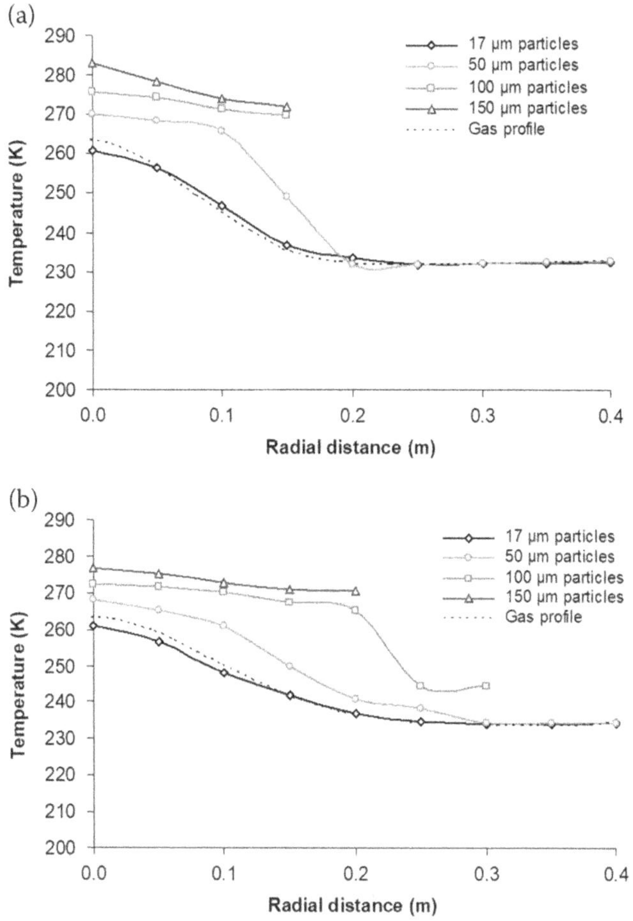

Figure 11.18 CFD simulated particle temperatures at axial distances of (a) z = 0.8 m and (b) z = 1.23 m (Anandharamakrishnan, 2008).

the minimum, maximum, and average particle RT were found to be 0.19 s, 20.9 s, and 1.3 s, respectively. Due to the influence of atomizer exit velocity on the speeding up transition through the chamber, the average RT was much lower than the gas residence time (18 s) (Figure 11.20).

The particle diameter and residence time were inversely related (Figure 11.21) and the smaller particles closely followed the gas flow as they were capable of penetrating outside the core region with low gas velocity and gas recirculation. Conversely, larger particles were dragged along the fast flowing center core region, such that they impact on the conical bottom or exit via the outlet of the spray-freezing chamber.

iv. **Particle impact positions**

Understanding the particle impact positions is vital in the design and operation of spray-freezing operation. Simulation results (Figure 11.22) which had a good agreement with the experimental results showed that 65% of particles reached the conical bottom of the spray-freezing chamber, 11% of particles struck the cylindrical part of the wall, with merely 22% of the particles that exited the chamber directly. Strikingly, none of the particles impacted the ceiling as gas

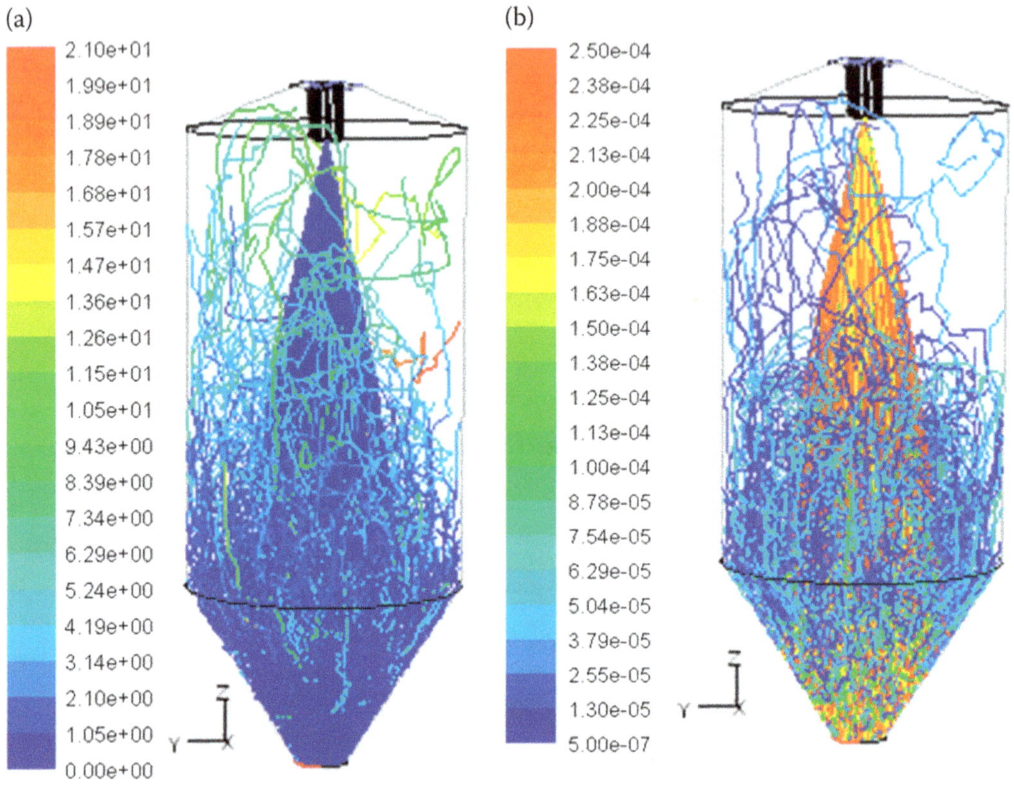

Figure 11.19 Particle trajectories: (a) colored by particle residence time (s) and (b) colored by particle diameter (m) (Anandharamakrishnan, 2008).

re-circulation was observed only at the bottom of the chamber. There were some 2% of incomplete particles (Figure 11.22), with diameter less than 1 μm. Unlike spray-drying, where the dried particles that had once hit the walls remain deposited there, during the spray-freezing operation, a frozen particle that strikes the wall, sticks and starts to build an icy layer. Subsequently, the icy layer may slide down gradually along the wall towards the main product outlet. Thus, a logically

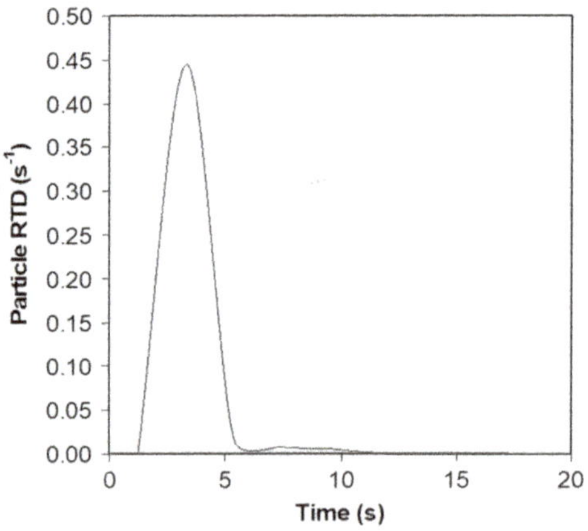

Figure 11.20 Overall particle residence time distribution (Anandharamakrishnan, 2008).

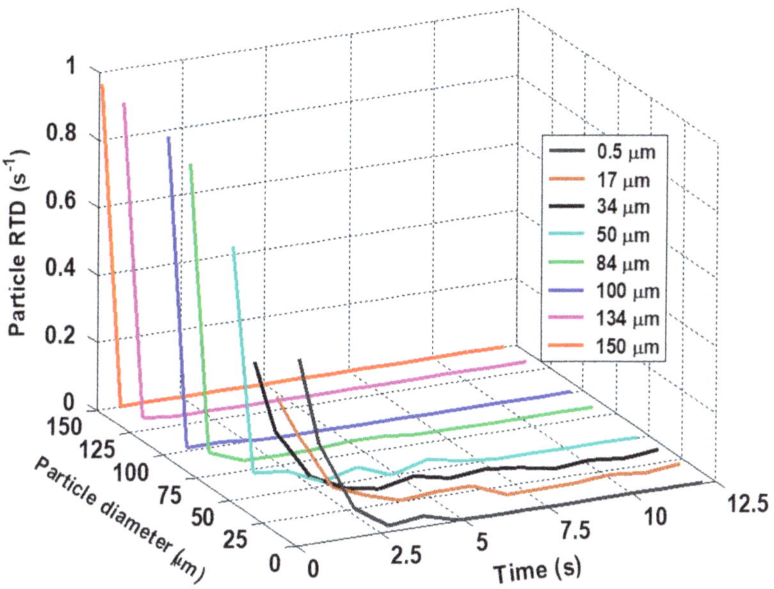

Figure 11.21 Predicted residence time distribution for different particle sizes (Anandharamakrishnan, 2008).

designed spray-freezing chamber can lead to more amounts of product exiting the outlet directly. This aspect is relevant from the purview of obtaining high product yield.

11.6.3.2.2 Case B: Hollow cone spray

In order to reduce the higher outlet temperature of particle observed with solid cone spray without altering the chamber design, CFD simulations with a hollow cone spray pattern was explored. Nevertheless, the boundary conditions of case B simulation were the same as case A, except for the change in spray pattern from a solid cone to a hollow cone. A comparison was sought between the simulation results of Case A and Case B to appreciate the influence of spray pattern on particle velocity, temperature, particle RTD and product collection efficiency.

11.6.3.2.2.1 Gas axial velocity and temperature While the axial velocity profiles resultant from using hollow cone spray (Figure 11.23[a]) were similar to those of solid cone spray, the temperature distributions were different. In the full cone spray, the gas was hot in the core region or center axial line of the chamber where the droplets were sprayed, due to a higher degree of heat transfer from the droplets to gas, Contrarily, the hollow cone spray was inherently typified by the absence of spray in the central axial line due to the spray pattern. As a result, a lower gas temperature was predicted in the simulations (Figure 11.23[b]).

11.6.3.2.2.2 Particle axial velocity The particle axial velocities exhibited similarities with gas velocities. Compared to solid cone spray, the hollow cone spray showed a wide range of particle sizes, with the larger ones penetrating outside the core region (0.1 m < r < 0.4 m). Consequently, the particle velocities and temperatures were reduced due to low gas velocity and increased residence time, respectively. Particularly, at z = 1.23 m, larger particles showed higher axial velocities than the smaller ones (Figure 11.24).

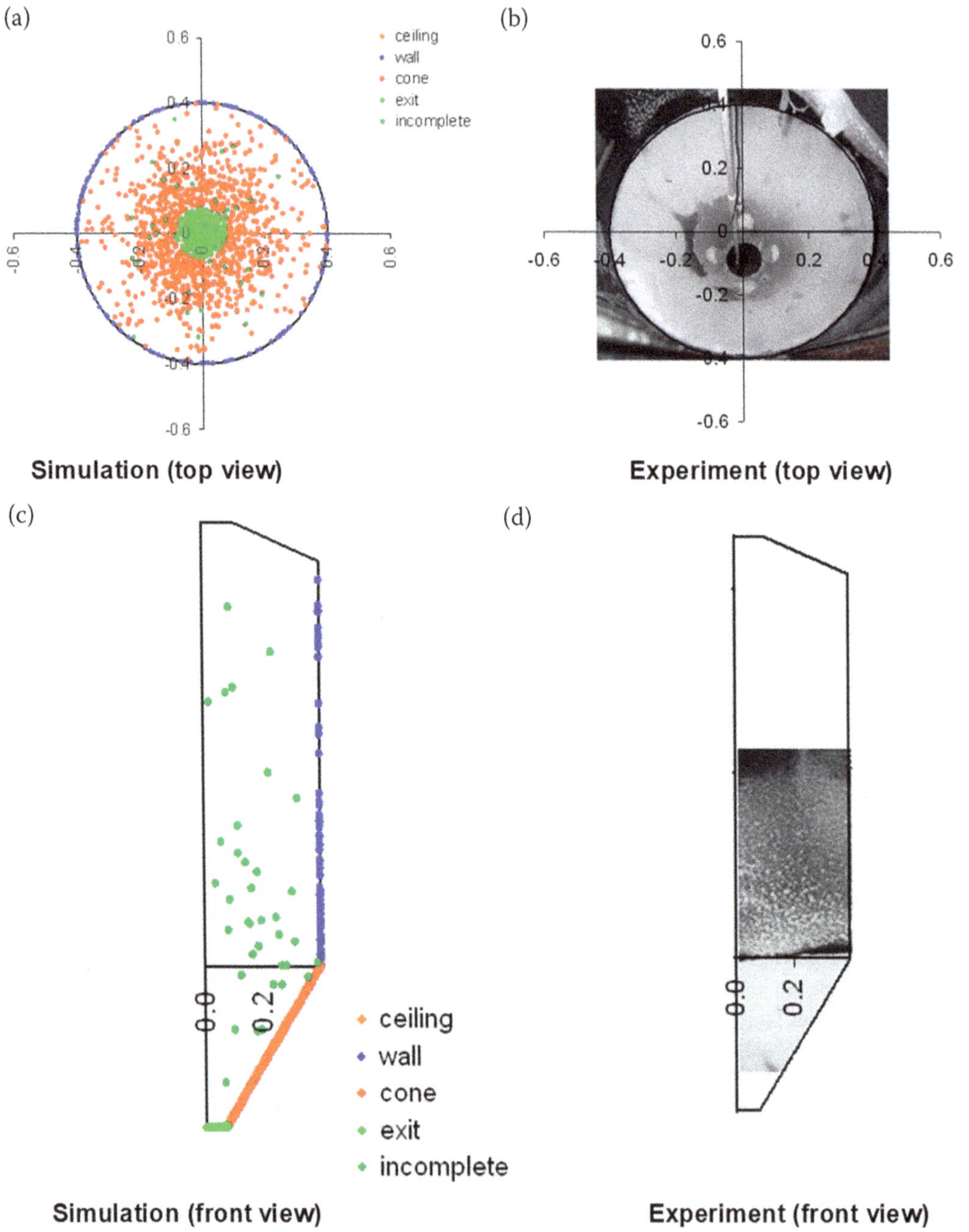

Figure 11.22 Results of the CFD simulation and experiments related to the particle impact position on the walls (Anandharamakrishnan, 2008).

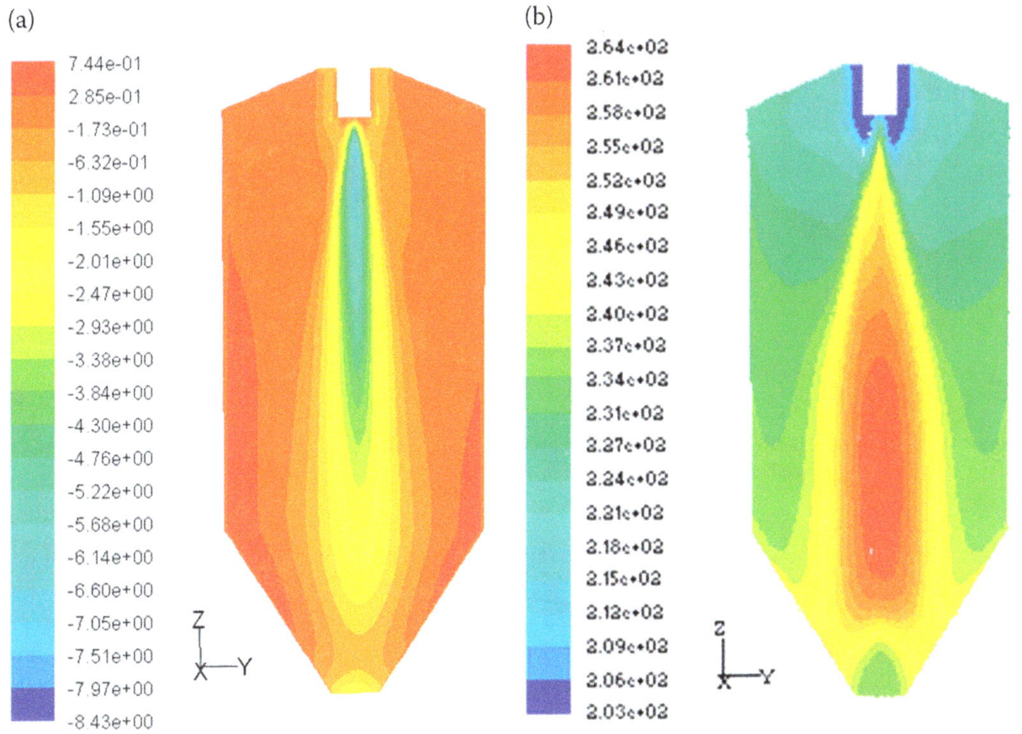

Figure 11.23 Gas profile contours with hollow cone spray: (a) axial velocity (m/s) and (b) temperature (K) (Anandharamakrishnan, 2008).

11.6.3.2.2.3 Particle temperature The radial profiles of particle temperatures (Figure 11.25) at an axial distance of z = 0.8 m and 1.23 m below the nozzle orifice established the higher temperature of larger particles than the smaller ones. The core region (0 < r < 0.2 m) showed constant particle temperatures. However, as the particles deviated from the core region (0.2 < r < 0.4 m), their temperatures began to drop. For instance, the temperature of a 50 μm particle at z = 0.8 m level (Figure 11.25) was ~270 K in the core region, which dropped to 235 K, as it moved outside. Thus, it is

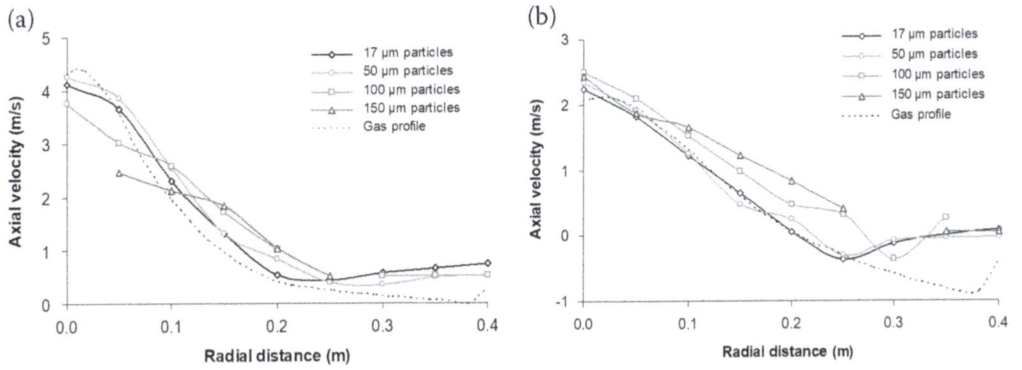

Figure 11.24 CFD simulated hollow cone spray particle axial velocities at axial position of (a) z = 0.8 m and (b) z = 1.23 m levels (Anandharamakrishnan, 2008).

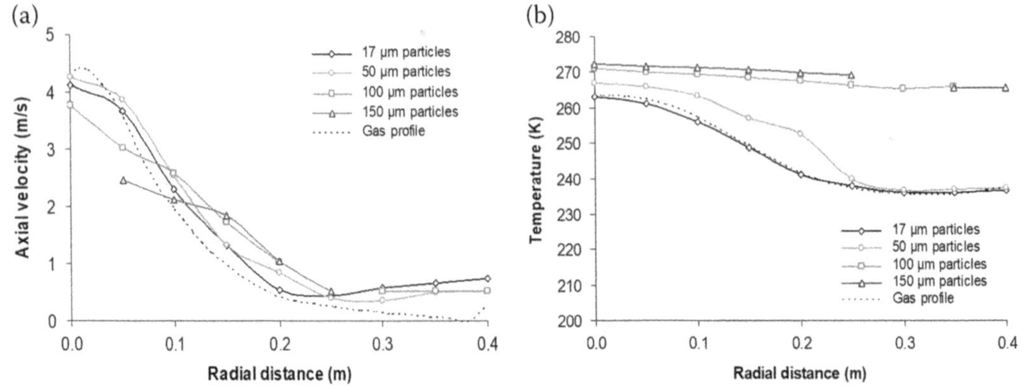

Figure 11.25 CFD simulated hollow cone spray particle temperatures at axial position of (a) z = 0.8 m and (b) z = 1.23 m levels (Anandharamakrishnan, 2008).

apparent that the droplets sprayed outside the core region attain a lower temperature on account of low gas axial velocity that prolongs the RT of particles and insulates them from other droplets (less warming effect). Also, as observed in solid cone spray, the larger diameter particles travelled faster and cooled more gradually than the smaller particles (Figure 11.26), due to their larger inertia and thermal capacity. Outlet temperature of the smaller particles was almost equivalent to the gas temperature. Thus, between the solid and hollow cone sprays, the difference in temperatures of larger particles (<2 K) were insignificant. But, the smaller particles in the hollow cone spray depicted significantly lower temperatures (~10 K) than those in the solid cone. The difference can be attributed to the low overall residence time distribution due to lower particle velocities in the hollow cone spray.

Thus, compared to solid cone spray, the lower particle velocities and the resultant cooler particle temperatures obtained from hollow cone spray is a clear advantage with respect to a low temperature process such as spray-freeze-drying. This is relevant as the high particle temperature (> 0°C) resultant from solid cone spray has a negative impact on the freeze-drying process. Particles at high temperature are susceptible to agglomeration during the sublimation stage, leading to final dried particles with larger diameter and irregular shape. This jeopardizes the whole rationale of the spray-freeze-drying process.

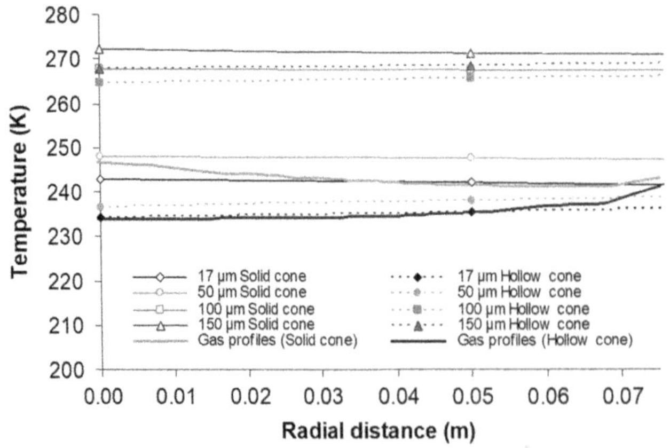

Figure 11.26 CFD simulated particles temperatures at the chamber outlet (Anandharamakrishnan, 2008).

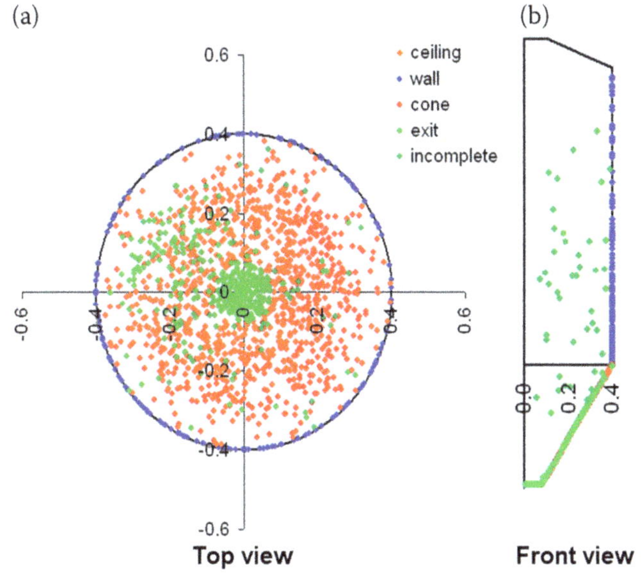

Figure 11.27 Predicted particle impact position on walls for case B (Anandharamakrishnan, 2008).

11.6.3.2.2.4 Particle impact position With the hollow cone spray, 73% of particles collided on the conical part of the chamber and 11% hit the cylindrical part of the wall. Only 13% of the particles exited the chamber directly, which is lesser than that observed in Case A. Similar to that observed with solid cone spray, 2% of incomplete particles were observed (Figure 11.27). A comparative analysis showed that, irrespective of the spray pattern, the particle impact was more than 60% in the bottom cone region and only 10% of the particles deposited on the cylindrical wall region. Though the particle collection efficiency was less than 20% in both the spray configurations, the solid cone spray pattern performed relatively better than the hollow cone spray due to its low overall RT (Figs. 11.28 & 11.29).

In terms of particle collection, spray-freezing operation is different from the spray drying operation, in which the dried particles are recovered from the wall and cone as they inherently slide

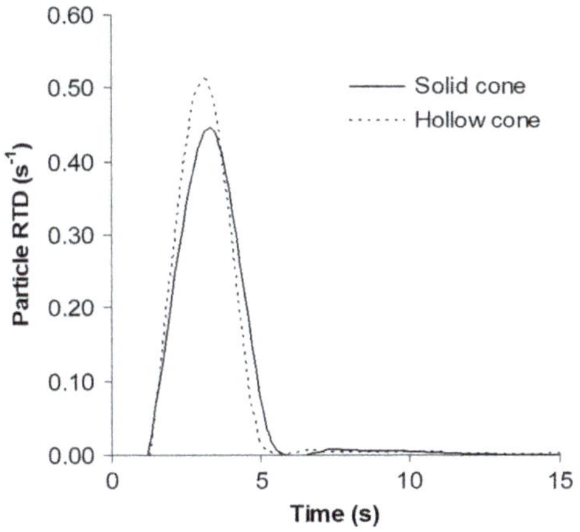

Figure 11.28 Comparison of solid and hollow cone spray particles overall primary RTD (Anandharamakrishnan, 2008).

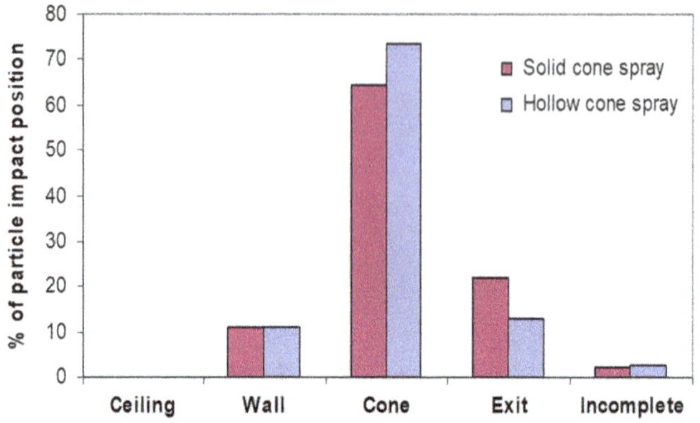

Figure 11.29 Comparison of solid and hollow cone spray particle impact position on walls (Anandharamakrishnan, 2008).

under the influence of outside hammer operations. Since the major applications of spray-freezing-drying process is for high-value food ingredients and biologicals, the product collection efficiency is central to its commercial viability. In this context, modifying the design of spray-freezing chamber by increasing the diameter of chamber outlet can improve the product collection efficiency.

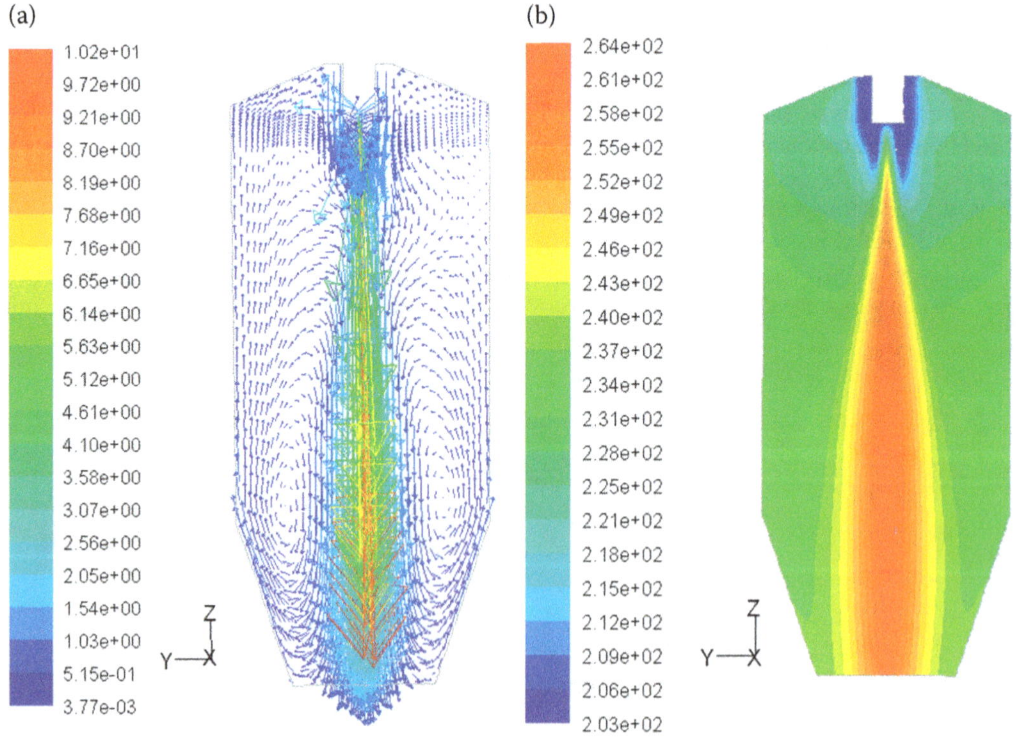

Figure 11.30 Gas profiles of (a) velocity magnitude vector (m/s); (b) contours of temperature (K) (Anandharamakrishnan, 2008).

11.6.3.2.3 Case C: Modified spray-freezing chamber design

In case C, the outlet diameter of chamber was increased from 0.15 m to 0.45 m. Boundary conditions were same as those of case A.

11.6.3.2.3.1 Axial velocity and temperature of gas The gas axial velocity and temperature profiles (Figure 11.30) were similar to those observed in case A. Trivial gas recirculation was observed near the wall in the conical region of the chamber (Figure 11.30[a]). But, the temperature profiles (Figure 11.30[b]) were slightly different near the outlet. The gas temperature was marginally higher (260 K) in case C than case A, as the narrow outlet cone in the latter caused higher gas recirculation and thereby reduced the gas temperature at the chamber outlet.

11.6.3.2.3.2 Particle axial velocity and temperature Interestingly, the particle axial velocity at the outlet in *case C* was lower than that of case A, as the increase in outlet diameter reduced the exit gas velocity (Figure 11.31). Yet, a mixed trend was seen in the temperature profiles (Figure 11.32). While the larger particles behaved similar to case A, smaller particles showed

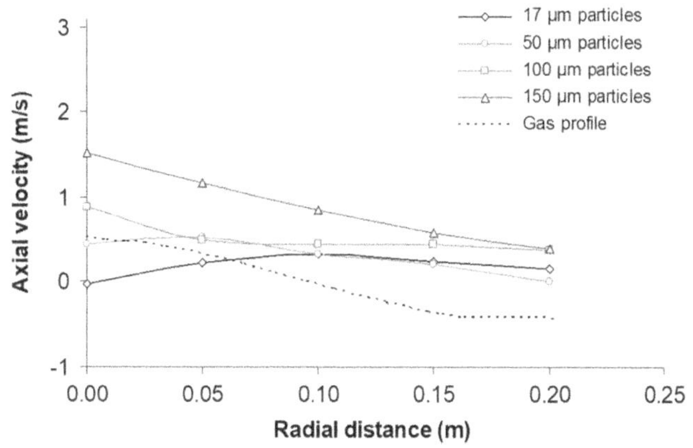

Figure 11.31 CFD simulated particle axial velocities at the chamber outlet (Anandharamakrishnan, 2008).

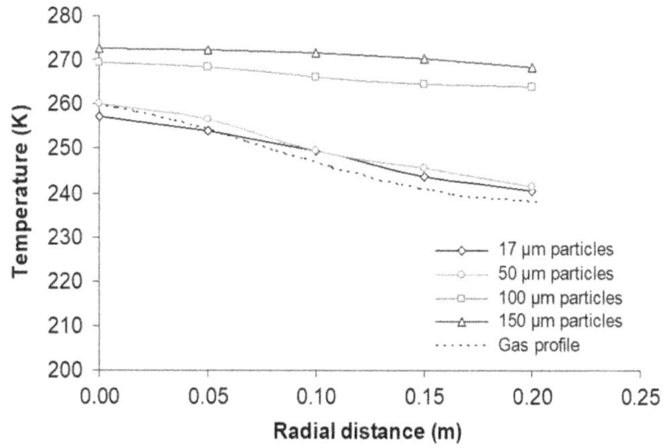

Figure 11.32 CFD simulated particle temperatures at the chamber outlet (Anandharamakrishnan, 2008).

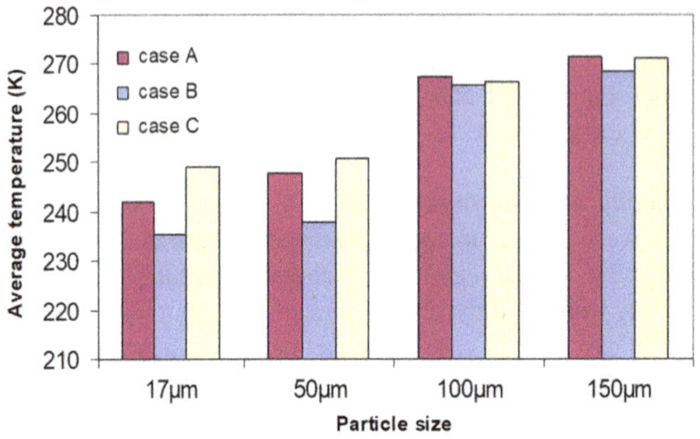

Figure 11.33 Average particle temperature at outlet (z = 1.73 m) (Anandharamakrishnan, 2008).

higher temperatures by nearly 5 K (Figure 11.33) due to the reduced degree of gas recirculation and shorter residence time. This is not very advantageous as higher temperatures can promote particle agglomeration during the freeze-drying stage and result in larger particle diameter and longer drying time.

11.6.3.2.3.3 Particle impact position As hypothesized, there was a substantial improvement in the particle collection efficiency (Figure 11.34) with 57% of the particles exiting the chamber directly, which was almost 3-fold higher than that obtained from case A and case B. Only 33% of particles impacted the conical part of the chamber and the remaining 10% hit the cylindrical part of the wall (Figure 11.35). Notably, incomplete particles were absent in case C as the wider outlet area reduced the

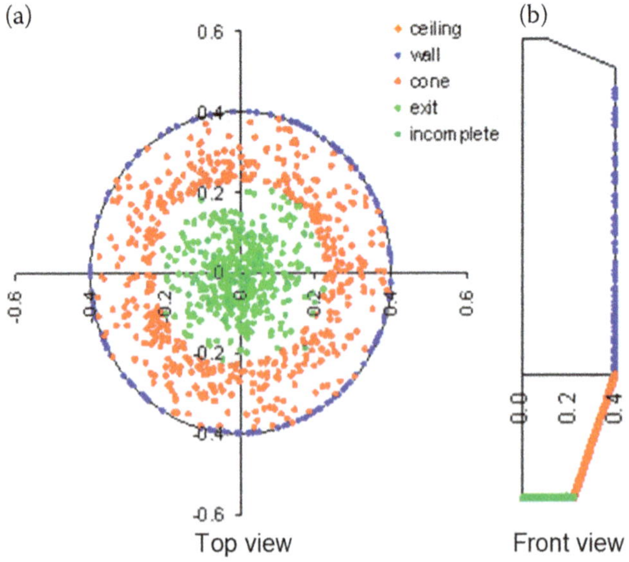

Figure 11.34 Particle impact position on walls for case C (Anandharamakrishnan, 2008).

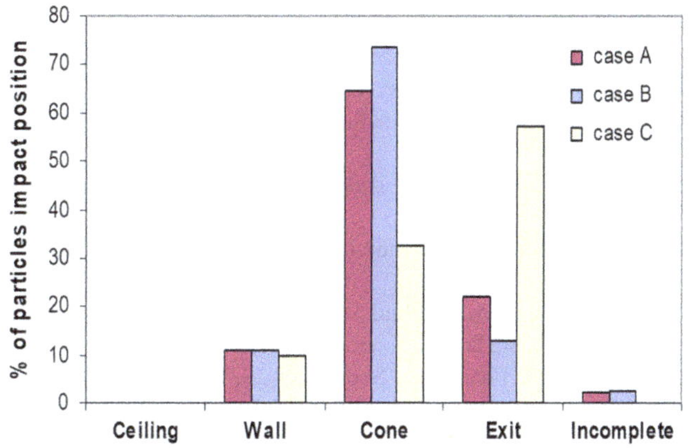

Figure 11.35 Comparison of all three cases of particle impact position on wall (Anandharamakrishnan, 2008).

gas recirculation. Therefore, modifying the design of spray-freezing design chamber led to a significant improvement in the outlet particle collection efficiency (Anandharamakrishnan, 2008).

CONCLUSIONS

This chapter clearly demonstrated the capability of a three-dimensional CFD model for spray-freezing to predict and evaluate the influence of spray patterns and chamber design on the gas and particle parameters (axial velocity, temperature profiles, particle residence time distributions) and the particle collection efficiency. Notably, the model included the latent heat effects during phase change. Collectively, the results of CFD modeling suggested that, for the same chamber design, the solid cone spray configuration would be a better option than the hollow cone spray in terms of product collection efficiency. Nevertheless, using a hollow cone spray in a spray-freezing chamber with increased outlet diameter may yield higher product recoveries with low particle temperatures. Thus, computational fluid dynamics modeling offers a flexible environment to alter the process parameters and study their influence on the performance of a complex process such as spray-freeze-drying, which is often tedious to be established with an experimental approach.

REFERENCES

Al-Hakim, K., Wigley, G., & Stapley, A. G. F. (2006). Phase Doppler anemometry studies of spray freezing. *Trans IChemE, Part A, Chemical Engineering Research and Design*, 84(A12), 1142–1151.

Anandharamakrishnan, C., & Ishwarya, S. P. (2019). Essentials and Applications of Food Engineering (1st ed.) (Chapter 16). CRC Press, Taylor & Francis Group, Boca Raton, FL.

Anandharamakrishnan, C. (2003). Computational fluid dynamics (CFD) – applications for the food industry. *Indian Food Industry* 22, 62–68.

Anandharamakrishnan, C. (2008). Experimental and computational fluid dynamics studies on spray-freeze-drying and spray-drying of proteins. *Ph.D. thesis*, Submitted to Loughborough University, United Kingdom.

Anandharamakrishnan, C. (2013). *Computational Fluid Dynamics Applications in Food Processing* (Springer Briefs in Food, Health, and Nutrition). Springer, New York.

Anandharamakrishnan, C., & Ishwarya, S. P. (2015). *Spray Drying Techniques for Food Ingredient Encapsulation*. John Wiley and Sons Ltd, United Kingdom.

Anandharamakrishnan, C., Gimbun, J., Stapley, A. G. F., & Rielly, C. D. (2010). Application of Computational Fluid Dynamics (CFD) Simulations to Spray-Freezing Operations. *Drying Technology*, 28, 94–102.

Anderson, J. D. (2013). *Computational fluid dynamics - The basics with applications*. McGraw-Hill Inc, New York.

Anderson, J. D. (1984). Computational fluid dynamics - The basics with applications. McGraw-Hill Inc., New York.

Anderson J. (2009). Governing Equations of Fluid Dynamics. In: Wendt J. F. (Ed.), Computational Fluid Dynamics. Springer, Berlin, Heidelberg.

ANSYS Fluent user guide (2006). © Fluent Inc. September 29, 2006. Available from: https://romeo.univ-reims.fr/documents/fluent/fluentUserGuide.pdf (Accessed on 14 August 2021).

Crowe, C. T., Sharma, M. P., & Stock, D. E. (1977). The particle source in cell (PSI-Cell) model for gas-droplet flows. *Journal of Fluid Engineering*, 9, 325–332.

Fletcher, A. J. (2000). *Computational techniques for fluid dynamics* (2nd edition). New York: Springer-Verlag.

Fluent (2005). Fluent manual. Available from: www.fluent.com.

Huang, L. , Kumar, K. , & Mujumdar, A. S. (2003). A parametric study of the gas flow patterns and drying performance of co-current spray dryer: Results of a computational fluid dynamics study. *Drying Technology*, 21, 957–978.

Jakobsen, H. A., Sannaes, B. H., Grevskott, S., & Svendsen, H. F. (1997). Modelling of vertical bubble-driven flows. *Industrial Engineering Chemistry Research*, 36, 4052–4074.

Lefebvre, A. H. (1989). *Atomization and Sprays*. Hemisphere Publishing, United States.

Madenci, E. & Guven, I. (2006). *The Finite Element Method and Applications in Engineering Using ANSYS®*. LLC, Springer Science+Business Media, USA.

Marshall, E. M. & Bakker, A. (2002). *Computational fluid mixing*. Fluent Inc, Lebanon.

Mostafa, A. A., & Mongia, H. C. (1987). On the modeling of turbulent evaporating sprays: Eulerian versus Lagrangian approach. *International Journal of Heat and Mass Transfer*, 30, 2583–2593.

Nijdam, J. J., Guo, B., Fletcher, D. F., & Langrish, T. A. G. (2006). Lagrangian and Eulerian models for simulating turbulent dispersion and coalescence of droplets within a spray. *Applied Mathematical Modelling*, 30, 1196–1211.

Papadakis, S. E. & King, C. J. (1988). Air temperature and humidity profiles in spray drying. 1. Features predicted by the particle source in cell model. *Industrial Engineering Chemistry Research*, 27, 2111–2116.

Ranz, W. , & Marshall, W. (1952). Evaporation from Drops. *Chemical Engineering Progress*, 48, 141–146.

Release 12.0 © ANSYS, Inc. (2009). Available from: https://www.afs.enea.it/project/neptunius/docs/fluent/html/rn/rel_notes_12.0.htm (Accessed on 14 August 2021).

Scott, G., & Richardson, P. (1997). The application of computational fluid dynamics in the food industry. *Trends in Food Science and Technology*, 8,119–124.

Tate, R. W. (1969). Sprays, Kirk-Othmer Encyclopedia of Chemical Technology, Vol. 18, 2nd edition. John Wiley & Sons, New York, pp. 634–654.

Xia, B. and Sun, D.- W. (2002). The application of computational fluid dynamics (CFD) in the food industry: a review. *Computers and Electronics in Agriculture*, 34, 5–24.

CHAPTER 12

Spray-freeze-drying: Challenges and Solutions in the Way Forward

As understood from the preceding chapters, it is evident that spray-freeze-drying (SFD) offers a multitude of advantages over its forerunner technologies. It yields mechanically robust particles with unique microstructure, instantaneous aqueous dissolution and greater retention of active ingredients. The synergistic benefits of atomization in combination with low-temperature processing are responsible for the outweighing merits of SFD process. Nevertheless, it is always imperative to understand the bottlenecks in an operation. Bottlenecks are limitations related to either the equipment or resource that reduces the throughput (Athimulam et al., 2006) and impedes the scalability of a process. Identification of bottlenecks will aid in overcoming them by devising the necessary technical solutions. This chapter intends to brainstorm on the bottlenecks of spray-freeze-drying and propose evidence-based and experience-based solutions for the same.

12.1 IDENTIFYING THE BOTTLENECKS OF SPRAY-FREEZE-DRYING PROCESS

12.1.1 Resource-related bottlenecks

The bottlenecks of spray-freeze-drying process can be classified as resource-related and process or equipment-related. Cryogenic liquid or gas and vacuum are the major resources for conducting a SFD trial. The requirements for cryogenic liquid/cold dry gas and vacuum are variable depending on the choice of spray-freezing (SFV, SFL or SFV/L) and freeze-drying (conventional or atmospheric freeze-drying) methods, respectively. Consequently, the aforementioned choices would also influence the energy intensity and process economics of the SFD process (Di Matteo, Donsi, & Ferrari, 2003; Wolff & Gibert, 1990). Besides, the given factors compel the batch operation of SFD process and renders it a time-consuming process. Although the process innovations in SFD have come a long way by imbibing several developments across the years, the development of a continuous spray-freeze-dryer for the production of pharmaceuticals or food ingredients is still at its nascent stage. A number of factors such as the process validation protocols, good manufacturing practices, operational cost and energy consumption must be considered while designing a continuous spray-freeze-dryer with fully integrated spray-freezing and freeze-drying units (Adali et al., 2020).

12.1.2 Process-related bottlenecks and relevant solutions

12.1.2.1 Challenges imposed by the spray-freezing step

Apart from the challenges related to the cost and logistics of handling resources, there are several other process-related challenges in SFD. For instance, biological products, especially

DOI: 10.1201/9781003019312-12

proteins are susceptible to instability during the spray-freezing-into-vapor over liquid (SFV/L) mode of spray-freezing. Majority of the protein aggregation occurs during the transit of droplets through the vapor phase rather than during the freeze-drying step (Costantino et al., 2000). Atomization of feed liquid in ambient gas causes protein adsorption and unfolding due to the large interfacial area available between the gas and liquid phases (Webb et al., 2002). On the other hand, instantaneous cooling of liquid droplets creates a large ice-liquid interface that turns out to be the site of protein denaturation (Costantino et al., 2000). Moreover, during SFV/L, when the solvent is frozen, the active ingredient (AI) turns supersaturated in the unfrozen zones of the droplet. This facilitates the nucleation and growth of AI crystals, which obstructs the rapidity of freezing, thereby leading to particle growth, larger particle sizes and lower specific surface areas (Gombotz et al., 1990; Gusman & Johnson, 1990). Adopting SFL as the mode of spray-freezing has been prescribed for protein-based active ingredients to avoid denaturation and aggregation. Using carbohydrate-based cryoprotectants (ex. lactose, trehalose, sucrose and inulin) and freezing adjuvants can also aid in preserving the native stability of proteins and other biological products during the SFD process.

Other practical challenges encountered during the SFL and SFV/L processes are the agglomeration of atomized feed inside the cryogenic liquid bath before being transferred to the shelves of freeze-dryer. Mixing impellers can be used to keep the contents of cryogenic liquid vessel under stirred conditions to avoid aggregation of the frozen particles. The container can be insulated and closed with a gas-permeable lid to reduce evaporation of cryogenic liquid. Further, care should be taken to replenish the product container with fresh cryogenic liquid frequently, as it evaporates at a rapid rate due to its ultra-low boiling point. Cryogenic liquid level sensors may be a potential option to sense liquid levels in containers and actuate automatic dispensing from a pressurized tank containing liquid nitrogen. Freezing-mediated blockage of feed within the nozzle is also encountered, which can be avoided by using heated or plastic nozzles, as discussed in earlier chapters.

12.1.2.2 Challenges associated with the freeze-drying stage and proposed solutions

In the atmospheric SFD (ASFD) process, feed formulations with their frozen matrices having low eutectic (T_e: temperature below which a mixture of substances can no more be in liquid state under one atmosphere pressure) or glass transition temperature (T_g: the temperature range within which a polymeric material transforms from a rigid glassy to a soft amorphous state) cause a drastic rise in the cost per kilogram of dry product. This is because of the requirement for low drying temperature and long processing time (Leuenberger, 2002). As discussed in the earlier chapters, use of excipients can aid in increasing the T_e and T_g of feed formulations. Hence, the application of ASFD is justified and its profitability can be assured only if the concerned product has a higher market value than the cost associated with excipients. The aforementioned apprehension is mainly associated with food-based feed substances that often exhibit low or imprecise eutectic and glass transition temperatures (Mennet, 1994). The low profit margins associated with spray-freeze-dried foods may be the reason for the greater prevalence of SFD in the pharmaceutical sector than in the food processing industry (Ishwarya, Anandharamakrishnan, & Stapley, 2015).

The uncertainties associated with T_e and T_g dismiss the scope of developing large-scale spray-freeze-driers for food powder production. Also, the equipment manufacturers would face limitations with the versatility of commercial-scale spray-freeze-dryers. Because, the prototypes of spray-freeze-dryers developed hitherto (Mennet, 1994; Mumenthaler & Leuenberger, 1991) were found to be too large for high-value pharmaceutical products and too small for the freeze-drying of foods (Leuenberger, 2002). Unlike the industry-friendly spray dryers, commercial-scale spray-freeze-dryers need to be customized based on their end-application. Similarly, particle elutriation and contamination on the wall are the major limitations associated with atmospheric spray fluidized bed freeze-drying (Ishwarya, Anandharamakrishnan, & Stapley, 2015).

12.2 CONTINUOUS SPRAY-FREEZE-DRYING PROCESS: A SOLUTION IN THE WAY FORWARD

Designing a continuous spray-freeze-drying process is apparently the one-stop solution to address the identified challenges. A continuous SFD process is operated under full containment and complete sterility from bulk liquid formulation through discharge of dried particles until powder filling into vials, syringes, blisters, inhalation systems or in bulk (IMA LYNFINITY, 2021; Sebastião et al., 2019). Given the larger surface area of frozen particles, a continuous spray-freeze-dryer can lead to high throughput and high energy efficiency, by leveraging the increased heat and mass transfer during the drying step (Adali et al., 2020). However, scaling-up of SFD process is an intricate task! Multiple factors must be accounted during the design phase. The forthcoming sections would present the recent efforts made towards devising a continuous and large-scale SFD process and equipment. These concepts can potentially lead to the installation of commercial-scale spray-freeze-dryers in the food and pharmaceutical industry in the near future.

12.2.1 Prototypes of continuous SFD systems

12.2.1.1 Rey's model

L. Rey (2010) proposed a prototype (Figure 12.1), in which the frozen microspheres of feed liquid are formed by dropping rather than atomization, followed by freezing in a counter-current cold air stream. Then, the frozen spheres are transported on an ultrasonic conveyor belt through an airlock towards a vacuum chamber, wherein the primary drying occurs. The latent heat of sublimation is provided by electrical heaters located at the roof of the drying chamber. Once primary drying is complete, the dry particles move through another airlock into the secondary drying chamber. Desorption of bound residual moisture during the secondary drying phase is effected by microwave and/or infrared radiation that is directed towards the particles along with the vacuum. Finally, the finished product moves to the collection section, wherein the vials are continuously filled with the powder and stoppered (Pisano et al., 2019; Rey, 2010).

12.2.1.2 Spray-freezing followed by dynamic freeze-drying

Another spray-freeze-drying system has been designed and manufactured by the Meridion Technologies, a German-based pharmaceutical company. The equipment has been designed to form frozen and spherical microparticles with diameter raging between 300–600 μm. In this novel design, the spray-freezing chamber is coupled to a cylindrical rotating drum (Figure 12.2). Unlike Rey's system, here the feed liquid atomization is brought about by a nozzle placed at the chamber's top. The droplets freeze as they fall through the double-walled cylindrical chamber, which is cooled to about −110°C by cryogenic liquid or gas (nitrogen). In order to ensure a sterile processing line, the spray-freezing column is sealed so that the liquefied nitrogen from a separate line is utilized as refrigerant fluid in the cooling jacket, which does not contact the product directly. The spray-freezing column is placed above and integrated with the rotary vacuum dryer under closed conditions in order to facilitate direct transfer of frozen particles by virtue of gravity without the need for manual handling. According to a recent report, the freezing column length decides the space requirement for an industrial SFD system. Accordingly, it can take up to two or three floor levels of a shop floor (Sebastião et al., 2019).

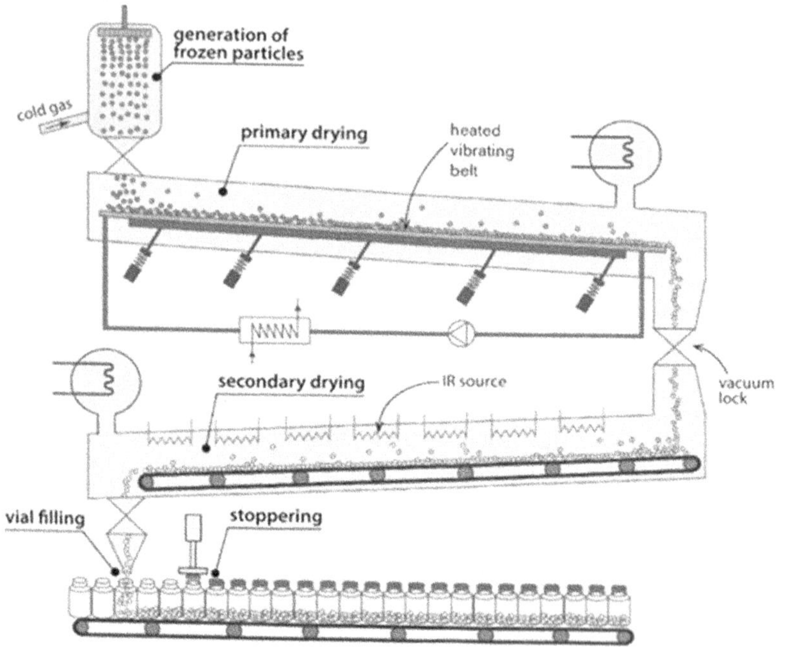

Figure 12.1 Schematic diagram of Rey's design of a continuous spray-freeze-dryer (Pisano et al., 2019).

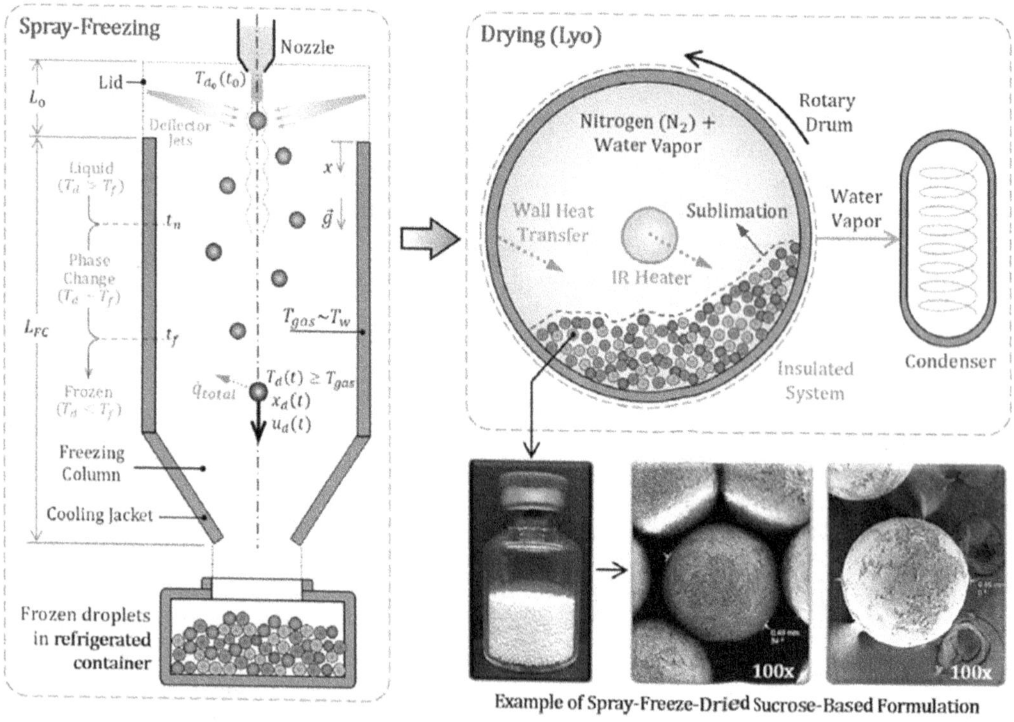

Figure 12.2 Scheme of the conceptual design of dynamic freeze-drying process with continuous removal of dried particle fraction (Sebastião et al., 2019).

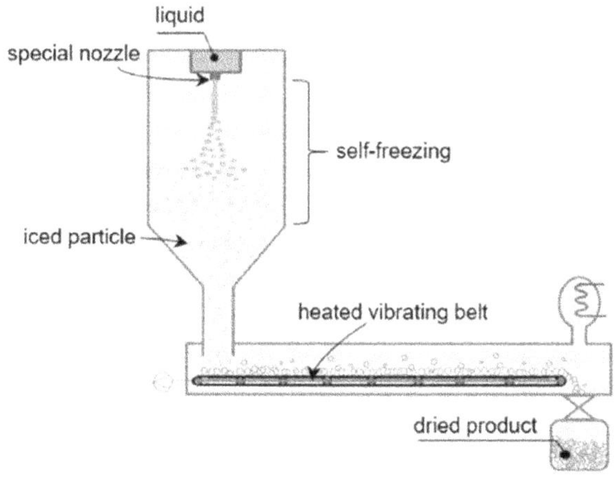

Figure 12.3 Schematic of the fine spray-freeze-drying system (Adali et al., 2020).

The subsequent stage constitutes the dynamic bulk freeze-drying, wherein the frozen microparticles are continuously released from the freezing chamber into the pre-cooled drum of the rotary dryer, inside which freeze-drying occurs under vacuum aided by constant and gentle mixing. Initially, the drum wall's temperature must be adequately low to prevent melting of the frozen particles. An infrared (IR) heater that is housed above the product bed's surface in the rotating drum supplies the necessary sublimation energy. In addition, conductive heat transfer via temperature-controlled walls of the drum functions as a secondary heat source (Sebastião et al., 2019). By using IR-based rotary freeze-dryers for sublimation, extremely less residual moisture content (<1% w/w) and high yield (>97%) can be achieved (Plitzko & Luy, unpublished data). The uniform bulk product obtained after freeze-drying is then filled in a container that can be connected to a sterile isolator for powder filling (Duralliu et al., 2020). Meridion's SFD system as described earlier is a break-through technology, which exhibits the potential to be transformed into a continuous process with the inclusion of a system for continuous particle production. Alternatively, multiple batch drum dryers may be installed in the spray-freezing unit, which operate simultaneously to result in a continuous flow of dry powder particles (Adali et al., 2020).

12.2.1.3 Fine-spray freeze-drying

Fine-spray freeze-drying is a fully-closed sterile system developed by the ULVAC Technologies, USA, for the microparticle production. Here, the feed liquid is atomized into a vacuum chamber, wherein, it is gradually dispersed to form uniform droplets. Then, the frozen particles are produced by a self-freezing mechanism when some vapor flashes off to cool the droplets that fall into the vacuum chamber (Figure 12.3). After atomization, the frozen particles are assembled on a heating/cooling belt that is heated to obtain the final dried powder (Bullich, 2015).

12.2.1.4 Stirred freeze-drying or active freeze drying

This system of continuous freeze-drying (Figure 12.4) was conceived by the Hosokawa Micron Corporation, Japan. In this design, after passing through the freezing medium or controlled cooling, the spray-frozen feed is freeze-dried under stirred conditions under low temperature and pressure within a conical and jacketed vacuum dryer. The design of stirring system must be such

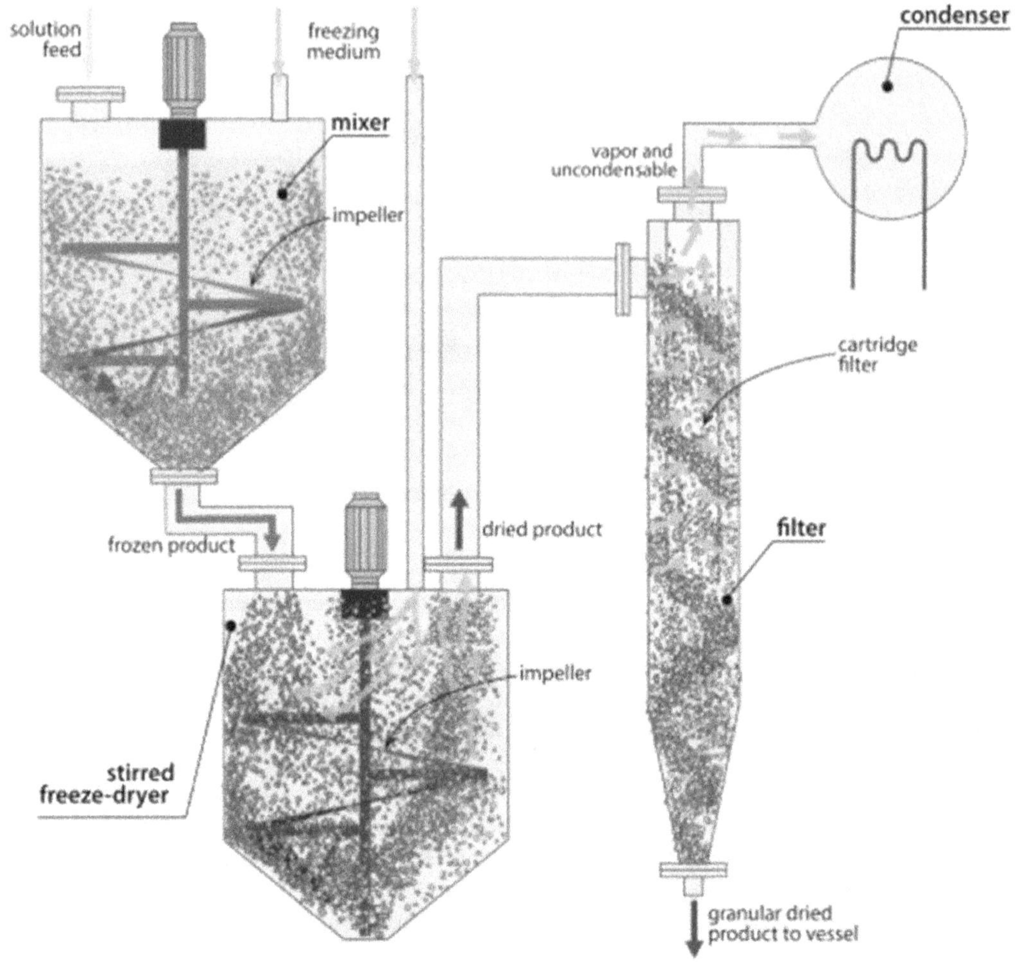

Figure 12.4 Schematic diagram of the design of stirred freeze-drying system (Pisano et al., 2019).

that the crushing is avoided. The thermal energy for sublimation is provided by the jacketed walls and disseminated by a stirrer that also assists in transporting the dry product to an integrated filter system. After the vacuum is cut-off, the free-flowing dry powder from the filter is collected in a sterile container. This stirred system outweighs the conventional SFD apparatuses owing to its shorter-duration and uniform drying and extremely fine particle size, all of which are facilitated by the continuous stirring operation (Bullich, 2015). Continuous process, high uniformity in drying and complete automation are the merits of this system (Adali et al., 2020).

12.2.1.5 Improved Rey's system for continuous processing of spray-freeze-dried products

An improved design over the Rey's SFD system has been proposed by the IMA Group (LYnfinity, Italy/USA) (Figure 12.5). Initially, the feed liquid is spray-frozen using a unique nozzle into a temperature-controlled freezing tower that is fixed on the roof of the freezing column and cooled by a cryogen. Nozzle frequency controls the homogeneity of droplet size distribution

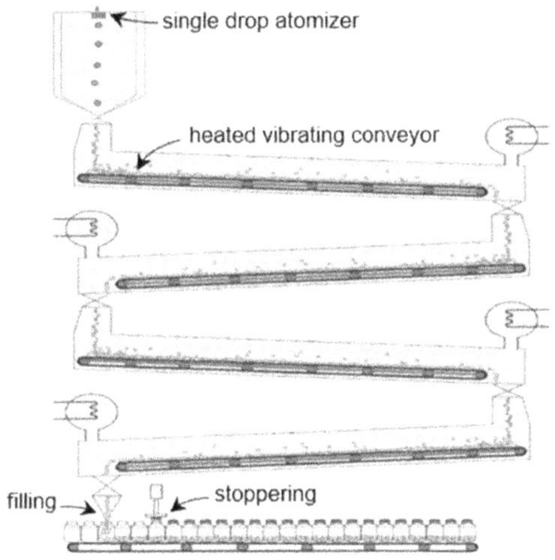

Figure 12.5 Schematic diagram of the LYnfinity continuous spray-freeze-dryer (Adali et al., 2020).

that facilitates consistent drying of wide-ranging product formulations. Post spray-freezing, the frozen particles are transferred to a heated and inclined conveyor in the drying chamber. The constant movement of particles on the shelf and the thermal contact between frozen particles and shelf promotes rapid sublimation and prevents particle agglomeration. Eventually, the powder is filled into containers for dosage or bulk storage. This system is proposed to have the following advantages of high degree of sterility, particle homogeneity, consistent drying and ease of powder filling enabled by the spraying technology (IMA, 2021).

A continuous and automated SFD process overcomes many limitations discussed in the introductory section of this chapter. Owing to the gentle drying process, the resultant products are expected to have enhanced uniformity and superior quality characteristics without losing their stability and inherent biological activity. Moreover, productivity can be increased and downtime be reduced owing to shortened drying time, direct heat transfer between the freeze-drying shelf and product and absence of resistance to heat and mass transfer since there is no build-up of cake thickness from the product. Since many continuous SFD systems employ conveyor belts for freeze-drying the frozen particles, proper control can allow adequate resident time for drying. The expenditure on cryogen shall be recouped from the lower cost of operation and utilities and reduced requirement for operators. Above all, higher energy transfer efficiency is possible with a continuous spray-freeze-drying process, due to the direct transmission of energy from the shelf to the surface of particles, where the drying front begins. This removes the major barrier to heat transfer such as the product cakes with poor thermal conductivity that are often encountered in conventional freeze-drying processes operated in batch mode (IMA LYNFINITY, 2021).

CONCLUSIONS

In the design of continuous spray-freeze-drying systems discussed earlier, additional sublimation energy is supplied by either a source of electromagnetic radiation or a temperature-controlled drum surface, which promotes rapid drying. This removes the commonly encountered heat transfer barriers in a usual batch freeze-drying and can even handle products with poor thermal conductivity (IMA LYNFINITY, 2021). Alternatively, spray-freezing followed by atmospheric freeze-drying in combination with a heat pump can lead to conservation of process energy and address the requirement for a large volume of cold dry gas. Nevertheless, the capital expenditure on heat pumps and shift in energy demand from thermal to electrical energy are some inevitable trade-offs in the given approach. Witnessing the active research on spray-freeze-drying, we can expect that the food and pharmaceutical manufacturing units would sooner have fully-automated and continuous spray-freeze-dryers with effective energy management systems in place.

REFERENCES

Adali, M. B., Barresi, A. A., Boccardo, G., & Pisano, R. (2020). Spray Freeze-Drying as a Solution to Continuous Manufacturing of Pharmaceutical Products in Bulk. *Processes*, 8(6), 709.

Athimulam, A., Kumaresan, S., Foo, D. C. Y., Sarmidi, M. R., & Aziz, R. A. (2006). Modelling and Optimization of *Eurycoma longifolia* Water Extract Production. *Food and Bioproducts Processing*, 84(2), 139–149.

Bullich, R. (2015). Telstar Industry Session: Continuous Freeze Drying. In: *Proceedings of the Innovation Forum in Pharmaceutical Process Professional*, Pharmaprocess Forum, Barcelona, Spain, 27–28 October 2015.

Costantino, H. R., Firouzabadian, L., Hogeland, K., Wu, C. C., Beganski, C., Carrasquillo, K. G., et al. (2000). Protein Spray-freeze Drying. Effect of Atomisation Conditions on Particle Size and Stability. *Pharmaceutical Research*, 17(11), 1374–1383.

Di Matteo, P., Donsi, G., & Ferrari, G. (2003). The Role of Heat and Mass Transfer Phenomena in Atmospheric Freeze-drying of Foods in a Fluidized Bed. *Journal of Food Engineering*, 59, 267–275.

Duralliu, A., Matejtschuk, P., Stickings, P., Hassall, L., Tierney, R. & Williams, D. R. (2020). The Influence of Moisture Content and Temperature on the Long-term Storage Stability of Freeze-dried High Concentration Immunoglobulin G (IgG). *Pharmaceutics*, 12, 303.

Gombotz, W. R., Healy, M. S., Brown, L. R., & Auer, H. E. (1990). *Process for Producing Small Particles of Biologically Active Pharmaceuticals.* WO/1990/013285.

Gusman, M. I., & Johnson, S. M. (1990). *Cryochemical Method of Preparing Ultrafine Particles of High-purity Superconducting Oxides.* Application (pp. 13). US: US, (SRI International, USA).

IMA (2021). IMA Group Web Page. Available from: https://ima.it/pharma/machine/lynfinity/ (Accessed on 13 August 2021).

IMA LYNFINITY (2021). https://ima.it/pharma/machine/lynfinity/ (Accessed on 13 August 2021).

Ishwarya S. P., Anandharamakrishnan C., & Stapley A. G. (2015). Spray-freeze-Drying: A Novel Process for the Drying of Foods and Bioproducts. *Trends in Food Science and Technology*, 41, 161–181.

Leuenberger, H. (2002). Spray Freeze Drying the Process of Choice for Low Water Soluble Drugs? *Journal of Nanoparticle Research*, 4, 111–119.

Mennet, H.-P. (1994). *Sprühgefriertrock-nung bei Atmosphärendruck Ein Beitrag zur Untersuchung des Prozesses und seiner Anwendungsmöglichkeiten.* Dissertation Basel.

Mumenthaler, M., & Leuenberger, H. (1991). Atmospheric Spray Freeze-drying: A Suitable Alternative in Freeze Drying Technology. *International Journal of Pharmaceutics*, 72, 97–110.

Pisano, R., Arsiccio, A., Capozzi, L. C., & Trout, B. L. (2019). Achieving Continuous Manufacturing in Lyophilization: Technologies and Approaches. *European Journal of Pharmaceutics and Biopharmaceutics*, 142, 265–279.

Rey, L. (2010). Glimpses into the Realm of Freeze-drying: Classical Issues and New Ventures. In: L. Rey & J.C. May (Eds.), *Freeze-Drying/Lyophilization of Pharmaceutical and Biological Products*, 3rd edition. CRC Press, New York, USA, pp. 1–32.

Sebastião, I. B., Bhatnagar, B., Tchessalov, S., Ohtake, S., Plitzko, M., Luy, B., & Alexeenko, A. (2019). Bulk Dynamic Spray Freeze-drying Part 1: Modeling of Droplet Cooling and Phase Change. *Journal of Pharmaceutical Sciences*, 108(6), 2063–2074.

Webb, S. D., Golledge, S. L., Cleland, J. L., Carpenter, J. F., & Randolph, T. W. (2002). Surface Adsorption of Recombinant Human Interferon-c in Lyophilized and Spray-lyophilized Formulations. *Journal of Pharmaceutical Sciences*, 91, 1474–1487.

Wolff, E., & Gibert, H. (1990). Atmospheric Freeze-drying Part 2: Modelling Drying Kinetics Using Adsorption Isotherms. *Drying Technology*, 8(2), 405–428.

Index

A

B

E

F

G

H

I

K

L

P

T

U

V

W

X

Z

For Product Safety Concerns and Information please contact our EU
representative GPSR@taylorandfrancis.com
Taylor & Francis Verlag GmbH, Kaufingerstraße 24, 80331 München, Germany

www.ingramcontent.com/pod-product-compliance
Lightning Source LLC
Chambersburg PA
CBHW080906220326
41598CB00034B/5490
* 9 7 8 1 0 3 2 2 0 7 7 5 9 *